Reviews of The Artificial Intelligence Papers

MW00837295

"James Stone has done it again: another masterful book that takes you straight to the heart of current thinking in artificial intelligence (AI) – and its foundations. From perceptrons in 1958 to generative pre-trained transformers (GPTs), this book scaffolds the history of AI with landmark papers that chart progress over the last half-century – as witnessed by the author. In short, this book represents an intellectual string of pearls that would complement the bookshelf of anyone invested in the forthcoming age of artificial intelligence."

Karl J Friston, MBBS, MA, MRCPsych, MAE, FMedSci, FRBS, FRS.
Scientific Director: Wellcome Centre for Human Neuroimaging.
Professor: Queen Square Institute of Neurology, University College London.
Honorary Consultant: The National Hospital for Neurology and Neurosurgery.

"I learned a lot from this collection of classic papers about the neural network approach to artificial intelligence. Spanning all the major advances from perceptrons to large language models (e.g. GPT), the collection is expertly curated and accompanied by insightful tutorials, along with intimate reminiscences from several of the pioneering researchers themselves."

Steven Strogatz, Professor of Mathematics, Cornell University, USA.
Author of Nonlinear Dynamics and Chaos, 2024.

"To define the future, one must study the past. Stone's book collects together the most significant papers on neural networks from the perceptron to GPT-2. Each paper is explained in modern terms and, in many cases, comments by the original authors are included. This book describes a riveting intellectual journey that is only just beginning."

Simon Prince, Honorary Professor of Computer Science, University of Bath, England.
Author of Understanding Deep Learning, 2023.

"Connectionist models of the brain date back to the work of Hebb in 1949, and the first faltering first steps towards practical applications followed soon after Rosenblatt's seminal 1958 paper on the perceptron. As of 2024, models firmly rooted in connectionism, from generative adversarial networks (GANs) to transformers, have heralded a renaissance in artificial intelligence that is revolutionising the nature of our digital age. This latest volume by James Stone collects the pivotal connectionist papers from 1958 right up to today's radical innovations, and provides an illuminating descriptive narrative charting the theoretical, technical, and application-based historical development in a lucid tutorial style. A welcome, much needed, and valuable addition to the current canon on artificial intelligence."

Mark A Girolami, FREng FRSE.
Chief Scientist: The Alan Turing Institute.
Sir Kirby Laing Professor of Civil Engineering, University of Cambridge, England.

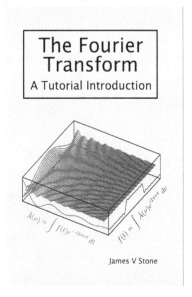

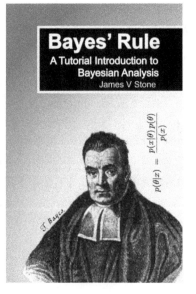

James V Stone is a Visiting Professor at the University of Sheffield, England.

The Artificial Intelligence Papers

Original Research Papers

With Tutorial Commentaries

James V Stone

Title: The Artificial Intelligence Papers
Author: James V Stone

©2024 Sebtel Press

First Edition, 2024
First Printing.
Typeset in LaTeX $\partial 2_\varepsilon$.

ISBN 9781068620003

For life, in all its endless forms most beautiful and most wonderful.

Contents

Preface

This book is intended to provide an account of key developments in modern artificial intelligence (AI) that is both informal and rigorous. By 'modern AI' I mean systems that rely heavily on *artificial neural networks*, in contrast to the purely *symbolic AI* that dominated AI research in the latter part of the 20th century. Each of these key developments is associated with a foundational research paper, which introduced new ideas to the field and acted as a stepping stone to further progress. As is always the case, many innovative and courageous papers presented novel ideas that ultimately did not advance the field, and these have been omitted.

Each foundational research paper included here is introduced in terms of its historical context and its specific contribution to the development of AI. This is followed by a technical summary of the paper, which includes a brief informal description (with diagrams) and a tutorial-style mathematical account of the paper. Most chapters have a technical summary containing several pages of mathematics, whereas other technical summaries consist of only a few paragraphs, depending on the particular demands of the paper in question.

The papers included here are inevitably biased by my own perspective. However, having been involved in neural network research over the past 40 years, I sincerely believe these papers to represent the most important contributions to the development of modern AI.

Note that the introductory material in several chapters has been adapted from my previous books on AI[122;123].

Who Should Read This Book? The material in this book should be accessible to anyone with an understanding of basic calculus and matrix algebra (for which a tutorial appendix is provided). The tutorial style adopted in the introduction to each paper ensures that the mathematics introduced in the early chapters acts as a foundation for the more sophisticated mathematics in later chapters. Thus, readers prepared to put in the effort will be amply rewarded with a solid grasp of the historically important neural networks that underpin modern AI systems.

Acknowledgements. Thanks to those authors who have provided an account of how their ideas came about, which is fascinating in terms of the history of science.

For debates about neural networks over many years, I am indebted to my friend Raymond Lister. For reading one or more chapters, I am very grateful to Alistair Bray, John Frisby, Nikki Hunkin, Llion Jones, Raymond Lister, Simon Prince, and Stephen Snow. Thanks to Alice Yew for meticulous copy-editing and proofreading. Finally, thanks to the authors and publishers for allowing their papers to be published in this book.

Corrections. Please email corrections to jvgfwstone@googlemail.com.
A list of corrections can be found at https://jamesstone.sites.sheffield.ac.uk/books/AIPapers.

About The Author

My education in artificial neural networks began in 1984 when I attended a research seminar given by Geoff Hinton on Boltzmann machine neural networks at Sussex University (UK), where I was studying for an MSc in Knowledge Based Systems. As a novice student, I was not confident enough to speak to Geoff Hinton after the lecture, but I was enormously excited by what I had seen (which stood in stark contrast to the symbolic AI systems taught on the MSc course). Despite initial reluctance from my supervisor (who, in fairness, had no experience of neural networks), Boltzmann machines became the topic of my MSc project (so thank you, Rudi Lutz). This project was based on Hinton and Sejnowski's technical report (reprinted in Chapter 4). After reading that report many times, I wrote a Boltzmann machine computer program and eventually succeeded in getting it to work, even though it crashed the PDP11 mainframe computer more than once as it ran overnight. In subsequent years, I enjoyed extended research visits to Hinton's laboratory in Toronto and to Sejnowski's laboratory in San Diego, funded by a Wellcome Trust Mathematical Biology Fellowship.

As an academic at Sheffield University in the UK, I have published papers on a range of relevant subjects, including connectionism[113], backprop[67], temporal backprop[114], optimisation[115], human memory[34;127], independent component analysis[116], evolution[117], and using neural networks to test principles of brain function[112;126]. I have also written several books: *A Brief Guide to Artificial Intelligence*[123], *Artificial Intelligence Engines*[122;123], *Neural Information Theory*[121], *Information Theory*[125], *Vision and Brain*[118], *Seeing*[26] (with J Frisby), and *The Quantum Menagerie*[124].

Once the machine thinking method has started, it would not take long to outstrip our feeble powers.
AM Turing, 1951.

Chapter 1

The Origins of Modern Artificial Intelligence

1.1. Introduction

There are two basic approaches to building machines intended to emulate human intelligence. The first assumes that the structure of the brain is not relevant and that intelligence requires only a purely symbolic or logic-based representation of the physical world. This approach, known as *symbolic AI* or *good old fashioned AI* (GOFAI), dominated AI research in the latter part of the 20th century. It achieved some successes on *toy world* models[139], but (amongst other drawbacks) its inability to learn was a major handicap.

The second approach, which is now the dominant form of AI, assumes that intelligent machines should be built on the same principles that govern the most intelligent entity known to exist: the human brain. These *artificial neural networks*, or simply neural networks as they are now known, are based on a simple hypothesis regarding learning in biological neurons, which can be traced back to the late 19th century:

> *When two impressions concur, or closely succeed one another, the nerve currents find some bridge or place of continuity. In the cells where the currents meet there is, in consequence of the meeting, a strengthened connexion.*
> Alexander Bain, 1873.

Similar ideas were expressed by Tanzi (1893) and Lugaro (1898), as quoted in Berlucchi and Buchtel (2009). To anyone who has ever considered how learning occurs in the brain, such ideas are all too familiar, but they were stunningly original in 1873. And anyone with even a superficial knowledge of modern AI will recognise that Bain's insight represents the basis of all neural network learning algorithms, an insight more usually attributed to the psychologist Donald Hebb, who wrote[36] in 1949:

> *When an axon of cell A is near enough to excite cell B and repeatedly or persistently takes part in firing it, some growth process or metabolic change takes place in one or both cells such that A's efficiency, as one of the cells firing B, is increased.*

Of course, there is a vast difference between formulating a general principle – based on Bain's 'strengthened connexion' (sic) or Hebb's increased 'efficiency' – and building an intelligent machine. Indeed, the gap between Bain's original proposal and intelligent machines is so vast that it has taken more than a century to begin to bridge that gap. But, as stated by the inventor of the perceptron (an early form of neural network), creating intelligent machines may provide intellectual rewards beyond the merely practical:

> *By the study of systems such as the perceptron, it is hoped that those fundamental laws of organization which are common to all information handling systems, machines and men included, may eventually be understood.*
> Frank Rosenblatt, 1958.

1.2. Turing on Computing Machinery and Intelligence

Alan Turing (1912–1954) single-handedly initiated at least two fields of scientific research, including mathematical models of morphogenesis and AI. Even though Turing did not explicitly discuss AI in terms of neural networks, his seminal 1950 paper[133], 'Computing machinery and intelligence', remains extraordinarily relevant to modern AI. His paper opens with a typically clear-eyed first sentence:

> *I propose to consider the question,* Can machines think?

In his paper, Turing proposed the following test to answer this question. A user is allowed to have a conversation via a computer keyboard with an entity, and the user's task is to assess whether the entity is human or not. If the user decides that the entity is human then that entity has passed the test, even if it is actually a computer. This is now known as the *Turing test*, which has generated much debate in the intervening years and remains a cornerstone of AI to this day.

Of particular relevance to modern AI, and especially neural networks, Turing considered an alternative to programming a computer to behave like a human – by allowing it to learn:

> *Instead of trying to produce a programme to simulate the adult mind, why not rather try to produce one which simulates the child's? If this were then subjected to an appropriate course of education one would obtain the adult brain.*

In so doing, Turing noted an important (but probably inevitable) side-effect of learning, which has become known as *black-box learning*:

> *An important feature of a learning machine is that its teacher will often be very largely ignorant of quite what is going on inside, although he may still be able to some extent to predict his pupil's behaviour.*

However, this is not so different from the situation with humans, who are (despite their protestations) archetypal examples of black-box learning systems. For example, when asked, 'How did you learn to ride a bike?', people usually answer with words that carry almost no useful information about how to do so, because the information required to learn to ride a bike cannot be expressed in language (not by humans, anyway). Indeed, in 1998, *reinforcement learning* (see Section 1.6) allowed a neural network to learn to ride a bike[90] – not by being given any explicit instructions but by being rewarded for remaining upright over long distances. This general approach is neatly expressed in a saying from the Tang dynasty of China (618–907 CE), which encapsulates the philosophy that underpins much of modern AI:

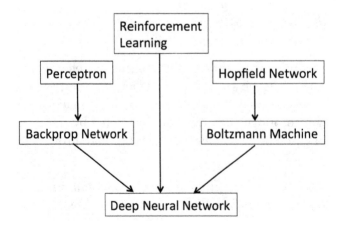

Figure 1.1. A simplified diagram of the evolution of deep neural networks and reinforcement learning, which underpin modern AI systems. Reproduced from Stone (2019).

Those who are good at archery learnt from the bow and not from Yi the Archer. Those who know how to manage boats learnt from the boats and not from Wo.
Anonymous.

In other words, the best lessons are not given by experts who are knowledgable about a particular aspect of Nature, but by Nature itself.

1.3. The Dartmouth Summer Research Project

By 1955, the idea of making a computer program imitate human behaviour was sufficiently familiar in academic circles that a project, *The Dartmouth Summer Research Project on Artificial Intelligence*, was proposed. This set of meetings, which spanned several weeks in the summer of 1956 in Dartmouth (New Hampshire, USA), is now acknowledged as the formal start of AI research. In this nascent period for AI, traditional AI methods in the guise of 'Turing machines' and 'neuron networks' appeared as explicit terms in the initial proposal for the project. Participants included people who made major contributions to research fields related to AI, including Warren McCulloch, Marvin Minsky, John Nash, Claude Shannon, Oliver Selfridge, Allen Newell, Herbert Simon and Bernard Widrow.

1.4. The Origins of Artificial Neural Networks

The development of modern artificial neural networks (see Figure 1.1) occurred in a series of step-wise dramatic improvements, interspersed with periods of stasis (often called *neural network winters*). As early as 1943, before commercial computers existed, McCulloch and Pitts[73] began to explore how small networks of artificial neurons or *units* could mimic brain-like processes. Crucially, they proved that a neural network is capable of *universal computation*, which means that it can perform any task achievable by a digital computer. Incidentally, the cross-disciplinary nature of this early research programme is evident from the fact that McCulloch and Pitts both contributed to a classic, and wonderfully titled, paper on the neurophysiology of vision ('What the frog's eye tells the frog's brain'[65]), published in 1959 in the *Proceedings of the Institute of Radio Engineers* (i.e. not in a biology journal, as might be expected nowadays).

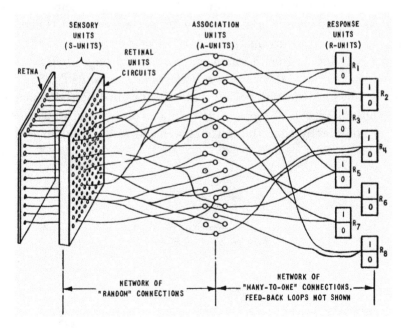

Figure 1.2. Perceptron architecture. From the *Mark I Perceptron Operators' Manual*[35].

In subsequent years, the increasing availability of computers allowed neural networks to be tested on simple pattern recognition tasks. But progress was slow, partly because most research funding was allocated to more conventional approaches that did not attempt to mimic the neural architecture of the brain, and partly because artificial neural network learning algorithms were limited.

A major landmark in the history of neural networks was Frank Rosenblatt's *perceptron* (1958), a network of interconnected units that was explicitly modelled on the neuronal structures in the brain (see Figure 1.2 and Chapter 2). The *perceptron learning algorithm* allowed a perceptron to learn to associate inputs with outputs in a manner apparently similar to learning in humans. Specifically, the perceptron could learn an association between an input image and a desired output, where this output indicated which one of several classes the image belonged to. Crucially, the perceptron could do so even if the input image was not an exact match to the image that had been learned. In effect, the perceptron was capable of *generalising* beyond the associations it had learned in the past.

Despite such human-like qualities, the perceptron was dealt a severe blow in 1969, when Minsky and Papert famously proved[75] that it could not learn to classify images correctly unless they were of a particularly simple kind. Specifically, the images in different classes had to be *linearly separable*. Consequently, a perceptron cannot solve the *exclusive OR* (XOR) problem, which requires a positive response when either of two inputs is presented to the network in isolation, but a negative response when both inputs are presented at the same time (see Chapter 2). This marked the beginning of the first neural network winter, during which most research funding went to symbolic AI, as opposed to neural networks. Despite the lack of funding, neural network research continued to be undertaken by a handful of scientists. Readers interested in the research conducted during this relatively fallow period can consult papers by Widrow and Hoff (1960), Longuett-Higgins et al. (1970), Kohonen (1972), Anderson (1968) and Sejnowski (1977).

1.5. Modern Neural Networks

The era of modern neural networks began in 1982 with the *Hopfield net*[44]. As shown in Figure 1.3, Hopfield nets are fully interconnected networks of units with bi-directional connection weights. They can learn to store particular patterns and to recall each of these when a subset of units is set to part of a pattern (see Chapter 3). Hopfield nets showed initial promise in solving classic examples of hard problems, such as the travelling salesman problem[46]. However, despite some valiant efforts to improve their performance, it was found that Hopfield nets were not very useful (but see Krotov (2023) for recent developments). At that juncture in the history of neural networks, the main contribution of Hopfield's work was to introduce a theoretical framework based on *statistical mechanics*, which laid the foundations for Ackley, Hinton and Sejnowki's *Boltzmann machine* in 1985 (see Chapter 4).

Unlike a Hopfield net, in which the *states* of all units are specified by the associations being learned, a Boltzmann machine has a reservoir of *hidden units*, which can be used to learn complex associations (see Figure 1.4). It is noteworthy that the authors went to some trouble to demonstrate that the

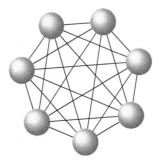

Figure 1.3. A small Hopfield network with seven units and 21 bi-directional weights.

Boltzmann machine could solve the (by now infamous) XOR problem (see Chapter 2), which had proved to be so catastrophic for neural network research in previous decades. The Boltzmann machine is also important because it facilitated a conceptual shift away from the idea of a neural network as a passive associative machine towards the view of a neural network as a *generative model*. The only problem is that Boltzmann machines learn at a rate best described as glacial. But on a practical level, the Boltzmann machine demonstrated that neural networks could learn to solve complex toy problems (e.g. XOR), which suggested that they could learn to solve almost any problem (at least in principle, and at least eventually).

The impetus supplied by the Boltzmann machine gave rise to a more tractable method devised in 1986 by Rumelhart, Hinton and Williams, the *backpropagation learning algorithm*[96] (see Chapter 5). The early backpropagation networks consisted of three layers of units: an *input layer*, which is connected with connection weights to a *hidden layer*, which in turn is connected to an *output layer*, as shown in Figure 1.4. The backpropagation algorithm is important because it demonstrated the potential of neural networks to learn sophisticated tasks in a human-like manner. Crucially, for the first time, a backpropagation neural network called NETtalk[102] learned to 'speak', inasmuch as it translated text to *phonemes* (the basic elements of speech), which a voice synthesizer then used to produce speech. A detailed account of the evolution of multi-layer neural networks can be found in Schmidhuber (2015).

Image Recognition. By 1989, LeCun had developed a backprop neural network which, to all intents and purposes, mimicked the micro-architecture of the neuronal structures within the human eye (see Chapter 7). The resultant *convolutional neural network* with several layers of units achieved an impressively low error rate of 5% in recognising handwritten digits. It is now standard practice to make use of convolution in neural networks designed for image recognition.

Before 2012, image recognition had been the exclusive reserve of the vision research community, which tended to rely on traditional image processing techniques. Performance of image recognition systems is gauged by their ability to classify standard test sets, such as the ImageNet collection of 1.2 million training images and a separate test set of 100,000 images. After 2012, attention shifted away from traditional techniques and towards neural networks. The reason was that, for the first time, a deep (seven-layer) convolutional neural network, now known as AlexNet, won first prize in the annual ImageNet competition (see Chapter 8). Consequently, conferences on vision research are no longer dominated by traditional vision methods, which have been superseded by convolutional neural networks. Today, the remarkable image manipulation capabilities of modern smartphone cameras rely heavily on convolutional neural networks. Because such networks have many layers of units, they are known as *deep neural networks*. These deep neural networks require enormous amounts of training data, which they learn using the backpropagation learning algorithm in a process referred to as *deep learning* (see Chapter 8).

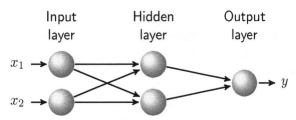

Figure 1.4. A backpropagation network with two input units, two hidden units, and one output unit. The term backpropagation refers to the fact that the difference between the actual output y and the correct output is propagated backwards through the network and is used to adjust the connection strengths between units in order to reduce that difference in future.

Generating Images. If a neural network can recognise images then it ought to possess (either explicitly or implicitly) a model of the statistical structure of those images. This, in turn, suggests that a network capable of recognising images should be able, at least in principle, to generate realistic images. Conversely, any network capable of generating images should be able, at least in principle, to recognise images. Thus, image recognition and generation are, to all intents and purposes, two sides of the same coin.

It is therefore unsurprising that networks designed to capture the underlying statistical structure of images can be used both for image recognition and for generating images. Three such networks are variational autoencoders[56] (VAEs, Chapter 9), generative adversarial networks[29] (GANs, Chapter 10), and diffusion models[41;111] (Chapter 11).

Language. Learning to deal with temporal sequences, such as language, has always posed a particularly hard challenge for neural networks. Early attempts to do so often transformed a temporal sequence into a static one-dimensional array, which could then be treated as a series of independent tokens (e.g. phonemes or words) and processed by a conventional neural network (i.e. one that is atemporal without internal dynamics).

Throughout the 1980s and 1990s, several *recurrent neural networks* (RNNs) with internal dynamics were devised, but (amongst other difficulties) they struggled to take account of events separated by long intervals. For example, in the sentence 'Julia got on the bus to go to the shop', two important words are 'Julia' and 'shop', which are separated by a long interval of eight intervening words. Given the difficulties posed by such long-range dependences for RNNs, it is unsurprising that the most successful system for processing language by the year 2000 treated language as a static one-dimensional array of linguistic tokens (see Chapter 13). By 2017, this system had been superseded by *transformers*, which can translate text between languages (see Chapter 14).

The rapid rise of AI in the popular imagination is due, in no small part, to the impressive linguistic abilities of several *large language models* (LLMs), such as the *generative pre-trained transformers* (GPT) first introduced in 2018 (see Chapter 15). As indicated by the name, GPTs are based on extended versions of transformer neural networks. It is worth noting that both transformers and LLMs consist of atemporal neural networks rather than RNNs. However, the atemporal approach used in transformers and LLMs and the temporal approach used in RNNs have recently been reconciled in the form of a proof[82] that transformers (and therefore LLMs, to some extent) are equivalent to RNNs.

1.6. Reinforcement Learning

In parallel with the evolution of neural networks, *reinforcement learning* was developed throughout the 1980s and 1990s, principally by Sutton and Barto (2018). Reinforcement learning is an inspired fusion of game playing by computers, as developed by Shannon (1950) and Samuel (1959), optimal control theory, and stimulus–response experiments in psychology. Early results showed that hard, albeit small-scale, problems such as balancing a pole can be solved using feedback in the form of simple reward signals[8] (see Chapter 6).

More recently, reinforcement learning has been combined with deep learning to produce impressive skill acquisition, such as in the case of a glider that learns to gain height on thermals[32], as well as world-class performance in games such as chess and Go[108]. Indeed, the two most important methods that underpin modern AI are deep learning and reinforcement learning.

Chapter 2

The Perceptron – 1958

Context

The perceptron was the first modern incarnation of an artificial neural network based explicitly on the micro-structure of the brain. Rosenblatt's perceptron paper is remarkable, not just for its detailed mathematical analysis but also for its *connectionist* philosophy, which emphasized the brain-like properties of the perceptron:

> *The memory of the perceptron is* distributed, *in the sense that any association may make use of a large proportion of the cells in the system, and the removal of a portion of the association system would not have an appreciable effect on the performance of any one discrimination or association, but would begin to show up as a general deficit in all learned associations.*
> Frank Rosenblatt, 1958.

Technical Summary

The original paper by Rosenblatt describes several variants, and sub-variants, of the perceptron. Because of the multiplicity of variants, combined with the somewhat opaque descriptions, it can be difficult to make sense of the paper as a whole. The following sections attempt to simplify the explanations by giving an account of the perceptron variant that seems most similar to the system described in textbooks.

Before we begin, it is vital to be aware of an important inaccuracy in most textbook accounts of perceptrons, which state that the connections between neurons get changed during learning. However, as described in the original paper, *the connection strengths between units are constant* and are not changed during the process of learning new associations. Instead, learning consists of changing the outputs in one particular set of units. As we shall see, this makes no difference to the behaviour of the network, but it requires a substantial shift in perspective when reading Rosenblatt's paper.

2.1. Architecture

The structure of the perceptron is shown in Figure 2.1. It consists of three or four sets of neurons or *units*, a *retina*, a *projection area*, an *association area*, and a set of *response units*. The only bi-directional connections are between the association area and the response units.

The retina consists of an array of binary photoreceptors, called *S-points*. The outputs of these *S*-points are transmitted to the *A*-units in the projection area A_I. The *S*-points connected to a particular *A*-unit are called the *origin points* of that *A*-unit. The connections from the retina to the projection area A_I can be positive, negative or zero (i.e. no connection).

Units in the projection area A_{I} transmit their outputs to the association area A_{II}. The connections between A_{I} and A_{II} are randomly distributed, and the values of the connection strengths can be positive, negative or zero (i.e. no connection).

Finally, units in the association area A_{II} transmit their outputs to a set of response units $(R_1, \ldots, R_n)$. The set of A-units connected to a particular response unit is called the *source-set* for that response.

The projection area is assumed to be omitted in the following, as shown in Figure 1.2, which is copied from the *Mark I Perceptron Operators' Manual*. For each architecture, there are three classes of learning rules, referred to as α, β and γ. We consider only the γ system here, because out of the three learning rules it is the closest to the modern incarnation of a Hebbian learning rule.

Note that Rosenblatt makes no attempt to distinguish between a unit and its state or output, so we are left to disambiguate terms from their context. For the sake of consistency, we follow Rosenblatt's notational convention, while identifying the role (e.g. state or type of unit) of each variable for clarity.

2.2. Activation

Consider a stimulus applied to the retinal S-points. If the net input to an association unit A_k from the S-points exceeds a threshold θ_k then A_k becomes active with a fixed output *value* V_k. Rosenblatt then describes two phases of activation: a *predominant* phase and a *postdominant* phase.

In the predominant phase, the outputs of the association units cause some of the response units to become active. Each response unit has excitatory feedback connections to the units in its own source-set and inhibitory feedback connections to the complement of its source-set (i.e. it tends to inhibit activity in any association units that do not connect to it). The feedback connections between the response units and the association units ensure that, after a short interval during which the system settles into a stable state, only one response unit R_j remains in an active state.

In this postdominant phase, the only active units are retinal units (S-points) activated by the stimulus image, the association units connected to those retinal units (but only if each association unit's threshold is reached), and a single response R_j. A learning rule can now be applied, which is intended to increase the probability that the same response R_j will be obtained if the same stimulus is applied to the retinal units.

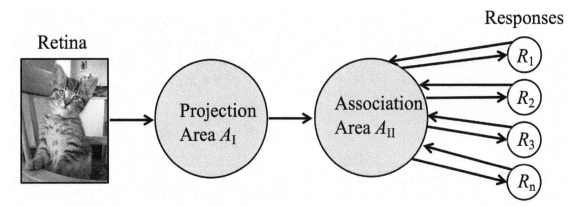

Figure 2.1. Basic structure of Rosenblatt's perceptron. Retinal units project to the projection area A_{I}, which projects to the association area A_{II}, which is connected to the response units. Only connections between the association area and the response units are bi-directional. All units are binary. All connections have fixed weights, and learning occurs by changing the permanent values of association units. The area A_{I} may be omitted so that the retina connects directly to A_{II}. Reproduced from Stone (2019).

2.3. Learning

The perceptron depends on *supervised learning* to correctly classify inputs, where the differences between the network outputs and the correct outputs are used to train the network.

As mentioned above, the perceptron learning rule is conventionally presented in terms of changing the connection weights between units. However, Rosenblatt used a different method, which has an equivalent effect but is very different in terms of conceptualisation and notation. In fact, Rosenblatt mentions modifying connections within the perceptron only once, and even that is presented as a choice between changing the connection weights *or* changing the values of association units as a way of learning associations (see p391, paragraph 1, of the paper). Having presented this choice, Rosenblatt decides to modify the values of the association units, rather than altering the connection weights between association units and response units. The equivalence between these two options can be understood by comparing a conventional Hebbian learning rule with Rosenblatt's update rule.

For readers who wish to skip the next two sections, here is a brief summary. In essence, using Rosenblatt's learning rule permanently changes the effective output of an association unit, and its effect on the inputs to response units changes accordingly. In contrast, a Hebbian learning rule changes each of the weights between an association unit and the response units to which it is connected in proportion to the product of their outputs, so the inputs to those response units change accordingly. Because the permanent change ΔV produced by Rosenblatt's learning rule is proportional to each of the weight changes Δw produced by a Hebbian learning rule, the two rules yield the same end result.

We do not present a complete proof of why this is true, but the brief outline below should convince the reader that such a proof is both plausible and attainable with a little extra (albeit tedious) calculation. In essence, an association unit A_k that projects to n response units can change the effect it has on the input to (and therefore the output of) each of those units via two learning routes.

A Hebbian Learning Rule. First, consider a network that uses a Hebbian learning rule. Typically, the input u_j to a response unit R_j is a weighted sum of values of units in its source-set, where the weight assigned to value V_k of association unit A_k is w_{kj}:

$$u_j \;=\; \sum_{k=1}^{K} w_{kj} V_k. \tag{2.1}$$

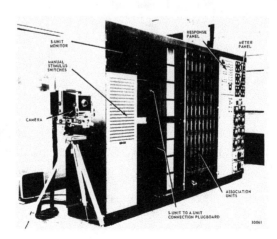

Figure 2.2. Photograph of the Mark I Perceptron with major components labelled. From the *Mark I Perceptron Operators' Manual*[35].

Using a Hebbian learning rule, the weight of each outgoing connection from an association unit A_k to a response unit R_j is changed in proportion to the product of their outputs:

$$\Delta w_{kj} = V_k R_j. \tag{2.2}$$

One variant of this rule, which is considered to be more robust, ignores the magnitude of the product $V_k R_j$ and replaces the product with its sign,

$$\Delta w_{kj} = \text{sign}(V_k R_j). \tag{2.3}$$

Consequently, a change δV_k in the output of A_k will change the input to the response unit R_j by

$$\Delta u_{kj} = \Delta w_{kj} \delta V_k \tag{2.4}$$
$$= \text{sign}(V_k R_j) \delta V_k. \tag{2.5}$$

Rosenblatt's Learning Rule. With Rosenblatt's learning rule γ, the association units A_k connected to the only active response unit R_j (in the postdominant phase) have their values V_k permanently changed. According to this rule, the permanent change in V_k caused by the states of the pair of connected units A_k and R_j is

$$\Delta V_{kj} = \begin{cases} +1 & \text{if } u_{kj} > \theta_k, \\ \dfrac{-N_{ar}}{N_{Ar} - N_{ar}} & \text{if } u_{kj} \leq \theta_k, \end{cases} \tag{2.6}$$

where N_{ar} is the number of active units and N_{Ar} is the total number of units in the source-set of response unit R_j, and u_{kj} is the input to response unit R_j from association unit A_k. In words, if the input u_{kj} is greater than the unit's threshold θ_k then the unit's value V_k is incremented by 1. Otherwise, the unit's value is decreased by an amount equal to the ratio of active to inactive units in the source-set of the response unit R_j. To simplify matters, if we assume that half of the units in the source-set are active then this ratio (and therefore the decrement in V_k) is equal to unity, so that

$$\Delta V_{kj} = \begin{cases} +1 & \text{if } u_{kj} > \theta_k, \\ -1 & \text{if } u_{kj} \leq \theta_k. \end{cases} \tag{2.7}$$

For simplicity we assume $\theta_k = 0$ here, so that Rosenblatt's learning rule can be approximated as

$$\Delta V_{kj} = \text{sign}(u_{kj}). \tag{2.8}$$

Because there is only one active unit, R_j, in the postdominant phase, we can write $u_{kj} = V_k R_j$ and therefore

$$\Delta V_{kj} = \text{sign}(V_k R_j). \tag{2.9}$$

If the output of A_k changes by an amount δV_k then the additional potency gained from the permanent change ΔV_{kj} in A_k's value will change the input u_{kj} to the response unit R_j by

$$\Delta u_{kj} = \Delta V_{kj} \delta V_k \tag{2.10}$$
$$= \text{sign}(V_k R_j) \delta V_k. \tag{2.11}$$

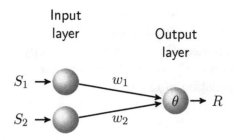

Figure 2.3. A perceptron with one output unit, which has a threshold θ.

Crucially, if we compare the Hebbbian learning rule in Equation 2.3 with Rosenblatt's learning rule in Equation 2.9, we see that they are identical. By implication, changes in the outputs of response units caused by learning will also be identical using both rules. This is confirmed by observing that the change in the input to the response unit resulting from Hebbian learning (Equation 2.5) is identical to the change resulting from Rosenblatt's learning rule (Equation 2.11).

The *perceptron convergence theorem*[80], proved by Novikoff in 1962, states that repeated application of the learning rule guarantees convergence to correct responses. However, the nemesis of the entire perceptron edifice appeared in the form of a theorem of Minsky and Papert (1969). In essence, this showed that perceptrons cannot solve an apparently simple problem, called exclusive OR, or XOR.

To understand why the perceptron cannot solve the XOR problem, we first need to understand how the perceptron works in more detail. Because we have already shown that the Rosenblatt and Hebbian learning rules are essentially equivalent, we will make use of the modern Hebbian approach, which involves modifying weights between units.

2.4. Hebbian Learning

We can consider a perceptron to be a neural network with many input units and one output unit, with a threshold applied to the state of the output unit (Figure 2.3). If the total input u to the output unit exceeds this threshold θ then the output unit's state is $R = +1$; otherwise it is $R = -1$. To see how classification might be achieved, let's start with the simplest possible example: classifying input vectors into two possible classes.

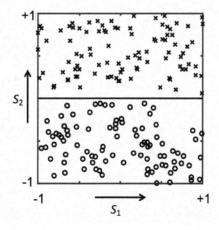

Figure 2.4. Defining two classes. Each data point $\mathbf{S} = (S_1, S_2)$ defines a location. All points (circles) on or below the horizontal line, defined by $S_2 \leq 0$, belong to class C_1, and all points (crosses) above the line belong to class C_2. Reproduced from Stone (2019).

For reasons that will become apparent, each input vector consists of just two elements, which can be interpreted as an image containing two pixels. However, images are rarely perfect, and they contain noise, which corrupts the original image. Consequently, we usually have a population of images, where each image contains a noisy version of the original noiseless image. Our objective is to train a neural network to classify the entire population of images correctly.

If an image $\mathbf{S} = (S_1, S_2)$ belongs to the class C_1 then we would like the neural network output R to be the target value $R_1^{\text{targ}} = -1$, and if $\mathbf{S}$ belongs to class C_2 then we would like the neural network output to be $R_1^{\text{targ}} = +1$. For simplicity, we define all input vectors $\mathbf{S}$ in which $S_2 \leq 0$ as belonging to class C_1 (i.e. $\mathbf{S} \in C_1$) and all input vectors in which $S_2 > 0$ as belonging to class C_2 (i.e. $\mathbf{S} \in C_2$), as shown in Figure 2.4. This is written succinctly as

$$\mathbf{S} \in \begin{cases} C_1 & \text{if } S_2 \leq 0, \\ C_2 & \text{if } S_2 > 0. \end{cases} \tag{2.12}$$

Because each image consists of just two pixels with values S_1 and S_2, we can represent each input vector as a point in a two-dimensional graph of S_1 versus S_2. For each class, the population of images defines a cluster of points, as shown in Figure 2.4; in this case the points on or below the line $S_2 = 0$ belong to class C_1, and the points above the line belong to C_2.

For a perceptron with a weight vector $\mathbf{w}$ and an input vector $\mathbf{S}$, the total input u to the output unit is given by the inner product

$$u = \mathbf{w} \cdot \mathbf{S}. \tag{2.13}$$

If this input exceeds the threshold θ then the output unit state is set to $R = +1$; otherwise it is set to $R = -1$. This arrangement is expressed succinctly as

$$R = \begin{cases} -1 & \text{if } u \leq \theta, \\ +1 & \text{if } u > \theta. \end{cases} \tag{2.14}$$

For simplicity, we assume $\theta = 0$ here.

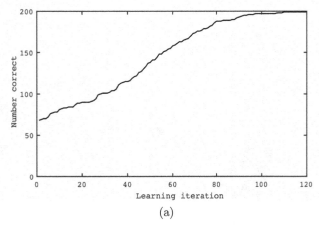
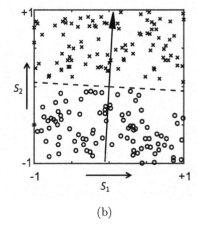

Figure 2.5. Learning in the perceptron in Figure 2.3. (a) Number of correctly classified input vectors during learning. (b) After training, the weight vector $\mathbf{w}$ (arrow) is nearly vertical, and the decision boundary (dashed line) is perpendicular to $\mathbf{w}$. If $|\mathbf{w}| = 1$ then the inner product $u_t = \mathbf{w} \cdot \mathbf{S}_t$ provides an orthogonal projection of each input $\mathbf{S}_t$ onto $\mathbf{w}$. All points above the decision boundary yield $u_t > 0$, whereas all points below the line yield $u_t < 0$, so all points are correctly classified. Reproduced from Stone (2019).

The Hebbian learning rule states that if an input vector is correctly classified then do nothing, but if it is misclassified then update the weights as follows: $\mathbf{w}_{\text{new}} \leftarrow \mathbf{w}_{\text{old}} + \Delta\mathbf{w}$, where

$$\Delta\mathbf{w} = \begin{cases} -\epsilon\,\mathbf{S} & \text{if } R^{\text{targ}} < 0 \text{ and } y > 0, \\ +\epsilon\,\mathbf{S} & \text{if } R^{\text{targ}} > 0 \text{ and } y < 0. \end{cases} \tag{2.15}$$

As an example, a set of $N = 200$ input vectors to be classified consists of $N_1 = 100$ vectors belonging to C_1 and $N_2 = 100$ belonging to C_2. During training, the number of correctly classified input vectors increases, as shown in Figure 2.5a.

When the neural network has finished learning, the final weight vector can be drawn as a line in the input space (the arrow in Figure 2.5b), and a line perpendicular to this vector defines a *decision boundary*. In this example, the decision boundary is a line that is approximately horizontal, as shown by the dashed line in Figure 2.5b. The network classifies input vectors below the decision boundary as belonging to C_1 and inputs above the boundary as belonging to C_2.

The point is that any weight vector defines a decision boundary, and in the perceptron this boundary is always a straight line. This matters because it means that any two classes that cannot be separated by a straight line cannot be classified by a perceptron. If two classes can be separated by a straight line then they are said to be *linearly separable*.

Now let's back up a little and work out why the decision boundary is perpendicular to the weight vector. To understand this, we need a geometric interpretation of the mapping from input to output. For this, we will consider a more general case in which the two classes are separated by a decision boundary that is not horizontal, as shown in Figure 2.6.

Given an input vector $\mathbf{S}$, the input u to the output unit is the inner product $u = \mathbf{w} \cdot \mathbf{S}$ (Equation 2.13). For convenience, we assume that the weight vector is maintained at unit length (i.e. $|\mathbf{w}| = 1$) by setting $\mathbf{w} \leftarrow \mathbf{w}/|\mathbf{w}|$, where $|\mathbf{w}| = \sqrt{w_1^2 + w_2^2}$. Geometrically, if $|\mathbf{w}| = 1$ then the inner product is the length of the *orthogonal projection* of $\mathbf{S}$ onto $\mathbf{w}$, as shown in Figure 2.6, so the input u is simply a length measured along $\mathbf{w}$. If $u \le \theta$ then $\mathbf{S}$ is classified as belonging to class C_1; otherwise, it is classified as C_2. Just as u is a length measured along $\mathbf{w}$, so θ is also a length along $\mathbf{w}$. From Figure 2.6, any input vector $\mathbf{S}$ that projects to a point on $\mathbf{w}$ such that $u = \mathbf{w} \cdot \mathbf{S} > \theta$ gets classified as C_2.

In Figure 2.5b, all points in class C_1 project to values u below $S_2 = 0$; similarly, all points in class C_2 project to values u above $S_2 = 0$. Once we have found the weight vector $\mathbf{w}$, the decision boundary is a line perpendicular to $\mathbf{w}$ that separates the two sets of projections onto $\mathbf{w}$.

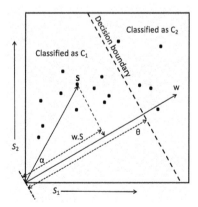

Figure 2.6. The decision boundary for the perceptron in Figure 2.3. The length of the orthogonal projection of a data point $\mathbf{S}$ onto a weight vector $\mathbf{w} = (0.89, 0.45)$ is given by the inner product $\mathbf{w} \cdot \mathbf{S} = |\mathbf{S}|\,|\mathbf{w}|\cos\alpha$, where α is the angle between $\mathbf{w}$ and $\mathbf{S}$, and $|\mathbf{w}| = 1$. If $\mathbf{w} \cdot \mathbf{S} \le \theta$ then $\mathbf{S}$ is classified into class C_1, or else it is classified into C_2; the threshold θ corresponds to a length along $\mathbf{w}$. Reproduced from Stone (2019).

2.5. The Perceptron's Nemesis: Exclusive OR

The *exclusive OR problem* (XOR) was almost single-handedly responsible for what is now called the first *neural network winter*. An initial flurry of excitement over perceptrons rapidly faded after Minsky and Papert (1969) proved that perceptrons cannot solve this apparently simple problem.

The XOR problem can be summarised as four associations, where each association comprises an input vector $\mathbf{S}_t$ and a corresponding target value R_t^{targ}:

$$\mathbf{S}_1 = (-1, -1) \rightarrow R_1^{\text{targ}} = -1, \tag{2.16}$$

$$\mathbf{S}_2 = (+1, +1) \rightarrow R_2^{\text{targ}} = -1, \tag{2.17}$$

$$\mathbf{S}_3 = (-1, +1) \rightarrow R_3^{\text{targ}} = +1, \tag{2.18}$$

$$\mathbf{S}_4 = (+1, -1) \rightarrow R_4^{\text{targ}} = +1. \tag{2.19}$$

This set of associations defines the *exclusive OR function*, depicted as four two-pixel images in Figure 2.7. It is called exclusive OR because the output unit has a state of $+1$ only if exactly one of the two input units has a state of $+1$; otherwise the output unit has a state of -1. When depicted geometrically, the XOR problem consists of classifying any pair of input vectors that lie in opposite quadrants as C_1, and input vectors that lie in the other pair of opposite quadrants as C_2.

However, whereas the classification examples in the previous section could be solved using a neural network with two input units and one output unit, this neural network cannot solve the XOR problem. The reason for this can be understood in terms of the decision boundaries possible with a perceptron.

We know that the only decision boundaries possible with a perceptron are straight lines. But the data points representing the class C_1 lie in two opposite quadrants of the input space defined by S_1 and S_2, as shown in Figure 2.7. Similarly, the data points representing the class C_2 lie in the other two opposite quadrants of the input space. This means that there is no straight line that can separate the data points in C_1 from those in C_2; and this means that it is impossible for a perceptron to classify these points correctly.

Why Exclusive OR Matters. The fact that a perceptron cannot solve the XOR problem may seem trivial. Surely, the XOR problem is of purely academic interest? In abstract terms, the exclusive OR problem can be expressed as discriminating between two pairs of features: one pair includes X and not Y, whereas the other pair includes Y but not X. These features can be pixel values, as in the example above, or whole sets of connected pixels that define a visual feature. The point is that if two classes of

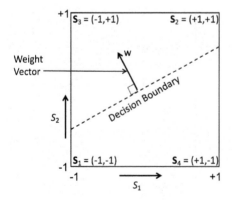

Figure 2.7. Geometric representation of the XOR problem, with each vector $\mathbf{S} = (S_1, S_2)$ in one corner. A weight vector $\mathbf{w}$ defines a decision boundary that splits the input space into two regions. If class $C_1 = \{\mathbf{S}_1, \mathbf{S}_2\}$ and class $C_2 = \{\mathbf{S}_3, \mathbf{S}_4\}$ then no straight-line decision boundary exists that can split the input space into C_1 and C_2. For example, the weight vector shown defines a decision boundary that fails to separate the two classes.

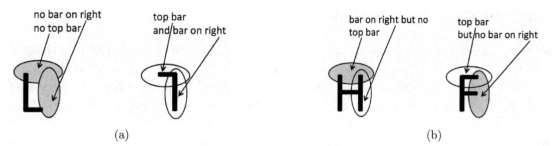

(a)   (b)

Figure 2.8. The exclusive OR (XOR) problem depicted with two sets of images.
(a) The letters L and ⌐ define the class $C_1 = \{$L, ⌐$\}$. (b) The letters F and H make up the class $C_2 = \{$F, H$\}$. Crucially, the two pixels at the arrow tips in (a) have values $\mathbf{S}_1 = (-1, -1)$ for L and $\mathbf{S}_2 = (+1, +1)$ for ⌐, whereas the same pixels in (b) have values $\mathbf{S}_3 = (-1, +1)$ for H and $\mathbf{S}_4 = (+1, -1)$ for F; so these two pixels alone represent an XOR problem, which implies that C_1 and C_2 are not linearly separable.

objects are classified as different because one class has a feature X and not Y, and the other class has a feature Y and not X, then these classes represent an XOR problem.

For example, the letter H has a vertical bar on the right but no horizontal bar at the top, whereas F has a horizontal top bar but no vertical right bar, as shown in Figure 2.8. In contrast, the letter L has no top bar and no bar on the right, whereas an inverted L (⌐) has both a top bar and a bar on the right. If we define the two classes

$$C_1 = \{\text{L, ⌐}\}, \qquad C_2 = \{\text{F, H}\}, \qquad (2.20)$$

then discriminating between these classes represents an XOR problem.

This classification task is not a fanciful vague analogy of the XOR problem, but an exact example of it. Consider two pixels located roughly at the arrow tips in Figure 2.8a. In an image of an L these two pixels are both white, $(-1, -1)$, whereas in the image of an ⌐ those same pixels are both black, $(+1, +1)$. Similarly, these two pixels are white and black, $(-1, +1)$, in the image of an H, but they are black and white, $(+1, -1)$, in the image of an F, as shown in Figure 2.8b.

For the two-pixel image considered above, a linear network defines a line through the 2D input space. Similarly, if a letter resides in an image of, say, $8 \times 8 = 64$ pixels, this defines a 64-dimensional space, and each image is represented by a point in this space. Additionally, just as a line separates the space of two-pixel images into two regions, so a 63-dimensional (hyper)plane separates images in the 64-dimensional space into two regions. We already know that if the image consisted only of the two pixels at the arrow tips in Figure 2.8a or Figure 2.8b, then a linear network could not divide the 2D space so that one region contains the two-pixel images $(-1, -1)$ and $(+1, +1)$ and the other contains the two-pixel images $(-1, +1)$ and $(+1, -1)$. It follows that a linear network cannot separate the 64-dimensional space into two regions such that one region contains the class $C_1 = \{$L, ⌐$\}$ and other contains $C_2 = \{$F, H$\}$. In other words, there is no linear decision boundary that can separate the class C_1 from C_2. Therefore, as a matter of principle, a perceptron cannot classify L and ⌐ into class C_1 and F and H into class C_2.

In summary, the ups and downs of the perceptron represent a microcosm of neural network history. Initially there is much excitement, as the neural network is compared to the human brain. Then there is disappointment, as the practical limitations of the neural network are gradually realised. Then, after a neural network winter lasting several years, a new kind of neural network is invented, accompanied by much excitement as the neural network is once again compared to the human brain. It is as hard to deny the truth of this pattern as it is to deny that, eventually, there will come a time when there will be no more winters.

2.6. List of Mathematical Symbols

origin points the set of units in the retina, called S-points, that are connected to a particular A-unit. For each A-unit, the origin points are clustered around a single central location on the retina, such that the density of connections falls as an exponential function of distance from that central location. The origin points correspond to a *receptive field* in modern terminology. Note that if the projection area A_I is not present then A-unit refers to a unit in the association area A_II.

projection area A_I the set of units, called A-units, to which retinal units (S-points) transmit impulses; A_I may be omitted from the network.

association area A_II the set of association units to which the units in A_I connect to. Connections between A_I and A_II are random. Up to A_II, all connections are forward only.

response units units $R_1, \ldots, R_n$ to which units in the association area transmit their outputs. Each response unit has excitatory feedback connections to the units in A_II that are connected to it (its source-set) and inhibitory connections to the complement of its source-set.

source-set the set of association units, or A-units, connected to a particular response unit.

N_A	number of A-units.
N_R	number of response units.
N_S	number of retinal S-points.
P_a	expected proportion of A-units activated by a retinal stimulus.
P_c	conditional probability that an A-unit responds to a stimulus S_2 given that it responds to the stimulus S_1.
P_g	probability of a correct generalisation response given two alternative responses; a correct response to a stimulus in a given class where that stimulus has not been seen before.
P_G	probability of a correct generalisation response given all possible alternative responses; a correct response to a stimulus in a given class where that stimulus has not been seen before.
P_r	probability of a correct response given two alternative responses.
P_R	probability of a correct response given all possible responses.
R	state of a binary response unit; also the proportion of S-points activated by the stimulus.
S	state of a binary retinal unit (S-point).
V	state of an A-unit.
x	number of excitatory connections per A-unit.
y	number of inhibitory connections to other A-units per A-unit.

α **system** learning rule whereby an active cell simply increments its value for every impulse and holds this gain indefinitely.

β **system** learning rule whereby each source-set is allowed a certain constant rate of gain, with the increments being apportioned among the cells of the source-set in proportion to their activity.

γ **system** learning rule whereby active cells increase their value at the expense of the inactive cells of their source-set, so that the total value of a source-set remains constant.

θ fixed threshold of an A-unit.

Research Paper:
The Perceptron: A Probabilistic Model for Information Storage and Organization in the Brain

Reference: Rosenblatt, F. (1958). The perceptron: a probabilistic model for information storage and organization in the brain. *Psychological Review*, 65(6):386–408. https://github.com/jgvfwstone/TheAIPapers/blob/main/perceptronPaperRosenblatt_1958.pdf This paper is in the public domain.

Psychological Review
Vol. 65, No. 6, 1958

THE PERCEPTRON: A PROBABILISTIC MODEL FOR INFORMATION STORAGE AND ORGANIZATION IN THE BRAIN [1]

F. ROSENBLATT

Cornell Aeronautical Laboratory

If we are eventually to understand the capability of higher organisms for perceptual recognition, generalization, recall, and thinking, we must first have answers to three fundamental questions:

1. How is information about the physical world sensed, or detected, by the biological system?
2. In what form is information stored, or remembered?
3. How does information contained in storage, or in memory, influence recognition and behavior?

The first of these questions is in the province of sensory physiology, and is the only one for which appreciable understanding has been achieved. This article will be concerned primarily with the second and third questions, which are still subject to a vast amount of speculation, and where the few relevant facts currently supplied by neurophysiology have not yet been integrated into an acceptable theory.

With regard to the second question, two alternative positions have been maintained. The first suggests that storage of sensory information is in the form of coded representations or images, with some sort of one-to-one mapping between the sensory stimulus and the stored pattern. According to this hypothesis, if one understood the code or "wiring diagram" of the nervous system, one should, in principle, be able to discover exactly what an organism remembers by reconstructing the original sensory patterns from the "memory traces" which they have left, much as we might develop a photographic negative, or translate the pattern of electrical charges in the "memory" of a digital computer. This hypothesis is appealing in its simplicity and ready intelligibility, and a large family of theoretical brain models has been developed around the idea of a coded, representational memory (**2, 3, 9, 14**). The alternative approach, which stems from the tradition of British empiricism, hazards the guess that the images of stimuli may never really be recorded at all, and that the central nervous system simply acts as an intricate switching network, where retention takes the form of new connections, or pathways, between centers of activity. In many of the more recent developments of this position (Hebb's "cell assembly," and Hull's "cortical anticipatory goal response," for example) the "responses" which are associated to stimuli may be entirely contained within the CNS itself. In this case the response represents an "idea" rather than an action. The important feature of this approach is that there is never any simple mapping of the stimulus into memory, according to some code which would permit its later reconstruction. Whatever in-

[1] The development of this theory has been carried out at the Cornell Aeronautical Laboratory, Inc., under the sponsorship of the Office of Naval Research, Contract Nonr-2381(00). This article is primarily an adaptation of material reported in Ref. 15, which constitutes the first full report on the program.

formation is retained must somehow be stored as a *preference for a particular response;* i.e., the information is contained in *connections* or *associations* rather than topographic representations. (The term *response*, for the remainder of this presentation, should be understood to mean any distinguishable state of the organism, which may or may not involve externally detectable muscular activity. The activation of some nucleus of cells in the central nervous system, for example, can constitute a response, according to this definition.)

Corresponding to these two positions on the method of information retention, there exist two hypotheses with regard to the third question, the manner in which stored information exerts its influence on current activity. The "coded memory theorists" are forced to conclude that recognition of any stimulus involves the matching or systematic comparison of the contents of storage with incoming sensory patterns, in order to determine whether the current stimulus has been seen before, and to determine the appropriate response from the organism. The theorists in the empiricist tradition, on the other hand, have essentially combined the answer to the third question with their answer to the second: since the stored information takes the form of new connections, or transmission channels in the nervous system (or the creation of conditions which are functionally equivalent to new connections), it follows that the new stimuli will make use of these new pathways which have been created, automatically activating the appropriate response without requiring any separate process for their recognition or identification.

The theory to be presented here takes the empiricist, or "connectionist" position with regard to these questions. The theory has been developed for a hypothetical nervous system, or machine, called a *perceptron*. The perceptron is designed to illustrate some of the fundamental properties of intelligent systems in general, without becoming too deeply enmeshed in the special, and frequently unknown, conditions which hold for particular biological organisms. The analogy between the perceptron and biological systems should be readily apparent to the reader.

During the last few decades, the development of symbolic logic, digital computers, and switching theory has impressed many theorists with the functional similarity between a neuron and the simple on-off units of which computers are constructed, and has provided the analytical methods necessary for representing highly complex logical functions in terms of such elements. The result has been a profusion of brain models which amount simply to logical contrivances for performing particular algorithms (representing "recall," stimulus comparison, transformation, and various kinds of analysis) in response to sequences of stimuli—e.g., Rashevsky (14), McCulloch (10), McCulloch & Pitts (11), Culbertson (2), Kleene (8), and Minsky (13). A relatively small number of theorists, like Ashby (1) and von Neumann (17, 18), have been concerned with the problems of how an imperfect neural network, containing many random connections, can be made to perform reliably those functions which might be represented by idealized wiring diagrams. Unfortunately, the language of symbolic logic and Boolean algebra is less well suited for such investigations. The need for a suitable language for the mathematical analysis of events in systems where only the gross organization can be characterized, and the precise structure is unknown, has led the author to formulate the current model in terms of probability theory rather than symbolic logic.

The theorists referred to above were chiefly concerned with the question of how such functions as perception and recall might be achieved by a deterministic physical system of any sort, rather than how this is actually done by the brain. The models which have been produced all fail in some important respects (absence of equipotentiality, lack of neuroeconomy, excessive specificity of connections and synchronization requirements, unrealistic specificity of stimuli sufficient for cell firing, postulation of variables or functional features with no known neurological correlates, etc.) to correspond to a biological system. The proponents of this line of approach have maintained that, once it has been shown how a physical system of any variety might be made to perceive and recognize stimuli, or perform other brainlike functions, it would require only a refinement or modification of existing principles to understand the working of a more realistic nervous system, and to eliminate the shortcomings mentioned above. The writer takes the position, on the other hand, that these shortcomings are such that a mere refinement or improvement of the principles already suggested can never account for biological intelligence; a *difference in principle* is clearly indicated. The theory of statistical separability (Cf. 15), which is to be summarized here, appears to offer a solution in principle to all of these difficulties.

Those theorists—Hebb (7), Milner (12), Eccles (4), Hayek (6)—who have been more directly concerned with the biological nervous system and its activity in a natural environment, rather than with formally anal-ogous machines, have generally been less exact in their formulations and far from rigorous in their analysis, so that it is frequently hard to assess whether or not the systems that they describe could actually work in a realistic nervous system, and what the necessary and sufficient conditions might be. Here again, the lack of an analytic language comparable in proficiency to the Boolean algebra of the network analysts has been one of the main obstacles. The contributions of this group should perhaps be considered as suggestions of what to look for and investigate, rather than as finished theoretical systems in their own right. Seen from this viewpoint, the most suggestive work, from the standpoint of the following theory, is that of Hebb and Hayek.

The position, elaborated by Hebb (7), Hayek (6), Uttley (16), and Ashby (1), in particular, upon which the theory of the perceptron is based, can be summarized by the following assumptions:

1. The physical connections of the nervous system which are involved in learning and recognition are not identical from one organism to another. At birth, the construction of the most important networks is largely random, subject to a minimum number of genetic constraints.

2. The original system of connected cells is capable of a certain amount of plasticity; after a period of neural activity, the probability that a stimulus applied to one set of cells will cause a response in some other set is likely to change, due to some relatively long-lasting changes in the neurons themselves.

3. Through exposure to a large sample of stimuli, those which are most "similar" (in some sense which must be defined in terms of the particular physical system) will tend

19

to form pathways to the same sets of responding cells. Those which are markedly "dissimilar" will tend to develop connections to different sets of responding cells.

4. The application of positive and/or negative reinforcement (or stimuli which serve this function) may facilitate or hinder whatever formation of connections is currently in progress.

5. *Similarity*, in such a system, is represented at some level of the nervous system by a tendency of similar stimuli to activate the same sets of cells. Similarity is not a necessary attribute of particular formal or geometrical classes of stimuli, but depends on the physical organization of the perceiving system, an organization which evolves through interaction with a given environment. The structure of the system, as well as the ecology of the stimulus-environment, will affect, and will largely determine, the classes of "things" into which the perceptual world is divided.

The Organization of a Perceptron

The organization of a typical photo-perceptron (a perceptron responding to optical patterns as stimuli) is shown in Fig. 1. The rules of its organization are as follows:

1. Stimuli impinge on a retina of sensory units (S-points), which are assumed to respond on an all-or-nothing basis, in some models, or with a pulse amplitude or frequency proportional to the stimulus intensity, in other models. In the models considered here, an all-or-nothing response will be assumed.

2. Impulses are transmitted to a set of association cells (A-units) in a "projection area" (A_I). This projection area may be omitted in some models, where the retina is connected directly to the association area (A_{II}).

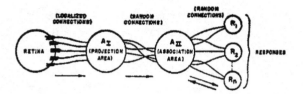

Fig. 1. Organization of a perceptron.

The cells in the projection area each receive a number of connections from the sensory points. The set of S-points transmitting impulses to a particular A-unit will be called the *origin points* of that A-unit. These origin points may be either *excitatory* or *inhibitory* in their effect on the A-unit. If the algebraic sum of excitatory and inhibitory impulse intensities is equal to or greater than the threshold (θ) of the A-unit, then the A-unit fires, again on an all-or-nothing basis (or, in some models, which will not be considered here, with a frequency which depends on the net value of the impulses received). The origin points of the A-units in the projection area tend to be clustered or focalized, about some central point, corresponding to each A-unit. The number of origin points falls off exponentially as the retinal distance from the central point for the A-unit in question increases. (Such a distribution seems to be supported by physiological evidence, and serves an important functional purpose in contour detection.)

3. Between the projection area and the association area (A_{II}), connections are assumed to be random. That is, each A-unit in the A_{II} set receives some number of fibers from origin points in the A_I set, but these origin points are scattered at random throughout the projection area. Apart from their connection distribution, the A_{II} units are identical with the A_I units, and respond under similar conditions.

4. The "responses," R_1, R_2, . . . , R_n are cells (or sets of cells) which

respond in much the same fashion as the A-units. Each response has a typically large number of origin points located at random in the A_{II} set. The set of A-units transmitting impulses to a particular response will be called the source-set for that response. (The source-set of a response is identical to its set of origin points in the A-system.) The arrows in Fig. 1 indicate the direction of transmission through the network. Note that up to A_{II} all connections are forward, and there is no feedback. When we come to the last set of connections, between A_{II} and the R-units, connections are established in both directions. The rule governing feedback connections, in most models of the perceptron, can be either of the following alternatives:

(a) Each response has excitatory feedback connections to the cells in its own source-set, or

(b) Each response has inhibitory feedback connections to the complement of its own source-set (i.e., it tends to prohibit activity in any association cells which do not transmit to it).

The first of these rules seems more plausible anatomically, since the R-units might be located in the same cortical area as their respective source-sets, making mutual excitation between the R-units and the A-units of the appropriate source-set highly probable. The alternative rule (b) leads to a more readily analyzed system, however, and will therefore be assumed for most of the systems to be evaluated here.

Figure 2 shows the organization of a simplified perceptron, which affords a convenient entry into the theory of statistical separability. After the theory has been developed for this simplified model, we will be in a better position to discuss the advantages of the system in Fig. 1. The feedback connections shown in Fig. 2 are inhibitory, and go to the complement of the source-set for the response from which they originate; consequently, this system is organized according to Rule b, above. The system shown here has only three stages, the first association stage having been eliminated. Each A-unit has a set of randomly located origin points in the retina. Such a system will form similarity concepts on the basis of *coincident areas* of stimuli, rather than by the similarity of contours or outlines. While such a system is at a disadvantage in many discrimination experiments, its capability is still quite impressive, as will be demonstrated presently. The system shown in Fig. 2 has only two responses, but there is clearly no limit on the number that might be included.

The responses in a system organized in this fashion are mutually exclusive. If R_1 occurs, it will tend to inhibit R_2, and will also inhibit the source-set for R_2. Likewise, if R_2 should occur, it will tend to inhibit R_1. If the total impulse received from all the A-units in one source-set is stronger or more frequent than the impulse received by the alternative (antagonistic) response, then the first response will

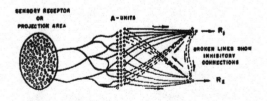

FIG. 2A. Schematic representation of connections in a simple perceptron.

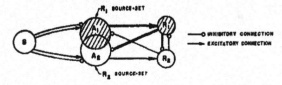

FIG. 2B. Venn diagram of the same perceptron (shading shows active sets for R_1 response).

tend to gain an advantage over the other, and will be the one which occurs. If such a system is to be capable of learning, then it must be possible to modify the A-units or their connections in such a way that stimuli of one class will tend to evoke a stronger impulse in the R_1 source-set than in the R_2 source-set, while stimuli of another (dissimilar) class will tend to evoke a stronger impulse in the R_2 source-set than in the R_1 source-set.

It will be assumed that the impulses delivered by each A-unit can be characterized by a value, V, which may be an amplitude, frequency, latency, or probability of completing transmission. If an A-unit has a high value, then all of its output impulses are considered to be more effective, more potent, or more likely to arrive at their endbulbs than impulses from an A-unit with a lower value. The value of an A-unit is considered to be a fairly stable characteristic, probably depending on the metabolic condition of the cell and the cell membrane, but it is not absolutely constant. It is assumed that, in general, periods of activity tend to increase a cell's value, while the value may decay (in some models) with inactivity. The most interesting models are those in which cells are assumed to compete for metabolic materials, the more active cells gaining at the expense of the less active cells. In such a system, if there is no activity, all cells will tend to remain in a relatively constant condition, and (regardless of activity) the net value of the system, taken in

TABLE 1

COMPARISON OF LOGICAL CHARACTERISTICS OF α, β, AND γ SYSTEMS

	α-System (Uncompensated Gain System)	β-System (Constant Feed System)	γ-System (Parasitic Gain System)
Total value-gain of source set per reinforcement	N_{a_r}	K	0
ΔV for A-units active for 1 unit of time	$+1$	K/N_{a_r}	$+1$
ΔV for inactive A-units outside of dominant set	0	K/N_{A_r}	0
ΔV for inactive A-units of dominant set	0	0	$\dfrac{-N_{a_r}}{N_{A_r}-N_{a_r}}$
Mean value of A-system	Increases with number of reinforcements	Increases with time	Constant
Difference between mean values of source-sets	Proportional to differences of reinforcement frequency $(n_{s_{r1}}-n_{s_{r2}})$	0	0

Note: In the β and γ systems, the total value-change for any A-unit will be the sum of the ΔV's for all source-sets of which it is a member.

N_{a_r} = Number of active units in source-set
N_{A_r} = Total number of units in source-set
$n_{s_{r_j}}$ = Number of stimuli associated to response r_j
K = Arbitrary constant

its entirety, will remain constant at all times. Three types of systems, which differ in their value dynamics, have been investigated quantitatively. Their principal logical features are compared in Table 1. In the alpha system, an active cell simply gains an increment of value for every impulse, and holds this gain indefinitely. In the beta system, each source-set is allowed a certain constant rate of gain, the increments being apportioned among the cells of the source-set in proportion to their activity. In the gamma system, active cells gain in value at the expense of the inactive cells of their source-set, so that the total value of a source-set is always constant.

For purposes of analysis, it is convenient to distinguish two phases in the response of the system to a stimulus (Fig. 3). In the *predominant phase*, some proportion of A-units (represented by solid dots in the figure) responds to the stimulus, but the R-units are still inactive. This phase is transient, and quickly gives way to the *postdominant phase*, in which one of the responses becomes active, inhibiting activity in the complement of its own source-set, and thus preventing the occurrence of any alternative response. The response which happens to become dominant is initially random, but if the A-units are reinforced (i.e., if the active units are allowed to gain in value), then when the same stimulus is presented again at a later time, the same response will have a stronger tendency to recur, and learning can be said to have taken place.

Analysis of the Predominant Phase

The perceptrons considered here will always assume a fixed threshold, θ, for the activation of the A-units. Such a system will be called a *fixed-threshold model*, in contrast to a *continuous transducer model*, where the response of the A-unit is some continuous function of the impinging stimulus energy.

In order to predict the learning curves of a fixed-threshold perceptron, two variables have been found to be of primary importance. They are defined as follows:

P_a = the expected proportion of A-units activated by a stimulus of a given size,

P_c = the conditional probability that an A-unit which responds to a given stimulus, S_1, will also respond to another given stimulus, S_2.

It can be shown (Rosenblatt, **15**) that as the size of the retina is increased, the number of S-points (N_s) quickly ceases to be a significant parameter, and the values of P_a and P_c approach the value that they would have for a retina with infinitely many points. For a large retina, therefore, the equations are as follows:

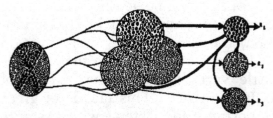

FIG. 3A. Predominant phase. Inhibitory connections are not shown. Solid black units are active.

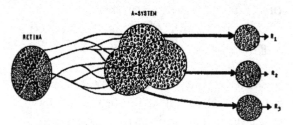

FIG. 3B. Postdominant phase. Dominant subset suppresses rival sets. Inhibitory connections shown only for R₁.

FIG. 3. Phases of response to a stimulus.

$$P_a = \sum_{e=\theta}^{x} \sum_{i=\theta}^{min \atop (y,\,e-\theta)} P(e,i) \qquad (1)$$

23

where

$$P(e,i) = \binom{x}{e} R^e (1 - R)^{x-e}$$
$$\times \binom{y}{i} R^i (1 - R)^{y-i}$$

and

R = proportion of S-points activated by the stimulus

x = number of excitatory connections to each A-unit

y = number of inhibitory connections to each A-unit

θ = threshold of A-units.

(The quantities e and i are the excitatory and inhibitory components of the excitation received by the A-unit from the stimulus. If the algebraic sum $\alpha = e + i$ is equal to or greater than θ, the A-unit is assumed to respond.)

$$P_c = \frac{1}{P_a} \sum_{e=\theta}^{x} \sum_{i=e-\theta}^{y} \sum_{l_e=0}^{e} \sum_{l_i=0}^{i}$$
$$\sum_{g_e=0}^{x-e} \sum_{g_i=0}^{y-i} P(e,i,l_e,l_i,g_e,g_i) \quad (2)$$

$$(e - i - l_e + l_i + g_e - g_i \geq \theta)$$

where

$P(e,i,l_e,l_i,g_e,g_i)$

$$= \binom{x}{e} R^e (1 - R)^{x-e}$$
$$\times \binom{y}{i} R^i (1 - R)^{y-i}$$
$$\times \binom{e}{l_e} L^{l_e} (1 - L)^{e-l_e}$$
$$\times \binom{i}{l_i} L^{l_i} (1 - L)^{i-l_i}$$
$$\times \binom{x - e}{g_e} G^{g_e} (1 - G)^{x-e-g_e}$$
$$\times \binom{y - i}{g_i} G^{g_i} (1 - G)^{y-i-g_i}$$

and

L = proportion of the S-points illuminated by the first stimulus, S_1, which are not illuminated by S_2

G = proportion of the residual S-set (left over from the first stimulus) which is included in the second stimulus (S_2).

The quantities R, L, and G specify the two stimuli and their retinal overlap. l_e and l_i are, respectively, the numbers of excitatory and inhibitory origin points "lost" by the A-unit when stimulus S_1 is replaced by S_2; g_e and g_i are the numbers of excitatory and inhibitory origin points "gained" when stimulus S_1 is replaced by S_2. The summations in Equation 2 are between the limits indicated, subject to the side condition $e - i - l_e + l_i + g_e - g_i \geq \theta$.

Some of the most important characteristics of P_a are illustrated in Fig. 4, which shows P_a as a function of the retinal area illuminated (R). Note that P_a can be reduced in magnitude by either increasing the threshold, θ, or by increasing the proportion of inhibitory connections (y). A comparison of Fig. 4b and 4c shows that if the excitation is about equal to the inhibition, the curves for P_a as a function of R are flattened out, so that there is little variation in P_a for stimuli of different sizes. This fact is of great importance for systems which require P_a to be close to an optimum value in order to perform properly.

The behavior of P_c is illustrated in Fig. 5 and 6. The curves in Fig. 5 can be compared with those for P_a in Fig. 4. Note that as the threshold is increased, there is an even sharper reduction in the value of P_c than was the case with P_a. P_c also decreases as the proportion of inhibitory connections increases, as does P_a. Fig. 5, which is

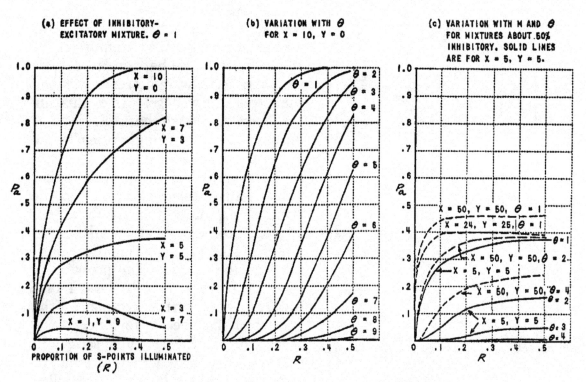

FIG. 4. P_a as function of retinal area illuminated.

calculated for nonoverlapping stimuli, illustrates the fact that P_c remains greater than zero even when the stimuli are completely disjunct, and illuminate no retinal points in common. In Fig. 6, the effect of varying amounts of overlap between the stimuli is shown. In all cases, the value of P_c goes to unity as the stimuli approach perfect identity. For smaller stimuli (broken line curves), the value of P_c is lower than for large stimuli. Similarly, the value is less for high thresholds than for low thresholds. The minimum value of P_c will be equal to

$$P_{cmin} = (1 - L)^x(1 - G)^y. \quad (3)$$

In Fig. 6, P_{cmin} corresponds to the curve for $\theta = 10$. Note that under these conditions the probability that the A-unit responds to both stimuli (P_c) is practically zero, except for stimuli which are quite close to identity. This condition can be of considerable help in discrimination learning.

MATHEMATICAL ANALYSIS OF LEARNING IN THE PERCEPTRON

The response of the perceptron in the predominant phase, where some fraction of the A-units (scattered throughout the system) responds to the stimulus, quickly gives way to the postdominant response, in which activity is limited to a single source-set, the other sets being suppressed. Two possible systems have been studied for the determination of the "dominant" response, in the postdominant phase. In one (the mean-discriminating system, or μ-system), the response whose inputs have the greatest mean value responds first, gaining a slight advantage over the others, so that it quickly becomes dominant. In the second case (the sum-discriminating system, or Σ-system), the response whose inputs have the greatest net value gains an advantage. In most cases, systems which respond to mean values have an advantage over systems which respond to sums, since the means are

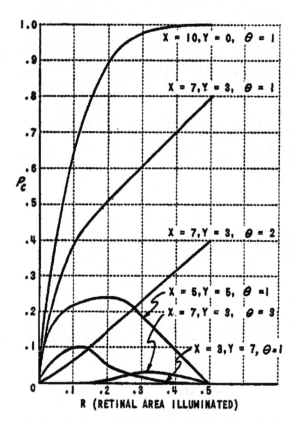

FIG. 5. P_c as a function of R, for nonoverlapping stimuli.

less influenced by random variations in P_a from one source-set to another. In the case of the γ-system (see Table 1), however, the performance of the μ-system and Σ-system become identical.

We have indicated that the perceptron is expected to learn, or to form associations, as a result of the changes in value that occur as a result of the activity of the association cells. In evaluating this learning, one of two types of hypothetical experiments can be considered. In the first case, the perceptron is exposed to some series of stimulus patterns (which might be presented in random positions on the retina) and is "forced" to give the desired response in each case. (This forcing of responses is assumed to be a prerogative of the experimenter. In experiments intended to evaluate trial-and-error learning, with more sophisticated perceptrons, the experimenter does not force the system to

respond in the desired fashion, but merely applies positive reinforcement when the response happens to be correct, and negative reinforcement when the response is wrong.) In evaluating the learning which has taken place during this "learning series," the perceptron is assumed to be "frozen" in its current condition, no further value changes being allowed, and the same series of stimuli is presented again in precisely the same fashion, so that the stimuli fall on identical positions on the retina. The probability that the perceptron will show a bias towards the "correct" response (the one which has been previously reinforced during the learning series) in preference to any given alternative response is called P_r, the probability of correct choice of response between two alternatives.

In the second type of experiment, a learning series is presented exactly as before, but instead of evaluating the perceptron's performance using the same series of stimuli which were shown before, a new series is presented, in which stimuli may be drawn from the same *classes* that were previ-

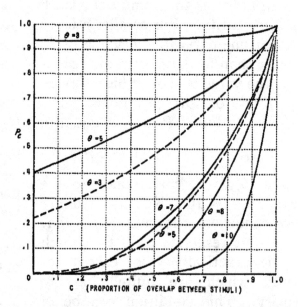

FIG. 6. P_c as a function of C. $X = 10$, $Y = 0$. Solid lines: R = .5; broken lines: R = .2.

ously experienced, but are not necessarily identical. This new test series is assumed to be composed of stimuli projected onto random retinal positions, which are chosen independently of the positions selected for the learning series. The stimuli of the test series may also differ in size or rotational position from the stimuli which were previously experienced. In this case, we are interested in the probability that the perceptron will give the correct response for the *class* of stimuli which is represented, regardless of whether the particular stimulus has been seen before or not. This probability is called P_g, the probability of correct generalization. As with P_r, P_g is actually the probability that a bias will be found in favor of the proper response rather than any one alternative; only one pair of responses at a time is considered, and the fact that the response bias is correct in one pair does not mean that there may not be other pairs in which the bias favors the wrong response. The probability that the correct response will be preferred over *all* alternatives is designated P_R or P_G.

In all cases investigated, a single general equation gives a close approximation to P_r and P_g, if the appropriate constants are substituted. This equation is of the form:

$$P = P(N_{a_r} > 0) \cdot \phi(Z) \qquad (4)$$

where

$$P(N_{a_r} > 0) = 1 - (1 - P_a)^{N_e}$$

$\phi(Z)$ = normal curve integral
     from $-\infty$ to Z

and

$$Z = \frac{c_1 n_{s_r} + c_2}{\sqrt{c_3 n_{s_r}^2 + c_4 n_{s_r}}}.$$

If R_1 is the "correct" response, and R_2 is the alternative response under consideration, Equation 4 is the probability that R_1 will be preferred over

R_2 after n_{s_r} stimuli have been shown for each of the two responses, during the learning period. N_e is the number of "effective" A-units in each source-set; that is, the number of A-units in either source-set which are not connected in common to both responses. Those units which are connected in common contribute equally to both sides of the value balance, and consequently do not affect the net bias towards one response or the other. N_{a_r} is the number of active units in a source-set, which respond to the test stimulus, S_t. $P(N_{a_r} > 0)$ is the probability that at least one of the N_e effective units in the source-set of the correct response (designated, by convention, as the R_1 response) will be activated by the test stimulus, S_t.

In the case of P_g, the constant c_2 is always equal to zero, the other three constants being the same as for P_r. The values of the four constants depend on the parameters of the physical nerve net (the perceptron) and also on the organization of the stimulus environment.

The simplest cases to analyze are those in which the perceptron is shown stimuli drawn from an "ideal environment," consisting of randomly placed points of illumination, where there is no attempt to classify stimuli according to intrinsic similarity. Thus, in a typical learning experiment, we might show the perceptron 1,000 stimuli made up of random collections of illuminated retinal points, and we might arbitrarily reinforce R_1 as the "correct" response for the first 500 of these, and R_2 for the remaining 500. This environment is "ideal" only in the sense that we speak of an ideal gas in physics; it is a convenient artifact for purposes of analysis, and does not lead to the best performance from the perceptron. In the ideal environment situation, the constant c_1 is always equal to zero, so that, in the case of P_g (where c_2 is also zero), the value of Z will be zero, and P_g can never be any better than the random expectation of 0.5. The evaluation of P_r for these conditions, however, throws some interesting light on the differences between the alpha, beta, and gamma systems (Table 1).

First consider the alpha system, which has the simplest dynamics of the three. In this system, whenever an A-unit is active for one unit of time, it gains one unit of value. We will assume an experiment, initially, in which N_{s_r} (the number of stimuli associated to each response) is constant for all responses. In this case, for the sum system,

$$\left.\begin{aligned}
c_1 &= 0 \\
c_2 &= (1 - P_a)N_e \\
c_3 &= 2P_a\omega \\
c_4 &\approx 0
\end{aligned}\right\} \quad (5)$$

where ω = the fraction of responses connected to each A-unit. If the source-sets are disjunct, $\omega = 1/N_R$, where N_R is the number of responses in the system. For the μ-system,

$$\left.\begin{aligned}
c_1 &= 0 \\
c_2 &= (1 - P_a)N_e \\
c_3 &= 0 \\
c_4 &= 2\omega
\end{aligned}\right\} \quad (6)$$

The reduction of c_3 to zero gives the μ-system a definite advantage over the Σ-system. Typical learning curves for these systems are compared in Fig. 7 and 8. Figure 9 shows the effect of variations in P_a upon the performance of the system.

If n_{s_r}, instead of being fixed, is treated as a random variable, so that the number of stimuli associated to each response is drawn separately from some distribution, then the per-

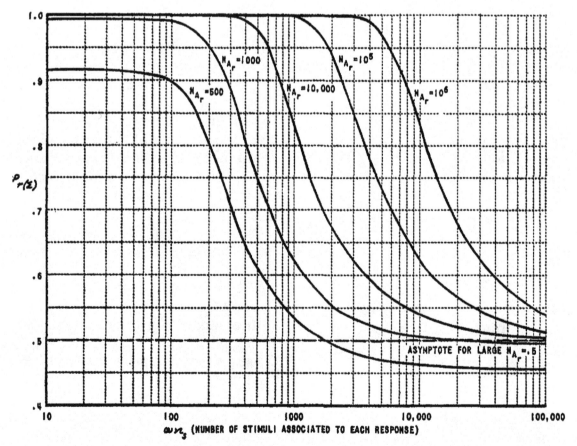

FIG. 7. $P_r(\Sigma)$ as function of ωn_s, for discrete subsets. ($\omega_e = 0$, $P_a = .005$. Ideal environment assumed.)

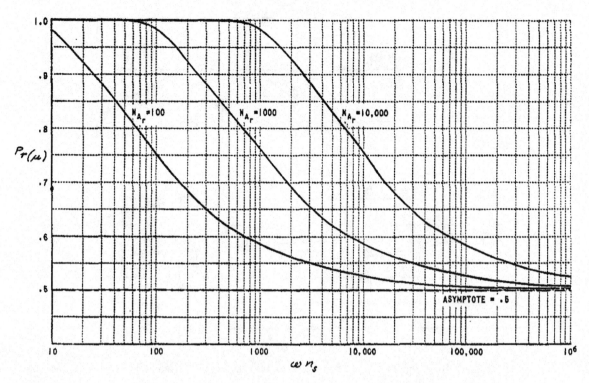

FIG. 8. $P_{r(\mu)}$ as function of ωn_s. (For $P_a = .07$, $\omega_c = 0$. Ideal environment assumed.)

formance of the α-system is considerably poorer than the above equations indicate. Under these conditions, the constants for the μ-system are

$$\left.\begin{aligned}
c_1 &= 0 \\
c_2 &= 1 - P_a \\
c_3 &= 2P_a^2 q^2 \left[\frac{(\omega N_R - 1)^2}{N_R - 2} + 1 \right] \\
c_4 &= \frac{2(1 - P_a)N_R}{(1 - \omega_c)N_A}
\end{aligned}\right\} \quad (7)$$

where

q = ratio of $\sigma_{n_{s_r}}$ to $\bar{n}_{s_r}$
N_R = number of responses in the system
N_A = number of A-units in the system
ω_c = proportion of A-units common to R_1 and R_2.

For this equation (and any others in which n_{s_r} is treated as a random variable), it is necessary to define n_{s_r} in Equation 4 as the expected value of this variable, over the set of all responses.

For the β-system, there is an even greater deficit in performance, due to the fact that the net value continues to grow regardless of what happens to the system. The large net values of the subsets activated by a stimulus tend to amplify small statistical differences, causing an unreliable performance. The constants in this case (again for the μ-system) are

$$\left.\begin{aligned}
c_1 &= 0 \\
c_2 &= (1 - P_a)N_e \\
c_3 &= 2(P_a N_e q\omega N_R^2)^2 \\
c_4 &= 2(1 - P_a)\omega N_R N_e
\end{aligned}\right\} \quad (8)$$

In both the alpha and beta systems, performance will be poorer for the sum-discriminating model than for the mean-discriminating case. In the gamma-system, however, it can be shown that $P_{r(\Sigma)} = P_{r(\mu)}$; i.e., it makes no difference in performance whether the Σ-system or μ-system is used. Moreover, the constants for the γ-system, with variable n_{s_r}, are identical to the constants for the alpha μ-sys-

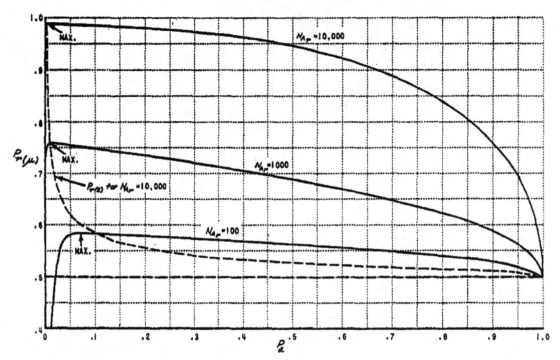

FIG. 9. $P_{r(\mu)}$ as function of P_a. (For $n_{s_r} = 1,000$, $\omega_o = 0$. Ideal environment assumed.)

tem, with n_{s_r} fixed (Equation 6). The performance of the three systems is compared in Fig. 10, which clearly demonstrates the advantage of the γ-system.

Let us now replace the "ideal en-

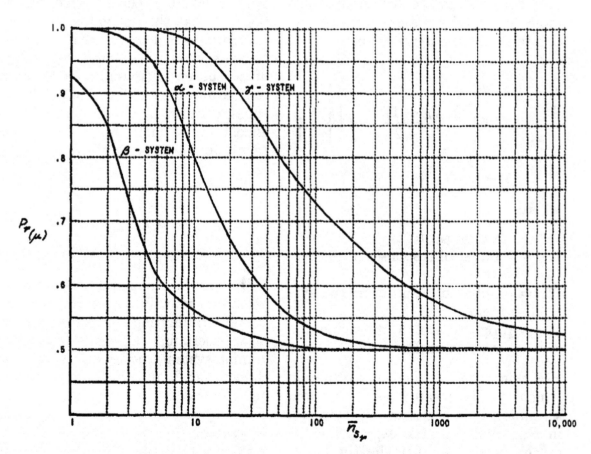

FIG. 10. Comparison of α, β, and γ systems, for variable n_{s_r} ($N_R = 100$, $\sigma_{n_{rs}} = .5\bar{n}_{s_i}$, $N_A = 10,000$, $P_a = .07$, $\omega = .2$).

vironment" assumptions with a model for a "differentiated environment," in which several distinguishable classes of stimuli are present (such as squares, circles, and triangles, or the letters of the alphabet). If we then design an experiment in which the stimuli associated to each response are drawn from a different class, then the learning curves of the perceptron are drastically altered. The most important difference is that the constant c_1 (the coefficient of n_{s_r} in the numerator of Z) is no longer equal to zero, so that Equation 4 now has a nonrandom asymptote. Moreover, in the form for P_g (the probability of correct generalization), where $c_2 = 0$, the quantity Z remains greater than zero, and P_g actually approaches the same asymptote as P_r. Thus the equation for the perceptron's performance after infinite experience with each class of stimuli is identical for P_r and P_g:

$$P_{r\infty} = P_{g\infty} = [1 - (1 - P_a)^{N_e}]$$
$$\times \phi\left(\frac{c_1}{\sqrt{c_3}}\right) \quad (9)$$

This means that *in the limit it makes no difference whether the perceptron has seen a particular test stimulus before or not; if the stimuli are drawn from a differentiated environment, the performance will be equally good in either case.*

In order to evaluate the performance of the system in a differentiated environment, it is necessary to define the quantity $P_{c\alpha\beta}$. This quantity is interpreted as the expected value of P_c between pairs of stimuli drawn at random from classes α and β. In particular, P_{c11} is the expected value of P_c between members of the same class, and P_{c12} is the expected value of P_c between an S_1 stimulus drawn from Class 1 and an S_2 stimulus drawn from Class 2. P_{c1x} is the expected value of P_c between members of Class 1 and

stimuli drawn at random from all other classes in the environment.

If $P_{c11} > P_a > P_{c12}$, the limiting performance of the perceptron ($P_{g\infty}$) will be better than chance, and learning of some response, R_1, as the proper "generalization response" for members of Class 1 should eventually occur. If the above inequality is not met, then improvement over chance performance may not occur, and the Class 2 response is likely to occur instead. It can be shown (15) that for most simple geometrical forms, which we ordinarily regard as "similar," the required inequality can be met, if the parameters of the system are properly chosen.

The equation for P_r, for the sum-discriminating version of an α-perceptron, in a differentiated environment where n_{s_r} is fixed for all responses, will have the following expressions for the four coefficients:

$$\left.\begin{aligned}
c_1 &= P_a N_e (P_{c11} - P_{c12}) \\
c_2 &= P_a N_e (1 - P_{c11}) \\
c_3 &= \sum_{r=1,2} P_a (1 - P_a) N_e \\
&\quad \times [P_{c^21r} + \sigma_s^2(P_{c1r}) \\
&\quad + \sigma_j^2(P_{c1r}) + (\omega N_R - 1)^2 \\
&\quad \times (P_{c1x} + \sigma_s^2(P_{c1x}) \\
&\quad + \sigma_j^2(P_{crx})) + 2(\omega N_R - 1) \\
&\quad (P_{c1r} P_{c1x})] + P_a^2 N_e^2 \\
&\quad \times [\sigma_s^2(P_{c1r}) + (\omega N_R - 1)^2 \\
&\quad \times \sigma_s^2(P_{c1x}) + 2(\omega N_R - 1)\epsilon] \\
c_4 &= \sum_{r=1,2} P_a N_e [P_{c1r} - P_{c1r}^2 \\
&\quad - \sigma_s^2(P_{c1r}) - \sigma_j^2(P_{c1r}) \\
&\quad + (\omega N_R - 1)(P_{c1x} - P_{c1x}^2 \\
&\quad \qquad - \sigma_j^2(P_{c1x}))]
\end{aligned}\right\} \quad (10)$$

where

$\sigma_s^2(P_{c1r})$ and $\sigma_s^2(P_{c1x})$ represent the variance of P_{c1r} and P_{c1x} measured over the set of possible test stimuli, S_t, and

$\sigma_j^2(P_{c1r})$ and $\sigma_j^2(P_{c1x})$ represent the variance of P_{c1r} and P_{c1x} measured over the set of all A-units, a_j.

ϵ = covariance of $P_{c1r}P_{c1x}$, which is assumed to be negligible.

The variances which appear in these expressions have not yielded, thus far, to a precise analysis, and can be treated as empirical variables to be determined for the classes of stimuli in question. If the sigma is set equal to half the expected value of the variable, in each case, a conservative estimate can be obtained. When the stimuli of a given class are all of the same shape, and uniformly distributed over the retina, the subscript s variances are equal to zero. $P_{g(\Sigma)}$ will be represented by the same set of coefficients, except for c_2, which is equal to zero, as usual.

For the mean-discriminating system, the coefficients are:

$$
\left.
\begin{aligned}
c_1 &= (P_{c11} - P_{c12}) \\
c_2 &= (1 - P_{c11}) \\
c_3 &= \sum_{r=1,2} \left[\frac{1}{P_a(N_e-1)} - \frac{1}{N_e-1} \right] \\
&\quad \times [\sigma_j^2(P_{c1r}) + (\omega N_R - 1)^2 \\
&\quad \times \sigma_j^2(P_{c1x})] + [\sigma_s^2(P_{c1r}) \\
&\quad + (\omega N_R - 1)^2 \sigma_s^2(P_{c1x})] \\
c_4 &= \sum_{r=1,2} \frac{1}{P_a N_e} [P_{c1r} - P_{c1r}^2 \\
&\quad - \sigma_s^2(P_{c1r}) - \sigma_j^2(P_{c1r}) \\
&\quad + (\omega N_R - 1)(P_{c1x} - P_{c1x}^2 \\
&\quad - \sigma_s^2(P_{c1x}) - \sigma_j^2(P_{c1x}))]
\end{aligned}
\right\} \quad (11)
$$

Some covariance terms, which are considered negligible, have been omitted here.

A set of typical learning curves for the differentiated environment model is shown in Fig. 11, for the mean-discriminating system. The parameters are based on measurements for a

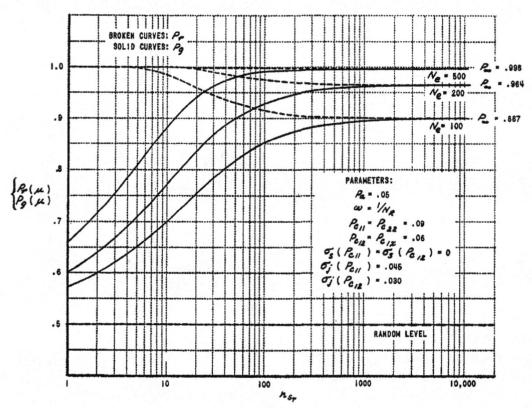

FIG. 11. P_r and P_g as function of n_{sr}. Parameters based on square-circle discrimination.

square-circle discrimination problem. Note that the curves for P_r and P_g both approach the same asymptotes, as predicted. The values of these asymptotes can be obtained by substituting the proper coefficients in Equation 9. As the number of association cells in the system increases, the asymptotic learning limit rapidly approaches unity, so that for a system of several thousand cells, the errors in performance should be negligible on a problem as simple as the one illustrated here.

As the number of responses in the system increases, the performance becomes progressively poorer, if every response is made mutually exclusive of all alternatives. One method of avoiding this deterioration (described in detail in Rosenblatt, 15) is through the binary coding of responses. In this case, instead of representing 100 different stimulus patterns by 100 distinct, mutually exclusive responses, a limited number of discriminating features is found, each of which can be independently recognized as being present or absent, and consequently can be represented by a single pair of mutually exclusive responses. Given an ideal set of binary characteristics (such as dark, light; tall, short; straight, curved; etc.), 100 stimulus classes could be distinguished by the proper configuration of only seven response pairs. In a further modification of the system, a single response is capable of denoting by its activity or inactivity the presence or absence of each binary characteristic. The efficiency of such coding depends on the number of independently recognizable "earmarks" that can be found to differentiate stimuli. If the stimulus can be identified only in its entirety and is not amenable to such analysis, then ultimately a separate binary response pair, or bit, is required to

denote the presence or absence of each stimulus class (e.g., "dog" or "not dog"), and nothing has been gained over a system where all responses are mutually exclusive.

BIVALENT SYSTEMS

In all of the systems analyzed up to this point, the increments of value gained by an active A-unit, as a result of reinforcement or experience, have always been *positive*, in the sense that an active unit has always *gained* in its power to activate the responses to which it is connected. In the gamma-system, it is true that some units lose value, but these are always the *inactive units*, the active ones gaining in proportion to their rate of activity. In a *bivalent system*, two types of reinforcement are possible (positive and negative), and an active unit may either gain or lose in value, depending on the momentary state of affairs in the system. If the positive and negative reinforcement can be controlled by the application of external stimuli, they become essentially equivalent to "reward" and "punishment," and can be used in this sense by the experimenter. Under these conditions, a perceptron appears to be capable of trial-and-error learning. A bivalent system need not necessarily involve the application of reward and punishment, however. If a binary-coded response system is so organized that there is a single response or response-pair to represent each "bit," or stimulus characteristic that is learned, with positive feedback to its own source-set if the response is "on," and negative feedback (in the sense that active A-units will lose rather than gain in value) if the response is "off," then the system is still bivalent in its characteristics. Such a bivalent system is particularly efficient in reducing some of the bias effects (preference for the wrong response due to greater size or frequency of its associated stimuli) which plague the alternative systems.

Several forms of bivalent systems have been considered (15, Chap. VII). The most efficient of these has the following logical characteristics.

If the system is under a state of positive reinforcement, then a positive ΔV is added to the values of all active A-units in the source-sets of "on" responses, while a negative ΔV is added to the active units in the source-sets of "off" responses. If the system is currently under negative reinforcement, then a negative ΔV is added to all active units in the source-set of an "on" response, and a positive ΔV is added to active units in an "off" source-set. If the source-sets are disjunct (which is essential for this system to work properly), the equation for a bivalent γ-system has the same coefficients as the monovalent α-system, for the μ-case (Equation 11).

The performance curves for this system are shown in Fig. 12, where the asymptotic generalization probability attainable by the system is plotted for the same stimulus parameters that were used in Fig. 11. This is the probability that all bits in an n-bit response pattern will be correct. Clearly, if a majority of correct responses is sufficient to identify a stimulus correctly, the performance will be better than these curves indicate.

In a form of bivalent system which utilizes more plausible biological assumptions, A-units may be either excitatory or inhibitory in their effect on connected responses. A positive ΔV in this system corresponds to the incrementing of an excitatory unit, while a negative ΔV corresponds to the incrementing of an inhibitory unit.

Such a system performs similarly to the one considered above, but can be shown to be less efficient.

Bivalent systems similar to those illustrated in Fig. 12 have been simulated in detail in a series of experiments with the IBM 704 computer at the Cornell Aeronautical Laboratory. The results have borne out the theory in all of its main predictions, and will be reported separately at a later time.

IMPROVED PERCEPTRONS AND SPONTANEOUS ORGANIZATION

The quantitative analysis of perceptron performance in the preceding sections has omitted any consideration of *time* as a stimulus dimension. A perceptron which has no capability for temporal pattern recognition is referred to as a "momentary stimulus perceptron." It can be shown (15) that the same principles of statistical separability will permit the perceptron to distinguish velocities, sound sequences, etc., provided the stimuli leave some temporarily persistent trace, such as an altered threshold,

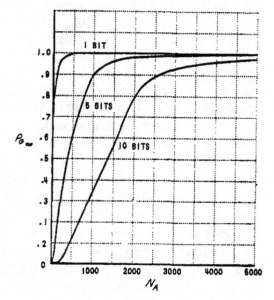

FIG. 12. P_{G_∞} for a bivalent binary system (same parameters as Fig. 11).

which causes the activity in the A-system at time t to depend to some degree on the activity at time $t - 1$.

It has also been assumed that the origin points of A-units are completely random. It can be shown that by a suitable organization of origin points, in which the spatial distribution is constrained (as in the projection area origins shown in Fig. 1), the A-units will become particularly sensitive to the location of contours, and performance will be improved.

In a recent development, which we hope to report in detail in the near future, it has been proven that if the values of the A-units are allowed to decay at a rate proportional to their magnitude, a striking new property emerges: the perceptron becomes capable of "spontaneous" concept formation. That is to say, if the system is exposed to a random series of stimuli from two "dissimilar" classes, and all of its responses are automatically reinforced without any regard to whether they are "right" or "wrong," the system will tend towards a stable terminal condition in which (for each binary response) the response will be "1" for members of one stimulus class, and "0" for members of the other class; i.e., the perceptron will spontaneously recognize the difference between the two classes. This phenomenon has been successfully demonstrated in simulation experiments, with the 704 computer.

A perceptron, even with a single logical level of A-units and response units, can be shown to have a number of interesting properties in the field of selective recall and selective attention. These properties generally depend on the intersection of the source sets for different responses, and are elsewhere discussed in detail (15). By combining audio and photo inputs, it is possible to associate sounds, or audi-

tory "names" to visual objects, and to get the perceptron to perform such selective responses as are designated by the command "Name the object on the left," or "Name the color of this stimulus."

The question may well be raised at this point of where the perceptron's capabilities actually stop. We have seen that the system described is sufficient for pattern recognition, associative learning, and such cognitive sets as are necessary for selective attention and selective recall. The system appears to be potentially capable of temporal pattern recognition, as well as spatial recognition, involving any sensory modality or combination of modalities. It can be shown that with proper reinforcement it will be capable of trial-and-error learning, and can learn to emit ordered sequences of responses, provided its own responses are fed back through sensory channels.

Does this mean that the perceptron is capable, without further modification in principle, of such higher order functions as are involved in human speech, communication, and thinking? Actually, the limit of the perceptron's capabilities seems to lie in the area of relative judgment, and the abstraction of relationships. In its "symbolic behavior," the perceptron shows some striking similarities to Goldstein's brain-damaged patients (5). Responses to definite, concrete stimuli can be learned, even when the proper response calls for the recognition of a number of simultaneous qualifying conditions (such as naming the color if the stimulus is on the left, the shape if it is on the right). As soon as the response calls for the recognition of a relationship between stimuli (such as "Name the object left of the square." or "Indicate the pattern that appeared before the circle."), however, the problem generally becomes excessively difficult for the perceptron. Statistical separability alone does not provide a sufficient basis for higher order abstraction. Some system, more advanced in principle than the perceptron, seems to be required at this point.

CONCLUSIONS AND EVALUATION

The main conclusions of the theoretical study of the perceptron can be summarized as follows:

1. In an environment of random stimuli, a system consisting of randomly connected units, subject to the parametric constraints discussed above, can learn to associate specific responses to specific stimuli. Even if many stimuli are associated to each response, they can still be recognized with a better-than-chance probability, although they may resemble one another closely and may activate many of the same sensory inputs to the system.

2. In such an "ideal environment," the probability of a correct response diminishes towards its original random level as the number of stimuli learned increases.

3. In such an environment, no basis for generalization exists.

4. In a "differentiated environment," where each response is associated to a distinct class of mutually correlated, or "similar" stimuli, the probability that a learned association of some specific stimulus will be correctly retained typically approaches a better-than-chance asymptote as the number of stimuli learned by the system increases. This asymptote can be made arbitrarily close to unity by increasing the number of association cells in the system.

5. In the differentiated environment, the probability that a stimulus *which has not been seen before* will be correctly recognized and associated to its appropriate class (the probability of correct generalization) approaches the same asymptote as the probability of a correct response to a previously reinforced stimulus. This asymptote will be better than chance if the inequality $P_{c12} < P_a < P_{c11}$ is met, for the stimulus classes in question.

6. The performance of the system can be improved by the use of a contour-sensitive projection area, and by the use of a binary response system, in which each response, or "bit," corresponds to some independent feature or attribute of the stimulus.

7. Trial-and-error learning is possible in bivalent reinforcement systems.

8. Temporal organizations of both stimulus patterns and responses can be learned by a system which uses only an extension of the original principles of statistical separability, without introducing any major complications in the organization of the system.

9. The memory of the perceptron is *distributed*, in the sense that any association may make use of a large proportion of the cells in the system, and the removal of a portion of the association system would not have an appreciable effect on the performance of any one discrimination or association, but would begin to show up as a general deficit in *all* learned associations.

10. Simple cognitive sets, selective recall, and spontaneous recognition of the classes present in a given environment are possible. The recognition of relationships in space and time, however, seems to represent a limit to the perceptron's ability to form cognitive abstractions.

Psychologists, and learning theorists in particular, may now ask: "What has the present theory accomplished, beyond what has already been done in the quantitative theories of Hull, Bush and Mosteller, etc., or physiological theories such as Hebb's?" The present theory is still too primitive, of course, to be considered as a full-fledged rival of existing theories of human learning. Nonetheless, as a first approximation, its chief accomplishment might be stated as follows:

For a given mode of organization (α, β, or γ; Σ or μ; monovalent or bivalent) the fundamental phenomena of *learning, perceptual discrimination, and generalization can be predicted entirely from six basic physical parameters*, namely:

x: the number of excitatory connections per A-unit,

y: the number of inhibitory connections per A-unit,

θ: the expected threshold of an A-unit,

ω: the proportion of R-units to which an A-unit is connected,

N_A: the number of A-units in the system, and

N_R: the number of R-units in the system.

N_s (the number of sensory units) becomes important if it is very small. It is assumed that the system begins with all units in a uniform state of value; otherwise the initial value distribution would also be required. *Each of the above parameters is a clearly defined physical variable, which is measurable in its own right, independently of the behavioral and perceptual phenomena which we are trying to predict.*

As a direct consequence of its foundation on physical variables, the present system goes far beyond existing learning and behavior theories in three main points: parsimony, verifiability, and explanatory power and generality. Let us consider each of these points in turn.

1. *Parsimony.* Essentially all of the basic variables and laws used in this system are already present in the structure of physical and biological science, so that we have found it necessary to postulate only one hypothetical variable (or construct) which we have called V, the "value" of an association cell; this is a variable which must conform to certain functional characteristics which can clearly be stated, and which is assumed to have a potentially measurable physical correlate.

2. *Verifiability.* Previous quantitative learning theories, apparently without exception, have had one important characteristic in common: they have all been based on measurements of *behavior*, in specified situations, using these measurements (after theoretical manipulation) to predict *behavior* in other situations. Such a procedure, in the last analysis, amounts to a process of curve fitting and extrapolation, in the hope that the constants which describe one set of curves will hold good for other curves in other situations. While such extrapolation is not necessarily circular, in the strict sense, it shares many of the logical difficulties of circularity, particularly when used as an "explanation" of behavior. Such extrapolation is difficult to justify in a new situation, and it has been shown that if the basic constants and parameters are to be derived anew for any situation in which they break down empirically (such as change from white rats to humans), then the basic "theory" is essentially irrefutable, just as any successful curve-fitting equation is irrefutable. It has, in fact, been widely conceded by psychologists that there is little point in trying to "disprove" any of the major learning theories in use today, since by extension, or a change in parameters, they

have all proved capable of adapting to any specific empirical data. This is epitomized in the increasingly common attitude that a choice of theoretical model is mostly a matter of personal aesthetic preference or prejudice, each scientist being entitled to a favorite model of his own. In considering this approach, one is reminded of a remark attributed to Kistiakowsky, that "given seven parameters, I could fit an elephant." This is clearly *not* the case with a system in which the independent variables, or parameters, can be measured *independently* of the predicted behavior. In such a system, it is not possible to "force" a fit to empirical data, if the parameters in current use should lead to improper results. In the current theory, a failure to fit a curve in a new situation would be a clear indication that either the theory or the empirical measurements are wrong. Consequently, if such a theory *does* hold up for repeated tests, we can be considerably more confident of its validity and of its generality than in the case of a theory which must be hand-tailored to meet each situation.

3. *Explanatory power and generality.* The present theory, being derived from basic physical variables, is not specific to any one organism or learning situation. It can be generalized in principle to cover any form of behavior in any system for which the physical parameters are known. A theory of learning, constructed on these foundations, should be considerably more powerful than any which has previously been proposed. It would not only tell us what behavior might occur in any known organism, but would permit the *synthesis* of behaving systems, to meet special requirements. Other learning theories tend to become increasingly qualitative as they are generalized.

Thus a set of equations describing the effects of reward on T-maze learning in a white rat reduces simply to a statement that rewarded behavior tends to occur with increasing probability, when we attempt to generalize it from any species and any situation. The theory which has been presented here loses none of its precision through generality.

The theory proposed by Donald Hebb (7) attempts to avoid these difficulties of behavior-based models by showing how psychological functioning might be derived from neurophysiological theory. In his attempt to achieve this, Hebb's philosophy of approach seems close to our own, and his work has been a source of inspiration for much of what has been proposed here. Hebb, however, has never actually achieved a model by which behavior (or any psychological data) can be *predicted* from the physiological system. His physiology is more a suggestion as to the *sort* of organic substrate which might underlie behavior, and an attempt to show the plausibility of a bridge between biophysics and psychology.

The present theory represents the first actual completion of such a bridge. Through the use of the equations in the preceding sections, it is possible to predict learning curves from neurological variables, and likewise, to predict neurological variables from learning curves. How well this bridge stands up to repeated crossings remains to be seen. In the meantime, the theory reported here clearly demonstrates the feasibility and fruitfulness of a quantitative statistical approach to the organization of cognitive systems. By the study of systems such as the perceptron, it is hoped that those fundamental laws of organization which are common to all information handling systems, machines and men included, may eventually be understood.

REFERENCES

1. ASHBY, W. R. *Design for a brain.* New York: Wiley, 1952.
2. CULBERTSON, J. T. *Consciousness and behavior.* Dubuque, Iowa: Wm. C. Brown, 1950.
3. CULBERTSON, J. T. Some uneconomical robots. In C. E. Shannon & J. McCarthy (Eds.), *Automata studies.* Princeton: Princeton Univer. Press, 1956. Pp. 99–116.
4. ECCLES, J. C. *The neurophysiological basis of mind.* Oxford: Clarendon, 1953.
5. GOLDSTEIN, K. *Human nature in the light of psychopathology.* Cambridge: Harvard Univer. Press, 1940.
6. HAYEK, F. A. *The sensory order.* Chicago: Univer. Chicago Press, 1952.
7. HEBB, D. O. *The organization of behavior.* New York: Wiley, 1949.
8. KLEENE, S. C. Representation of events in nerve nets and finite automata. In C. E. Shannon & J. McCarthy (Eds.), *Automata studies.* Princeton: Princeton Univer. Press, 1956. Pp. 3–41.
9. KÖHLER, W. Relational determination in perception. In L. A. Jeffress (Ed.), *Cerebral mechanisms in behavior.* New York: Wiley, 1951. Pp. 200–243.
10. MCCULLOCH, W. S. Why the mind is in the head. In L. A. Jeffress (Ed.), *Cerebral mechanisms in behavior.* New York: Wiley, 1951. Pp. 42–111.
11. MCCULLOCH, W. S., & PITTS, W. A logical calculus of the ideas immanent in nervous activity. *Bull. math. Biophysics*, 1943, **5**, 115–133.
12. MILNER, P. M. The cell assembly: Mark II. *Psychol. Rev.*, 1957, **64**, 242–252.
13. MINSKY, M. L. Some universal elements for finite automata. In C. E. Shannon & J. McCarthy (Eds.), *Automata studies.* Princeton: Princeton Univer. Press, 1956. Pp. 117–128.
14. RASHEVSKY, N. *Mathematical biophysics.* Chicago: Univer. Chicago Press, 1938.
15. ROSENBLATT, F. *The perceptron: A theory of statistical separability in cognitive systems.* Buffalo: Cornell Aeronautical Laboratory, Inc. Rep. No. VG-1196-G-1, 1958.
16. UTTLEY, A. M. Conditional probability machines and conditioned reflexes. In C. E. Shannon & J. McCarthy (Eds.), *Automata studies.* Princeton: Princeton Univer. Press, 1956. Pp. 253–275.
17. VON NEUMANN, J. The general and logical theory of automata. In L. A. Jeffress (Ed.), *Cerebral mechanisms in behavior.* New York: Wiley, 1951. Pp. 1–41.
18. VON NEUMANN, J. Probabilistic logics and the synthesis of reliable organisms from unreliable components. In C. E. Shannon & J. McCarthy (Eds.), *Automata studies.* Princeton: Princeton Univer. Press, 1956. Pp. 43–98.

(Received April 23, 1958)

Chapter 3

Hopfield Nets – 1982

Context

Hopfield nets are important because they harness the mathematical machinery of *statistical mechanics*, which enabled learning to be interpreted in terms of *energy functions*. Hopfield nets led directly to *Boltzmann machines* (Chapter 4), which represent an important stepping stone to modern AI systems. Today, Hopfield nets and Boltzmann machines are collectively classified as *energy-based networks*.

Hopfield nets were almost entirely responsible for ending a neural network winter. By identifying the states of neurons with the states of elements in a physical system, the whole mathematical framework of statistical mechanics, constructed piece by piece over the previous century, could be harnessed for learning in artificial neural networks. Hopfield's epiphany heralded a step-change in the perception of neural networks, which had been viewed with considerable scepticism by the more traditional members of the AI community. Probably because Hopfield nets were formulated within the ultra-respectable framework of statistical mechanics, neural networks attracted the attention of physicists and gradually gained acceptance amongst the more mathematically inclined AI researchers. And even though Hopfield nets turned out to be of limited practical use[†], there is little doubt that they laid the foundations upon which the modern era of neural networks was built.

Technical Summary

A Hopfield net consists of N binary units, or *neurons*, with *symmetric, bi-directional* connection weights between them, which means that each neuron is connected to all other neurons (Figure 3.1). Each neuron has a binary state $V^{s'}$, such that $V^{s'} = 0$ or $V^{s'} = 1$. The superscript s' indicates that this refers to the state of a neuron. In contrast, the state that is to be learned is represented by V^s, with superscript s.

The State Update Rule. The total input x_i to a neuron i depends only on the current state of each of the other $N - 1$ neurons:

$$x_i = \sum_{j=1}^{N} T_{ij} V_j^{s'}. \tag{3.1}$$

Here T_{ij} is the weight of the connection between neurons i and j. Note that there is no connection from a neuron to itself, so if $i = j$ then $T_{ij} = 0$. Each neuron has the same activation function as a unit in the perceptron, so the state of a neuron is given by Equation 1 in Hopfield's paper as

$$V_i^{s'} = f\left(\sum_{j=1}^{N} T_{ij} V_j^{s'}\right), \tag{3.2}$$

[†]But see Krotov (2023) for recent developments.

where f is a step function,

$$f(x_i) \;=\; \begin{cases} 0 & \text{if } x_i < 0, \\ 1 & \text{if } x_i > 0. \end{cases} \tag{3.3}$$

In order for the network to settle into a stable state, neuron states are updated asynchronously (i.e. one at a time) and in random order. Hopfield also introduced a threshold U_i for each neuron, but in Equation 3.2 this is set to $U_i = 0$, so we ignore it here.

3.1. Learning

For a network with N neurons, the state of the network can be represented as a vector of binary values

$$\mathbf{V}^{s'} \;=\; (V_1^{s'}, \dots, V_N^{s'}). \tag{3.4}$$

Suppose we wish the network to memorise a particular set of values, represented by the *training vector* of N binary values

$$\mathbf{V}^{s} \;=\; (V_1^{s}, \dots, V_N^{s}). \tag{3.5}$$

Stable States. If we can find a set of weights such that the network state $\mathbf{V}^{s'}$ matches the training vector $\mathbf{V}^{s}$ then $\mathbf{V}^{s'}$ will be a *stable state* of the network. Specifically, suppose we set the weights so that the state of each neuron $V_i^{s'} = V_i^{s}$, where $V_i^{s'}$ is determined by the $N - 1$ states $(V_1^{s}, \dots, V_{i-1}^{s}, V_{i+1}^{s}, \dots, V_N^{s})$ of the training vector:

$$V_i^{s'} \;=\; f\left(\sum_{j=1}^{N} T_{ij}\, V_j^{s} \right). \tag{3.6}$$

In this case, no further state changes will occur, so the network will stay in the state $\mathbf{V}^{s'} = \mathbf{V}^{s}$. Thus, if Equation 3.6 holds then $\mathbf{V}^{s'} = \mathbf{V}^{s}$ is a stable state of the network.

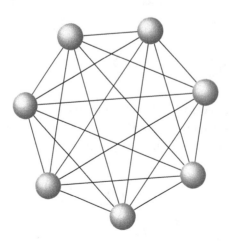

Figure 3.1. A Hopfield network with seven units and $(7 \times (7 - 1))/2 = 21$ connections. From Stone (2019).

The Learning Rule. To set the weights correctly, we use the learning rule defined in Equation 2 of Hopfield's paper:

$$T_{ij} = \sum_{s=1}^{n}(2V_i^s - 1)(2V_j^s - 1). \tag{3.7}$$

Notice that the weight T_{ij} is a measure of the correlation between the states of neurons i and j (T_{ij} is actually the n times the correlation). From Equation 3.1, the total input to neuron i is

$$x_i = \sum_{j=1}^{N}T_{ij}V_j^{s'}. \tag{3.8}$$

Because T_{ij} is a measure of the correlation, it ensures that if the states of neurons i and j are positively correlated, and if a neuron j has state $V_j^{s'} = 1$, then neuron i to which it is connected will receive a positive value, whereas if $V_j^{s'} = 0$ then neuron i will receive a zero value. Substituting T_{ij} from Equation 3.7, the total input to neuron i becomes

$$\sum_j T_{ij}V_j^{s'} = \sum_{j=1}^{N}\left[\sum_{s=1}^{n}(2V_i^s - 1)(2V_j^s - 1)\right]V_j^{s'}. \tag{3.9}$$

Rearranging this yields

$$\sum_j T_{ij}V_j^{s'} = \sum_{s=1}^{n}(2V_i^s - 1)\left[\sum_{j=1}^{N}(2V_j^s - 1)V_j^{s'}\right]. \tag{3.10}$$

For convenience, we define the term in square brackets as

$$z^{s'} = \sum_{j=1}^{N}(2V_j^s - 1)V_j^{s'}. \tag{3.11}$$

Notice that $z^{s'}$ is a measure of the correlation between the network states $V_j^{s'}$ in $\mathbf{V}^{s'}$ and the corresponding states V_j^s in the training vector $\mathbf{V}^s$. We now write

$$x_i = \sum_{s=1}^{n}(2V_i^s - 1)z^{s'}. \tag{3.12}$$

As we shall see, when $s' = s$ (Scenario 1 below), $z^{s'} \approx N/2$, and when $s' \neq s$ (Scenario 2 below), $z^{s'} = 0$. Hopfield refers to this as *pseudoorthogonality*, and it ensures that only vectors that were part of the training set yield non-zero inputs to neurons. Specifically, if the states of neurons are set to m arbitrary vectors, none of which are in the set of n training vectors, then $x_i = 0$ for each of the m arbitrary vectors. But if a vector $\mathbf{V}^{s'}$ is an element of the training set then we obtain Hopfield's Equation 4:

$$x_i \approx (2V_i^{s'} - 1)N/2, \tag{3.13}$$

so that $x_i \approx \pm N/2$.

Next, we explain the two scenarios mentioned above: $s = s'$ and $s \neq s'$.

Scenario 1 ($s = s'$). If $s = s'$ then the network state $\mathbf{V}^{s'}$ matches the training vector $\mathbf{V}^s$. In this case, for each neuron j we have that either

$$V_j^s = V_j^{s'} = 0 \tag{3.14}$$

or

$$V_j^s = V_j^{s'} = 1. \tag{3.15}$$

If $V_j^s = V_j^{s'} = 0$ then each term in the sum in Equation 3.11 evaluates to

$$(2V_j^s - 1)V_j^{s'} = 0, \tag{3.16}$$

and if $V_j^s = V_j^{s'} = 1$ then each term evaluates to

$$(2V_j^s - 1)V_j^{s'} = 1. \tag{3.17}$$

Therefore, if we assume that approximately half of the neurons in each network state are 'on' (i.e. $V_j^{s'} = 1$) then, from Equation 3.11,

$$z^{s'} \approx N/2. \tag{3.18}$$

Scenario 2 ($s \neq s'$). If $s \neq s'$ then the network state $\mathbf{V}^{s'}$ does not match the training vector $\mathbf{V}^s$. In this case, for each neuron j there are two possible states, $V_j^{s'} = 0$ and $V_j^{s'} = 1$, and two possible desired states, $V_j^s = 0$ and $V_j^s = 1$. Thus, the four possible pairs of states and desired states for each neuron are

$$(V_j^s, V_j^{s'}) = (0,0), \tag{3.19}$$
$$(V_j^s, V_j^{s'}) = (0,1), \tag{3.20}$$
$$(V_j^s, V_j^{s'}) = (1,0), \tag{3.21}$$
$$(V_j^s, V_j^{s'}) = (1,1), \tag{3.22}$$

and the corresponding values of $(2V_j^s - 1)V_j^{s'}$ are $(0, -1, 0, +1)$:

$$(2V_j^s - 1)V_j^{s'} = (2 \times 0 - 1) \times 0 = 0, \tag{3.23}$$
$$(2V_j^s - 1)V_j^{s'} = (2 \times 0 - 1) \times 1 = -1, \tag{3.24}$$
$$(2V_j^s - 1)V_j^{s'} = (2 \times 1 - 1) \times 0 = 0, \tag{3.25}$$
$$(2V_j^s - 1)V_j^{s'} = (2 \times 1 - 1) \times 1 = +1. \tag{3.26}$$

If we assume that for a given training vector $\mathbf{V}^{s'}$ each of the four possible pairs $(V_j^s, V_j^{s'})$ occurs approximately equally often (i.e. at a frequency of $N/4$) then

$$z^{s'} = \left[\sum_{j=1}^{N} (2V_j^s - 1)V_j^{s'} \right] \tag{3.27}$$
$$\approx 0. \tag{3.28}$$

3.2. Recall: Content Addressable Memory

The pseudoorthogonality property described in Section 3.1 ensures that Hopfield nets implement *content addressable memory*. This can be understood by examining the network in Figure 3.1. This network has $N = 7$ neurons, and we can represent the collective state of the network as a vector

$$\mathbf{V}^{s'} = (V_1, \ldots, V_7). \tag{3.29}$$

Suppose this network has learned a vector of $+1$s,

$$\mathbf{V}^s = (+1, +1, +1, +1, +1, +1, +1), \tag{3.30}$$

and that $\mathbf{V}^{s'} = \mathbf{V}^s$. If we change the state of the seventh neuron then this defines a corrupted network state

$$\mathbf{V}^{s'}_{\text{corrupted}} = (+1, +1, +1, +1, +1, +1, 0), \tag{3.31}$$

so that the initial network state is $\mathbf{V}^{s'} = \mathbf{V}^{s'}_{\text{corrupted}}$.

Note that each neuron receives input from six other neurons (i.e. all neurons excluding itself). The settling process defined in Equation 3.2 means that neuron $i = 7$ receives a total input x_7 that will switch it from a corrupted state $V_7^{s'} = 0$ to an uncorrupted state $V_7^{s'} = +1$. Thus, if it so happens that neuron $i = 7$ is updated before any other then the learned state is recalled almost immediately, so that $\mathbf{V}^{s'} = \mathbf{V}^s$. But what if the neuron $i = 7$ is not updated before any other neuron?

In this case, the contribution from neuron $i = 7$ to the inputs of the other six neurons $j = 1, \ldots, 6$ is $x_i = 0$, which will have little or no effect on their states (depending on how an input of exactly zero is interpreted). Irrespective of how an input $x_i = 0$ is interpreted, each of those six neurons $j = 1, \ldots, 6$ also receives inputs from five other neurons with uncorrupted states (e.g. $j = 1$ receives uncorrupted input from five neurons, $j = 2, \ldots, 6$). Thus, to all intents and purposes, the effect of the seventh neuron's corrupted state V_7 on the other six neurons $j = 1, \ldots, 6$ will be out-voted by the 'majority' effect that those six uncorrupted neurons have on each other. This majority voting effect is quite literal, so that content addressable memory works provided more than half of the neuron states are set to a state that has been learned previously.

3.3. Tolerance to Damage

Hopfield nets can tolerate damage to their connection weights. Just as a corrupted neuron state pushes all other neuron states towards incorrect states, so a corrupted weight can push one pair of neurons towards incorrect states. However, just as the effect of a corrupted neuron state can be out-voted by the collective effect of uncorrupted neuron states, so the effect of a corrupted weight can be compensated for by the presence of uncorrupted weights. Both content addressable memory and tolerance to damage yield results like that shown in Figure 3.2.

3.4. The Energy Function

The defining property of the Hopfield net energy function is that its value always decreases (or stays the same) as the network evolves according to its dynamics (Equation 3.2). This, in turn, means that the system converges to a stable state. The energy function is derived from statistical mechanics, where it is called the *Hamiltonian*; in optimisation theory it is called a *Lyapunov function*.

The state update rule (Equation 3.2) ensures that the state of the network 'rolls' downhill on an *energy landscape* or *energy function*. This is hard to visualise for the binary neurons considered here, but if the neuron states are continuous then the energy landscape can be represented as in Figure 3.3 (with neuron state along each horizontal axis). The energy landscape consists of many minima, some of which correspond to learned vectors. In effect, setting the weights to particular values in the energy function defines a particular energy landscape. Ideally, the process of learning defines an energy landscape in which the largest minima correspond to learned vectors. If the network state is initialised in a state that corresponds to the side of a *basin of attraction*, then the state update rule guarantees that the network state will end up in a stable state that corresponds to the bottom of this basin. We will not consider the continuous version of the Hopfield net here (Hopfield, 1984); suffice it to say that the continuous version has all of the key properties that the binary Hopfield net has. Indeed, the binary Hopfield net is really a special case of the continuous Hopfield net.

Given a set of weight values, the value of the energy function is defined for every possible set of neuron states by Equation 7 in Hopfield's paper:

$$E = -\frac{1}{2}\sum_{i=1}^{N}\sum_{j=1;i\neq j}^{N} T_{ij} V_i V_j. \tag{3.32}$$

The factor of $1/2$ is included to take account of the fact that $T_{ij} = T_{ji}$, so every term is counted twice in the double summation (except for terms with $i = j$, which are not counted).

What happens to E if we change V_i according to the update rule in Equation 3.3? Taking the derivative of E with respect to V_i gives the value of the slope,

$$\frac{\partial E}{\partial V_i} = -\frac{1}{2}\sum_{i=1}^{N}\sum_{j=1;i\neq j}^{N} T_{ij} V_j \frac{dV_i}{dV_i}, \tag{3.33}$$

and since $dV_i/dV_i = 1$, we have

$$\frac{\partial E}{\partial V_i} = -\frac{1}{2}\sum_{i=1}^{N}\sum_{j=1;i\neq j}^{N} T_{ij} V_j. \tag{3.34}$$

The double counting implicit in the summations over i and j means that every term is counted twice, but if we use the factor of $1/2$ to multiply the sums then we obtain

$$\frac{\partial E}{\partial V_i} = -\sum_{j=1}^{N} T_{ij} V_j. \tag{3.35}$$

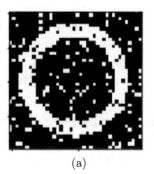 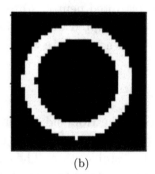

(a)   (b)

Figure 3.2. (a) Input image, which is a corrupted version of a learned network state. (b) Recalled image, obtained after the settling process during recall, which is identical to the image learned. Reproduced with permission from Tom Stafford.

From Equation 3.1, we can see that this is equal to $-x_i$, i.e. the negative of the input to neuron i. For a finite change in state ΔV_i, we obtain a corresponding change in E, which is Hopfield's Equation 8,

$$\Delta E = -\Delta V_i \sum_{j=1}^{N} T_{ij} V_j, \tag{3.36}$$

and can be rewritten as

$$\Delta E = -\Delta V_i x_i. \tag{3.37}$$

We now prove that the update rule in Equation 3.3 reduces E or leaves its value unchanged.

Consider two scenarios: (a) $x_i > 0$ and (b) $x_i < 0$. For scenario (a), the input $x_i > 0$, and if initially $V_i = 0$ then the update rule implies that we should set $V_i = 1$, so that $\Delta V_i = 1 - 0 = 1$. In this case, Equation 3.37 implies that $\Delta E < 0$. Therefore, changing $V_i^{s'}$ from 0 to 1 when $x_i > 0$ ensures that E will get smaller. On the other hand, if $V_i^{s'} = 1$ then the update rule implies that $V_i^{s'}$ should be set to 1, so no change is required and therefore $\Delta E = 0$. Thus, if $x_i > 0$ then the update rule either reduces E or leaves E unchanged. Similarly, for scenario (b), the input $x_i < 0$, and a similar line of reasoning yields that the update rule also either reduces E or leaves E unchanged.

In summary, for a given set of weights, if the network state $\mathbf{V}^{s'}$ is set according to the update rule in Equation 3.3 then the network will settle into a state that minimises the energy E, which implies that the network is in a stable state.

It is noteworthy that all contemporary accounts of Hopfield nets define the binary neuron states as -1 and $+1$, rather than as the states 0 and $+1$ used in Hopfield's paper. This is because using $V = \pm 1$ simplifies the mathematics considerably; however, Hopfield's original notation has been retained here for the sake of authenticity.

3.5. Results

Interestingly, the simulations reported in the paper removed the restriction that weights should be symmetric. Most results were obtained with $N = 30$ neurons, because computers had limited capability in 1982. The most striking result was that recall was found to be reliable provided that the number n of encoded network states was less than $0.15N$.

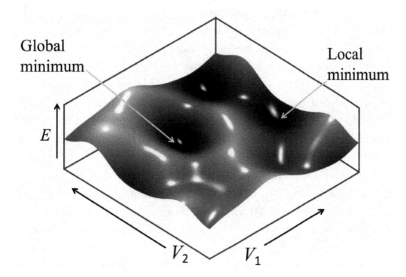

Figure 3.3. Schematic diagram of local and global minima of the energy function E for a network with two neurons whose continuous states are represented along the two horizontal axes.

3.6. Comments by the Paper's Author: JJ Hopfield

Here is John Hopfield's account of how he came to write the 1982 paper (reproduced with permission):

1977: Searching for THE PROBLEM. I spent the long winter of 1977 at the Bohr Institute/Nordita in Copenhagen, as part of its sporadic but historical outreach toward biology. My nominal task was to run a broad 'modern biology for physicists' seminar series, with distinguished speakers from across Europe. I managed to recruit a powerful set of speakers with the help of the Bohr name. In that era, everyone still remembered Niels Bohr's interest in biology. Arriving in Copenhagen after a vacation in 1932, young theoretical physicist Max Delbruck rushed to the opening of the International Congress on Light Therapy where Bohr was to give a lecture entitled 'Light and Life'. It was a transformative experience for Delbruck, setting him on the road from theoretical physics to a Nobel Prize in Physiology and Medicine. However, my distinguished experts emphasized what they understand, and skirted around all lacunae in their conceptual frameworks and research paradigms. They certainly did not describe biological science as being in need of help from (or hospitable to) theoretical physicists. No transformative experience. Slightly disappointed, I returned to Princeton with no glimmer of a new problem for myself.

One can always construct variations and extensions on a theme of a previous paper, whether your own or that of others. The science research literature is overwhelmingly dominated by such work. But I was looking for A PROBLEM, not a problem. The distinction between the two? In the early 1970's I served as the PhD supervisor of Bill Topp, a highly imaginative but somewhat erratic chemistry student. Theoretical chemistry degree in hand, Topp next became an experimental post-doc in the laboratory of James D. Watson at Cold Springs Harbor. I saw Jim about nine months later. Learning that Bill had not yet settled down to work on something, I made appropriate apologetic noises. Jim cut me off, remarking "That's all right. When I first met Francis he was 35, and had not yet found a problem." At that point in his career Crick had written 10 publishable papers, chiefly on x-ray crystallography, but what Watson meant was that Crick was working on problems, not A PROBLEM. Whereas what Watson had already grasped, and which he was pitching to Crick, was the outlines of A PROBLEM in biology. In mature hindsight, I was looking for A PROBLEM in biology for which my unusual background was unique and appropriate preparation.

My office telephone rang very shortly before my morning traverse of the Princeton campus, from Jadwin Laboratory (physics) to Frick (chemistry) where in the fall of 1977 I was giving a course in physical biochemistry. That a chemistry department could permit me, who had never taken a course in either organic chemistry or biochemistry, to teach such a course was a consequence of University mismanagement of the interface between chemistry and biology at the highest administrative levels. That I would rashly attempt to do so was part of my looking for a significant new research project. Expecting a call from a tennis buddy, I hastily answered, only to find Francis O. Schmidt on the line. He said only that he ran the Neuroscience Research Program at MIT, would be transiting Princeton next Wednesday, and would greatly appreciate a half hour (no more) of my time. I had never heard either of Schmidt or the NRP, needed to head off for lecture, thought 'what's 30 minutes' and agreed to meet with no further conversation.

The following week, Francis O. Schmidt descended on me. He described an entity called the Neuroscience Research Program (NRP), which chiefly held small meetings in Boston attended by 20 regular members of the program and 20 visitors. The visitors were broadly chosen, but generally selected with an emphasis on the special topic under consideration at that particular meeting. Schmidt invited me to talk at the next meeting. I suggested that a physics graduate student named Terry Sejnowski who was attending my physical biochemistry course and who had written a mathematical paper on neural coding might give a talk of greater interest, a suggestion that Schmidt quickly rejected. I told him I knew nothing of neuroscience (a word Schmidt had coined some years earlier). He said that it didn't matter, "just speak on what interests you", so I talked about kinetic proofreading and the general issue of accuracy in the cellular biosynthesis of large molecules.

The audience – neurologists, neuroendocrinologists, psychologists, immunologists, electrophysiologists, neuroanatomists, biochemists – understood little of what I said. It didn't matter. Frank wanted to add a physicist member to the group, hoping to bring someone with diverse science experience to interact with his subject and perhaps help it to become more integrated and a more predictive science. Frank was a believer with a zealot's faith that somehow, sometime, science would be able to bridge the gaps between molecules, brains, minds, and behavior. He had gotten my name from Princeton relativist John A. Wheeler, who (for reasons that I have never grasped) had always been one of my staunch supporters. (Wheeler had also chaired the search committee that brought me to Princeton as professor of physics in 1964 because of my research in solid state physics.) It was a put-up job. Following Frank's leadership, the group voted to make me a member.

I was captivated by the talks at the meeting. How mind emerges from brain is to me the deepest question posed by our humanity. Definitely A PROBLEM. It was being pursued in narrow slices – as problems – by this NRP club of scientists with diverse talents and great enthusiasm. But it appeared to me that this group of scientists would never possibly address THE PROBLEM because the solution can be expressed only in an appropriate mathematical language and structure. None then involved with the NRP moved easily in this sphere. So I joined the group, hoping to define, construct, or discover something I could usefully do in this field. The fact that Frank knew how to run a meeting with enough style and grace to attract the best of an international community into his orbit was also influential.

My basic education in neurobiology came from attending the semiannual NRP meetings, sitting next to world experts in their fields, who would patiently explain to me their interpretation of what was going on. Although Schmidt did his best to try to get experts to lecture broadly on their topics, and to describe neuroscience in an integrative fashion, he was generally defeated in this enterprise. So my introduction consisted of a set of disjoint expert views of experimental neuroscience, with interpretive commentary by an expert in some other corner of neuroscience. These commentaries were often impatient with the details studied by those in other subfields, but science is ever thus, and it was in any event not detail that I was seeking. I had tasked myself with finding an integrative view, of trying to somehow rise above the disjoint details of lectures ranging over primate neuroanatomy, insect flight behavior, electrophysiology in aplasia, learning in rat hippocampus, Alzheimer's disease, potassium channels, human language processing, ... to find a project to work on with the tools of theoretical physics.

Cellular automata had their brief moment in the sun in the late 1970's. I first heard of them in reading about Conway's 'game of life' in *Scientific American*, and surmised that a generalization or modification of that basic idea might be useful in understanding how a brain operates. Cellular automata are constructed from an array of equivalent 'cells'. Each 'cell' has an internal state that changes in time in a deterministic fashion according to rules that involve the internal state of that 'cell' and the state of its neighboring 'cell'. All cells are equivalent, and change their internal states simultaneously. I speculated that if the cell state transition rules were made less rigidly structured, more like the less systematic synapse connections that provide inputs between neurons, and less rigid in time to reflect neural delays in signal propagation and processing, it might be possible to bridge the chasm between digital computers and neural systems.

Simple digital computers obtained their answers by starting at an initial state of the machine, implicitly described by a program and some data. They then change state again and again according to simple rules built into the hardware chips of the machine, until finally the state change stops. An end-state where the rules generate no further state change has been reached. The answer to the programmed problem is the information now contained in a few specific memory registers. In the fall of 1978 I began playing with corruptions of the 'game of life' to make it slightly more like neurobiology, hoping to see it 'compute' by following the state trajectory to an answer. Unfortunately I was unable to carry out the mathematics necessary to follow the trajectory of the changing state for any such model. I needed to program a digital computer to simulate such a system, and to do computer experiments to gain insight into a diversity of such models.

It is difficult these days to imagine the primitive state of computers and computer laboratories in universities 37 years ago[¶]. Machines were slow, machine time was expensive, input to computers was chiefly through punched cards, output written on massive printers, TV-type display terminals rare. Computer power as measured by the number of transistors in a microprocessor, has followed 'Moore's Law' for 50 years, roughly doubling every two years. That makes 18 doublings between 1978 and now. Thus I was stuck with computing power $(1/2)^{18} = 1/250,000$ of present day systems. With a very few notable exceptions (such as the AI lab at MIT) computers were used to produce numerical results from reliable programs and costly data. If you computer-simulated a model, it was because you had a well-based belief that the model was in good correspondence with reality. You did not guess or speculate about possible models – it was too expensive, a waste of a valuable resource. There was no emphasis on easy-to-program languages, and the computationally efficient languages were ungainly to use.

Princeton general computing and the computers of Princeton's high-energy physics group (the only departmental computer in physics) were run in a number-crunching mode. The idea of guessing about models, quickly and easily exploring the consequences of these guesses on a digital machine, and hoping to find interesting behavior, was foreign to the computing facilities and environments of Princeton and Bell Laboratories, where I also was affiliated.

Given my computing environment, I made little progress. The basic idea was that any computer, whether digital machine or a brain, operates by following a dynamical trajectory from a starting point (program and data) to an end-point, and that the trajectory needed stability against perturbations to get to the answer reliably in spite of noise and system imperfections. I did give an NRP talk on this idea of neurobiology as a system that computed with dynamical attractors. But there were neither computer simulations nor mathematics to give credence to this view. One Young Turk visitor came up to me afterward to tell me it was a beautiful talk but unfortunately had nothing to do with neurobiology. The others ignored it. I wryly note that my 2015 Swartz Prize (in computational neuroscience) from the Society of Neuroscience is really for this basic idea. But of course, the very existence of the term computational neuroscience implies that there are many mathematically sophisticated scientists now in the field, and these were an extreme rarity in 1979.

In 1978 Harold Brown, the recently appointed Caltech president, resigned to become Secretary of Defense, and Caltech was again in the market for a physicist-president. They turned to Marvin (Murph) Goldberger, an eminent theoretical physicist who had chaired Princeton physics. Caltech, with Delbruck on the faculty of the Biology Division, had been making an effort to have more of a link between biology and physics. Goldberger had seen me struggling to do biological physics within the Princeton physics department. So during his honeymoon phase as Caltech president, he talked his faculty into offering me an endowed professorship jointly between Chemistry and Biology. Caltech Physics, dominated by the mindsets of Murray Gell-Mann and Richard Feynman, had no interest in appointing in such a direction.

What was their effect on the Princeton Physics Department? I had never given a physics colloquium or seminar on either of my most interesting biology-related papers (on kinetic proofreading and on biological electron transfer). The general attitude was that I was probably doing something interesting, but it involved too many details for Princeton Physics. When in October 1979 I went to see Physics Chair Val Fitch to tell him about the offer from Caltech, there was no counteroffer. Val said it would be best for both of us for me to go – best for me scientifically, and a simplification of his problem of departmental focus.

[¶]This account was written in 2018.

1980. The quantum chemistry computing facility at Caltech was a splendid computing environment for trying out models. It supported multi-user real-time computing, with CRT displays and direct keyboard input, and no compilation delays. My research was a perversion of its intended purpose, but no one was looking. With this help, it rapidly became apparent that the previous year's speculation on a possible relationship to conventional cellular automata was useless.

It is surprisingly difficult to give up on a wrong idea that has been nurtured for a year. So rather than being abandoned, the cellular automata became perverted into a random quasi-neural network. The periodic nature of cellular automata was abandoned in favor of randomly chosen connections. The complex set of logical rules for state transitions was replaced with a rule inspired by biology. After a year of simulations and mathematics, I finally gave up on random networks. Instead, why not try a network with a particular structure chosen to accomplish some simple but profound task that neurobiology does rapidly, and that seems natural to biology but not to computers. The conceptually simplest such task, and one that fit naturally into the basic computing paradigm of computing through dynamical system attractors, is associative memory.

Associative memory is reciprocal – seeing someone reminds you of their name (or at least did when I was younger), and hearing their name reminds you of what they look like. That fact can be expressed in network structure by making connections that are reciprocal. The mathematics of such networks is closely related to the mathematics of 'spin' systems that are responsible for all the complex forms of magnetism in solids. I knew something of these systems through my connections with theoretical physics at Bell Labs. Suddenly there was a connection between neurobiology and physics systems I understood (thanks to a lifetime of interaction with P.W. Anderson). A month later I was writing a paper.

I had previously agreed to attend a symposium entitled 'From Physics to Biology' of the Institut de la Vie at Versailles in the summer of 1981. This was an unusual gathering organized by Maurice Marois, a medical doctor who had dreams of strengthened connections between diverse scientists. He was persuasive with sponsors and flattering to Nobel Prize winners, and ran a posh, pretentious meeting with the conference meetings themselves in the Hall of Mirrors in the Palace of Versailles, and the speakers staying at the Trianon Palace Hotel next to the Chateau. I had happily (if slightly corruptly) accepted the invitation for such an all-expenses paid trip to Paris. I discarded my previously chosen topic in favor of a talk based on the recent work, making this talk at Versailles the first public talk on this subject. I have never met anyone who remembered hearing the talk.

The conference was to involve a book. As a result, the first manuscript I wrote on this research was a short but wide-ranging account of my recent research and its intellectual setting. This manuscript was submitted, but no conference proceedings ever appeared. With this draft in hand, I set about writing an article. I had two target audiences, physicists and neurobiologists, so thought immediately of *PNAS*. Neurobiologists read *PNAS*, and might see the article. And although few physicists regularly read *PNAS* in that era, at least *PNAS* was typically available in physics libraries. Not ideal, but the best I could come up with. A simple choice, for as an Academy member I could publish such a paper without review. Also a difficult choice, for there was an absolute 5-page limit to article length and there was much to say.

This *PNAS* paper is the first publication in which I use the word 'neuron'. It was to provide an entryway to working in neuroscience for many physicists and computer scientists. Further work connected these networks to many significant applications far beyond associative memory. It is the most cited paper I have ever written (6800 citations). Even AT&T was pleased (I had a part-time affiliation with Bell Labs all during this epoch), for the research also generated a very frequently referenced patent [http://www.google.com/patents/US4660166] for their patent pool, as well as strengthened links between neural biophysics and condensed matter physics at the Bell Labs.

Concerning the writing of non-fiction, Ernest Hemingway remarked, "If a writer of prose knows enough about what he is writing about he may omit things that he knows and the reader, if the writer is writing truly enough, will have a feeling of those things as strongly as though the writer had stated them."** The *PNAS* length limitation forced me to be highly selective in what was said – and what was omitted. Had Hemingway been a physicist, he would have recognized the style. In hindsight, the omission of the almost obvious probably increased the impact of the paper. The unstated became an invitation for others to add to the subject, and thus encouraged a community of contributors to work on such network models. Successful science is always a community enterprise.

Research Paper:
Neural Networks and Physical Systems with Emergent Collective Computational Abilities

Reference: Hopfield, J. J. (1982). Neural networks and physical systems with emergent collective computational abilities. *Proc. Nat. Acad. Sci. USA*, 79(8):2554–2558.
https://www.pnas.org/doi/10.1073/pnas.79.8.2554
Reproduced with permission.

** E Hemingway, *Death in the Afternoon* (1932).

Neural networks and physical systems with emergent collective computational abilities

(associative memory/parallel processing/categorization/content-addressable memory/fail-soft devices)

J. J. HOPFIELD

Division of Chemistry and Biology, California Institute of Technology, Pasadena, California 91125; and Bell Laboratories, Murray Hill, New Jersey 07974

Contributed by John J. Hopfield, January 15, 1982

ABSTRACT   Computational properties of use to biological organisms or to the construction of computers can emerge as collective properties of systems having a large number of simple equivalent components (or neurons). The physical meaning of content-addressable memory is described by an appropriate phase space flow of the state of a system. A model of such a system is given, based on aspects of neurobiology but readily adapted to integrated circuits. The collective properties of this model produce a content-addressable memory which correctly yields an entire memory from any subpart of sufficient size. The algorithm for the time evolution of the state of the system is based on asynchronous parallel processing. Additional emergent collective properties include some capacity for generalization, familiarity recognition, categorization, error correction, and time sequence retention. The collective properties are only weakly sensitive to details of the modeling or the failure of individual devices.

Given the dynamical electrochemical properties of neurons and their interconnections (synapses), we readily understand schemes that use a few neurons to obtain elementary useful biological behavior (1–3). Our understanding of such simple circuits in electronics allows us to plan larger and more complex circuits which are essential to large computers. Because evolution has no such plan, it becomes relevant to ask whether the ability of large collections of neurons to perform "computational" tasks may in part be a spontaneous collective consequence of having a large number of interacting simple neurons.

In physical systems made from a large number of simple elements, interactions among large numbers of elementary components yield collective phenomena such as the stable magnetic orientations and domains in a magnetic system or the vortex patterns in fluid flow. Do analogous collective phenomena in a system of simple interacting neurons have useful "computational" correlates? For example, are the stability of memories, the construction of categories of generalization, or time-sequential memory also emergent properties and collective in origin? This paper examines a new modeling of this old and fundamental question (4–8) and shows that important computational properties spontaneously arise.

All modeling is based on details, and the details of neuroanatomy and neural function are both myriad and incompletely known (9). In many physical systems, the nature of the emergent collective properties is insensitive to the details inserted in the model (e.g., collisions are essential to generate sound waves, but any reasonable interatomic force law will yield appropriate collisions). In the same spirit, I will seek collective properties that are robust against change in the model details.

The model could be readily implemented by integrated circuit hardware. The conclusions suggest the design of a delo-

The publication costs of this article were defrayed in part by page charge payment. This article must therefore be hereby marked "*advertisement*" in accordance with 18 U. S. C. §1734 solely to indicate this fact.

calized content-addressable memory or categorizer using extensive asynchronous parallel processing.

The general content-addressable memory of a physical system

Suppose that an item stored in memory is "H. A. Kramers & G. H. Wannier *Phys. Rev.* **60**, 252 (1941)." A general content-addressable memory would be capable of retrieving this entire memory item on the basis of sufficient partial information. The input "& Wannier, (1941)" might suffice. An ideal memory could deal with errors and retrieve this reference even from the input "Vannier, (1941)". In computers, only relatively simple forms of content-addressable memory have been made in hardware (10, 11). Sophisticated ideas like error correction in accessing information are usually introduced as software (10).

There are classes of physical systems whose spontaneous behavior can be used as a form of general (and error-correcting) content-addressable memory. Consider the time evolution of a physical system that can be described by a set of general coordinates. A point in state space then represents the instantaneous condition of the system. This state space may be either continuous or discrete (as in the case of N Ising spins).

The equations of motion of the system describe a flow in state space. Various classes of flow patterns are possible, but the systems of use for memory particularly include those that flow toward locally stable points from anywhere within regions around those points. A particle with frictional damping moving in a potential well with two minima exemplifies such a dynamics.

If the flow is not completely deterministic, the description is more complicated. In the two-well problems above, if the frictional force is characterized by a temperature, it must also produce a random driving force. The limit points become small limiting regions, and the stability becomes not absolute. But as long as the stochastic effects are small, the essence of local stable points remains.

Consider a physical system described by many coordinates $X_1 \cdots X_N$, the components of a state vector X. Let the system have locally stable limit points $X_a, X_b, \cdots$. Then, if the system is started sufficiently near any X_a, as at $X = X_a + \Delta$, it will proceed in time until $X \approx X_a$. We can regard the information stored in the system as the vectors $X_a, X_b, \cdots$. The starting point $X = X_a + \Delta$ represents a partial knowledge of the item X_a, and the system then generates the total information X_a.

Any physical system whose dynamics in phase space is dominated by a substantial number of locally stable states to which it is attracted can therefore be regarded as a general content-addressable memory. The physical system will be a potentially useful memory if, in addition, any prescribed set of states can readily be made the stable states of the system.

The model system

The processing devices will be called neurons. Each neuron i has two states like those of McCullough and Pitts (12): $V_i = 0$

("not firing") and $V_i = 1$ ("firing at maximum rate"). When neuron i has a connection made to it from neuron j, the strength of connection is defined as T_{ij}. (Nonconnected neurons have $T_{ij} \equiv 0$.) The instantaneous state of the system is specified by listing the N values of V_i, so it is represented by a binary word of N bits.

The state changes in time according to the following algorithm. For each neuron i there is a fixed threshold U_i. Each neuron i readjusts its state randomly in time but with a mean attempt rate W, setting

$$\begin{matrix} V_i \to 1 \\ V_i \to 0 \end{matrix} \quad \text{if} \quad \sum_{j \neq i} T_{ij} V_j \quad \begin{matrix} > U_i \\ < U_i \end{matrix} \quad . \qquad [1]$$

Thus, each neuron randomly and asynchronously evaluates whether it is above or below threshold and readjusts accordingly. (Unless otherwise stated, we choose $U_i = 0$.)

Although this model has superficial similarities to the Perceptron (13, 14) the essential differences are responsible for the new results. First, Perceptrons were modeled chiefly with neural connections in a "forward" direction $A \to B \to C \to D$. The analysis of networks with strong backward coupling $A \rightleftarrows B \rightleftarrows C$ proved intractable. All our interesting results arise as consequences of the strong back-coupling. Second, Perceptron studies usually made a random net of neurons deal directly with a real physical world and did not ask the questions essential to finding the more abstract emergent computational properties. Finally, Perceptron modeling required synchronous neurons like a conventional digital computer. There is no evidence for such global synchrony and, given the delays of nerve signal propagation, there would be no way to use global synchrony effectively. Chiefly computational properties which can exist in spite of asynchrony have interesting implications in biology.

The information storage algorithm

Suppose we wish to store the set of states V^s, $s = 1 \cdots n$. We use the storage prescription (15, 16)

$$T_{ij} = \sum_s (2V_i^s - 1)(2V_j^s - 1) \qquad [2]$$

but with $T_{ii} = 0$. From this definition

$$\sum_j T_{ij} V_j^{s'} = \sum_s (2V_i^s - 1) \left[\sum_j V_j^{s'} (2V_j^s - 1) \right] \equiv H_i^{s'} . \qquad [3]$$

The mean value of the bracketed term in Eq. 3 is 0 unless $s = s'$, for which the mean is $N/2$. This pseudoorthogonality yields

$$\sum_j T_{ij} V_j^{s'} \equiv \langle H_i^{s'} \rangle \approx (2V_i^{s'} - 1) N/2 \qquad [4]$$

and is positive if $V_i^{s'} = 1$ and negative if $V_i^{s'} = 0$. Except for the noise coming from the $s \neq s'$ terms, the stored state would always be stable under our processing algorithm.

Such matrices T_{ij} have been used in theories of linear associative nets (15–19) to produce an output pattern from a paired input stimulus, $S_1 \to O_1$. A second association $S_2 \to O_2$ can be simultaneously stored in the same network. But the confusing simulus $0.6\, S_1 + 0.4\, S_2$ will produce a generally meaningless mixed output $0.6\, O_1 + 0.4\, O_2$. Our model, in contrast, will use its strong nonlinearity to make choices, produce categories, and regenerate information and, with high probability, will generate the output O_1 from such a confusing mixed stimulus.

A linear associative net must be connected in a complex way with an external nonlinear logic processor in order to yield true computation (20, 21). Complex circuitry is easy to plan but more difficult to discuss in evolutionary terms. In contrast, our model obtains its emergent computational properties from simple properties of many cells rather than circuitry.

The biological interpretation of the model

Most neurons are capable of generating a train of action potentials—propagating pulses of electrochemical activity—when the average potential across their membrane is held well above its normal resting value. The mean rate at which action potentials are generated is a smooth function of the mean membrane potential, having the general form shown in Fig. 1.

The biological information sent to other neurons often lies in a short-time average of the firing rate (22). When this is so, one can neglect the details of individual action potentials and regard Fig. 1 as a smooth input–output relationship. [Parallel pathways carrying the same information would enhance the ability of the system to extract a short-term average firing rate (23, 24).]

A study of emergent collective effects and spontaneous computation must necessarily focus on the nonlinearity of the input–output relationship. The essence of computation is nonlinear logical operations. The particle interactions that produce true collective effects in particle dynamics come from a nonlinear dependence of forces on positions of the particles. Whereas linear associative networks have emphasized the linear central region (14–19) of Fig. 1, we will replace the input–output relationship by the dot-dash step. Those neurons whose operation is dominantly linear merely provide a pathway of communication between nonlinear neurons. Thus, we consider a network of "on or off" neurons, granting that some of the interconnections may be by way of neurons operating in the linear regime.

Delays in synaptic transmission (of partially stochastic character) and in the transmission of impulses along axons and dendrites produce a delay between the input of a neuron and the generation of an effective output. All such delays have been modeled by a single parameter, the stochastic mean processing time $1/W$.

The input to a particular neuron arises from the current leaks of the synapses to that neuron, which influence the cell mean potential. The synapses are activated by arriving action potentials. The input signal to a cell i can be taken to be

$$\sum_j T_{ij} V_j \qquad [5]$$

where T_{ij} represents the effectiveness of a synapse. Fig. 1 thus

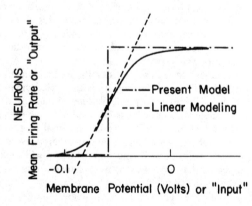

FIG. 1. Firing rate versus membrane voltage for a typical neuron (solid line), dropping to 0 for large negative potentials and saturating for positive potentials. The broken lines show approximations used in modeling.

becomes an input–output relationship for a neuron.

Little, Shaw, and Roney (8, 25, 26) have developed ideas on the collective functioning of neural nets based on "on/off" neurons and synchronous processing. However, in their model the relative timing of action potential spikes was central and resulted in reverberating action potential trains. Our model and theirs have limited formal similarity, although there may be connections at a deeper level.

Most modeling of neural learning networks has been based on synapses of a general type described by Hebb (27) and Eccles (28). The essential ingredient is the modification of T_{ij} by correlations like

$$\Delta T_{ij} = [V_i(t)V_j(t)]_{\text{average}} \qquad [6]$$

where the average is some appropriate calculation over past history. Decay in time and effects of $[V_i(t)]_{\text{avg}}$ or $[V_j(t)]_{\text{avg}}$ are also allowed. Model networks with such synapses (16, 20, 21) can construct the associative T_{ij} of Eq. 2. We will therefore initially assume that such a T_{ij} has been produced by previous experience (or inheritance). The Hebbian property need not reside in single synapses; small groups of cells which produce such a net effect would suffice.

The network of cells we describe performs an abstract calculation and, for applications, the inputs should be appropriately coded. In visual processing, for example, feature extraction should previously have been done. The present modeling might then be related to how an entity or *Gestalt* is remembered or categorized on the basis of inputs representing a collection of its features.

Studies of the collective behaviors of the model

The model has stable limit points. Consider the special case $T_{ij} = T_{ji}$, and define

$$E = -\frac{1}{2} \sum_{i \neq j} \sum T_{ij} V_i V_j \ . \qquad [7]$$

ΔE due to ΔV_i is given by

$$\Delta E = -\Delta V_i \sum_{j \neq i'} T_{ij} V_j \ . \qquad [8]$$

Thus, the algorithm for altering V_i causes E to be a monotonically decreasing function. State changes will continue until a least (local) E is reached. This case is isomorphic with an Ising model. T_{ij} provides the role of the exchange coupling, and there is also an external local field at each site. When T_{ij} is symmetric but has a random character (the spin glass) there are known to be many (locally) stable states (29).

Monte Carlo calculations were made on systems of $N = 30$ and $N = 100$, to examine the effect of removing the $T_{ij} = T_{ji}$ restriction. Each element of T_{ij} was chosen as a random number between -1 and 1. The neural architecture of typical cortical regions (30, 31) and also of simple ganglia of invertebrates (32) suggests the importance of 100–10,000 cells with intense mutual interconnections in elementary processing, so our scale of N is slightly small.

The dynamics algorithm was initiated from randomly chosen initial starting configurations. For $N = 30$ the system never displayed an ergodic wandering through state space. Within a time of about $4/W$ it settled into limiting behaviors, the commonest being a stable state. When 50 trials were examined for a particular such random matrix, all would result in one of two or three end states. A few stable states thus collect the flow from most of the initial state space. A simple cycle also occurred occasionally—for example, $\cdots A \to B \to A \to B \cdots$.

The third behavior seen was chaotic wandering in a small region of state space. The Hamming distance between two binary states A and B is defined as the number of places in which the digits are different. The chaotic wandering occurred within a short Hamming distance of one particular state. Statistics were done on the probability p_i of the occurrence of a state in a time of wandering around this minimum, and an entropic measure of the available states M was taken

$$\ln M = -\sum p_i \ln p_i \ . \qquad [9]$$

A value of $M = 25$ was found for $N = 30$. *The flow in phase space produced by this model algorithm has the properties necessary for a physical content-addressable memory* whether or not T_{ij} is symmetric.

Simulations with $N = 100$ were much slower and not quantitatively pursued. They showed qualitative similarity to $N = 30$.

Why should stable limit points or regions persist when $T_{ij} \neq T_{ji}$? If the algorithm at some time changes V_i from 0 to 1 or vice versa, the change of the energy defined in Eq. 7 can be split into two terms, one of which is always negative. The second is identical if T_{ij} is symmetric and is "stochastic" with mean 0 if T_{ij} and T_{ji} are randomly chosen. The algorithm for $T_{ij} \neq T_{ji}$ therefore changes E in a fashion similar to the way E would change in time for a symmetric T_{ij} but with an algorithm corresponding to a finite temperature.

About 0.15 N states can be simultaneously remembered before error in recall is severe. Computer modeling of memory storage according to Eq. 2 was carried out for $N = 30$ and $N = 100$. n random memory states were chosen and the corresponding T_{ij} was generated. If a nervous system preprocessed signals for efficient storage, the preprocessed information would appear random (e.g., the coding sequences of DNA have a random character). The random memory vectors thus simulate efficiently encoded real information, as well as representing our ignorance. The system was started at each assigned nominal memory state, and the state was allowed to evolve until stationary.

Typical results are shown in Fig. 2. The statistics are averages over both the states in a given matrix and different matrices. With $n = 5$, the assigned memory states are almost always stable (and exactly recallable). For $n = 15$, about half of the nominally remembered states evolved to stable states with less than 5 errors, but the rest evolved to states quite different from the starting points.

These results can be understood from an analysis of the effect of the noise terms. In Eq. 3, $H_i^{s'}$ is the "effective field" on neuron i when the state of the system is s', one of the nominal memory states. The expectation value of this sum, Eq. 4, is $\pm N/2$ as appropriate. The $s \neq s'$ summation in Eq. 2 contributes no mean, but has a rms noise of $[(n - 1)N/2]^{1/2} \equiv \sigma$. For nN large, this noise is approximately Gaussian and the probability of an error in a single particular bit of a particular memory will be

$$P = \frac{1}{\sqrt{2\pi\sigma^2}} \int_{N/2}^{\infty} e^{-x^2/2\sigma^2} \, dx \ . \qquad [10]$$

For the case $n = 10$, $N = 100$, $P = 0.0091$, the probability that a state had no errors in its 100 bits should be about $e^{-0.91} \approx 0.40$. In the simulation of Fig. 2, the experimental number was 0.6.

The theoretical scaling of n with N at fixed P was demonstrated in the simulations going between $N = 30$ and $N = 100$. The experimental results of half the memories being well retained at $n = 0.15\,N$ and the rest badly retained is expected to

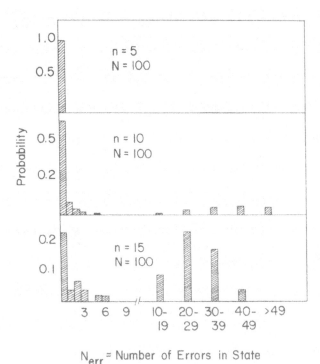

N_{err} = Number of Errors in State

FIG. 2. The probability distribution of the occurrence of errors in the location of the stable states obtained from nominally assigned memories.

be true for all large N. The information storage at a given level of accuracy can be increased by a factor of 2 by a judicious choice of individual neuron thresholds. This choice is equivalent to using variables $\mu_i = \pm 1$, $T_{ij} = \Sigma_s \mu_i^s \mu_j^s$, and a threshold level of 0.

Given some arbitrary starting state, what is the resulting final state (or statistically, states)? To study this, evolutions from randomly chosen initial states were tabulated for $N = 30$ and $n = 5$. From the (inessential) symmetry of the algorithm, if $(101110\cdots)$ is an assigned stable state, $(010001\cdots)$ is also stable. Therefore, the matrices had 10 nominal stable states. Approximately 85% of the trials ended in assigned memories, and 10% ended in stable states of no obvious meaning. An ambiguous 5% landed in stable states very near assigned memories. There was a range of a factor of 20 of the likelihood of finding these 10 states.

The algorithm leads to memories near the starting state. For $N = 30$, $n = 5$, partially random starting states were generated by random modification of known memories. The probability that the final state was that closest to the initial state was studied as a function of the distance between the initial state and the nearest memory state. For distance ≤ 5, the nearest state was reached more than 90% of the time. Beyond that distance, the probability fell off smoothly, dropping to a level of 0.2 (2 times random chance) for a distance of 12.

The phase space flow is apparently dominated by attractors which are the nominally assigned memories, each of which dominates a substantial region around it. The flow is not entirely deterministic, and *the system responds to an ambiguous starting state by a statistical choice* between the memory states it most resembles.

Were it desired to use such a system in an Si-based content-addressable memory, the algorithm should be used and modified to hold the known bits of information while letting the others adjust.

The model was studied by using a "clipped" T_{ij}, replacing T_{ij} in Eq. 3 by ± 1, the algebraic sign of T_{ij}. The purposes were to examine the necessity of a linear synapse supposition (by making a highly nonlinear one) and to examine the efficiency of storage. Only $N(N/2)$ bits of information can possibly be stored in this symmetric matrix. Experimentally, for $N = 100$, $n = 9$, the level of errors was similar to that for the ordinary algorithm at $n = 12$. The signal-to-noise ratio can be evaluated analytically for this clipped algorithm and is reduced by a factor of $(2/\pi)^{1/2}$ compared with the unclipped case. For a fixed error probability, the number of memories must be reduced by $2/\pi$.

With the μ algorithm and the clipped T_{ij}, both analysis and modeling showed that the maximal information stored for $N = 100$ occurred at about $n = 13$. Some errors were present, and the Shannon information stored corresponded to about $N(N/8)$ bits.

New memories can be continually added to T_{ij}. The addition of new memories beyond the capacity overloads the system and makes all memory states irretrievable unless there is a provision for forgetting old memories (16, 27, 28).

The saturation of the possible size of T_{ij} will itself cause forgetting. Let the possible values of T_{ij} be 0, ± 1, ± 2, ± 3, and T_{ij} be freely incremented within this range. If $T_{ij} = 3$, a next increment of $+1$ would be ignored and a next increment of -1 would reduce T_{ij} to 2. When T_{ij} is so constructed, only the recent memory states are retained, with a slightly increased noise level. Memories from the distant past are no longer stable. How far into the past are states remembered depends on the digitizing depth of T_{ij}, and $0, \cdots, \pm 3$ is an appropriate level for $N = 100$. Other schemes can be used to keep too many memories from being simultaneously written, but this particular one is attractive because it requires no delicate balances and is a consequence of natural hardware.

Real neurons need not make synapses both of $i \rightarrow j$ and $j \rightarrow i$. Particular synapses are restricted to one sign of output. We therefore asked whether $T_{ij} = T_{ji}$ is important. Simulations were carried out with only one ij connection: if $T_{ij} \neq 0$, $T_{ji} = 0$. The probability of making errors increased, but the algorithm continued to generate stable minima. A Gaussian noise description of the error rate shows that the signal-to-noise ratio for given n and N should be decreased by the factor $1/\sqrt{2}$, and the simulations were consistent with such a factor. This same analysis shows that the system generally fails in a "soft" fashion, with signal-to-noise ratio and error rate increasing slowly as more synapses fail.

Memories too close to each other are confused and tend to merge. For $N = 100$, a pair of random memories should be separated by 50 ± 5 Hamming units. The case $N = 100$, $n = 8$, was studied with seven random memories and the eighth made up a Hamming distance of only 30, 20, or 10 from one of the other seven memories. At a distance of 30, both similar memories were usually stable. At a distance of 20, the minima were usually distinct but displaced. At a distance of 10, the minima were often fused.

The algorithm categorizes initial states according to the similarity to memory states. With a threshold of 0, the system behaves as a forced categorizer.

The state $00000\cdots$ is always stable. For a threshold of 0, this stable state is much higher in energy than the stored memory states and very seldom occurs. Adding a uniform threshold in the algorithm is equivalent to raising the effective energy of the stored memories compared to the 0000 state, and 0000 also becomes a likely stable state. The 0000 state is then generated by any initial state that does not resemble adequately closely one of the assigned memories and represents positive recognition that the starting state is not familiar.

Familiarity can be recognized by other means when the memory is drastically overloaded. We examined the case $N = 100$, $n = 500$, in which there is a memory overload of a factor of 25. None of the memory states assigned were stable. The initial rate of processing of a starting state is defined as the number of neuron state readjustments that occur in a time $1/2W$. Familiar and unfamiliar states were distinguishable most of the time at this level of overload on the basis of the initial processing rate, which was faster for unfamiliar states. This kind of familiarity can only be read out of the system by a class of neurons or devices abstracting average properties of the processing group.

For the cases so far considered, the expectation value of T_{ij} was 0 for $i \neq j$. A set of memories can be stored with average correlations, and $\overline{T}_{ij} = C_{ij} \neq 0$ because there is a consistent internal correlation in the memories. If now a partial new state X is stored

$$\Delta T_{ij} = (2X_i - 1)(2X_j - 1) \quad i,j \leq k < N \quad [11]$$

using only k of the neurons rather than N, an attempt to reconstruct it will generate a stable point for all N neurons. The values of $X_{k+1} \cdots X_N$ that result will be determined primarily from the sign of

$$\sum_{j=1}^{k} c_{ij} x_j \quad [12]$$

and X is completed according to the mean correlations of the other memories. The most effective implementation of this capacity stores a large number of correlated matrices weakly followed by a normal storage of X.

A nonsymmetric T_{ij} can lead to the possibility that a minimum will be only metastable and will be replaced in time by another minimum. Additional nonsymmetric terms which could be easily generated by a minor modification of Hebb synapses

$$\Delta T_{ij} = A \sum_{s} (2V_i^{s+1} - 1)(2V_j^s - 1) \quad [13]$$

were added to T_{ij}. When A was judiciously adjusted, the system would spend a while near V_s and then leave and go to a point near V_{s+1}. But sequences longer than four states proved impossible to generate, and even these were not faithfully followed.

Discussion

In the model network each "neuron" has elementary properties, and the network has little structure. Nonetheless, collective computational properties spontaneously arose. Memories are retained as stable entities or *Gestalts* and can be correctly recalled from any reasonably sized subpart. Ambiguities are resolved on a statistical basis. Some capacity for generalization is present, and time ordering of memories can also be encoded. These properties follow from the nature of the flow in phase space produced by the processing algorithm, which does not appear to be strongly dependent on precise details of the modeling. This robustness suggests that similar effects will obtain even when more neurobiological details are added.

Much of the architecture of regions of the brains of higher animals must be made from a proliferation of simple local circuits with well-defined functions. The bridge between simple circuits and the complex computational properties of higher nervous systems may be the spontaneous emergence of new computational capabilities from the collective behavior of large numbers of simple processing elements.

Implementation of a similar model by using integrated circuits would lead to chips which are much less sensitive to element failure and soft-failure than are normal circuits. Such chips would be wasteful of gates but could be made many times larger than standard designs at a given yield. Their asynchronous parallel processing capability would provide rapid solutions to some special classes of computational problems.

The work at California Institute of Technology was supported in part by National Science Foundation Grant DMR-8107494. This is contribution no. 6580 from the Division of Chemistry and Chemical Engineering.

1. Willows, A. O. D., Dorsett, D. A. & Hoyle, G. (1973) *J. Neurobiol.* 4, 207–237, 255–285.
2. Kristan, W. B. (1980) in *Information Processing in the Nervous System*, eds. Pinsker, H. M. & Willis, W. D. (Raven, New York), 241–261.
3. Knight, B. W. (1975) *Lect. Math. Life Sci.* 5, 111–144.
4. Smith, D. R. & Davidson, C. H. (1962) *J. Assoc. Comput. Mach.* 9, 268–279.
5. Harmon, L. D. (1964) in *Neural Theory and Modeling*, ed. Reiss, R. F. (Stanford Univ. Press, Stanford, CA), pp. 23–24.
6. Amari, S.-I. (1977) *Biol. Cybern.* 26, 175–185.
7. Amari, S.-I. & Akikazu, T. (1978) *Biol. Cybern.* 29, 127–136.
8. Little, W. A. (1974) *Math. Biosci.* 19, 101–120.
9. Marr, J. (1969) *J. Physiol.* 202, 437–470.
10. Kohonen, T. (1980) *Content Addressable Memories* (Springer, New York).
11. Palm, G. (1980) *Biol. Cybern.* 36, 19–31.
12. McCulloch, W. S. & Pitts, W. (1943) *Bull. Math Biophys.* 5, 115–133.
13. Minsky, M. & Papert, S. (1969) *Perceptrons: An Introduction to Computational Geometry* (MIT Press, Cambridge, MA).
14. Rosenblatt, F. (1962) *Principles of Perceptrons* (Spartan, Washington, DC).
15. Cooper, L. N. (1973) in *Proceedings of the Nobel Symposium on Collective Properties of Physical Systems*, eds. Lundqvist, B. & Lundqvist, S. (Academic, New York), 252–264.
16. Cooper, L. N., Liberman, F. & Oja, E. (1979) *Biol. Cybern.* 33, 9–28.
17. Longuet-Higgins, J. C. (1968) *Proc. Roy. Soc. London Ser. B* 171, 327–334.
18. Longuet-Higgins, J. C. (1968) *Nature (London)* 217, 104–105.
19. Kohonen, T. (1977) *Associative Memory—A System-Theoretic Approach* (Springer, New York).
20. Willwacher, G. (1976) *Biol. Cybern.* 24, 181–198.
21. Anderson, J. A. (1977) *Psych. Rev.* 84, 413–451.
22. Perkel, D. H. & Bullock, T. H. (1969) *Neurosci. Res. Symp. Summ.* 3, 405–527.
23. John, E. R. (1972) *Science* 177, 850–864.
24. Roney, K. J., Scheibel, A. B. & Shaw, G. L. (1979) *Brain Res. Rev.* 1, 225–271.
25. Little, W. A. & Shaw, G. L. (1978) *Math. Biosci.* 39, 281–289.
26. Shaw, G. L. & Roney, K. J. (1979) *Phys. Rev. Lett.* 74, 146–150.
27. Hebb, D. O. (1949) *The Organization of Behavior* (Wiley, New York).
28. Eccles, J. G. (1953) *The Neurophysiological Basis of Mind* (Clarendon, Oxford).
29. Kirkpatrick, S. & Sherrington, D. (1978) *Phys. Rev.* 17, 4384–4403.
30. Mountcastle, V. B. (1978) in *The Mindful Brain*, eds. Edelman, G. M. & Mountcastle, V. B. (MIT Press, Cambridge, MA), pp. 36–41.
31. Goldman, P. S. & Nauta, W. J. H. (1977) *Brain Res.* 122, 393–413.
32. Kandel, E. R. (1979) *Sci. Am.* 241, 61–70.

Chapter 4

Boltzmann Machines – 1984

Context

Invented by Ackley, Hinton and Sejnowski in 1984, the Boltzmann machine represents a transition between Hopfield nets and backprop networks[96], which were introduced by Rumelhart, Hinton and Williams in 1986. The Boltzmann machine is important because it facilitated a conceptual shift away from the idea of a neural network as a passive associative machine towards the view of a neural network as a *generative model*. This is based on the insight that a neural network should be capable of generating unit states that have the same statistical structure as its inputs. Accordingly, a fundamental assumption of generative models is that the correct interpretation of input data can only be achieved if the model has learned the underlying statistical structure of those data.

Learning in generative models can be considered an (extremely loose) analogy to dreaming in humans. If a human grew up in a forest, it is likely that their dreams would contain trees and animals rather than cars and TVs. The ability of the human to generate (dream) data that exist in their environment can be taken as evidence that they have learned a model of that environment. In a similar manner, if a network can be shown to generate data that are similar to its 'environment', then it may be inferred that the network has learned about that environment.

The only problem is that Boltzmann machines learn at a rate that is best described as glacial. But on a practical level, the Boltzmann machine demonstrated that neural networks could learn to solve complex toy (i.e. small-scale) problems, which suggested that they could learn to solve almost any problem (at least in principle, and at least eventually).

The statistical mechanics framework introduced by Hopfield nets allowed Boltzmann machines to learn interesting toy problems, but their slow learning rate effectively barred their application to more realistic problems. It is worth noting that a relatively small modification yielded the *restricted Boltzmann machine*[39], which learned more quickly and performed well on real-world problems such as handwritten digit recognition.

Technical Summary

A Boltzmann machine is essentially a Hopfield net, but with the addition of *hidden units*, as shown in Figure 4.1. Units that receive inputs from the environment (i.e. from the training set) are called *visible units*, whereas hidden units receive inputs only from visible units. Visible and hidden units are binary, with states that are either 0 or 1.

Learning consists of finding weights that force the visible units of the network to accurately mirror the states of those units when they are exposed to the physical environment. Thus, the problem is to find a set of weights which generates visible unit states that are similar to vectors in the training set, which is derived from the environment.

The strategy for solving this problem is implemented as two nested loops, where the inner loop is used to gather statistics that are then used to adjust the weights in the outer loop, as shown in

Figure 4.4. Specifically, the inner loop functions with a fixed set of weights and has two components, commonly called the *wake* and *sleep* components (although this terminology became popular some time after the publication of the paper). The wake inner loop component is used to estimate the correlation between the states of connected units (e.g. how often both are on) when the visible units are *clamped* (i.e. fixed) to each of the vectors in the training set. The sleep inner loop component is used to estimate the correlation between the states of connected units when the visible units are unclamped from the environment (i.e. free to vary according to the current weight values). Both inner loop components involve *simulated annealing* (described below), which is used to settle the network so that it attains *thermal equilibrium* before correlations are measured.

In the outer loop, for each pair of connected units, the difference between their clamped and unclamped correlations is used to adjust the weight between them. Specifically, each weight is adjusted to reduce the difference between the correlations measured in the clamped and unclamped conditions. When learning is complete, these correlations should be similar, so that the states of visible units in the unclamped network will be similar to the states of visible units when the network is clamped to vectors in the training set.

The reason this works is because all connections are symmetric. Therefore, the hidden unit states induced by training vectors are precisely the hidden unit states most likely to generate states resembling the training vectors when the visible units are unclamped. This procedure is described in more detail in the following sections.

4.1. The Boltzmann Machine Energy Function

The input x_j to the unit U_j is a weighted sum of the states s_i of units U_i that are connected to U_j, minus a threshold:

$$x_j = \sum_{i=1}^{N} w_{ij} s_i - \theta_j, \tag{4.1}$$

where N is the total number of units in the network and θ_j is the threshold of unit U_j. Note that $w_{ij} = 0$ if units U_i and U_j are not connected or if $i = j$ (i.e. no unit is connected to itself). The binary

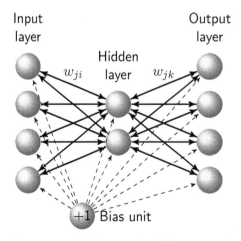

Figure 4.1. A 4-2-4 autoencoder architecture, with bias unit. All units within each layer are also connected to each other (not shown). The bias unit has a fixed state of $s = +1$. The input and output units are visible units.

output state of a unit is defined as

$$
s_j = \begin{cases} 0 & \text{if } \sum_{i=1}^{N} w_{ij} s_i < \theta_j, \\ 1 & \text{if } \sum_{i=1}^{N} w_{ij} s_i > \theta_j, \end{cases}
\tag{4.2}
$$

or equivalently

$$
s_j = \begin{cases} 0 & \text{if } x_j < 0, \\ 1 & \text{if } x_j > 0. \end{cases}
\tag{4.3}
$$

The authors simplify the notation in Equation 4.1 by adding a *bias unit* U_{N+1}, which always has state $s_{N+1} = +1$, as shown in Figure 4.1. Thus, if the threshold term is defined as $\theta_j = -w_{N+1,j}$ then Equation 4.1 can be written as

$$
x_j = \sum_{i=1}^{N+1} w_{ij} s_i.
\tag{4.4}
$$

Using this simplified notation, the threshold of each unit acts like an additional weight, which can be learned like any other weight.

The state of each unit is updated using a stochastic update rule. The probability P_j that the state of unit U_j will be set to $s_j = 1$ is defined by the sigmoidal activation function of the unit's input x_j:

$$
P_j = \frac{1}{1 + e^{-x_j/T}},
\tag{4.5}
$$

where T is a temperature parameter that determines the steepness of the activation function, as shown in Figure 4.2a. Temperature plays a vital role in finding stable states of the network.

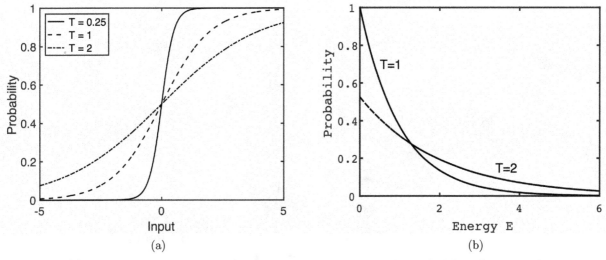

Figure 4.2. (a) The sigmoidal activation function used to determine the probability that a unit's state is $s = 1$ (Equation 4.5). As the temperature T increases, decisions regarding s become increasingly noisy. (b) The Boltzmann distribution at temperatures $T = 1$ and $T = 2$. As the energy associated with a state increases, the probability of observing that state decreases exponentially, as defined in Equation 4.8. High-energy states become more probable at high temperatures. Reproduced from Stone (2019).

The states of the N network units are represented as a vector

$$S = (s_1, \ldots, s_N). \tag{4.6}$$

For a given set of weights and a particular state vector S, the Boltzmann machine has an energy defined by the *energy function*

$$E(S) = -\sum_{i=1}^{N+1} \sum_{j=i+1}^{N+1} w_{ij} s_i s_j. \tag{4.7}$$

The probability $p(S)$ that a Boltzmann machine adopts a state S is defined in terms of a *Boltzmann distribution*,

$$p(S) = \frac{1}{Z} e^{-E(S)/T}, \tag{4.8}$$

as shown in Figure 4.2b. The term Z is the *partition function*

$$Z = \sum_{k=1}^{2^N} e^{-E/T}, \tag{4.9}$$

where the sum is taken over all 2^N possible combinations of binary unit states of the entire network. The partition function ensures that values of $p(S)$ sum to unity. Unless N is small, calculating Z is impractical; it is therefore estimated using *simulated annealing*, as described in the next section.

Equation 4.8 is borrowed from *statistical mechanics*. It states that, once the network has reached *thermal equilibrium* at temperature T, the probability that the network is in state S falls exponentially as the energy associated with that state increases. In other words, almost all states observed at thermal equilibrium are low-energy states. From a practical perspective, the temperature term T allows us to find low-energy states using simulated annealing.

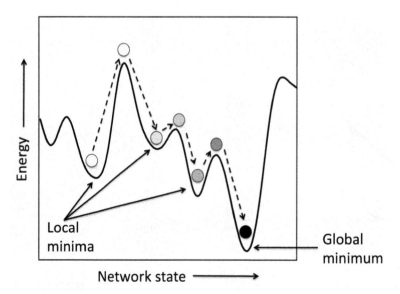

Figure 4.3. Simulated annealing. The network state is represented as a disc, and the colour of the disc represents temperature (white = hot, black = cool). At high temperatures, the network often chooses states that increase energy, allowing escape from shallow minima to deep minima. As temperature decreases, the network is more likely to choose states that decrease energy. Reproduced from Stone (2019).

4.2. Simulated Annealing

If white-hot iron is allowed to cool slowly then the atomic structure forms a low-energy, stable matrix. Conversely, if iron is cooled quickly by quenching it in cold water, then the atomic structure forms small islands of stability but the overall structure is disorganised, resulting in a brittle metal.

By analogy, if the temperature term in a Boltzmann machine is reduced slowly, this encourages the formation of low-energy stable states. More precisely, at low temperatures, Equation 4.5 dictates that a unit's state will be chosen so that it almost always decreases energy. In contrast, at high temperatures, the decision regarding a unit's state is very noisy, so its chosen state frequently increases energy. This may sound like a bad idea, but increasing energy can allow the state to escape from local minima in the energy function, as shown in Figure 4.3. If the temperature is decreased gradually, then the state can hop between successively lower local minima, so that it almost certainly arrives at the global minimum[27]. Formally, simulated annealing is a type of *Markov chain Monte Carlo* (MCMC) method.

4.3. Learning by Sculpting Distributions

We can decompose the network of N unit states S into a vector of h hidden unit states

$$H_\beta = (s_1, \ldots, s_h) \tag{4.10}$$

and a vector of v visible unit states

$$V_\alpha = (s_{h+1}, \ldots, s_{h+v}), \tag{4.11}$$

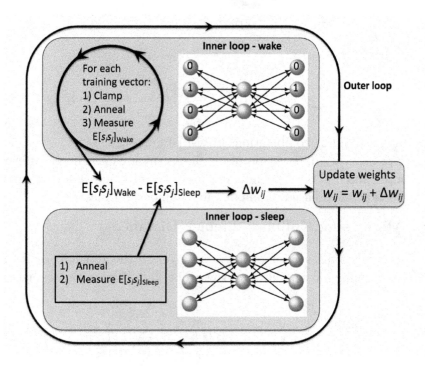

Figure 4.4. Learning in a Boltzmann machine comprises two nested loops. On each iteration of the inner loop, the correlation between unit states s_i and s_j is measured under two conditions: 1) when visible units are clamped to training vectors, which yields $\mathbb{E}[s_i s_j]_{\text{Wake}}$; and 2) when visible units are unclamped, which yields $\mathbb{E}[s_i s_j]_{\text{Sleep}}$. On each iteration of the outer loop, the weights are adjusted so that after learning, $\mathbb{E}[s_i s_j]_{\text{Wake}} \approx \mathbb{E}[s_i s_j]_{\text{Sleep}}$. Reproduced from Stone (2019).

where V_α includes both input and output unit states. Now the entire set of N units in the network can be represented as a vector

$$S_{\alpha\beta} = (H_\beta, V_\alpha). \tag{4.12}$$

Using these definitions, we can express formally how a network can generate visible unit states with a probability distribution that matches the probability distribution of visible unit states in the environment: for each training vector V_t, modify the weights so that the joint distribution $P(H_\beta, V_t)$ of the network units when the visible units are clamped to the training vector V_t is the same as the joint distribution $P'(H_\beta, V_\alpha)$ when the visible units are free to vary.

4.4. Learning in Boltzmann Machines

We wish to learn the set of J training vectors

$$\{V\} = (V_1, \ldots, V_J). \tag{4.13}$$

In essence, if the network weights are adjusted so that the probability distribution P' of visible unit states *generated by the network* matches the probability distribution $P(V)$ of the training vectors, then the network can be said to have learned the training vectors.

Accordingly, the learning rule for Boltzmann machines is obtained by forcing the distribution P' of visible unit states generated by the network when the visible units are unclamped to be similar to the distribution $P(V)$ of visible unit states when the visible units are clamped to the training set V.

So, considered over all visible unit states when unclamped, we would like the probability of each state to be the same as when the visible units are clamped. The extent to which this is not true is measured using the Kullback–Leibler divergence (KL) between $P'(V)$ and $P(V)$:

$$G = \sum_{k=1}^{2^v} P(V_k) \log \frac{P(V_k)}{P'(V_k)} \tag{4.14}$$

$$= \sum_{k=1}^{2^v} P(V_k) \log P(V_k) - \sum_{k=1}^{2^v} P(V_k) \log P'(V_k), \tag{4.15}$$

where each summation is over all 2^v combinations of visible unit states.

At this juncture, we present two different proofs of the learning rule. The proof in Section 4.5 is an exact copy of the proof presented in the Appendix of the paper, with additional explanatory comments. The simpler proof in Section 4.6 is based on maximum likelihood estimation.

4.5. Learning by Minimising Kullback–Liebler Distance

The following is a slightly rewritten version of the proof given in the appendix of the original paper.

When a network is free-running at equilibrium, the probability distribution over the visible units is

$$P'(V_\alpha) = \sum_\beta P'(V_\alpha \wedge H_\beta) = \frac{\sum_\beta e^{-E_{\alpha\beta}/T}}{\sum_{\lambda\mu} e^{-E_{\lambda\mu}/T}}, \tag{4.16}$$

where $E_{\alpha\beta}$ is the energy of the network in state $V_\alpha \wedge H_\beta$,

$$E_{\alpha\beta} = -\sum_{i<j} w_{ij} s_i^{\alpha\beta} s_j^{\alpha\beta}, \tag{4.17}$$

with $s_i^{\alpha\beta}$ representing the state of the ith unit in the global state indexed by $\alpha\beta$ when the network is free-running at equilibrium. Hence,

$$\frac{\partial e^{-E_{\alpha\beta}/T}}{\partial w_{ij}} = \frac{1}{T} s_i^{\alpha\beta} s_j^{\alpha\beta} e^{-E_{\alpha\beta}/T}. \tag{4.18}$$

Differentiating Equation 4.16 then yields

$$\frac{\partial P'(V_\alpha)}{\partial w_{ij}} = \frac{\frac{1}{T}\sum_\beta s_i^{\alpha\beta} s_j^{\alpha\beta} e^{-E_{\alpha\beta}/T}}{\sum_{\lambda\mu} e^{-E_{\lambda\mu}/T}} - \frac{\sum_\beta e^{-E_{\alpha\beta}/T} \frac{1}{T}\sum_{\lambda\mu} s_i^{\lambda\mu} s_j^{\lambda\mu} e^{-E_{\lambda\mu}/T}}{\left(\sum_{\lambda\mu} e^{-E_{\lambda\mu}/T}\right)^2} \tag{4.19}$$

$$= \frac{1}{T}\left[\sum_\beta P'(V_\alpha \wedge H_\beta) s_i^{\alpha\beta} s_j^{\alpha\beta} - P'(V_\alpha)\sum_{\lambda\mu} P'(V_\lambda \wedge H_\mu) s_i^{\lambda\mu} s_j^{\lambda\mu}\right]. \tag{4.20}$$

The derivative in Equation 4.20 is used to compute the gradient of the G-measure

$$G = \sum_\alpha P(V_\alpha) \ln \frac{P(V_\alpha)}{P'(V_\alpha)}, \tag{4.21}$$

where $P(V_\alpha)$ is the clamped probability distribution over the visible units and is independent of w_{ij}. Differentiating Equation 4.21 gives

$$\frac{\partial G}{\partial w_{ij}} = -\sum_\alpha \frac{P(V_\alpha)}{P'(V_\alpha)} \frac{\partial P'(V_\alpha)}{\partial w_{ij}}. \tag{4.22}$$

Using Equation 4.20, this becomes

$$\frac{\partial G}{\partial w_{ij}} = -\frac{1}{T}\sum_\alpha \frac{P(V_\alpha)}{P'(V_\alpha)}\left[\sum_\beta P'(V_\alpha \wedge H_\beta) s_i^{\alpha\beta} s_j^{\alpha\beta} - P'(V_\alpha)\sum_{\lambda\mu} P'(V_\lambda \wedge H_\mu) s_i^{\lambda\mu} s_j^{\lambda\mu}\right]. \tag{4.23}$$

Moving the ratio $\frac{P(V_\alpha)}{P'(V_\alpha)}$ inside the square brackets of Equation 4.23,

$$\frac{\partial G}{\partial w_{ij}} = -\frac{1}{T}\sum_\alpha\left[\sum_\beta P'(V_\alpha \wedge H_\beta) \frac{P(V_\alpha)}{P'(V_\alpha)} s_i^{\alpha\beta} s_j^{\alpha\beta} - \frac{P(V_\alpha)}{P'(V_\alpha)} P'(V_\alpha)\sum_{\lambda\mu} P'(V_\lambda \wedge H_\mu) s_i^{\lambda\mu} s_j^{\lambda\mu}\right]. \tag{4.24}$$

We now find an expression for the ratio $P'(V_\alpha \wedge H_\beta)\frac{P(V_\alpha)}{P'(V_\alpha)}$. Using *Bayes' rule* (see Appendix D),

$$P(V_\alpha \wedge H_\beta) = P(H_\beta|V_\alpha)P(V_\alpha), \tag{4.25}$$
$$P'(V_\alpha \wedge H_\beta) = P'(H_\beta|V_\alpha)P'(V_\alpha), \tag{4.26}$$

so

$$P(V_\alpha) = \frac{P(V_\alpha \wedge H_\beta)}{P(H_\beta|V_\alpha)}, \tag{4.27}$$

$$P'(V_\alpha) = \frac{P'(V_\alpha \wedge H_\beta)}{P'(H_\beta|V_\alpha)}, \tag{4.28}$$

and therefore

$$\frac{P(V_\alpha)}{P'(V_\alpha)} = \frac{P(V_\alpha \wedge H_\beta)/P(H_\beta|V_\alpha)}{P'(V_\alpha \wedge H_\beta)/P'(H_\beta|V_\alpha)}. \tag{4.29}$$

But because the probability of a hidden state given some visible state must be the same in equilibrium whether the visible units were clamped in that state or arrived there by free-running, it follows that

$$P'(H_\beta|V_\alpha) = P(H_\beta|V_\alpha), \tag{4.30}$$

and therefore Equation 4.29 becomes

$$\frac{P(V_\alpha)}{P'(V_\alpha)} = \frac{P(V_\alpha \wedge H_\beta)}{P'(V_\alpha \wedge H_\beta)}, \tag{4.31}$$

so that

$$P'(V_\alpha \wedge H_\beta)\frac{P(V_\alpha)}{P'(V_\alpha)} = P(V_a \wedge H_\beta). \tag{4.32}$$

Substituting Equation 4.32 into Equation 4.24 gives

$$\frac{\partial G}{\partial w_{ij}} = -\frac{1}{T}\sum_\alpha\left[\sum_\beta P(V_\alpha \wedge H_\beta)s_i^{\alpha\beta}s_j^{\alpha\beta} - P(V_\alpha)\sum_{\lambda\mu}P'(V_\lambda \wedge H_\mu)s_i^{\lambda\mu}s_j^{\lambda\mu}\right]. \tag{4.33}$$

Moving the summation over α inside the brackets (which can now be omitted), we have

$$\frac{\partial G}{\partial w_{ij}} = -\frac{1}{T}\sum_{\alpha\beta}P(V_\alpha \wedge H_\beta)s_i^{\alpha\beta}s_j^{\alpha\beta} - \frac{1}{T}\sum_\alpha P(V_\alpha)\sum_{\lambda\mu}P'(V_\lambda \wedge H_\mu)s_i^{\lambda\mu}s_j^{\lambda\mu}. \tag{4.34}$$

Note that

$$\sum_\alpha P(V_\alpha) = 1, \tag{4.35}$$

so

$$\frac{\partial G}{\partial w_{ij}} = -\frac{1}{T}\sum_{\alpha\beta}P(V_\alpha \wedge H_\beta)s_i^{\alpha\beta}s_j^{\alpha\beta} - \frac{1}{T}\sum_{\lambda\mu}P'(V_\lambda \wedge H_\mu)s_i^{\lambda\mu}s_j^{\lambda\mu}. \tag{4.36}$$

Define

$$p_{ij} = \sum_{\alpha\beta}P(V_\alpha \wedge H_\beta)s_i^{\alpha\beta}s_j^{\alpha\beta}, \tag{4.37}$$

$$p'_{ij} = \sum_{\lambda\mu}P'(V_\lambda \wedge H_\mu)s_i^{\lambda\mu}s_j^{\lambda\mu}. \tag{4.38}$$

Then we can write

$$\frac{\partial G}{\partial w_{ij}} = -\frac{1}{T}[p_{ij} - p'_{ij}]. \tag{4.39}$$

The term p_{ij} is a measure of the correlation between unit states s_i and s_j when the visible units are clamped to a training vector, whereas the term p'_{ij} is a measure of the correlation between unit states s_i and s_j when the network is free-running. The difference $p_{ij} - p'_{ij}$ is a measure of how well the probability distribution of free-running states matches the probability distribution of the network when the visible units are clamped to training vectors. Changing the weights in the direction of the gradient $\partial G/\partial w_{ij}$ ensures that this difference shrinks over learning iterations.

4.6. Learning by Maximising Likelihood

A simpler derivation of the learning rule than that given in the original paper can be obtained using maximum likelihood estimation. Note that G (Equation 4.14) achieves its minimum value of zero when $P(V_k) = P'(V_k)$. The distribution of training vectors $P(V_k)$ is fixed by the environment, so the first term on the right-hand side of Equation 4.15 is constant. Therefore, instead of minimising G we can minimise the second term, the *cross-entropy*

$$g \;=\; \sum_{k=1}^{2^v} P(V_k) \log \frac{1}{P'(V_k)}. \tag{4.40}$$

If the number J of training vectors is large, then the number n_k of occurrences of the training vector V_k is approximately $P(V_k) \times J$, so that

$$P(V_k) \;\approx\; n_k/J. \tag{4.41}$$

Substituting Equation 4.41 into Equation 4.40 yields

$$g \;\approx\; \frac{1}{J} \sum_{k=1}^{2^v} n_k \log \frac{1}{P'(V_k)}. \tag{4.42}$$

Given that each vector V_k occurs n_k times in the training set, g can be estimated by summing over the J vectors in the training set:

$$g \;\approx\; \frac{1}{J} \sum_{k=1}^{J} \log \frac{1}{P'(V_k)}. \tag{4.43}$$

In practice, we ignore the factor of $1/J$ because it has no effect on the weight values $\mathbf{w}$ that minimise g. Finally, rather than minimising g, we can choose to maximise $-g \times J$ (and drop the approximation symbol), which defines

$$L(\mathbf{w}) \;=\; \sum_{k=1}^{J} \log P'(V_k), \tag{4.44}$$

where $L(\mathbf{w})$ is standard notation for the *log likelihood* of $\mathbf{w}$. We are primarily interested in the set of training vectors $\{V\}$, defined over the visible units. The Boltzmann machine energy function was defined in Equation 4.7, which can be written succinctly as

$$E \;=\; -\sum_{i,j} w_{ij} s_i s_j. \tag{4.45}$$

As a reminder, the probability that the network is in the state S is given by Equation 4.8,

$$p(S) \;=\; \frac{1}{Z} e^{-\gamma E(S)}, \tag{4.46}$$

where we have simplified the notation by defining $\gamma = 1/T$, and where (Equation 4.9)

$$Z = \sum_{k,j} e^{-\gamma E(H_j, V_k)}. \tag{4.47}$$

Throughout this section, the subscript k indexes the 2^v combinations of binary visible unit states, and the subscript j indexes the 2^h combinations of binary hidden unit states. Given that $p(S) = p(H, V)$, the probability of observing the visible unit state V_k can be found by *marginalising* over the hidden unit states:

$$P'(V_k) = \sum_{j=1}^{2^h} p(H_j, V_k). \tag{4.48}$$

Assuming that all J training vectors are mutually independent, the probability of the training set $\{V\}$ given the weight vector $\mathbf{w}$ is

$$P'(\{V\}|\mathbf{w}) = \prod_{k=1}^{J} P'(V_k), \tag{4.49}$$

$$= \prod_{k=1}^{J} \sum_{j=1}^{2^h} p(H_j, V_k), \tag{4.50}$$

which is called the *likelihood* of the weight vector $\mathbf{w}$. Because the logarithm is a monotonic function, the weight values that maximise the likelihood also maximise the logarithm of the likelihood, so it does not matter whether we choose to maximise the likelihood or the log likelihood. In practice, it is usually easier to maximise the log likelihood

$$L(\mathbf{w}) = \log P'(\{V\}|\mathbf{w}) \tag{4.51}$$

$$= \sum_{k=1}^{J} \log P'(V_k), \tag{4.52}$$

which tallies with Equation 4.44. Thus, in the process of learning weights that maximise $L(\mathbf{w})$, we implement *maximum likelihood estimation* (see Appendix C). Substituting Equation 4.48 into Equation 4.52 gives

$$L(\mathbf{w}) = \sum_{k} \log \sum_{j} p(H_j, V_k), \tag{4.53}$$

where (from Equation 4.46)

$$p(H_j, V_k) = \frac{1}{Z} e^{-\gamma E(H_j, V_k)}. \tag{4.54}$$

Substituting Equation 4.54 into Equation 4.53 yields

$$L(\mathbf{w}) = \sum_k \log\left(\frac{1}{Z}\sum_j e^{-\gamma E(H_j, V_k)}\right) \quad (4.55)$$

$$= \sum_k \log\left(\sum_j e^{-\gamma E(H_j, V_k)}\right) - \sum_k \log Z. \quad (4.56)$$

Substituting Equation 4.47 into Equation 4.56 then gives

$$L(\mathbf{w}) = \sum_k \log \sum_j e^{-\gamma E(H_j, V_k)} - \sum_k \log \sum_{k,j} e^{-\gamma E(H_j, V_k)}. \quad (4.57)$$

Using Equation 4.45, this can be rewritten as

$$L(\mathbf{w}) = \sum_k \log \overset{\text{Wake}}{\sum_j} \exp\left(\gamma \sum_{i,j} w_{ij} s_i s_j\right) - \sum_k \log \overset{\text{Sleep}}{\sum_{k,j}} \exp\left(\gamma \sum_{i,j} w_{ij} s_i s_j\right). \quad (4.58)$$

However, the transition from Equation 4.57 to Equation 4.58 should be interpreted with care. Specifically, comparison of Equations 4.57 and 4.58 implies that all visible unit states in the sum marked with 'Wake' are set to an element of the current training vector. In contrast, all unit states in the sum marked with 'Sleep' are independent of the training vectors. The terms 'wake' and 'sleep' are in common usage, by analogy with the waking and sleeping states of brains.

In order to maximise the log likelihood using gradient ascent, we need an expression for the derivative of $L(\mathbf{w})$ with respect to the weights. It can be shown that the derivative for a single weight w_{ij} is

$$\frac{\partial L(\mathbf{w})}{\partial w_{ij}} = \gamma \sum_k \mathbb{E}_k[s_i s_j]_{\text{Wake}} - \gamma \sum_k \mathbb{E}_k[s_i s_j]_{\text{Sleep}} \quad (4.59)$$

$$= \gamma\left(\mathbb{E}[s_i s_j]_{\text{Wake}} - \mathbb{E}[s_i s_j]_{\text{Sleep}}\right), \quad (4.60)$$

where $\mathbb{E}[\cdot]$ represents *expected value*. The two terms on the right are often called the *wake* and *sleep* terms. The wake term refers to the co-occurrence of the states s_i and s_j weighted by the probability of those states when the visible units are clamped to a vector V_k from the training set. In contrast, the sleep term refers to the co-occurrence of the states s_i and s_j under the probability distribution $p(H_h, V_v)$ in which hidden and visible units are free to vary. The learning rule is therefore

$$\Delta w_{ij} = \epsilon\gamma\left(\mathbb{E}[s_i s_j]_{\text{Wake}} - \mathbb{E}[s_i s_j]_{\text{Sleep}}\right), \quad (4.61)$$

where ϵ is the learning rate. This can be used to perform gradient ascent using simulated annealing, as follows. For each of a sequence of temperatures from high to low, clamp each of the J training vectors to the visible units. Then repeatedly update the hidden unit states to find low-energy states at the lowest temperature, and then estimate the correlations $\mathbb{E}[s_i s_j]_{\text{Wake}}$ between all connected units. Next, unclamp the visible units and, for each of a sequence of temperatures from high to low, repeatedly sample the hidden and visible unit states to find low-energy states at the lowest temperature, and estimate the correlations $\mathbb{E}[s_i s_j]_{\text{Sleep}}$ between all connected units. Finally, use these correlations to adjust all weights in the direction of the gradient of $L(\mathbf{w})$ with respect to $\mathbf{w}$, as defined in Equation 4.61.

4.7. Results: Autoencoders and Exclusive OR

The main ideas implicit in the Boltzmann machine can be exemplified with the *autoencoder* architecture, shown in Figure 4.1. Training consists of associating each input vector with itself, so the input and output layers have the same number of units. However, because the hidden layer has fewer units than the input layer, the states of the units in the hidden layer represent a compressed version of the input. In terms of *information theory*[125], the hidden units should represent the same amount of information as the input units. For example, the autoencoder in Figure 4.1 was trained to associate each of four binary vectors with itself:

$$(0001) \rightarrow (0001), \tag{4.62}$$
$$(0010) \rightarrow (0010), \tag{4.63}$$
$$(0100) \rightarrow (0100), \tag{4.64}$$
$$(1000) \rightarrow (1000), \tag{4.65}$$

where (input vector) → (output vector). After training, each input vector induces one of four unique states (codes) in the two hidden units (i.e. 00, 01, 10 or 11), which allows the input vector to be reproduced at the output layer[2]. The paper also demonstrated that a Boltzmann machine can solve more complex encoders, such as the 8-3-8 and 40-10-40 encoders.

The autoencoder is important for three main reasons. First, it provided one of the earliest demonstrations of a network finding a code (implicit in the states of the hidden units) to represent its inputs. Second, even though Boltzmann machines have fallen out of favour, the general idea of forcing a network to compress its input into a hidden layer with minimal loss of information has re-emerged several times, and is now a fundamental part of neural network methodology using the backprop algorithm. Third, the development of *variational autoencoders*[3;56] promises to provide a more efficient method for training autoencoder networks.

Crucially, on page 27 of the paper, it is shown that the Boltzmann machine can solve the XOR problem (see Section 2.5), which had been the Achilles heel of earlier networks, such as the perceptron.

4.8. List of Mathematical Symbols

α index for the visible unit states of the network.
β index for the hidden unit states of the network.
γ $1/T$ where T is temperature.
ϵ learning rate.
θ_i bias of unit i.
G Kullback–Liebler distance between the distributions $P'(V_\alpha)$ and $P(V_\alpha)$.
h the number of hidden units.
H_α the αth hidden unit state.
p_k probability that the kth unit will adopt output state $s_k = 1$.
$P(V_\alpha)$ probability of the visible unit states V_α as determined by the environment.
$P'(V_\alpha)$ probability of the visible unit states V_α when the network is running freely.
p'_{ij} probability that units i and j are both on when the network is at equilibrium.
p_{ij} probability that units i and j are both on when the network is clamped to the environment.
s_i binary state of unit i.
T temperature.
V_α the αth visible unit state.
v the number of visible units.
x_j the total input to the jth unit.

4.9. Comments by the Paper's Author: T Sejnowski

The following is an edited transcript of comments by T Sejnowski made in an impromptu talk at the NeurIPS 2023 conference. Reproduced with permission from `https://neurips.cc/virtual/2023/workshop/66524` (starting at 00:36min).

John Hopfield gave a talk[†], and I was sitting next to Geoff Hinton, and Hopfield was saying, "We could do optimisation here", ... and that was something we were interested in. I had just read an article by Scott Kirkpatrick on simulated annealing[*], and I said, "Geoff, we could heat it up, we could heat up the Hopfield network by adding a temperature", so instead of always going downhill sometimes you pop up and with that, if you slowly cool it we can find the optimal solution. ... It turns out something magical happens when you heat up a Hopfield network: it becomes capable of learning the weights to a multi-layer network; which is the first time that had been done.

Research Paper:
Boltzmann Machines: Constraint Satisfaction Networks That Learn

Reference: Hinton, G. E., Sejnowski, T. J., and Ackley, D. H. (1984). Boltzmann machines: Constraint satisfaction networks that learn. Technical Report CMU-CS-84-119, Department of Computer Science, Carnegie-Mellon University. `https://www.cs.toronto.edu/~hinton/absps/bmtr.pdf`
Ackley, D. H., Hinton, G. E., and Sejnowski, T. J. (1985). A learning algorithm for Boltzmann machines. *Cognitive Science*, 9:147–169.
The 1985 journal paper[2] forms the first half of the more comprehensive 1984 technical report[40] reproduced here, with permission.

[†]Presumably in 1982.
[*]This is probably a reference to a technical report, published later as Kirkpatrick, Gelat and Vecchi (1983).

Boltzmann Machines:
Constraint Satisfaction Networks that Learn [*]

Geoffrey E. Hinton
Department of Computer Science
Carnegie-Mellon University, Pittsburgh, PA 15213

Terrence J. Sejnowski
Department of Biophysics
The Johns Hopkins University, Baltimore, MD 21218

David H. Ackley
Department of Computer Science
Carnegie-Mellon University, Pittsburgh, PA 15213

May, 1984

Abstract

The computational power of massively parallel networks of simple processing elements resides in the communication bandwidth provided by the hardware connections between elements. These connections can allow a significant fraction of the knowledge of the system to be applied to an instance of a problem in a very short time. One kind of computation for which massively parallel networks appear to be well suited is large constraint satisfaction searches, but to use the connections efficiently two conditions must be met: First, a search technique that is suitable for parallel networks must be found. Second, there must be some way of choosing internal representations which allow the pre-existing hardware connections to be used efficiently for encoding the constraints in the domain being searched. We describe a general parallel search method, based on statistical mechanics, and we show how it leads to a general learning rule for modifying the connection strengths so as to incorporate knowledge about a task domain in an efficient way. We describe some simple examples in which the learning algorithm creates internal representations that are demonstrably the most efficient way of using the pre-existing connectivity structure.

[*] The research reported here was supported by grants from the System Development Foundation. We thank Peter Brown, Francis Crick, Mark Derthick, Scott Fahlman, Jerry Feldman, Stuart Geman, Gail Gong, John Hopfield, Jay McClelland, Barak Pearlmutter, Harry Printz, Dave Rumelhart, Tim Shallice, Paul Smolensky, Rick Szeliski, and Venkataraman Venkatasubramanian for helpful discussions.

Table of Contents

The first 5 sections of this technical report together with the conclusion and appendix will appear as a paper in the journal **Cognitive Science**. The reference is:

Ackley, D. H., Hinton, G. E., and Sejnowski, T. J. A learning algorithm for Boltzmann Machines. **Cognitive Science** (in press).

1 Introduction

Evidence about the architecture of the brain and the potential of the new VLSI technology have led to a resurgence of interest in "connectionist" systems (Hinton & Anderson, 1981; Feldman & Ballard, 1982) that store their long term knowledge as the strengths of the connections between simple neuron-like processing elements. These networks are clearly suited to tasks like vision that can be performed efficiently in parallel networks which have physical connections in just the places where processes need to communicate. For problems like surface interpolation from sparse depth data (Grimson, 1981; Terzopoulos, 1984) where the necessary decision units and communication paths can be determined in advance, it is relatively easy to see how to make good use of massive parallelism. The more difficult problem is to discover parallel organizations that do not require so much problem-dependent information to be built into the architecture of the network. Ideally, such a system would adapt a given structure of processors and communication paths to whatever problem it was faced with.

This paper presents a type of parallel constraint satisfaction network which we call a "Boltzmann Machine" that is capable of learning the underlying constraints that characterize a domain simply by being shown examples from the domain. The network modifies the strengths of its connections so as to construct an internal *generative* model that produces examples with the same probability distribution as the examples it is shown. Then, when shown any particular example, the network can "interpret" it by finding values of the variables in the internal model that would generate the example. When shown a partial example, the network can complete it by finding internal variable values that generate the partial example and using them to generate the remainder. At present, we have an interesting mathematical result that guarantees that a certain learning procedure will build internal representations which allow the connection strengths to capture the underlying constraints that are implicit in a large ensemble of examples taken from a domain. We also have simulations that show that the theory works for some simple cases, but the current version of the learning algorithm is very slow.

The search for general principles that allow parallel networks to learn the structure of their environment has often begun with the assumption that networks are randomly wired. This seems to us to be just as wrong as the view that *all* knowledge is innate. If there are connectivity structures that are good for particular tasks that the network will have to perform, it is much more efficient to build these in at the start. However, not all tasks can be foreseen, and even for ones that can, fine-tuning may still be helpful.

Another common belief is that a general connectionist learning rule would make sequential "rule-based" models unnecessary. We believe that this view stems from a misunderstanding of the need for multiple levels of description of large systems, which can be usefully viewed as either parallel or serial depending on the grain of the analysis. Most of the key issues and questions that have been studied in the context of sequential models do not magically disappear in connectionist models. It is still necessary to perform searches for good solutions to problems or good interpretations of perceptual input, and to create complex internal representations. Ultimately it will be necessary to bridge the gap between hardware-oriented connectionist descriptions and the more abstract symbol manipulation models that have proved to be an extremely powerful and pervasive way of describing human information processing (Newell & Simon, 1972).

2 The Boltzmann Machine

The Boltzmann Machine is a parallel computational organization that is well suited to constraint satisfaction tasks involving large numbers of "weak" constraints. Constraint-satisfaction searches (e.g. Waltz, 1975; Winston, 1984) normally use "strong" constraints that *must* be satisfied by any solution. In problem domains such as games and puzzles, for example, the goal criteria often have this character, so strong constraints are the rule.[1] In some problem domains, such as finding the most plausible interpretation of an image, many of the criteria are not all-or-none, and frequently even the best possible solution violates some constraints (Hinton, 1977). A variation that is more appropriate for such domains uses weak constraints that incur a cost when violated. The quality of a solution is then determined by the total cost of all the constraints that it violates. In a perceptual interpretation task, for example, this total cost should reflect the implausibility of the interpretation.

The machine is composed of primitive computing elements called *units* that are connected to each other by bidirectional *links*. A unit is always in one of two states, *on* or *off*, and it adopts these states as a probabilistic function of the states of its neighboring units and the *weights* on its links to them. The weights can take on real values of either sign. A unit being on or off is taken to mean that the system currently accepts or rejects some elemental hypothesis about the the domain. The weight on a link represents a weak pairwise constraint between two hypotheses. A positive weight indicates that the two hypotheses tend to support one another; if one is currently accepted, accepting the other should be more likely. Conversely, a negative weight suggests, other things being equal, that the two hypotheses should not both be accepted. Link weights are *symmetric*, having the same strength in both directions (Hinton & Sejnowski, 1983).[2]

---

[1] But see (Berliner & Ackley, 1982) for argument that, even in such domains, strong constraints must be used only where absolutely necessary for legal play, and in particular must not propagate into the determination of *good* play.

[2] Requiring the weights to be symmetric may seem to restrict the constraints that can be represented. Although a constraint on boolean variables A and B such as "$A \equiv B$ with a penalty of 2 points for violation" is obviously symmetric in A and B, "$A \Rightarrow B$ with a penalty of 2 points for violation" appears to be fundamentally asymmetric. Nevertheless, this constraint can be represented by the combination of a constraint on A alone and a symmetric pairwise constraint as follows: "Lose 2 points if A is true" and "Win 2 points if both A and B are true."

The resulting structure is related to a system described by Hopfield (1982), and as in his system, each global state of the network can be assigned a single number called the "energy" of that state. With the right assumptions, the individual units can be made to act so as to *minimize the global energy*. If *some* of the units are externally forced or "clamped" into particular states to represent a particular input, the system will then find the minimum energy configuration that is compatible with that input. The energy of a configuration can be interpreted as the extent to which that combination of hypotheses violates the constraints implicit in the problem domain, so in minimizing energy the system evolves towards "interpretations" of that input that increasingly satisfy the constraints of the problem domain.

The energy of a global configuration is defined as

$$E = -\sum_{i<j} w_{ij} s_i s_j + \sum_i \theta_i s_i \tag{1}$$

where w_{ij} is the strength of connection between units i and j, s_i is 1 if unit i is on and 0 otherwise, and θ_i is a threshold.

2.1 Minimizing energy

A simple algorithm for finding a combination of truth values that is a *local* minimum is to switch each hypothesis into whichever of its two states yields the lower total energy given the current states of the other hypotheses. If hardware units make their decisions asynchronously, and if transmission times are negligible, then the system always settles into a local energy minimum (Hopfield, 1982). Because the connections are symmetric, the difference between the energy of the whole system with the kth hypothesis rejected and its energy with the kth hypothesis accepted can be determined locally by the kth unit, and this "energy gap" is just

$$\Delta E_k = \sum_i w_{ki} s_i - \theta_k \tag{2}$$

Therefore, the rule for minimizing the energy contributed by a unit is to adopt the *on* state if its total input from the other units and from outside the system exceeds its threshold. This is the familiar rule for binary threshold units.

The threshold terms can be eliminated from Eqs. (1) and (2) by making the following observation: the effect of θ_i on the global energy or on the energy gap of an individual unit is identical to the effect of a link with strength $-\theta_i$ between unit i and a special unit that is by definition always held in the *on* state. This "true unit" need have no physical reality, but it simplifies the computations by allowing the threshold of a unit to be treated in the same manner as the links. The value $-\theta_i$ is called the *bias* of unit i. If a permanently active "true unit" is assumed to be part of every network, then Eqs. (1) and (2) can be written as:

$$E = -\sum_{i<j} w_{ij} s_i s_j \tag{3}$$

$$\Delta E_k = \sum_i w_{ki} s_i \tag{4}$$

2.2 Using noise to escape from local minima

The simple, deterministic algorithm suffers from the standard weakness of gradient descent methods: It gets stuck in *local* minima that are not globally optimal. This is not a problem in Hopfield's system because the local energy minima of his network are used to store "items": If the system is started near some local minimum, the desired behavior is to fall into that minimum, not to find the global minimum. For constraint satisfaction tasks, however, the system must try to escape from local minima in order to find the configuration that is the global minimum given the current input.

A simple way to get out of local minima is to occasionally allow jumps to configurations of higher energy. An algorithm with this property was introduced by Metropolis et. al. (1953) to study average properties of thermodynamic systems (Binder, 1978) and has recently been applied to problems of constraint satisfaction (Kirkpatrick, Gelatt, & Vecchi, 1983). We adopt a form of the Metropolis algorithm that is suitable for parallel computation: If the energy gap between the *on* and *off* states of the kth unit is ΔE_k then regardless of the previous state set $s_k = 1$ with probability

$$p_k = \frac{1}{(1 + e^{-\Delta E_k/T})} \tag{5}$$

where T is a parameter that acts like temperature (see Figure 1).

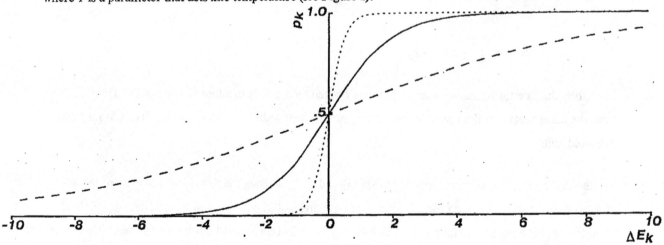

Figure 1: Eq. (5) at $T = 1.0$ *(solid)*, $T = 4.0$ *(dashed)*, and $T = 0.25$ *(dotted)*.

The decision rule in Eq. (5) is the same as that for a particle which has two energy states. A system of such particles in contact with a heat bath at a given temperature will eventually reach thermal equilibrium and the probability of finding the system in any global state will then obey a Boltzmann distribution. Similarly, a network of units obeying this decision rule will eventually reach "thermal equilibrium" and the relative probability of two global states will follow the Boltzmann distribution:

$$\frac{P_\alpha}{P_\beta} = e^{-(E_\alpha - E_\beta)/T} \tag{6}$$

where P_α is the probability of being in the αth global state, and E_α is the energy of that state.

The Boltzmann distribution has some beautiful mathematical properties and it is intimately related to information theory. In particular, the difference in the log probabilities of two global states is just their energy difference (at a temperature of 1). The simplicity of this relationship and the fact that the equilibrium distribution is independent of the path followed in reaching equilibrium are what make Boltzmann machines interesting.

At low temperatures there is a strong bias in favor of states with low energy, but the time required to reach equilibrium may be long. At higher temperatures the bias is not so favorable but equilibrium is reached faster. A good way to beat this trade-off is to start at a high temperature and gradually reduce it. This corresponds to annealing a physical system (Kirkpatrick et. al. 1983). At high temperatures, the network will ignore small energy differences and will approach equilibrium rapidly. In doing so it will perform a search of the coarse overall structure of the space of global states, and will find a good minimum at that coarse level. As the temperature is lowered, it will begin to respond to smaller energy differences and will find one of the better minima within the coarse-scale minimum it discovered at high temperature. Kirkpatrick et. al. have shown that this way of searching the coarse structure before the fine is very effective for combinatorial problems like graph partitioning, and we believe it will also prove useful when trying to satisfy multiple weak constraints, even though it will clearly fail in cases where the best solution corresponds to a minimum that is deep, narrow and isolated.

3 A learning algorithm

Perhaps the most interesting aspect of the Boltzmann Machine formulation is that it leads to a domain-independent learning algorithm that modifies the connection strengths between units in such a way that the whole network develops an internal model which captures the underlying structure of its environment. There has been a long history of failure in the search for such algorithms (Newell, 1982), and many people (particularly in Artificial Intelligence) now believe that no such algorithms exist. The major technical stumbling block which prevented the generalization of simple learning algorithms to more complex networks was this: To be capable of interesting computations, a network must contain non-linear elements that are not directly constrained by the input, and when such a network does the wrong thing it appears to be impossible to decide which of the many connection strengths is at fault. This "credit-assignment" problem was what led to the demise of Perceptrons (Rosenblatt, 1961; Minsky & Papert, 1968). The perceptron convergence theorem guarantees that the weights of a single layer of decision units can be trained, but it could not be generalized to networks of such units when the task did not directly specify how to use all the units in the network.

This version of the credit-assignment problem can be solved within the Boltzmann Machine formulation. By using the right stochastic decision rule, and by running the network until it reaches "thermal equilibrium" at some finite temperature, we achieve a mathematically simple relationship between the probability of a global state and its energy. For a network that is running freely without any input from the environment, this relationship is given by Eq. (6). Because the energy is a *linear* function of the weights (Eq. 1) this leads to a remarkably simple relationship between the log probabilities of global states and the individual connection strengths:

$$\frac{\partial \ln P_\alpha}{\partial w_{ij}} = \frac{1}{T} [s_i^\alpha s_j^\alpha - p'_{ij}] \tag{7}$$

where s_i^α is the state of the ith unit in the αth global state (so $s_i^\alpha s_j^\alpha$ is 1 only if units i and j are both on in state α), and p'_{ij} is just the probability of finding the two units i and j on at the same time when the system is at equilibrium.

Given Eq. (7), it is possible to manipulate the log probabilities of global states. If the environment directly specifies the required probabilities P_α for each global state α, there is a straightforward way of converging on a set of weights that achieve those probabilities, provided any such set exists (See Hinton & Sejnowski, 1983a for details). However, this is not a particularly interesting kind of learning because the system has to be given the required probabilities of *complete* global states. This means that the central question of what internal representation should be used has already been decided by the environment. The interesting problem arises when the environment implicitly contains high-order constraints and the network must choose internal representations that allow these constraints to be expressed efficiently.

3.1 Modeling the underlying structure of an environment

The units of a Boltzmann Machine partition into two functional groups, a non-empty set of *visible* units and a possibly empty set of *hidden* units. The visible units are the interface between the network and the environment; during training all the visible units are clamped into specific states by the environment; when testing for completion ability any subset of the visible units may be clamped. The hidden units, if any, are never clamped by the environment and can be used to "explain" underlying constraints in the ensemble of input vectors that cannot be represented by pairwise constraints among the visible units. A hidden unit would be needed, for example, if the environment demanded that the states of three visible units should have even parity — a regularity that cannot be enforced by pairwise interactions alone. Using hidden units to represent more complex hypotheses about the states of the visible units, such higher-order constraints among the visible units can be reduced to first and second-order constraints among the whole set of units.

We assume that each of the environmental input vectors persists for long enough to allow the network to approach thermal equilibrium, and we ignore any structure that may exist in the *sequence* of environmental vectors. The structure of an environment can then be specified by giving the probability distribution over all 2^v states of the v visible units. The network will be said to have a perfect model of the environment if it achieves exactly the same probability distribution over these 2^v states when it is running freely at thermal equilibrium with all units unclamped so there is no environmental input.

Unless the number of hidden units is exponentially large compared to the number of visible units, it will be impossible to achieve a *perfect* model because even if the network is totally connected the $(v+h-1)(v+h)/2$ weights and $(v+h)$ biases among the v visible and h hidden units will be insufficient to model the 2^v probabilities of the states of the visible units specified by the environment. However, if there are regularities in the environment, and if the network uses its hidden units to capture these regularities, it may achieve a good match to the environmental probabilities.

An information-theoretic measure of the discrepancy between the network's internal model and the environment is

$$G = \sum_\alpha P(V_\alpha) \ln \frac{P(V_\alpha)}{P'(V_\alpha)} \tag{8}$$

where $P(V_\alpha)$ is the probability of the αth state of the visible units when their states are determined by the environment, and $P'(V_\alpha)$ is the corresponding probability when the network is running freely with no environmental input. The G metric, sometimes called the asymmetric divergence or information gain (Kullback, 1959; Renyi, 1962), is a measure of the distance from the distribution given by the $P'(V_\alpha)$ to the distribution given by the $P(V_\alpha)$. G is zero if and only if the distributions are identical; otherwise it is positive.

The term $P'(V_\alpha)$ depends on the weights, and so G can be altered by changing them. To perform gradient descent in G, it is necessary to know the partial derivative of G with respect to each individual weight. In most cross-coupled non-linear networks it is very hard to derive this quantity, but because of the simple relationships that hold at thermal equilibrium, the partial derivative of G is straightforward to derive for our networks. The probabilities of global states are determined by their energies (Eq. 6) and the energies are determined by the weights (Eq. 1). Using these equations the partial derivative of G (see the appendix) is:

$$\frac{\partial G}{\partial w_{ij}} = -\frac{1}{T}(p_{ij} - p'_{ij}) \tag{9}$$

where p_{ij} is the average probability of two units both being in the *on* state when the environment is clamping the states of the visible units, and p'_{ij}, as in Eq. (7), is the corresponding probability when the environmental input is not present and the network is running freely. (Both these probabilities must be measured at equilibrium). Note the similarity between this equation and Eq. (7), which shows how changing a weight affects the log probability of a single state.

To minimize G, it is therefore sufficient to observe p_{ij} and p'_{ij} when the network is at thermal equilibrium, and to change each weight by an amount proportional to the difference between these two probabilities:

$$\Delta w_{ij} = \varepsilon(p_{ij} - p'_{ij}) \tag{10}$$

where ε scales the size of each weight change.

A surprising feature of this rule is that it uses only *locally available* information. The change in a weight depends only on the behavior of the two units it connects, even though the change optimizes a global measure, and the best value for each weight depends on the values of all the other weights. If there are no hidden units, it can be shown that G-space is concave (when viewed from above) so that simple gradient descent will not get trapped at poor local minima. With hidden units, however, there can be local minima that correspond to different ways of using the hidden units to represent the higher-order constraints that are implicit in the probability distribution of environmental vectors. Some techniques for handling these more complex G-spaces are discussed in the next section.

Once G has been minimized the network will have captured as well as possible the regularities in the environment, and these regularities will be enforced when performing completion. An alternative view is that the network, in minimizing G, is finding the set of weights that is most likely to have generated the set of environmental vectors. It can be shown that maximizing this likelihood is mathematically equivalent to minimizing G (Peter Brown, personal communication).

3.2 Controlling the learning

There are a number of free parameters and possible variations in the learning algorithm presented above. As well as the size of ε, which determines the size of each step taken for gradient descent, the lengths of time over which p_{ij} and p'_{ij} are estimated have a significant impact on the learning process. The values employed for the simulations presented here were selected primarily on the basis of empirical observations.

A practical system which estimates p_{ij} and p'_{ij} will necessarily have some noise in the estimates, leading to occasional "uphill steps" in the value of G. Since hidden units in a network can create local minima in G, this is not necessarily a liability. The effect of the noise in the estimates can be reduced, if desired, by using a small value for ε or by collecting statistics for a longer time, and so it is relatively easy to implement an annealing search for the minimum of G.

The objective function G is a metric that specifies how well two probability distributions match. Problems arise if an environment specifies that only a small subset of the possible patterns over the visible units ever occur. By default, the unmentioned patterns must occur with probability zero, and the only way a Boltzmann Machine running at a non-zero temperature can guarantee that certain configurations *never* occur is to give those configurations infinitely high energy, which requires infinitely large weights.

One way to avoid this implicit demand for infinite weights is to occasionally provide "noisy" input vectors. This can be done by filtering the "correct" input vectors through a process that has a small probability of reversing each of the bits. These noisy vectors are then clamped on the visible units. If the noise is small, the correct vectors will dominate the statistics, but every vector will have some chance of occurring and so infinite energies will not be needed. This "noisy clamping" technique was used for all the examples presented here. It works quite well, but we are not entirely satisfied with it and have been investigating other methods of preventing the weights from growing too large when only a few of the possible input vectors ever occur.

The simulations presented in the next section employed a modification of the obvious steepest descent method implied by Eq. (10). Instead of changing w_{ij} by an amount proportional to $p_{ij} - p'_{ij}$, it is simply incremented by a fixed "weight-step" if $p_{ij} > p'_{ij}$ and decremented by the same amount if $p_{ij} < p'_{ij}$. The advantage of this method over steepest descent is that it can cope with wide variations in the first and second derivatives of G. It can make significant progress on dimensions where G changes gently without taking very large divergent steps on dimensions where G falls rapidly and then rises rapidly again. There is no suitable value for the ε in Eq. (10) in such cases. Any value large enough to allow progress along the gently sloping floor of a ravine will cause divergent oscillations up and down the steep sides of the ravine.[3]

4 The encoder problem

The "encoder problem" (suggested to us by Sanjaya Addanki) is a simple abstraction of the recurring task of communicating information among various components of a parallel network. We have used this problem to test out the learning algorithm because it is clear what the optimal solution is like and it is non-trivial to discover it. Two groups of visible units, designated V_1 and V_2, represent two systems that wish to communicate their states. Each group has v units. In the simple formulation we consider here, each group has only one unit on at a time, so there are only v different states of each group. V_1 and V_2 are not connected directly but both are connected to a group of h hidden units H, with $h < v$ so H may act as a limited capacity bottleneck through which information about the states of V_1 and V_2 must be squeezed. Since all simulations began with all weights set to zero, finding a solution to such a problem requires that the two visible groups come to agree upon the meanings of a set of codes without any *a priori* conventions for communication through H.

To permit perfect communication between the visible groups, it must be the case that $h \geq \log_2 v$. We investigated minimal cases in which $h = \log_2 v$, and cases when h was somewhat larger than $\log_2 v$. In all cases, the environment for the network consisted of v equiprobable vectors of length $2v$ which specified that one unit in V_1 and the corresponding unit in V_2 should be on together with all other units off. Each visible group is completely connected internally and each is completely connected to H, but the units in H are not connected to each other.

Because of the severe speed limitation of simulation on a sequential machine, and because the learning requires many annealings, we have primarily experimented with small versions of the encoder problem. For example, Figure 2 shows a good solution to a "4-2-4" encoder problem in which $v = 4$ and $h = 2$. The interconnections between the visible groups and H have developed a binary coding — each visible unit causes a different pattern of *on* and *off* states in the units of H, and corresponding units in V_1 and V_2 support identical patterns in H. Note how the bias of the second unit of V_1 and V_2 is positive to compensate for the fact that the code which represents that unit has all the H units turned off.

---

[3]The problem of finding a suitable value for ε disappears if one performs a line search for the lowest value of G along the current direction of steepest descent, but line searches are inapplicable in this case. *Only* the local gradient is available. There are bounds on the second derivative that can be used to pick conservative values of ε (Mark Derthick, personal communication), and methods of this kind are currently under investigation.

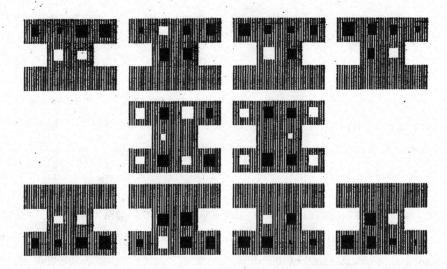

Figure 2: A solution to an encoder problem. The link weights are displayed using a recursive notation. Each unit is represented by a shaded I-shaped box; from top to bottom the rows of boxes represent groups V_1, H, and V_2. Within each box, the black or white rectangles show the strengths of that unit's connections to the other units. The *size* of a rectangle indicates the magnitude of the weight; white rectangles represent positive weights and black rectangles represent negative weights. Within the box representing a unit, the *relative position* of a rectangle indicates which other unit is involved in the connection. For example, the large white rectangle in the third unit along in the top row represents a positive connection to the first unit in the second row. All connections between units appear twice in the diagram, once in the box for each of the two units being connected. For example, the connection described above also appears in the first box of the second row of units, but it occupies a different relative position within this box. To give an idea of the scale of the weights, this connection has a weight of 26. In the position that would correspond to a unit connecting to itself (the second position in the top row of the second unit in the top row, for example), the bias is displayed.

4.1 The 4-2-4 encoder

The experiments on networks with $v = 4$ and $h = 2$ were performed using the following learning cycle:

1. *Estimation of p_{ij}:* Each environmental vector in turn was clamped over the visible units. For each environmental vector, the network was allowed to reach equilibrium twice. Statistics about how often pairs of units were both on together were gathered at equilibrium. To prevent the weights from growing too large we used the "noisy" clamping technique described in Section 3.2. Each *on* bit of a clamped vector was set to *off* with a probability of 0.15 and each *off* bit was set to *on* with a probability of 0.05.

2. *Estimation of p'_{ij}:* The network was completely unclamped and allowed to reach equilibrium at a temperature of 10. Statistics about co-occurrences were then gathered for as many annealings as were used to estimate p_{ij}.

3. *Updating the weights:* All weights in the network were incremented or decremented by a fixed weight-step of 2, with the sign of the increment being determined by the sign of $p_{ij} - p'_{ij}$.

When a settling to equilibrium was required, all the unclamped units were randomized with equal probability on or off (corresponding to raising the temperature to infinity), and then the network was allowed to run for the following times at the following temperatures: [2@20, 2@15, 2@12, 4@10].[4] After this annealing schedule it was assumed that the network had reached equilibrium, and statistics were collected at a temperature of 10 for 10 units of time.

We observed three main phases in the search for the global minimum of G, and found that the occurrence of these phases was relatively insensitive to the precise parameters used. The first phase begins with all the weights set to zero, and is characterized by the development of negative weights throughout most of the network, implementing two winner-take-all networks that model the simplest aspect of the environmental structure — only one unit in each visible group is normally active at a time. In a 4-2-4 encoder, for example, the number of possible patterns over the visible units is 2^8. By implementing a winner-take-all network among each group of four this can be reduced to 4 x 4 low energy patterns. Only the final reduction from 2^4 to 2^2 low energy patterns requires the hidden units to be used for communicating between the two visible groups. Figure 3a shows a 4-2-4 encoder network after 4 learning cycles.

Although the hidden units are exploited for inhibition in the first phase, the lateral inhibition task can be handled by the connections within the visible groups alone. In the second phase, the hidden units begin to develop positive weights to some of the units in the visible groups, and they tend to maintain symmetry between the sign and approximate magnitude of a connection to a unit in V_1 and the corresponding unit in V_2. The second phase finishes when every hidden unit has significant connection weights to each unit in V_1 and analogous weights to each unit in V_2, and most of the different codes are being used, but there are some codes that are used more than once and some not at all. Figure 3b shows the same network after 60 learning cycles.

Occasionally all the codes are being used at the end of the second phase in which case the problem is solved. Usually, however, there is a third and longest phase during which the learning algorithm sorts out the remaining conflicts and finds a global minimum. There are two basic mechanisms involved in the sorting out process. Consider the conflict between the 1st and 4th units in Figure 3b, which are both employing the code $\langle -, + \rangle$. When the system is running without environmental input, the two units will be on together quite frequently. Consequently, $p'_{1,4}$ will be higher than $p_{1,4}$ because the environmental input tends to prevent the two units from being on together. Hence the learning algorithm keeps decreasing the weight of the

---

[4]One unit of time is defined as the time required for each unit to be given, on average, one chance to change its state. This means that if there are n unclamped units, a time period of 1 involves n random probes in which some unit is given a chance to change its state.

connection between the first and fourth units in each group, and they come to inhibit each other strongly. (This effect explains the variations in inhibitory weights in Figure 2. Visible units with similar codes are the ones that inhibit each other strongly.) Visible units thus compete for "territory" in the space of possible codes, and this repulsion effect causes codes to migrate away from similar neighbors. In addition to the repulsion effect, we observed another process that tends to eventually bring the unused codes adjacent (in terms of hamming distance) to codes that are involved in a conflict. The mechanics of this process are somewhat subtle and we do not take the time to expand on them here.

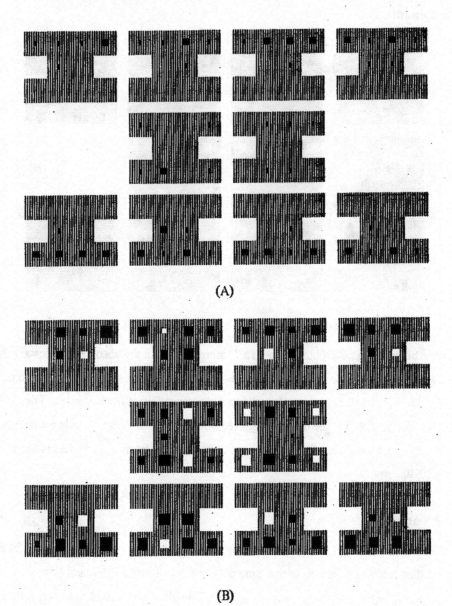

(A)

(B)

Figure 3: Two phases in the development of the perfect binary encoding shown in Figure 2. The weights are shown (A) after 4 learning trials and (B) after 60 learning trials.

The third phase finishes when all the codes are being used, and the weights then tend to increase so that the solution locks in and remains stable against the fluctuations caused by random variations in the cooccurrence statistics. (Figure 2 is the same network shown in Figure 3, after 120 learning cycles.)

In 250 different tests of the 4-2-4 encoder, it always found one of the global minima, and once there it remained there. The median time required to discover four different codes was 110 learning cycles. The longest time was 1810 learning cycles.

4.2 The 4-3-4 encoder

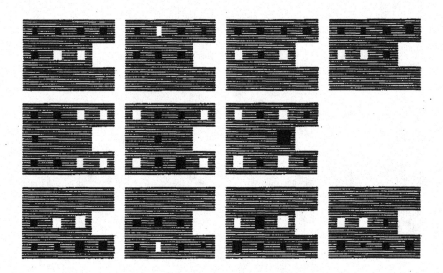

Figure 4:   A 4-3-4 encoder that has developed optimally spaced codes.

A variation on the binary encoder problem is to give H more units than are absolutely necessary for encoding the patterns in V_1 and V_2. A simple example is the 4-3-4 encoder which was run with the same parameters as the 4-2-4 encoder. In this case the learning algorithm quickly finds four different codes. Then it always goes on to modify the codes so that they are optimally spaced out and no pair differ by only a single bit, as shown in Figure 4. The median time to find four well-spaced codes was 270 learning cycles and the maximum time in 200 trials was 1090.

4.3 The 8-3-8 encoder

With $v=8$ and $h=3$ it took many more learning cycles to find all 8 three-bit codes. We did 20 simulations, running each for 4000 learning cycles using the same parameters as for the 4-2-4 case (but with a probability of 0.02 of reversing each *off* unit during noisy clamping). The algorithm found all 8 codes in 16 out of 20 simulations and found 7 codes in the rest. The median time to find 7 codes was 210 learning cycles and the median time to find all 8 was 1570 cycles.

The difficulty of finding all 8 codes is not surprising since the fraction of the weight space that counts as a solution is much smaller than in the 4-2-4 case. Sets of weights that use 7 of the 8 different codes are found fairly rapidly and they constitute local minima which are far more numerous than the global minima and have almost as good a value of G. In this type of G-space, the learning algorithm must be carefully tuned to achieve a global minimum, and even then it is very slow. We believe that the G-spaces for which the algorithm is well-suited are ones where there are a great many possible solutions and it is not essential to get the very best one. For large networks to learn in a reasonable time, it may be necessary to have enough units and weights and a liberal enough specification of the task so that no single unit or weight is essential. The next example illustrates the advantages of having some spare capacity.

4.4 The 40-10-40 encoder

A somewhat larger example is the 40-10-40 encoder. The 10 units in H is almost twice the theoretical minimum, but H still acts as a limited bandwidth bottleneck. The learning algorithm works well on this problem. Figure 5 displays the resulting network. Figure 6 shows its performance when given a pattern in V_1 and required to settle to the corresponding pattern in V_2. Each learning cycle involved annealing once with each of the 40 environmental vectors clamped, and the same number of times without clamping. The codes that the network selected to represent the patterns in V_1 and V_2 were all separated by a hamming distance of at least 2, which is very unlikely to happen by chance. As a test, we compared the weights of the connections between visible and hidden units. Each visible unit has 10 weights connecting it to the hidden units, and to avoid errors, the 10 dimensional weight vectors for two different visible units should not be too similar. The cosine of the angle between two vectors was used as a measure of similarity, and no two codes had a similarity greater than 0.73, whereas many pairs had similarities of 0.8 or higher when the same weights were randomly rearranged to provide a control group for comparison.

To achieve good performance on the completion tests, it was necessary to use a very gentle annealing schedule during testing. The schedule spent twice as long at each temperature and went down to half the final temperature of the schedule used during learning. As the annealing was made faster, the error rate increased, thus giving a very natural speed/accuracy trade-off. We have not pursued this issue any further, but it may prove fruitful because some of the better current models of the speed/accuracy trade-off in human reaction time experiments involve the idea of a biased random walk (Ratcliff 1978) and the annealing search gives rise to similar underlying mathematics.

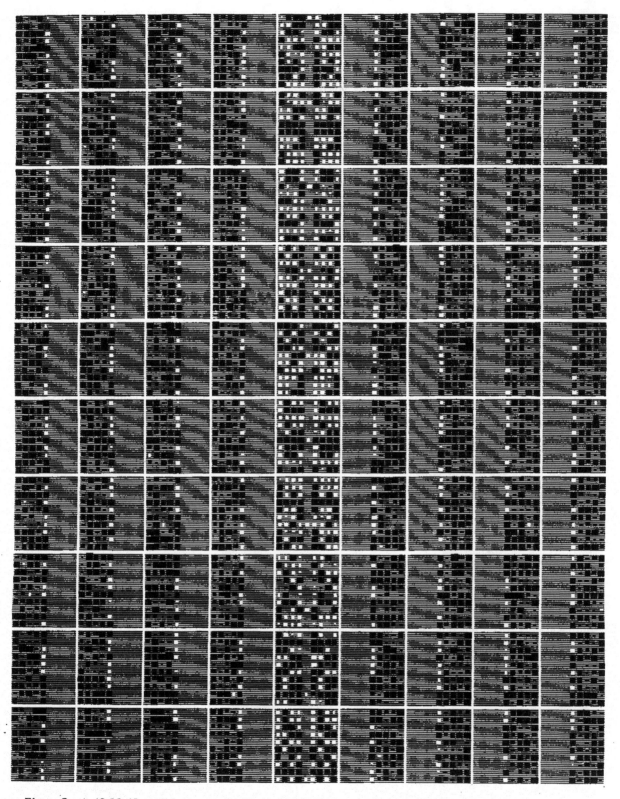

Figure 5: A 40-10-40 encoder network. The center column is group H, and the left and right four columns are groups V_1 and V_2.

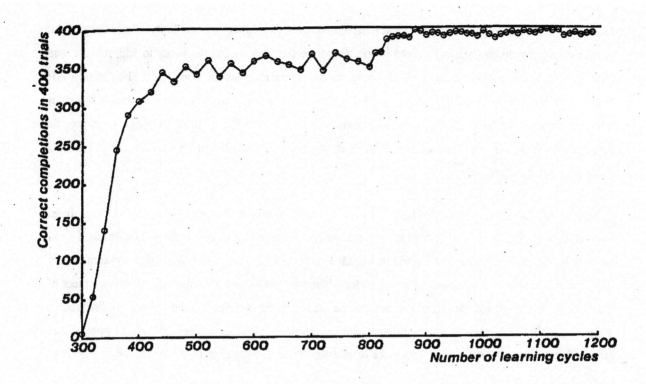

Figure 6: Completion accuracy of a 40-10-40 encoder during learning. The network was tested by clamping the states of the units in V_1 and letting the remainder of the network reach equilibrium. If just the correct unit was on in V_2, the test was successful. This was repeated 10 times for each of the 40 units in V_1. For the first 300 learning cycles the network was run without connecting up the hidden units. This ensured that each group of 40 visible units developed enough lateral inhibition to implement an effective winner-take-all network. The hidden units were then connected up and for the next 500 learning cycles we used "noisy" clamping, switching *on* bits to *off* with a probability of 0.1 and *off* bits to *on* with a probability of 0.0025. After this we removed the noise and this explains the sharp rise in performance after 800 cycles. The final performance asymptotes at 98.6% correct.

5 Representation in parallel networks

So far, we have avoided the issue of how complex concepts would be represented in a Boltzmann machine. The individual units stand for "hypotheses", but what is the relationship between these hypotheses and the kinds of concepts for which we have words? Some workers suggest that a concept should be represented in an essentially "local" fashion: the activation of one or a few computing units is the representation for a concept (Feldman & Ballard, 1982), while others view concepts as "distributed" entities: a particular pattern of activity over a large group of units represents a concept, and different concepts correspond to *alternative* patterns of activity over the same group of units (Hinton, 1981a).

One of the better arguments in favor of local representations is their inherent modularity. Knowledge about relationships between concepts is localized in specific connections and is therefore easy to add, remove, and modify, if some reasonable scheme for forming hardware connections can be found (Feldman, 1981; Fahlman, 1980). With distributed representations, however, the knowledge is diffuse. This is good for tolerance to local hardware damage, but it appears to make the design of modules to perform specific functions much harder. It is particularly difficult to see how new distributed representations of concepts could originate spontaneously.

In a Boltzmann machine, a distributed representation corresponds to an energy minimum, and so the problem of creating a good collection of distributed representations is equivalent to the problem of creating a good "energy landscape." The learning algorithm we have presented is capable of solving this problem, and it therefore makes distributed representations considerably more plausible. The diffuseness of any one piece of knowledge is no longer a serious objection, because the mathematical simplicity of the Boltzmann distribution makes it possible to manipulate all the diffuse local weights in a coherent way on the basis of purely local information. The formation of a simple set of distributed representations is illustrated by the encoder problems.

5.1 Communicating information between modules

The encoder problem examples also suggest a method for communicating symbols between various components of a parallel computational network. Feldman & Ballard (1982) present sketches of two implementations for this task, using the example of the transmission of the concept "wormy apple" from where it is recognized in the perceptual system to where the phrase "wormy apple" can be generated by the speech system. They argue that there appear to be only two ways that this could be accomplished. In the first method, the perceptual information is encoded into a set of symbols that are then transmitted as messages to the speech system, where they are decoded into a form suitable for utterance. In this case, there would be a set of general-purpose communication lines, analogous to a bus in a conventional computer, that would be used as the medium for all such messages from the visual system to the speech system. Feldman & Ballard describe the problems with such a system as:

- Complex messages would presumably have to be transmitted sequentially over the communications lines.

- Both sender and receiver would have to learn the common code for each new concept.

- The method seems biologically implausible as a mechanism for the brain.

The alternative implementation they suggest requires an individual, dedicated hardware pathway for each concept that is communicated from the perceptual system to the speech system. The idea is that the simultaneous activation of "apple" and "worm" in the perceptual system can be transmitted over private links to their counterparts in the speech system. The critical issues for such an implementation are having the necessary connections available between concepts, and being able to establish new connection pathways as new concepts are learned in the two systems. The main point of this approach is that the links between the computing units carry simple, non-symbolic information such as a single activation level.

The behavior of the Boltzmann machine when presented with an encoder problem demonstrates a way of communicating concepts that largely combines the best of the two implementations mentioned above. Like the second approach, the computing units are small, the links carry a simple numeric value, and the computational and connection requirements are within the range of biological plausibility. Like the first approach, the architecture is such that many different concepts can be transmitted over the same communication lines, allowing for effective use of limited connections. The learning of new codes to represent new concepts emerges automatically as a cooperative process from the G-minimization learning algorithm.

6 The Shifter Problem

The encoder example is a good test case for the learning algorithm because it is clear how the hidden units must be used. However, the encoder example is very untypical in an important way: the only reason the hidden units are needed at all is that the visible units do not have the appropriate direct connections. For most interesting tasks, hidden units are needed even if each visible unit is connected to every other because the sum of a lot of pairwise interactions is incapable of capturing any higher-order statistical structure that exists in the ensemble of patterns which the environment clamps on the visible units.

A simple example which can only be solved by capturing the higher-order structure is the shifter problem. The visible units are divided into three groups. Group V_1 is a one-dimensional array of 8 units each of which is clamped on or off at random with a probability of 0.3 of being on. Group V_2 also contains 8 units and their states are determined by shifting and copying the states of the units in group V_1. The only shifts allowed are one to the left, one to the right, or no shift. Wrap-around is used so that when there is a right shift, the state of the right-most unit in V_1 determines the state of the left-most unit in V_2. The three shifts are chosen at random with equal probabilities. Group V_3 contains three units to represent the three possible shifts, so at any one time one of them is clamped on and the others are clamped off.

The problem is to "recognize" the shift — i.e. to complete a partial input vector in which the states of V_1 and V_2 are clamped but the units in V_3 are left free. It is fairly easy to see why this problem cannot possibly be solved by just adding together a lot of pairwise interactions between units in V_1, V_2, and V_3. If you know that

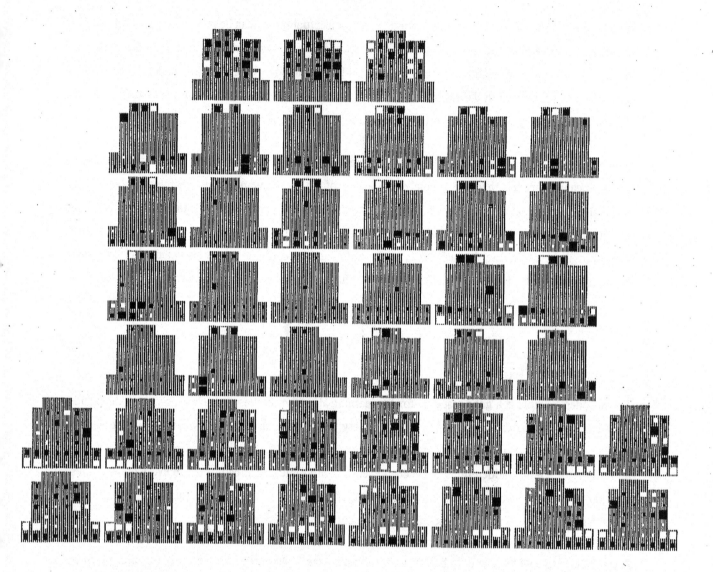

Figure 7: A shifter network. The bottom two rows of eight units are groups V_1 and V_2. The three units of group V_3 appear in the top row. The activation of the first unit of V_3 indicates that a left shift applied to V_1 produces V_2, the second unit indicates no shift, and the third unit indicates a right shift. The central four rows are the 24 hidden units. The connection strengths appearing at the bottom of a hidden unit show its "receptive field" over V_1 and V_2, and those at the top show how it affects the units in V_3.

a particular unit in V_1 is on, it tells you nothing whatsoever about what the shift is. It is only by finding *combinations* of active units in V_1 and V_2 that it is possible to predict the shift, so the information required is of at least third-order.

The obvious way to recognize the shift is to have extra units which detect informative features such as an active unit in V_1 and an active unit one place to the right in V_2 and then support the unit in V_3 that represents a right shift. The empirical question is whether the learning algorithm is capable of turning some hidden units into feature detectors of this kind, and whether it will generate a set of detectors that work well together rather than duplicating the same detector. The set of weights that minimizes G defines the *optimal* set of detectors but it is not at all obvious what these detectors are, nor is it obvious that the learning algorithm is capable of finding them.

Figure 7 shows the result of running a variation of the standard learning algorithm. Of the 24 hidden units, 5 seem to be doing very little but the remainder are sensible looking detectors and most of them have become spatially localised. One type of detector which occurs several times consists of two large negative weights, one above the other, flanked by smaller excitatory weights on each side. This is a more discriminating detector of no-shift than simply having two positive weights, one above the other. It is interesting to note that the various instances of this feature type all have different locations in V_1 and V_2, even though the hidden units are not connected to each other. The pressure for the feature detectors to be different from each other comes from the gradient of G, rather than from the kind of lateral inhibition among the feature detectors that is used in "competitive learning" paradigms (Rumelhart & Zipser, 1984; Fukushima, 1980).

The shifter problem is encouraging because it is a clear example of the kind of learning of higher-order structure that was beyond perceptrons, but it also illustrates several weaknesses in the current approach to learning:

1. The weights are fairly clearly not optimal because of the 5 hidden units that appear to do nothing useful, and further learning did not fix this. Also, the performance is not as good as it could be. When the states of the units in V_1 and V_2 are clamped and the network is annealed gently to half the final temperature used during learning, the shift units quite frequently adopt the wrong states. If the number of *on* units in V_1 is 1, 2, 3, 4, 5, 6, 7 the percentage of correctly recognized shifts is 50% 71% 81% 86% 89% 82% 66% respectively. The wide variation in the number of active units in V_1 naturally makes the task harder to learn than if a constant proportion of the units were active.

2. When the "standard" version of the learning algorithm described above was applied to the shifter problem there was a pronounced tendency for some of the hidden units to develop weights that caused them to be permanently off. This "suicide" effect is hard to recover from because if a unit is always off, its co-occurence statistics will always be the same (zero) in both the clamped and the free-running phases, and so the estimated derivative of G with respect to any of its weights will be zero. Some of the reasons for the suicide effect are discussed by Derthick (1984). To avoid it we used a variation of the learning algorithm that was suggested and tested by Barak Pearlmutter. This variation is described in Section 6.1.

3. The learning was very slow. It required 9000 learning cycles each of which involved reaching equilibrium 20 times with clamped input vectors and the same number of times without any clamping. The speed of learning is discussed further in Section 7.

6.1 Keeping the weights small

The learning algorithm presupposes that the network reaches thermal equilibrium, and it uses the co-occurence statistics measured at equilibrium to create an energy landscape that models the structure of the ensemble of vectors produced by the environment. Unfortunately there is nothing in the learning algorithm to prevent it from creating an energy landscape that contains large energy barriers which prevent the network from reaching equilibrium. If this happens, the network may perform badly, and more importantly, the statistics that are collected will not be equilibrium statistics so there is no guarantee that the changes in the weights will improve G.

One way to avoid the large weights that tend to make it hard to achieve equilibrium is to keep all the weights small. A simple way of doing this is to redefine the quantity to be minimized as:

$$G + h \sum_{ij} (w_{ij})^2$$

where h is a coefficient that determines the relative importance of minimizing G and keeping the weights small. The effect of the extra term is to make all the weights decay towards zero by an amount proportional to their current magnitude. This ensures that large weights which are not important for achieving a low value of G tend to shrink. For the shifter example shown in Figure 7 the value of h was 0.0005.

7 The speed of learning

The examples presented here took a long time to learn even though they are very small-scale. The time taken is not just a consequence of using serial machines to simulate parallel networks. Even in a truly parallel machine[5] the learning would be slow because the gradient descent requires a great many annealings with different input vectors. This slowness raises several questions:

1. How does the learning time scale with the size of the problem?

2. Can the learning algorithm be generalized to exhibit the kind of "one-shot" learning in which a person is told a fact once and then remembers it for a long time?

3. How much faster is the learning when the connectivity of the network and the initial values of the weights are approximately correct for the task at hand (as they might be for parts of the visual or motor system that have had time to evolve an appropriate architecture)?

4. Do good solutions generally have a particular statistical structure? If so, it may be possible to impose strong *a priori* domain-independent constraints on the values of the weights or the connectivity that will constrain the search for a good set of weights to a subspace. For example, it

---

[5]A truly parallel machine requires the annealing process to be tolerant of time delays. When one unit makes its decision it will be "unaware" of the states of other units that changed very recently, and this must not prevent the network from reaching equilibrium. See Section 11 for a discussion of time delays.

may be that for many problems a layered network with lateral inhibition within a layer and excitation between layers is a good solution. If so, the search can be confined to this subspace of the weight space.

We do not yet have good answers to any of these questions, but some comments on "one-shot" learning and on the way learning-time scales with the size of the problem may be helpful.

7.1 One-shot learning versus regularity detection

In the shifter problem there are 19 visible units and hence 2^{19} possible vectors that can be clamped on the visible units. Of these, only 3×2^8 actually occur. This is far too many for each pattern to be learned as an independent fact. The only way to learn the task is to capture the relatively low-order underlying regularities, and this is what the hidden units do. It is not surprising that the learning is slow. To reach the weights shown in Figure 7 required 20 x 9000 samples, but this does not seem all that excessive when the problem is expressed as learning a particular subset of 768 out of half a million patterns. A lot of samples must be taken before this subset is revealed.

It is hard for people to realise the difficulty of the task because Figure 7 is laid out in such a way that our existing notions of spatial locality make it easy to express the task. If the visible units were randomly reordered, the learning algorithm would do just as well because it starts from scratch, but people would find the task much harder. When it creates localised feature detectors among the hidden units, the learning algorithm is actually *constructing* spatial locality.

Learning a single new fact that can be expressed in terms of familiar concepts is very different from learning from scratch, particularly if the fact is one which is plausible given the existing knowledge that the system has. In the Boltzmann machine, the existing knowledge is an energy landscape, and to learn a new fact is to create a new energy minimum. For a plausible fact this means taking a state which had an energy somewhat higher than the energy of the known facts and lowering its energy. For successful one-shot learning it is important that creating this new minimum does not disturb the existing minima too much. If we use local representations in which one unit is dedicated to each fact, only the connections to this unit need to be modified, and so it is obvious that a new fact can be incorporated without interfering with existing facts. If we use distributed representations it is less obvious, because each weight will be involved in many minima, and so changing the weights to lower one minimum will interfere with many others.

The following reasoning shows that the interference effect can be made very small by using patterns of activity in which the fraction, f, of the units which are on at any one time is small. To keep the mathematics simple we assume a version of the learning algorithm that performs gradient descent by changing each weight by $\varepsilon\,(p_{ij} - p'_{ij})$. We also assume that all the global states which occur have the same fraction of *on* bits. This means that the total number of pairs of *on* units remains constant. Hence, the mean values of p_{ij} and p'_{ij} averaged over all connections must be the same and so the total of all the weights must remain constant.

Suppose that the units in the network are totally connected.[6] Suppose also that a set of global states are used to estimate p_{ij} before changing the weights. Now consider the effect of adding a single additional global state, α, to this set. This will have two effects:

1. The weights between units which are both *on* in α will get a bigger increment than before. Define δ to be the size of this increase in the increment.

2. To keep the sum of the weights constant, the remaining weights must have a reduction of $f^2\delta/(1-f^2)$ in their expected increment.

So the global state α will have its energy lowered by $N\delta$ where N is the number of pairs of *on* units in a global state. A global state, β, that is randomly related to α will have about Nf^2 of its weights incremented by δ more than before and the remainder incremented by $f^2\delta/(1-f^2)$ less than before. The net effect of the occurence of α on the energy of β will therefore be about 0, and this effect will be composed of two conflicting effects each of which is only about f^2 of the change in the energy of α. So provided $f \ll 1$ there will be very little unwanted transfer of learning. Only patterns which are significantly similar to α will have their energies significantly effected, and this is just what is need to achieve generalization. Hence, one-shot learning is possible without significantly disrupting the existing knowledge.

7.2 How the learning-time scales

The question of how the learning time depends on the size of the problem is very important, but it is also very complex because many other factors have to be scaled at the same time and it is not at all obvious how they should be scaled. The factors include:

1. The ratio of hidden to visible units.

2. The number of connections per unit.

3. The number of constraints in which each visible unit is involved.

4. The order of the underlying constraints: If the constraints among the visible units become higher order as the problem gets bigger, the problem may get much harder.

5. The compatibility of the constraints: If different constraints typically conflict with one another, the problem may become much harder as its size increases because many constraints must be

---

[6]The use of sparse connectivity does not affect the argument.

traded off against each other and this means that the system will take a long time to reach equilibrium and will only perform well if the various weak constraints have just the right relative strengths. If, on the other hand, the constraints typically agree with each other the global minimum will be over-determined.[7]. It will be possible to reach equilibrium rapidly, and the learning algorithm will only need to find a subset of the constraints in order to achieve good performance.

The encoder problem illustrates some of these points about scaling. The 8-3-8 encoder takes much longer to achieve 8 different codes than the 4-2-4 takes to achieve 4 different codes. However, the 40-10-40 encoder achieves almost perfect performance in *fewer* annealings than the 8-3-8 encoder.

There are an exponential number of valid constraints that the hidden units can capture in the encoder problem. These constraints have the form: If the active unit in V_1 is in subset S, then the active unit in V_2 must be in the equivalent subset within V_2. If the subset, S, is small the hidden unit will rarely be active and so it will convey little information on average. The best solution is for each hidden unit to dedicate itself to representing a different subset of about half the alternatives. The subsets should be chosen so that when a unit in V_1 is activated, the represented subsets that contain the unit are sufficient to encode which unit it is. For the 4-2-4 and 8-3-8 cases this can only be done by choosing orthogonal subsets each of which contains exactly half the alternatives. For the 40-10-40 case there is much more freedom in choosing the particular constraints that the hidden units should represent, and that is why the learning is relatively easier.

7.3 Degeneracy and the speed of learning

The small examples presented here require each and every visible unit to behave correctly. This may be unreasonable for larger problems. It may well be sufficient to have rather broad, degenerate energy minima in which many of the visible units are not strongly constrained to be on or off. In large networks it will probably be much easier to construct and modify these degenerate minima than to construct deep narrow minima in which the state of every unit is crucial. Narrow deep minima require large weights because changing the states of a few units must make a big change in the energy. Broad minima can be made deep without any of the weights being large.

A further interesting property of broad minima is that they suggest a way in which one concept can be differentiated into several more refined ones. By varying the weights between units that are not firmly on or off within a minimum it is possible to modify the shape of the floor of the minimum and thus to differentiate one large minimum into several closely related minima which are only separated by small energy barriers.

---

[7]The problem of labeling a line-drawing in computer vision has this property. Simple drawings without shadows are often highly ambiguous. More complex drawing with shadows can be *easier* to label because more constraints conspire together to rule out all but the correct labeling (Waltz, 1975)

This kind of differentiation could be used to model what happens when a concept like "dog" gets refined into two more specific concepts like "big nasty dog" and "little nice dog." The small barriers between concepts within the same general class should make it easy for the network to "side-track" to a neighboring, similar concept when the current concept does not quite match the data, as suggested by Minsky (1975).

When a neural model does not work as well as was hoped, it is not uncommon for its creators to claim that if only they had a *really big* network everything would be fine. Despite this unfortunate tendency, we feel that the degeneracy of large broad minima may be very important for the speed of learning and the ease of search, and so large networks may be essential because degeneracy effects disappear in very small networks.

8 Reducing higher-order constraints

An important way of distinguishing between constraints is by their "order." A first-order constraint involves only a single variable; a system consisting of only first-order constraints is trivial, since the constraints can be considered independently. Second-order constraints, involving two variables, are the simplest constraints that can require a combinatoric search. Constraints which *necessarily* involve more than two variables will be called "higher-order." When no strong constraints are involved, weak higher-order constraints among three or more given variables can always be approximated by introducing further variables and using only weak first and second-order constraints among the larger set of variables. This means that a general architecture for solving problems involving weak constraints need only implement first and second-order constraints directly, provided it can automatically reduce higher-order constraints to lower-order ones among a larger set of variables.

Some weak constraints on a set of variables that appear to be higher-order can be expressed as collections of first and second-order constraints without any extra variables. Consider, for example, the constraint on n boolean variables "If exactly one variable is true, win one point, otherwise lose at least one point". This appears to an nth-order constraint, but it can actually be expressed by the following collection of first- and second-order constraints: For each variable "Win one point if the variable is true", and for each pair of variables "Lose two points if the variables are both true".

On the other hand, some third-order constraints cannot be reduced to first- and second-order ones without introducing extra variables. The prototypical example of such a constraint is the exclusive-OR function: C should be true if A is true or B is true, but not both. The difficulty is that only the *combination* of the states of A and B constrains the state of C; the state of A or B alone says nothing about the state of C. Section 8.1 discusses this key example at length and demonstrates how the addition of an extra variable makes the reduction to second-order possible.

In general, there may be an exponentially large number of higher-order constraints among a given set of variables. A practical system will have only a limited number of extra variables at its disposal for doing the reduction to second order, so it must make compromises in selecting the best reduction. Determining which higher-order constraints to reduce, and what to reduce them to, is a central problem for efficient constraint-based knowledge representation. The general purpose learning algorithm presented in Section 3 can be viewed as an automatic method of reducing higher-order constraints among a set of variables to second-order constraints among a larger set. One can view the set of states of the visible units on which the machine is trained as a single, very high-order, disjunctive constraint. To perform search efficiently, the machine must reduce this constraint to a large set of first and second-order constraints, and to do this it must typically use extra "hidden" units that are not mentioned in the task specification.

8.1 An example

The approach and the key issues can be made clearer by considering in some detail a very simple example. Suppose an environment consisted of the following equiprobable vectors: $\{\langle 0,0,0\rangle,\langle 0,1,1\rangle,\langle 1,0,1\rangle,\langle 1,1,0\rangle\}$. In this case, there is an obvious rule that characterizes the set: the third element is the exclusive-OR of the first two. That is not the only such rule; it could also be phrased as the second being the exclusive-OR of the first and the third, or as the set of all triples of bits with even parity. The difference between these rules is in how the instances of the problem are specified. For the first rule, an instance would provide constraints on the first and second elements, and in finding a solution state that satisfied those constraints, the system could naturally be viewed as computing the exclusive-OR of the first two elements and representing the result in the third. With the third rule, any one or two of the elements might be constrained by an instance, and the system could be viewed as finding an even parity triple that satisfies the constraints of the instance.

This example is simple because the set of solution states is so small that a learning system of any significant size could simply produce an internal list of the solution states and enumerate them to find a solution state that fits any particular partial pattern. Interesting constraint satisfaction problems are rarely so compact. To learn to be an effective problem-solver, a system must have the ability to extract rules from a presentation of solution states; to extract regularities that tend to characterize membership in the set of solution states.

On the other hand, this example is non-trivial for one important reason. The state of any one element, by itself, provides *no information* about whether any other element should be one or zero. For each pair of elements, the solution set contains all four combinations of ones and zeros. It is for precisely this reason that Perceptrons were unable to compute the exclusive-OR function (Rosenblatt, 1961; Minsky & Papert, 1968). A single-level decision unit is unable to capture the distinguishing feature of all characterizations of this solution set: the states of two of the units, *taken together*, determines the third.

For the same reason, a Boltzmann Machine with only three interconnected units is unable to represent this solution set. Hypothesizing that the first element is a 1, by itself, does not constrain hypotheses about the second or the third sufficiently to solve the problem. However, notice the following decomposition of the set of solution states: $S_1 = \{\langle 0,0,0\rangle, \langle 0,1,1\rangle, \langle 1,0,1\rangle\}, S_2 = \{\langle 1,1,0\rangle\}$. S_1 alone is a partial description of the *inclusive*-OR function, and for that set there *are* constraints between pairs of units.[8] The hypothesis that the first unit is a 1 constrains the third unit also to be a one, and analogously for the second unit. S_1 can be represented by linking the first and second units to the third with positive weights, and providing the third unit with a negative bias, so that in absence of input from either of the other units, the third unit will tend to be off.

The network will now correctly find the elements of S_1, but fails on S_2, preferring $\langle 1,1,1\rangle$ over the correct state. By adding a hidden unit, however, this preference can be overridden. The hidden unit is given positive weights to the first two units, a negative weight to the third unit, and a bias low enough that both of the first two units must be on for the hidden unit to be likely to come on. The magnitudes of the weights are chosen so that the negative weight between the hidden unit and the third unit overrides the positive weights between the third unit and the first two. Figure 8 shows the resulting network.

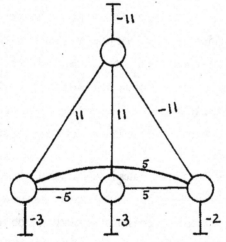

Figure 8: A Boltzmann Machine that computes the exclusive-OR function

It is worth emphasizing that connections in a Boltzmann Machine are *symmetric*; inside the machine there is no concept of "input lines" or "output lines." For example, it can cause confusion to think of the hidden unit in the exclusive-OR network as taking the states of the first two units as "input", and affecting the state of the third unit as "output." There is no preferred direction of causal flow through a Boltzmann Machine, there are only states and energies associated with those states. Units function as "inputs" when they are clamped by the

---

[8]There are other decompositions which work equally well.

environment and therefore cannot be changed by the network, and they function as "outputs" when they are not clamped, so with the pattern completion approach that we use in this paper, different instances of the same problem can completely redefine the notions of input and output.

The exclusive-OR example is a very simple illustration of a very hard general problem that a learning system must face: the task of discovering and representing the higher-order regularities of the solution set, those that do not manifest themselves as constraints on individual units or constraints between pairs of visible units. Hidden units are not directly constrained by the instances in a solution set and are thus available for reducing these higher-order constraints. Based on the environment and the structure of the network, a learning algorithm must discover distinctions that are helpful for solving the problem, as was the distinction between S_1 and S_2 above.

9 Bayesian inference

Bayesian inference suggests a general paradigm for perceptual interpretation problems. Suppose the probability associated with one unit represents the probability that a particular hypothesis, h, is correct. Suppose, also, that the "true" state of another unit is used to represent the existence of some evidence, e. Bayes theorem prescribes a way of updating the probability of the hypothesis $p(h)$ given the existence of new evidence e:

$$p(h|e) = \frac{p(h)\,p(e|h)}{p(h)\,p(e|h) + p(\bar{h})\,p(e|\bar{h})}$$

$$= 1/\left(1 + \frac{p(\bar{h})\,p(e|\bar{h})}{p(h)\,p(e|h)}\right)$$

$$= 1/\left(1 + e^{-\left(\ln\frac{p(h)}{p(\bar{h})} + \ln\frac{p(e|h)}{p(e|\bar{h})}\right)}\right)$$

where $\bar{h}$ is the negation of h.

The Bayes rule has the same form as the decision rule in (5) if we identify the probability of the unit with the probability of the hypothesis. The bias of the unit implements the *a priori* likelihood ratio and the weight implements the effect of the evidence provided by the state of the other unit (assuming the temperature is fixed at 1):

$$bias_h = \ln \frac{p(h)}{p(\bar{h})} \qquad\qquad w_{he} = \ln \frac{p(e|h)}{p(e|\bar{h})}$$

Bayesian inference with one piece of evidence can therefore be implemented by units of the type we have been considering. There are, however, several difficulties with this simple formulation.

1. It provides no way for the negation of the evidence e to affect the probability of h.

2. It does not lead to symmetrical weights when two units affect each other since $p(e|h)/p(e|\bar{h})$ is generally not equal to $p(h|e)/p(h|\bar{e})$.

3. Although it can easily be generalized to cases where there are many *independent* pieces of evidence, it is much harder to generalize to cases where the pieces of evidence not independent of each other.

A diagrammatic representation of the way to solve the first difficulty is shown in Figure 9.

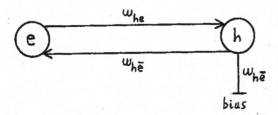

$$\text{where } w_{h\bar{e}} = \ln \frac{p(\bar{e}|h)}{p(\bar{e}|\bar{h})}$$

Figure 9: How to implement the required effect of an *off* unit. The effect of $\bar{e}$ is implemented by putting it into the bias term for h, and by subtracting an equal amount from the weighting coefficient from e, so that when e is in the true state the effect of the bias term on h is cancelled out.

The combined weight from e in Figure 9 is:

$$w_{total} = w_{he} - w_{h\bar{e}}$$

$$= \ln \frac{p(e|h)}{p(e|\bar{h})} - \ln \frac{p(\bar{e}|h)}{p(\bar{e}|\bar{h})}$$

$$= \ln \frac{p(e \wedge h)\,[1 - p(e) - p(h) + p(e \wedge h)]}{[p(e) - p(e \wedge h)] \cdot [p(h) - p(e \wedge h)]}$$

This equation is symmetrical in e and h, so in solving the problem of how to make the negation of e have the correct effect on h we have also solved the second problem — the required weights are now symmetrical. The more complicated weight in the resultant equation does not alter the fact that the probability of a hypothesis has the form of the Boltzmann distribution for a unit with two energy states.

Systems which use Bayesian inference often make the assumption that pieces of evidence are independent. This assumption is motivated by the belief that it would be too difficult to discover all the dependencies and too expensive to store them even if they were known. Unfortunately, the independence assumption is hard to justify and it is typically a poor approximation in systems with many mutually interdependent hypotheses. This has led people working in the domain of medical diagnosis, for example, to reject Bayesian statistics in favor of more discrete, symbolic systems that use underlying causal models. Recently, however, Charniak (1983) has argued that the two approaches can be reconciled. It is possible to use Bayesian statistics and to do much better than the independence assumption by using a relatively small set of additional "causal factors" which capture the most significant non-independencies. These causal factors are analogous to our hidden units because they are not part of the definition of the task — which is to relate diseases to symptoms — but they are useful because they allow the higher-order structure of the relationships to be expressed.

The analogy can be taken further. Any practical medical diagnosis system that attempted to extend the applicability of Bayesian statistics by using underlying causal factors would probably get the appropriate factors from medical experts. It is tempting, however, to ask if there is an automatic procedure for deriving the factors from the data, or for fine-tuning the factors provided by experts. This is just what our learning algorithm does. Given some hidden units, it adjusts the strengths of their connections to the visible units and to each other so as to turn them into a useful set of "causal factors."

To summarize: The independence assumption keeps things simple, but it is likely to be wrong for complex problems. The full set of inter-dependencies (higher-order statistics) is much too large to acquire or to store. A good compromise is to focus on the most significant violations of independence and to use extra "causal" factors to express them. The hidden units of a Boltzmann machine are just like these causal factors, and the learning algorithm can discover effective ways of using them.

10 Sequences

There is a serious obstacle which appears to prevent Boltzmann machines from modeling sequential symbol processing: At thermal equilibrium, there are no consistent sequences, and so a Boltzmann machine of the kind we have described is unable to produce sequential behavior. It can only respond to environmental changes, it cannot generate internal sequences.

It is tempting, when confronted with this problem to abandon the symmetry assumption and to use asymmetrical weights to encode sequential behavior. Sequences then correspond to continuous paths through phase-space. This is certainly a possibility (Hopfield, 1983), but it is incompatible with the central idea of performing constraint-satisfaction searches by relaxing to equilibrium under the influence of sustained input.[9] If *systematic* asymmetries are used to encode sequential information, the whole power of the search technique and of the resultant learning algorithm is destroyed.

What is at issue here is not just a minor modification. Systems with symmetric weights form a very interesting class of computational device because their dynamics is governed by an energy function.[10] This is what makes it possible to analyze their behavior and to use them for iterative constraint-satisfaction. They form an important natural class precisely because they *lack* the ability to go through regular sequences. In their influential exploration of Perceptrons, Minsky & Papert (1968; p 231) concluded that: "Multilayer machines with loops clearly open up all the questions of the general theory of automata." This statement is very plausible but recent developments suggest that it may be misleading because it ignores the symmetric case and it seems to have led to the general belief that it would be impossible to find an interesting generalization of perceptrons, and in particular a generalized perceptron convergence theorem for multi-layered networks.

An alternative to simply mixing in some asymmetric links in order to allow sequences is to separate out the symmetric and asymmetric components. A system could be composed of a number of internally symmetric modules that are asymmetrically connected to one another. Each module could then perform constraint-

---

[9]It is, of course, possible to *represent* sequences in a symmetric Boltzmann machine. If several different groups of units are dedicated to the different temporal pieces of a sequence, the states of these groups can represent the "fillers" of these temporal "slots," and the weights can be set so as to create energy minima for particular combinations of temporal slot fillers. It is then possible to recall the rest of a sequence from some fraction of it. However, this ability to recreate a *static* "spatial" representation of a sequence is quite different from the ability to proceed through the sequence. It does not require any temporal regularity in the internal states of the machine.

[10]One can easily write down a similar energy function for asymmetric networks, but this energy function does not govern the behavior of the network when the links are given their normal causal interpretation.

satisfaction searches, with the asymmetric inputs to the module acting as boundary conditions that determine the particular problem it has to solve at any moment. Given an architecture of this kind, it might be appropriate to use very different kinds of description at different time-scales. In the short term, the modules would perform parallel iterative constraint-satisfaction searches. In the longer term, the result of each search could be viewed as a single step in a strictly serial process, with each search setting up the boundary conditions for the next.

The idea of a coarse-grained, sequential description that is implemented by a series of parallel constraint-satisfaction searches seems to fit quite well with the idea of a production system architecture in which all the heavy computational work is done by a parallel "recognition process" that decides which rule best fits the current state of working memory.

One of the major difficulties in implementing this kind of matching is to ensure that there is a consistent set of bindings between the variables in a rule and the constants and variables of the instances in working memory. Discovering a consistent set of bindings is called unification and it requires a rather flexible matching process. Newell (1978) has suggested that this kind of flexible matching acts as a sequential bottleneck in people. Hinton (1981) describes a parallel network that is capable of performing the equivalent process in object-recognition.[11] This network has the interesting property that it can only settle on one match at a time. The settling process performs a large parallel search among alternative "rules" and "bindings" but the only way to ensure consistency of the bindings is to settle on a single match.

There are clearly many difficult problems in implementing production systems in a collection of Boltzmann machine modules, but we feel that this "discrete symbolic" approach may be more fruitful than trying to model sequences by continuous paths through state space, and it may help to explain how a collection of massively parallel, rather noisy modules can behave like a machine that manipulates discrete symbols according to formal rules.

11 The relationship to the brain

One of the main reasons for studying Boltzmann machines is that they bear some resemblances to brains. The hope is that by studying a simple and idealized machine that is in the same general class of computational device as the brain, we can gain insight into the principles that underlie biological computation (especially the kinds of computation that occur in mammalian neocortex). This is clearly a vain hope if there are

---

[11]In object recognition the problem is to find the viewpoint-independent object-model that best fits the current collection of viewpoint-dependent features. The object-models are like rules and the viewer-centered features are like instances in working memory. The viewing transform that relates viewer-centered and object-centered features is like the set of variable bindings.

irreconcilable differences between the cortex and Boltzmann machines that are crucial to the way Boltzmann machines compute. In this section we discuss some of the most obvious differences.

11.1 Binary states and action potentials

Neurons are complex biochemical entities and the simple binary units studied here are not meant to be literal models of cortical neurons. However, two of our key assumptions, that the binary units change state asynchronously and that they use a probabilistic decision rule, seem closer to reality than a model with synchronous, deterministic updating.

The energy gap for a binary unit has a role similar to that played by the membrane potential for a neuron: both are the sum of the excitatory and inhibitory inputs and both are used to determine the output state. However, neurons produce action potentials — brief spikes that propagate down axons — rather than binary outputs. When an action potential reaches a synapse, the signal it produces in the postsynaptic neuron rises to a maximum and then exponentially decays with the time constant of the membrane (typically around 5 msec for neurons in cerebral cortex). The effect of a single spike on the postsynaptic cell body may be further broadened by electrotonic transmission down the dendrite to the cell body.

The energy gap represents the summed input from all the recently active binary units. If the average time between updates is identified with the average duration of a postsynaptic potential then the binary pulse between updates can be considered an approximation to the postsynaptic potential. Although the shape of a single binary pulse differs significantly from a postsynaptic potential, the sum of a large number of stochastic pulses is independent of the shape of the individual pulses and depends only on their amplitudes and durations. Thus for large networks having the large fan-ins typical of cerebral cortex (around 10,000) the binary approximation may not be too bad.

11.2 Implementing temperature in neurons

What significance could the probabilistic decision rule in Eq. (5) have for neurons, and in particular, what does the temperature correspond to and how can it be controlled? The membrane potential of a neuron is graded, but if it exceeds a fairly sharp threshold an action potential is produced followed by a refractory period lasting several msec during which another action potential cannot be elicited. If Gaussian noise is added to the membrane potential, then even if the total synaptic input is below threshold, there is a finite probability that the membrane potential will reach threshold.

The amplitude of the Gaussian noise will determine the width of the sigmoidal probability distribution for the neuron to fire during a short time interval, and it therefore plays the role of temperature in the model.

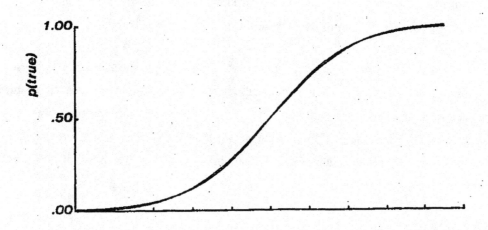

Figure 10: Two superimposed curves. One is the sigmoid function of Figure 1. The other is the cumulative area under a Gaussian, which gives the probability that the threshold of a unit will be exceeded by the sum of its energy gap and some Gaussian noise with mean 0. If the standard deviation of the Gaussian is chosen appropriately, the two curves never differ by more than about 1%.

Surprisingly, a cumulative Gaussian is a very good approximation to the required probability distribution (Figure 10). Intracellular recordings from neurons show that there is stochastic variability in the membrane potential of most neurons, in part due to fluctuations in the transmitter released by presynaptic terminals. Other sources of noise may also be present and could be controlled by cellular mechanisms.

In the visual cortex of primates single neurons respond to the same visual stimulus with different sequences of action potentials on each trial (Sejnowski, 1981). In order to measure a repeatable response the spike trains are typically averaged over 10 trials. The result, called the post-stimulus time histogram, gives the probability for a spike to occur as a function of the time after the onset of the stimulus. However, this averaging procedure throws out all information about the variance of the noise, so that there is no way to determine whether the noise varies systematically during the response to the stimulus or perhaps on a longer time scale while the stimulus is being attended.

Unlike the bulk of the brain, which is composed of many morphologically different nuclei, the cerebral cortex is relatively uniform in structure. Different areas of cerbral cortex are specialized for processing information from different sensory modalities, such as visual cortex, auditory cortex, and somatosensory cortex, and other areas are specialized for motor functions; however, all of these cortical areas have a similar anatomical organization and are more similar to each other in cytoarchitecture than they are to any other part of the brain. Many problems in vision, speech recognition, associative recall, and motor control can be formulated as searches. The similarity between different areas of cerebral cortex suggests that the same kind of massively parallel searches may be performed in many different cortical areas.

It may be difficult to arrange for all neurons in a network to have the same amplitude of noise. How sensitive are our statistical results to the assumption that all binary units are making decisions at the same temperature? The effect of variations in the temperature was tested in large-scale simulations of a problem reported in (Sejnowski, Hinton, Kienker, & Schumacher, 1984). Random variations in the temperature from unit to unit of up to 25% did not significantly affect the annealing and equilibrium solutions. The effect of variations in temperature on learning has not been studied.

11.3 Asymmetry and time-delays

In a Boltzmann Machine of the kind we have presented all connections are symmetrical. It is very unlikely that this assumption is strictly true of neurons in cerebral cortex. However, if the constraints of a problem are inherently symmetrical and if the network on average approximates the required symmetrical connectivity, then random asymmetries in a large network will be reflected as an increase in the Gaussian noise in each unit. Systematic asymmetries may have other purposes. For example, some interneurons are thought to have only inhibitory links to other neurons, and these could serve as automatic gain controls to keep a network within a narrow operating range of firing rates.

To see why random asymmetry acts like gaussian noise, consider a symmetrical network in which pairs of units are linked by two equal one-way connections, one in each direction. Now, perform the following operation on all pairs of these one-way connections: Remove one of the connections and double the strength of the other. Provided the choice of which connection to remove is made randomly, this operation will not alter the *expected* value of the input to a unit from the other units. On average, it will "see" half as many other units, but with twice the weight. So if a unit has a large fan-in it will be able to make a good unbiased estimate of what its total input would have been if the links had not been cut. However, the use of fewer, larger weights will increase the variance of the energy gap, and will thus act like added noise.

The idea that a large fan-in is needed to reduce the effects of random asymmetries has an interesting consequence for artificially produced systems of this kind. There may be a trade-off between symmetry and fan-in. In systems like the brain where connections are *grown* it may be hard to ensure symmetry, so a large fan-in may be essential (Cortical pyramidal cells typically receive between 10^3 and 10^5 input connections). In artificial systems it may be hard to achieve very high fan-ins, but this may not be essential provided the connections are symmetrical.

The analysis of the effects of time-delays is somewhat more complex, but simulations suggest that time-delays act like added noise, and preliminary mathematical results (Venkataraman Venkatasubramanian, personal communication) show that this is true to first order provided the fan-in is large and the individual weights are small compared with the energy gaps.

The main difficulty in the treatment of both asymmetry and time-delays is that it is hard to know how they will interact with the learning algorithm. It is quite possible, for example, that a network which starts with random asymmetry will develop systematic asymmetry.

12 Conclusion

The application of statistical mechanics to constraint-satisfaction searches raises a great many issues that we have only mentioned in passing. Some of these issues are discussed in greater detail elsewhere: Hinton and Sejnowski (1983) describe the relation to more conventional relaxation techniques; Fahlman, Hinton and Sejnowski (1983) compare Boltzmann machines with some alternative parallel schemes, and discuss some of the knowledge representation issues; Geman and Geman (1983) describe a very similar model, developed independently, and discuss its relationship to Markov random fields.

The two main ideas that led to the Boltzmann Machine are that noise can help with search, and that Boltzmann distributions make it possible to assign credit on the basis of *local* information in a non-linear network. It is interesting that a similar approach can be derived from entirely different considerations. While investigating how to perform computation reliably with unreliable components, Von Neumann was led to the following conclusion:

> All of this will lead to theories [of computation] which are much less rigidly of an all-or-none nature than past and present formal logic. They will be of a much less combinatorial, and much more analytical, character. In fact, there are numerous indications to make us believe that this new system of formal logic will move closer to another discipline which has been little linked in the past with logic. This is thermodynamics, primarily in the form it was received from Boltzmann, and is that part of theoretical physics which comes nearest in some of its aspects to manipulating and measuring information.
> — John Von Neumann, *Collected Works* Vol 5, pg 304

References

Berliner, H.J., & Ackley, D.H. The QBKG system: Generating explanations from a non-discrete knowledge representation. *Proceedings of the National Conference on Artificial Intelligence AAAI-82*, Pittsburgh, PA, August 1982, 213-216.

Binder, K. (Ed.) *The Monte-Carlo Method in Statistical Physics.* New York: Springer-Verlag, 1978.

Charniak, E. The Bayesian basis of common sense medical diagnosis. *Proceedings of the National Conference on Artificial Intelligence AAAI-83*, Washington, DC, August 1983, 70-73.

Derthick, M. Learning in Boltzmann Machines and why it's slow. Technical report CMU-CS-84-120, Pittsburgh, PA: Carnegie-Mellon University, 1982.

Fahlman, S.E. The Hashnet Interconnection Scheme. Technical report CMU-CS-80-125, Carnegie-Mellon University, June 1980.

Fahlman, S.E., Hinton, G.E., & Sejnowski, T.J. Massively parallel architectures for AI: NETL, Thistle, and Boltzmann Machines. *Proceedings of the National Conference on Artificial Intelligence AAAI-83*, Washington, DC, August 1983, 109-113.

Feldman, J.A. A connectionist model of visual memory. In G.E. Hinton & J.A. Anderson (Eds.) *Parallel Models of Associative Memory.* Hillsdale, NJ: Erlbaum, 1981.

Feldman, J.A., & Ballard, D.H. Connectionist models and their properties. *Cognitive Science*, 1982, *6*, 205-254.

Fukushima, K. Neocognitron: A self-organizing neural network model for a mechanism of pattern recognition unaffected by shift in position. *Biological Cybernetics*, 1980, *36*, 193-202.

Geman, S. & Geman, D. Stochastic relaxation, Gibbs distributions, and the Bayesian restoration of images. Unpublished manuscript, 1983.

Grimson, W.E.L. *From Images to Surfaces.* Cambridge, MA: MIT Press, 1981.

Hinton, G.E. Relaxation and its role it vision. PhD Thesis, University of Edinburgh, 1977; Described in: *Computer Vision*, D.H. Ballard and C.M. Brown (Eds.) Englewood Cliffs, NJ: Prentice-Hall, 1982, 408-430.

Hinton, G.E. Implementing semantic networks in parallel hardware. In G.E. Hinton & J.A. Anderson (Eds.) *Parallel Models of Associative Memory.* Hillsdale, NJ: Erlbaum, 1981a.

Hinton, G. E. A parallel computation that assigns canonical object-based frames of reference. In *Proceedings of the Seventh International Joint Conference on Artificial Intelligence*, Vol 2. Vancouver BC, Canada. August 1981b.

Hinton, G.E., & Anderson, J.A. *Parallel Models of Associative Memory.* Hillsdale, NJ: Erlbaum, 1981.

Hinton, G.E., & Sejnowski, T.J. Analyzing cooperative computation. *Proceedings of the Fifth Annual Conference of the Cognitive Science Society.* Rochester, NY, May 1983. (a)

Hinton, G.E., & Sejnowski, T.J. Optimal perceptual inference. *Proceedings of the IEEE Computer Society Conference on Computer Vision and Pattern Recognition.* Washington, DC, June 1983, 448-453. (b)

Hinton, G.E., Sejnowski, T.J., & Ackley, D.H. Boltzmann Machines: Constraint satisfaction networks that learn. Technical report CMU-CS-84-119, Carnegie-Mellon University, May 1984.

Hopfield, J.J. Neural networks and physical systems with emergent collective computational abilities. *Proceedings of the National Academy of Sciences USA,* 1982, *79,* 2554-2558.

Hopfield, J.J. Collective processing and neural states. To appear in: *Modeling and analysis in Biomedicine.*

Kirkpatrick, S., Gelatt, C.D., & Vecchi, M.P. Optimization by simulated annealing. *Science,* 1983, *220,* 671-680.

Kullback, S. *Information Theory and Statistics.* New York: Wiley, 1959.

Metropolis, N., Rosenbluth, A., Rosenbluth, M., Teller, A. & Teller, E. *Journal of Chemical Physics,* 1953, *6,* 1087.

Minsky, M. L. A framework for representing knowledge. In *The Psychology of Computer Vision,* P. Winston (Ed.) NY: McGraw-Hill, 1975.

Minsky, M., & Papert, S. *Perceptrons.* Cambridge, MA: MIT Press, 1968.

Newell, A. Harpy, Production Systems and Human Cognition. Technical report CMU-CS-78-140, Pittsburgh, PA: Carnegie-Mellon University, 1978.

Newell, A. Intellectual issues in the history of artificial intelligence. Technical report CMU-CS-82-142, Pittsburgh, PA: Carnegie-Mellon University, 1982.

Newell, A. & Simon, H.A. *Human Problem Solving.* Englewood Cliffs, NJ: Prentice-Hall, 1972.

Ratcliff, R. A theory of memory retrieval. *Psychological Review,* 1978, *85*(2), 59-108.

Renyi, A. *Probability Theory.* Amsterdam: North-Holland, 1962.

Rosenblatt, F. *Principles of neurodynamics: Perceptrons and the theory of brain mechanisms.* Washington, DC: Spartan, 1961.

Rumelhart, D. E. and Zipser, D. Competitive Learning. *Cognitive Science,* (in press).

Sejnowski, T. J., Hinton, G. E., Kienker, P. & Schumacher, L. (In preparation).

Sejnowski, T. J., Skeleton filters in the brain. In: G. E. Hinton & J. A. Anderson (Eds.) *Parallel Models of Associative Memory.* Hillsdale, NJ: Erlbaum, 1981.

Smolensky, P., Schema selection and stochastic inference in modular environments. *Proceedings of the National Conference on Artificial Intelligence AAAI-83*, Washington, DC, August 1983, 109-113.

Terzopoulos, D., Multiresolution computation of visible-surface representations. PhD. Thesis, Cambridge, MA: MIT, 1984.

Waltz, D.L. Understanding line drawings of scenes with shadows. In *The Psychology of Computer Vision*, P. Winston (Ed.) NY: McGraw-Hill, 1975.

Winston, P.H., *Artificial Intelligence.* (2nd edition) Reading, MA: Addison-Wesley, 1984.

Appendix: Derivation of the learning algorithm

When a network is free-running at equilibrium the probability distribution over the visible units is given by

$$P'(V_\alpha) = \sum_\beta P'(V_\alpha \wedge H_\beta) = \frac{\sum_\beta e^{-E_{\alpha\beta}/T}}{\sum_{\lambda\mu} e^{-E_{\lambda\mu}/T}} \tag{11}$$

where V_α is a vector of states of the visible units, H_β is a vector of states of the hidden units, and $E_{\alpha\beta}$ is the energy of the system in state $V_\alpha \wedge H_\beta$

$$E_{\alpha\beta} = -\sum_{i<j} w_{ij} s_i^{\alpha\beta} s_j^{\alpha\beta}$$

Hence,

$$\frac{\partial e^{-E_{\alpha\beta}/T}}{\partial w_{ij}} = \frac{1}{T} s_i^{\alpha\beta} s_j^{\alpha\beta} e^{-E_{\alpha\beta}/T}$$

Differentiating (11) then yields

$$\frac{\partial P'(V_\alpha)}{\partial w_{ij}} = \frac{\frac{1}{T}\sum_\beta e^{-E_{\alpha\beta}/T} s_i^{\alpha\beta} s_j^{\alpha\beta}}{\sum_{\lambda\mu} e^{-E_{\lambda\mu}/T}} - \frac{\sum_\beta e^{-E_{\alpha\beta}/T} \frac{1}{T}\sum_{\lambda\mu} e^{-E_{\lambda\mu}/T} s_i^{\lambda\mu} s_j^{\lambda\mu}}{\left(\sum_{\lambda\mu} e^{-E_{\lambda\mu}/T}\right)^2}$$

$$= \frac{1}{T}\left[\sum_\beta P'(V_\alpha \wedge H_\beta) s_i^{\alpha\beta} s_j^{\alpha\beta} - P'(V_\alpha)\sum_{\lambda\mu} P'(V_\lambda \wedge H_\mu) s_i^{\lambda\mu} s_j^{\lambda\mu}\right]$$

This derivative is used to compute the gradient of the G-measure

$$G = \sum_\alpha P(V_\alpha) \ln \frac{P(V_\alpha)}{P'(V_\alpha)}$$

where $P(V_\alpha)$ is the clamped probability distribution over the visible units and is independent of w_{ij}. So

$$\frac{\partial G}{\partial w_{ij}} = -\sum_\alpha \frac{P(V_\alpha)}{P'(V_\alpha)} \frac{\partial P'(V_\alpha)}{\partial w_{ij}}$$

$$= -\frac{1}{T}\sum_\alpha \frac{P(V_\alpha)}{P'(V_\alpha)}\left[\sum_\beta P'(V_\alpha \wedge H_\beta) s_i^{\alpha\beta} s_j^{\alpha\beta} - P'(V_\alpha)\sum_{\lambda\mu} P'(V_\lambda \wedge H_\mu) s_i^{\lambda\mu} s_j^{\lambda\mu}\right]$$

Now,

$$P(V_\alpha \wedge H_\beta) = P(H_\beta|V_\alpha)P(V_\alpha),$$
$$P'(V_\alpha \wedge H_\beta) = P'(H_\beta|V_\alpha)P'(V_\alpha),$$

and

$$P'(H_\beta|V_\alpha) = P(H_\beta|V_\alpha) \tag{12}$$

Equation (12) holds because the probability of a hidden state given some visible state must be the same in equilibrium whether the visible units were clamped in that state or arrived there by free-running. Hence,

$$P'(V_\alpha \wedge H_\beta) \frac{P(V_\alpha)}{P'(V_\alpha)} = P(V_\alpha \wedge H_\beta)$$

Also,

$$\sum_\alpha P(V_\alpha) = 1$$

Therefore,

$$\frac{\partial G}{\partial w_{ij}} = -\frac{1}{T}[p_{ij} - p'_{ij}]$$

where

$$p_{ij} \stackrel{\text{def}}{=} \sum_{\alpha\beta} P(V_\alpha \wedge H_\beta) s_i^{\alpha\beta} s_j^{\alpha\beta}$$

and

$$p'_{ij} \stackrel{\text{def}}{=} \sum_{\lambda\mu} P'(V_\lambda \wedge H_\mu) s_i^{\lambda\mu} s_j^{\lambda\mu}$$

as given in (9).

The Boltzmann Machine learning algorithm can also be formulated as an input-output model. The visible units are divided into an input set I and an output set O, and an environment specifies a set of conditional probabilities of the form $P(O_\beta|I_\alpha)$. During the "training" phase the environment clamps both the input and output units, and p_{ij}s are estimated. During the "testing" phase the input units are clamped and the output units and hidden units free-run, and p'_{ij}s are estimated. The appropriate G measure in this case is

$$G = \sum_{\alpha\beta} P(I_\alpha \wedge O_\beta) \ln \frac{P(O_\beta|I_\alpha)}{P'(O_\beta|I_\alpha)}$$

Similar mathematics apply in this formulation and $\partial G/\partial w_{ij}$ is the same as before.

Chapter 5

Backpropagation Networks – 1985

Context

The backpropagation or *backprop* algorithm is widely recognised as the work-horse of modern neural networks, and its promise was apparent even from its inception.

An early pattern recognition application from 1988 involved using underwater sonar signals to discriminate between rocks and metal cylinders, on which a performance of 90% was achieved[30]. The idea of driverless cars is now commonplace, but in 1989 one of the first attempts to use a backprop network for this task involved learning to steer a car using only a sequence of images as training data[85]. In the domain of games, as early as 1995 a backprop network was used in combination with a *reinforcement learning algorithm* to beat the best human players at backgammon[131]. Finally, one of the first successful attempts to train networks to recognise handwritten digits was achieved using a backprop network[63] in 1989, as described in Chapter 7.

The achievement of Sejnowski and Rosenberg's NETtalk[102] in 1987 has already been mentioned in Section 1.5 and is described more fully in Section 12.2. As Sejnowski and Rosenberg noted, a crucial property of backprop networks is that they are multi-layered:

> *Until recently, learning in multilayered networks was an unsolved problem and considered by some impossible.*

Moreover, because each layer utilised units with nonlinear (e.g. sigmoidal) input/output functions, it seemed likely that such a network should be able to learn a wide variety of tasks requiring nonlinear transformations between its input and output layers.

Indeed, one reason for the rapid take-up of backprop networks is the existence of several reassuring theorems, of which Cybenko's remarkable *universal approximation theorem*[22] is an early example from 1989. This states that the single layer of nonlinear hidden units in a backprop network is sufficient to approximate any mapping from input to output units. The only problem is that Cybenko's theorem does not say how many hidden units are required to learn a given mapping, nor how long it should take for the network to learn that mapping. For a problem such as XOR, it is a relatively simple matter to discover by trial and error that two hidden units suffice. However, for more complex problems, the number of hidden units required is not known in general.

Technical Summary

The backprop algorithm was brought to prominence by Rumelhart, Hinton and Williams (1986), although it had been discovered on previous occasions (e.g. Werbos, 1974).

It might seem obvious that the performance of a perceptron could be improved by increasing the number of layers of units. The problem is knowing how to train such a *multi-layer perceptron*. If we used a method based on gradient descent then this would involve the derivative of the perceptron unit activation function. However, because the perceptron unit activation function is a step function, its derivative is infinite (which effectively precludes performing gradient descent). Another strategy would be to increase the number of layers in a linear associative network, where each unit has a linear activation function. However, it can be shown that the composite input/output function of any multi-layer network in which all units have linear activation functions is also linear, so adding more layers adds no functionality.

A compromise would be to add extra layers with units that have activation functions that lie somewhere between the step function of a perceptron and the linear activation function of a linear associative network. One such compromise is represented by a *sigmoidal* (S-shaped) activation function, often referred to as a semi-linear activation function (Figure 5.1). A typical sigmoidal activation function is

$$
\begin{aligned}
O_{pj} &= f(net_{pj}) & (5.1)\\
&= 1/(1 + e^{-net_{pj}}), & (5.2)
\end{aligned}
$$

where net_{pj} is the total input to the jth unit and O_{pj} is the output of the unit. The derivative of this activation function is

$$
\frac{dO_{pj}}{dnet_{pj}} = O_{pj}(1 - O_{pj}). \qquad (5.3)
$$

Unlike a step function, a sigmoidal activation function has a finite derivative for all input values. Additionally, in contrast to a network of linear units, the overall input/output function computed by a network with three (or more) layers of sigmoidal units cannot be computed by any network with just two layers of sigmoidal units; so adding layers here does add functionality.

In historical terms, the compromise represented by a sigmoidal activation function was not deduced in the straightforward manner suggested above, but rather it is the solution implicit in backprop networks. Moreover, backprop networks only became well known after an interlude of several years, during which Hopfield nets and Boltzmann machines dominated the field. Even though there was a considerable interval between the introduction of perceptrons and the development of backprop networks, the *supervised learning* algorithms they employ are qualitatively similar.

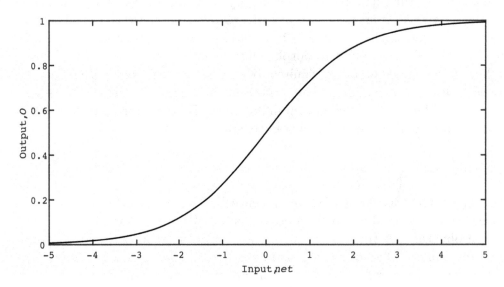

Figure 5.1. Sigmoidal activation function, Equation 5.2.

5.1. The Backpropagation Algorithm: Summary

Given a set of N_p associations to be learned, where each association is a pair of input and output *target* patterns, the backprop algorithm comprises three phases:

1) A forward phase, in which the states of units are propagated from the input layer to the output layer, for each of the N_p associations. The difference between the network output and the target output is calculated, and this is used to compute a *delta term* for each association.

2) A backward phase, in which the delta term of each unit in the output layer is recursively propagated back to the input layer, resulting in a delta term for every unit in the network.

3) A weight update phase, in which the accumulated N_p delta terms for each unit are used to change the weights between that unit and the units to which it projects. Each weight update reduces a measure of the overall difference between the network outputs and the target outputs.

Notation. The paper considers backprop networks with three layers of units, an example of which is shown in Figure 5.2. The output or state of the ith unit in the input layer is represented as i_i, the state of the jth unit in the hidden layer is represented as O_j, and the state of the kth unit in the output layer is represented as O_k (although this notation is not used consistently throughout the paper). The numbers of units in the input, hidden and output layers are I, J and K, respectively, and the total input to a unit in the input, hidden and output layers is denoted by net_i, net_j and net_k, respectively.

Because the activation function of each input unit is the identity function, the state of the input layer is the same as the current input (training) vector. In contrast, the hidden and output layers have sigmoidal activation functions (even though this is not strictly necessary for the output layer). Here, we refer to a network with I input units, J hidden units and K output units as an I-J-K network. For example, a 2-2-1 network is shown in Figure 5.2.

5.2. Forward Propagation of Input States

For a network with a single output unit (i.e. $K = 1$), each of the $p = 1, \ldots, N_p$ associations to be learned comprises an input vector $\mathbf{V}_p = (i_{p1}, \ldots, i_{pI})$ and a target output t_p. For the pth association, the total input to the jth hidden unit is a weighted sum of the I unit states in the input layer:

$$net_{pj} = \theta_j + \sum_{i=1}^{I} w_{ji} i_{pi}, \qquad (5.4)$$

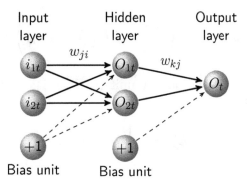

Figure 5.2. A 2-2-1 network showing how bias weights for the hidden layer can be implemented by adding a bias unit to the input layer. Similarly, the bias weight for the output layer can be implemented by adding a bias unit to the hidden layer.

where w_{ji} is the weight of the connection from the ith input unit to the jth hidden unit and θ_j is the bias of the jth hidden unit. The bias term reflects the hidden unit's propensity to be in a state $O_{pj} = 1$ even when that unit receives no input from other units. To simplify notation, θ_j can be implemented as the weight $w_{j,I+1}$ of a connection from a *bias unit* that has a constant output $i_{I+1} = +1$,

$$net_{pj} \;\; = \;\; w_{j,I+1}i_{I+1} + \sum_{i=1}^{I} w_{ji}\, i_{pi}, \tag{5.5}$$

which can now be written succinctly as

$$net_{pj} \;\; = \;\; \sum_{i=1}^{I+1} w_{ji}\, i_{pi}. \tag{5.6}$$

Using this notation, the bias term of each unit can be learned just like every other weight in the network. The bias terms for all units in a given layer can be represented by a single bias unit in the preceding layer, as shown in Figure 5.2. Alternatively, the bias unit for the entire network can be represented by a single global bias unit.

The activation function of units in the input layer is usually the identity function, whereas the activation functions f_j of units in the hidden layer are sigmoidal,

$$O_{pj} \;\; = \;\; f_j(net_{pj}) \tag{5.7}$$
$$\;\; = \;\; 1/(1 + e^{-net_{pj}}), \tag{5.8}$$

as shown in Figure 5.1. Similarly, the kth output unit receives a total input of

$$net_{pk} \;\; = \;\; \sum_{j=1}^{J+1} w_{kj}\, O_{pj}, \tag{5.9}$$

where w_{kj} is the weight of the connection from the jth hidden unit to the kth output unit and the $(J+1)$th unit in the hidden layer is a bias unit.

As the output layer activation function is also sigmoidal, the state of the kth output unit is

$$O_{pk} \;\; = \;\; f_k(net_{pk}) \tag{5.10}$$
$$\;\; = \;\; 1/(1 + e^{-net_{pk}}). \tag{5.11}$$

In summary, the state of the kth output unit is

$$O_{pk} \;\; = \;\; f_k\!\left(\sum_{j=1}^{J+1} w_{kj}\, f_j\!\left(\sum_{i=1}^{I+1} w_{ji}\, i_{pi} \right) \right). \tag{5.12}$$

If we consider a network with only one output unit (so $k = 1$), given a target state t_p for the pth association, the error is defined as

$$E_p \;\; = \;\; \frac{1}{2}(O_p - t_p)^2. \tag{5.13}$$

Taken over all N_p associations, the error function is

$$E = \frac{1}{2}\sum_{p=1}^{N_p}(O_p - t_p)^2. \tag{5.14}$$

5.3. Backward Propagation of Errors

There is a kind of anti-symmetry between the calculation of unit states and the adaptive change in weight values. We have seen how the input to an output unit is a weighted sum of states of units in the previous hidden layer, and that the input to a hidden unit is a weighted sum of states of units in the previous input layer. When adapting weights, rather than working from the input layer to the output layer, we work backwards from the output layer to the input layer. To achieve this, we define a *delta term* for each unit. Specifically, we shall see that the changes in weight between an input unit and a hidden unit depend on the delta term of the hidden unit, and the delta term of the hidden unit is a weighted sum of delta terms of units in the output layer. We can express this formally as follows.

For the pth association, the change in the weight between the jth hidden unit and the kth output unit is

$$\Delta w_{pkj} = -\epsilon \frac{\partial E_p}{\partial w_{kj}}, \tag{5.15}$$

where ϵ is the learning rate. When considered over all N_p training vectors, the change in the weight between the jth hidden unit and the kth output unit is

$$\Delta w_{kj} = -\epsilon \sum_{p=1}^{N_p} \frac{\partial E_p}{\partial w_{kj}}. \tag{5.16}$$

The gradient of E_p with respect to w_{kj} can be expressed using the chain rule as

$$\frac{\partial E_p}{\partial w_{kj}} = \frac{\partial E_p}{\partial net_{pk}} \frac{\partial net_{pk}}{\partial w_{kj}}, \tag{5.17}$$

where net_{pk} is the input to the kth output unit for the pth association.

It will prove useful to define the delta term of the kth output unit as the first factor on the right-hand side of Equation 5.17, which is the derivative of E_p with respect to the input to that unit:

$$\delta_{pk} = \frac{\partial E_p}{\partial net_{pk}}. \tag{5.18}$$

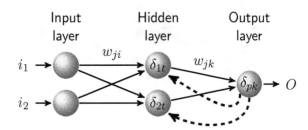

Figure 5.3. Backward propagation of delta term δ_{pk} of the output unit through connection weights in a 2-2-1 network. Bias unit not shown. Reproduced from Stone (2019).

121

Given Equation 5.9, the second term on the right-hand side of Equation 5.17 is

$$\frac{\partial net_{pk}}{\partial w_{kj}} = O_{pj}, \tag{5.19}$$

so we can write Equation 5.15 as

$$\Delta w_{pkj} = -\epsilon\, \delta_{pk}\, O_{pj}. \tag{5.20}$$

This is a general recipe for updating weights in a backprop network. The only problem lies in calculating the delta terms of hidden layers. Fortunately, just as the state of each unit in the output layer is calculated by propagating unit states forward from the input layer, so the delta term of each unit in the hidden layer can be calculated by propagating delta values backward from the output layer. To do this, we first need to evaluate the delta term of each output unit.

The Delta Term of an Output Unit. For each output unit, we can use the chain rule to express the delta term in Equation 5.18 as

$$\delta_{pk} = \frac{\partial E_p}{\partial O_{pk}} \frac{dO_{pk}}{dnet_{pk}}, \tag{5.21}$$

where (from Equations 5.3 and 5.13)

$$\frac{\partial E_p}{\partial O_{pk}} = (O_{pk} - t_{pk}), \tag{5.22}$$

$$\frac{dO_{pk}}{dnet_{pk}} = O_{pk}(1 - O_{pk}). \tag{5.23}$$

So Equation 5.21 can be written as

$$\delta_{pk} = (O_{pk} - t_{pk})O_{pk}(1 - O_{pk}). \tag{5.24}$$

Considered over all N_p training vectors (and using Equation 5.20), the change in the weight w_{kj} is

$$\Delta w_{kj} = -\epsilon \sum_{p=1}^{N_p} \delta_{pk}\, O_{pj}. \tag{5.25}$$

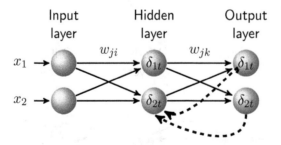

Figure 5.4. Backward propagation of delta terms of output units through connection weights in a 2-2-2 network. Note that the delta terms of *all* output units contribute to the delta term of *each* hidden unit. Reproduced from Stone (2019).

The Delta Term of a Hidden Unit. The gradient of E_p with respect to the weight w_{ji} of a connection between the ith input unit and the jth hidden unit can be obtained using the chain rule:

$$\frac{\partial E_p}{\partial w_{ji}} = \frac{\partial E_p}{\partial net_{pj}} \frac{\partial net_{pj}}{\partial w_{ji}}. \tag{5.26}$$

Just as the input to each output unit depends on every hidden unit state, so the delta term of each hidden unit depends on the delta term of every output unit (see Figure 5.3). We can recognise the first term on the right of Equation 5.26 as the delta term of the jth hidden unit,

$$\delta_{pj} = \frac{\partial E_p}{\partial net_{pj}}, \tag{5.27}$$

which we evaluate below. Using Equation 5.6, the second term on the right of Equation 5.26 evaluates to

$$\frac{\partial net_{pj}}{\partial w_{ji}} = i_{pi}, \tag{5.28}$$

so that

$$\frac{\partial E_p}{\partial w_{ji}} = \delta_{pj} i_{pi}, \tag{5.29}$$

which fits the general recipe defined in Equation 5.20. The hidden unit delta term can be evaluated using the chain rule:

$$\delta_{pj} = \frac{\partial E_p}{\partial O_{pj}} \frac{d O_{pj}}{d net_{pj}}. \tag{5.30}$$

Because the error E_p includes all output units, and because the state O_{pj} of the jth hidden unit affects all output unit states, O_{pj} must contribute to the error of every output unit. The extent to which the error of each output unit is affected by each hidden unit state O_{pj} clearly depends on the value of O_{pj}. Accordingly, we proceed by evaluating the first term on the right of Equation 5.30 as

$$\frac{\partial E_p}{\partial O_{pj}} = \sum_{k=1}^{K} \frac{\partial E_p}{\partial net_{pk}} \frac{\partial net_{pk}}{\partial O_{pj}}. \tag{5.31}$$

We can recognise the delta term $\delta_{pk} = \partial E_p / \partial net_{pk}$ of each output unit. Also, $\partial net_{pk} / \partial O_{pj} = w_{kj}$ from Equation 5.9 so that

$$\frac{\partial E_p}{\partial O_{pj}} = \sum_{k=1}^{K} \delta_{pk} w_{kj}. \tag{5.32}$$

Substituting Equations 5.31 and 5.32 into Equation 5.30 yields

$$\delta_{pj} = \frac{d O_{pj}}{d net_{pj}} \sum_{k=1}^{K} \delta_{pk} w_{kj}. \tag{5.33}$$

Now we can see that the delta term of a hidden unit is proportional to a weighted sum of the delta terms of units in the output layer.

From Equation 5.3 (repeated here),

$$\frac{dO_{pj}}{dnet_{pj}} = O_{pj}(1 - O_{pj}), \qquad (5.34)$$

so the delta term of a hidden unit is

$$\delta_{pj} = O_{pj}(1 - O_{pj}) \sum_{k=1}^{K} \delta_{pk} \, w_{kj}, \qquad (5.35)$$

and therefore Equation 5.29 can be written as

$$\frac{\partial E_p}{\partial w_{ji}} = \left(O_{pj} (1 - O_{pj}) \sum_{k=1}^{K} \delta_{pk} \, w_{kj} \right) i_{pi}. \qquad (5.36)$$

Over all N_p associations, the gradient of the total error E with respect to w_{ji} is

$$\frac{\partial E}{\partial w_{ji}} = \sum_{p=1}^{N_p} \delta_{pj} \, i_{pi} \qquad (5.37)$$

so that

$$\Delta w_{ji} = -\epsilon \sum_{p=1}^{N_p} \delta_{pj} \, i_{pi}. \qquad (5.38)$$

At this point, we have obtained an expression for the weight change applied to every weight in the neural network.

5.4. Weights as Vectors

For practical purposes, we define the $I + 1$ weights (including bias) between the input layer and the jth hidden unit as a weight vector

$$\mathbf{w}_j^{\text{hid}} = (w_{j1}, \dots, w_{jI}, w_{j,I+1}) \qquad (5.39)$$

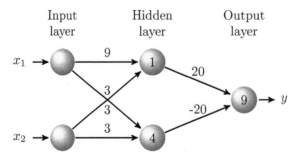

Figure 5.5. Solving the XOR problem. Each unit is labelled with a bias weight, which gets subtracted from its input. Binary vectors of the XOR problem are assumed to consist of 0s and 1s, and the hidden and output units have sigmoidal activation functions. Reproduced from Stone (2019).

and collect the connection weights to all hidden units in the hidden layer weight vector

$$\mathbf{w}^{\text{hid}} = (\mathbf{w}_1^{\text{hid}}, \ldots, \mathbf{w}_J^{\text{hid}}), \tag{5.40}$$

which consists of $N^{\text{hid}} = (I + 1) \times J$ weights. Similarly, we define the $J + 1$ weights from the hidden layer to the kth output unit as

$$\mathbf{w}_k^{\text{out}} = (w_{k1}, \ldots, w_{kJ}, w_{k,J+1}) \tag{5.41}$$

and the entire set of weights to all output units as

$$\mathbf{w}^{\text{out}} = (\mathbf{w}_1^{\text{out}}, \ldots, \mathbf{w}_K^{\text{out}}), \tag{5.42}$$

which consists of $N^{\text{out}} = (J + 1) \times K$ weights. We then concatenate the weights to the hidden and output units into a single vector

$$\mathbf{w} = (\mathbf{w}^{\text{hid}}, \mathbf{w}^{\text{out}}) = (w_1, \ldots, w_N), \tag{5.43}$$

which consists of $N = N^{\text{hid}} + N^{\text{out}}$ weights.

The Vector Gradient. We can then write the gradient with respect to $\mathbf{w}$ as a vector,

$$\nabla E = \left(\frac{\partial E}{\partial w_1}, \ldots, \frac{\partial E}{\partial w_N} \right). \tag{5.44}$$

The change in the weight vector in the direction of steepest descent is

$$\Delta \mathbf{w} = -\epsilon \nabla E, \tag{5.45}$$

and the rule for changing the weight vector is

$$\mathbf{w}_{\text{new}} = \mathbf{w}_{\text{old}} + \Delta \mathbf{w}. \tag{5.46}$$

5.5. Results: Exclusive OR and Autoencoders

Backprop can be used to solve the XOR problem using the 2-2-1 neural network shown in Figure 5.5 or Figure 2 in the paper. If the activation function of the hidden and output units is sigmoidal then the XOR problem is solved in under 40 learning iterations. However, the final weights produced by backprop can be inscrutable. Accordingly, using input values and target outputs of 0 or 1, idealised weights are shown in Figure 5.5.

As described in Section 4.7, the autoencoder problem provides an interesting benchmark for discovering useful codes. This paper demonstrates that a backprop network can solve the autoencoder problem using a network with 8 input units, 3 hidden units and 8 output units (i.e. with the minimum number, $\log_2 8$, of hidden units).

5.6. List of Mathematical Symbols

E error.

I number of units in the input layer.

i index of units in the input layer.

i_{pi} value of the ith component of the pth input (training) vector.

J number of units in the hidden layer.

j index of units in the hidden layer.

K number of units in the output layer.

k index of units in the output layer.

N_p number of training vectors, i.e. number of associations to be learned.

net_j total or net input to the jth unit.

O_k output or state of the kth unit.

p index of the set of associations to be learned.

t_k target value of the kth output unit.

w_{ji} weight of the connection from the ith input unit to the jth hidden unit.

w_{kj} weight of the connection from the jth hidden unit to the kth output unit.

δ_{pj} delta term of the jth unit for the pth training vector.

ϵ learning rate.

θ_j bias term of the jth unit.

Research Paper:
Learning Internal Representations by Error Propagation

Reference: Rumelhart, D. E., Hinton, G. E., and Williams, R. J. (1985). Learning internal representations by error propagation. Technical Report ICS 8506, Institute for Cognitive Science, University of California, San Diego. `https://apps.dtic.mil/sti/tr/pdf/ADA164453.pdf` Reproduced with permission.

Rumelhart, D. E., Hinton, G. E., and Williams, R. J. (1986). Learning representations by back-propagating errors. *Nature*, 323:533–536.

The original three-page backprop paper[96] published in *Nature* in 1986 is essentially a compressed version of the following 14-page extract of a technical report[95] from 1985. This technical report is reproduced here because its additional length provides a more insightful account than the *Nature* paper. However, due to the poor condition of the photocopied technical report available online, the following is a reformatted version, with much of the lengthy discussion omitted.

Learning Internal Representations by Error Propagation

David E. Rumelhart, Geoffrey E. Hinton, and Ronald J. Williams

THE PROBLEM

We now have a rather good understanding of simple two-layer associative networks in which a set of input patterns arriving at an input layer are mapped directly to a set of output patterns at an output layer. Such networks have no *hidden* units. They involve only *input* and *output* units. In these cases there is no *internal representation*. The coding provided by the external world must suffice. Perhaps the essential character of such networks is that they map similar input patterns to similar output patterns. This is what allows these networks to make reasonable generalizations and perform reasonably on patterns that have never before been presented. The similarity of patterns in a PDP[†] system is determined by their overlap. The overlap in such networks is determined outside the learning system itself—by whatever produces the patterns.

The constraint that similar input patterns lead to similar outputs can lead to an inability of the system to learn certain mappings from input to output. Whenever the representation provided by the outside world is such that the similarity structure of the input and output patterns are very different, a network without internal representations (i.e., a network without hidden units) will be unable to perform the necessary mapping. A classic example of this case is the *exclusive-or* (XOR) problem illustrated in Table 1. Here we see that those patterns which overlap least are supposed to generate identical output values. This problem and many others like it cannot be performed by networks without hidden units with which to create their own internal representations of the input patterns. It is interesting to note that had the input patterns contained a third input taking the value 1 whenever the first two have value 1 as shown in Table 2, a two-layer system would be able to solve the problem.

Minsky and Papert (1969) have provided a very careful analysis of conditions under which such systems are capable of carrying out the required mappings. They show that in a large number of interesting cases, networks of this kind are incapable of solving the problems. On the other hand, as Minsky and Papert also pointed out, if there is a layer of simple perceptron like hidden units, as shown in Figure 1, with which the original input pattern can be augmented, there is always a recoding (i.e., an internal representation) of the input patterns in the hidden units in which the similarity of the patterns among the hidden units can support any required mapping from the input to the output units. Thus, if we have the right connections from the input units to a large enough set of hidden units, we can always find a representation that will

---

[†]PDP = parallel distributed processing

Table 1:

Input Patterns		Output Patterns
00	→	0
01	→	1
10	→	1
11	→	0

Table 2:

Input Patterns		Output Patterns
000	→	0
010	→	1
100	→	1
111	→	0

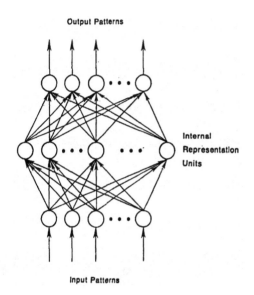

Figure 1: A multilayer network. In this case the information coming to the input units is *recoded* into an internal representation and the outputs are generated by the internal representation rather than by the original pattern. Input patterns can always be encoded, if there are enough hidden units, in a form so that the appropriate output pattern can be generated from any input pattern.

perform any mapping from input to output through these hidden units. In the case of the XOR problem, the addition of a feature that detects the conjunction of the input units changes the similarity structure of the patterns sufficiently to allow the solution to be learned. As illustrated in Figure 2, this can be done with a single hidden unit. The numbers on the arrows represent the strengths of the connections among the units. The numbers written in the circles represent the thresholds of the units. The value of +1.5 for the threshold of the hidden unit insures that it will be turned on only when both input units are on. value 0.5 for the output unit insures that it will turn on only when it receives a net positive input greater than 0.5. The weight of −2 from the hidden unit to the output unit insures that the output unit will not come on when both input units are on. Note that from the point of view of the output unit, the hidden unit is treated as simply another input unit. It is as if the input patterns consisted of three rather than two units.

The existence of networks such as this illustrate the potential power of hidden units and internal representations. The problem, as noted by Minsky and Papert, is that whereas there is a very simple guaranteed learning rule for all problems that can be solved without hidden units, namely, the perceptron convergence procedure (or the variation due originally to Widrow and Hoff, 1960, which we call the delta rule; see Chapter[*] 11), there is no equally powerful rule for learning in networks with hidden units. There have been three basic responses to this lack.

---

[*]This paper was copied from a technical report that later became a book chapter; the references to other

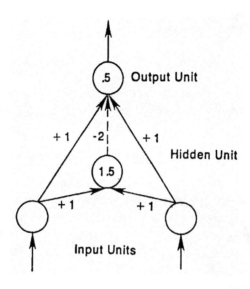

Figure 2: A simple XOR network with one hidden unit. See text for explanation.

One response is represented by competitive learning (Chapter* 5) in which simple *unsupervised* learning rules are employed so that useful hidden units develop. Although these approaches are promising, there is no external force to *insure* that hidden units appropriate for the required mapping are developed. The second response is to simply *assume* an internal representation that, on some a priori grounds, seems reasonable. This is the tack taken in the chapter on verb learning (Chapter* 18) and in the interactive activation model of word perception (McClelland & Rumelhart, 1981; Rumelhatt & McClelland, 1982). The third approach is to attempt to *develop* a learning procedure capable of learning an internal representation adequate for performing the task at hand. One such development is presented in the discussion of Boltzmann machines in Chapter* 7. As we have seen, this procedure involves the use of stochastic units, requires the network to reach equilibrium in two different phases, and is limited to symmetric networks. Another recent approach, also employing stochastic units, has been developed by Barto (1985) and various of his colleagues (cf. Barto & Anandan, 1985). In this chapter we present another alternative that works with deterministic units, that involves only local computations, and that is a clear generalization of the delta rule. We call this the *generalized delta rule*. From other considerations, Parker (1985) has independently derived a similar generalization, which he calls *learning-logic*. Le Cun (1985) has also studied a roughly similar learning scheme. In the remainder of this chapter we first derive the generalized delta rule, then we illustrate its use by providing some results of our simulations, and finally we indicate some further generalizations of the basic idea.

THE GENERALIZED DELTA RULE

The learning procedure we propose involves the presentation of a set of pairs of input and output patterns. The system first uses the input vector to produce its own output vector and then compares this with the *desired output*, or *target* vector. If there is no difference, no learning takes place. Otherwise the weights are changed to reduce the difference. In this case, with no hidden units, this generates the standard delta rule as described in Chapters* 2 and 11. The

chapters refer to that book: *Parallel Distributed Processing, Volume 1: Explorations in the Microstructure of Cognition: Foundations.* Cambridge MA: Bradford Books/MIT Press, 1986.

rule for changing weights following presentation of input/output pair p is given by

$$\Delta_p w_{ji} \;=\; \eta(t_{pj} - O_{pj})i_{pi} \;=\; \eta\delta_{pj}i_{pi} \tag{1}$$

where t_{pj} is the target input for jth component of the output pattern for pattern p, O_{pj} is the jth element of the actual output pattern produced by the presentation of input pattern p, i_{pi} is the value of the ith element of the input pattern $\delta_{pj} = p_j - O_{pj}$, and $\Delta_p w_{ji}$ is the change to be made to the weight from the ith to the jth unit following presentation of pattern p.

The delta rule and gradient descent. There are many ways of deriving this rule. For present purposes, it is useful to see that for linear units it minimizes the squares of the differences between the actual and the desired output values summed over the output units and all pairs of input/output vectors. One way to show this is to show that the derivative of the error measure with respect to each weight is proportional to the weight change dictated by the delta rule, with negative constant of proportionality. This corresponds to performing steepest descent on a surface in weight space whose height at any point in weight space is equal to the error measure. (Note that some of the following sections are written in italics. These sections constitute informal derivations of the claims made in the surrounding text and can be omitted by the reader who finds such derivations tedious.)

To more specific, then, let

$$E_p \;=\; \frac{1}{2}\sum_j (t_{pj} - O_{pj})^2 \tag{2}$$

be our measure of the error on input/output pattern p and let $E = \sum E_p$ be our overall measure of the error. We wish to show that the delta rule implements a gradient descent in E when the units are linear. We will proceed by simply showing that

$$-\frac{\partial E_p}{\partial w_{ji}} \;=\; \delta_{pj}i_{pi},$$

which is proportional to $\Delta_p w_{ji}$ as prescribed by the delta rule. When there are no hidden units it is straightforward to compute the relevant derivative. For this purpose we use the chain rule to write the derivative as the product of two parts: the derivative of the error with respect to the output of the unit times the derivative of the output with respect to the weight,

$$\frac{\partial E_p}{\partial w_{ji}} \;=\; \frac{\partial E_p}{\partial O_{pj}}\frac{\partial O_{pj}}{\partial w_{ji}}. \tag{3}$$

The first part tells how the error changes with the output of the jth unit and the second part tells how much changing w_{ji} changes that output. Now, the derivatives are easy to compute. First, from Equation 2

$$\frac{\partial E_p}{\partial O_{pj}} = -(t_{pj} - O_{pj}) = -\delta_{pj}. \tag{4}$$

Not surprisingly, the contribution of unit u_j to the error is simply proportional to δ_{pj}. Moreover, since we have linear units,

$$O_{pj} \;=\; \sum_i w_{ji}i_{pi}, \tag{5}$$

from which we conclude that

$$\frac{\partial O_{pj}}{\partial w_{ji}} \;=\; i_{pi}.$$

Thus, substituting back into Equation 3, we see that

$$\frac{\partial E_p}{\partial w_{ji}} = -\delta_{pj}i_{pi} \qquad (6)$$

as desired. Now, combining this with the observation that

$$\frac{\partial E}{\partial w_{ji}} = \sum_p \frac{\partial E_p}{\partial w_{ji}}$$

should lead us to conclude that the net change in w_{ji} after one complete cycle of pattern presentations is proportional to this derivative and hence that the delta rule implements a gradient descent in E. In fact, this is strictly true only if the values of the weights are not changed during this cycle. By changing the weights after each pattern is presented we depart to some extent from a true gradient descent in E. Nevertheless, provided the learning rate (i.e., the constant of proportionality) is sufficiently small, this departure will be negligible and the delta rule will implement a very close approximation to gradient descent in sum-squared error. In particular, with small enough learning rate, the delta rule will find a set of weights minimizing this error function.

The delta rule for semilinear activation functions in feedforward networks. We have shown how the standard delta rule essentially implements gradient descent in sum-squared error for linear activation functions. In this case, without hidden units, the error surface is shaped like a bowl with only one minimum, so gradient descent is guaranteed to find the best set of weights. With hidden units, however, it is not so obvious how to compute the derivatives, and the error surface is not concave upwards, so there is the danger of getting stuck in local minima. The main theoretical contribution of this chapter is to show that there is an efficient way of computing the derivatives. The main empirical contribution is to show that the apparently fatal problem of local minima is irrelevant in a wide variety of learning tasks.

At the end of the chapter we show how the generalized delta rule can be applied to arbitrary networks, but, to begin with, we confine ourselves to *layered feedforward* networks. In these networks, the input units arc the bottom layer and the output units are the top layer. There can be many layers of hidden units in between, but every unit must send its output to higher layers than its own and must receive its input from lower layers than its own. Given an input vector, the output vector is computed by a forward pass which computes the activity levels of each layer in turn using the already computed activity levels in the earlier layers.

Since we are primarily in extending this result to the case with hidden units and since, for reasons outlined in Chapter* 2, hidden units with linear activation functions provide no advantage, we begin by generalizing our analysis to the set of nonlinear activation functions which we call *semilinear* (see Chapter* 2). A semilinear activation function is one in which the output of a unit is a differentiable function of the net total input,

$$net_{pj} = \sum_i w_{ji}O_{pi}, \qquad (7)$$

where $O_i = i_i$ if unit i is an input unit. Thus, a semilinear activation function is one in which

$$O_{pj} = f_j(net_{pj}) \qquad (8)$$

and f is differentiable. The generalized delta rule works if the network consists of units having semilinear activation functions. Notice that linear threshold units do not satisfy the requirement because their derivative is infinite at the threshold and zero elsewhere.

To get the correct generalization the delta rule, we must set

$$\Delta_p w_{ji} \quad \propto \quad -\frac{\partial E_p}{\partial w_{ji}},$$

where E is the same sum-squared function defined earlier. As in the standard delta rule it is again useful to see this derivative as resulting from the product of two parts: one reflecting the change in error as a function of the change in the net input to the unit and one part representing the effect of changing a particular weight on the net input. Thus we can write

$$\frac{\partial E_p}{\partial w_{ji}} \quad = \quad \frac{\partial E_p}{\partial net_{pj}}\frac{\partial net_{pj}}{\partial w_{ji}}. \qquad (9)$$

By Equation 7 we see that the second factor is

$$\frac{\partial net_{pj}}{\partial w_{ji}} \quad = \quad \frac{\partial}{w_{ji}}\sum_k w_{jk}O_{pk} \quad = \quad O_{pi}. \qquad (10)$$

Now let us define

$$\delta_{pj} \quad = \quad -\frac{\partial E_p}{\partial net_{pj}}.$$

(By comparing this to Equation 4, note that this is consistent with the definition of δ_{pj} used in the original delta rule for linear units since $O_{pj} = net_{pj}$ when unit u_j is linear.) Equation 9 thus has the equivalent form

$$-\frac{\partial E_p}{\partial w_{ji}} \quad = \quad \delta_{pj}O_{pi}.$$

This says that to implement gradient descent in E we should make our weight changes according to

$$\Delta_p w_{ji} \quad = \quad \eta\delta_{pj}O_{pi}, \qquad (11)$$

just as in the standard delta rule. The trick is to figure out what δ_{pj} should be for each unit u_j in the network. The interesting result, which we now derive, is that there is a simple recursive computation of these δ's which can be implemented by propagating error signals backward through the network.

To compute $\delta_{pj} = -\dfrac{\partial E_p}{\partial net_{pj}}$, we apply the chain rule to write this partial derivative as the product of two factors, one factor reflecting the change in error as a function of the output of the unit and one reflecting the change in the output as a function of changes in the input. Thus, we have

$$\delta_{pj} \quad = \quad -\frac{\partial E_p}{\partial net_{pj}} \quad = \quad -\frac{\partial E_p}{\partial O_{pj}}\frac{\partial O_{pj}}{\partial net_{pj}}.$$

Let us compute the second factor. By Equation 8 we see that

$$\frac{\partial O_{pj}}{\partial net_{pj}} \quad = \quad f_j'(net_{pj}), \qquad (12)$$

which is simply the derivative of the squashing function f_j for the jth unit, evaluated at the net input net_{pj} to that unit. To compute the first factor, we consider two cases. First, assume that *unit u_j is an output unit of the network. In this case, it follows from the definition of E_p that*

$$\frac{\partial E_p}{\partial O_{pj}} \;=\; -(t_{pj} - O_{pj}),$$

which is the same result as we obtained with the standard delta rule. Substituting for the two factors in Equation 12, we get

$$\delta_{pj} \;=\; (t_{pj} - O_{pj})f'_j(net_{pj}) \tag{13}$$

for any output unit u_j. If u_j is not an output unit we use the chain rule to write

$$\sum_k \frac{\partial E_p}{\partial net_{pk}}\frac{\partial net_{pk}}{O_{pj}} = \sum_k \frac{\partial E_p}{\partial net_{pk}}\frac{\partial}{\partial O_{pj}}\sum_i w_{ki}O_{pi} = \sum_k \frac{\partial E_p}{\partial net_{pk}}w_{kj} = \sum_k \delta_{pk}w_{kj}.$$

In this case, substituting for the two factors in Equation 12 yields

$$\delta_{pj} \;=\; f'_j(net_{pj})\sum_k \delta_{pk}w_{kj} \tag{14}$$

whenever u_j is not an output unit. Equations 13 and 14 give a recursive procedure for computing the δ's for all units in the network, which are then used to compute the weight changes in the network according to Equation 11. This procedure constitutes the generalized delta rule for a feedforward network of semilinear units.

These results can be summarized in three equations. First, the generalized delta rule has exactly the same form as the standard delta rule of Equation 1. The weight on each line should be changed by an amount proportional to the product of an error signal, δ, available to the unit receiving input along that line and the output of the unit sending activation along that line. In symbols,

$$\Delta_p w_{ji} \;=\; \eta \delta_{pj}O_{pi}.$$

The other two equations specify the error signal. Essentially, the determination of the error signal is a recursive process which starts with the output units. If a unit is an output unit, its error signal is very similar to the standard delta rule. It is given by

$$\delta_{pj} \;=\; (t_{pj} - O_{pj})f'_j(net_{pj})$$

where $f'(net_{pj})$ is the derivative of the semilinear activation function which maps the total input to the unit to an output value. Finally, the error signal for hidden units for which there is no specified target is determined recursively in terms of the error signals of the units to which it directly connects and the weights of those connections. That is,

$$\delta_{pj} \;=\; f'_j(net_{pj})\sum_k \delta_{pk}w_{kj}$$

whenever the unit is not an output unit.

The application of the generalized delta rule, thus, involves two phases: During the first phase the input is presented and propagated forward through the network to compute the output value O_{pj} for each unit. This output is then compared with the targets, resulting in an error signal δ_{pj} for each output unit. The second phase involves a backward pass through the network (analogous to the initial forward pass) during which the error signal is passed to each unit in the network and the appropriate weight changes are made. This second, backward pass allows the recursive computation of δ as indicated above. The first step is to compute δ for each of the output units. This is simply the difference between the actual and desired output values times the derivative of the squashing function. We can then compute weight changes for all connections that feed into the final layer. After this is done, then compute δ's for all units in the penultimate layer. This propagates the errors back one layer, and the same process can be repeated for every layer. The backward pass has the same computational complexity as the forward pass, and so it is not unduly expensive.

We have now generated a gradient descent method for finding weights in any feedforward network with semilinear units. Before reporting our results with these networks, it is useful to note some further observations. It is interesting that not all weights need be variable. Any number of weights in the network can be fixed. In this case, error is still propagated as before; the fixed weights are simply not modified. It should also be noted that there is no reason why some output units might not receive inputs from other output units in earlier layers. In this case, those units receive two different kinds of error: that from the direct comparison with the target and that passed through the other output units whose activation it affects. In this case, the correct procedure is to simply add the weight changes dictated by the direct comparison to that propagated back from the other output units.

SIMULATION RESULTS

We now have a learning procedure which could, in principle, evolve a set of weights to produce an arbitrary mapping from input to output. However, the procedure we have produced is a gradient descent procedure and, as such, is bound by all of the problems of any hill climbing procedure—namely, the problem of local maxima or (in our case) minima. Moreover, there is a question of how long it might take a system to learn. Even if we could guarantee that it would eventually find a solution, there is the question of whether our procedure could learn in a reasonable period of time. It is interesting to ask what hidden units the system actually develops in the solution of particular problems. This is the question of what kinds of internal representations the system actually creates. We do not yet have definitive answers to these questions. However, we have carried out many simulations which lead us to be optimistic about the local minima and time questions and to be surprised by the kinds of representations our learning mechanism discovers. Before proceeding with our results, we must describe our simulation system in more detail. In particular, we must specify an activation function and show how the system can compute the derivative of this function.

A useful activation function. In our above derivations the derivative of the activation function of unit u_j, $f_j'(net_j)$, always played a role. This implies that we need an activation function for which a derivative exists. It is interesting to note that the linear threshold function, on which the perceptron is based, is discontinuous and hence will not suffice for the generalized delta rule. Similarly, since a linear system achieves no advantage from hidden units, a linear activation function will not suffice either. Thus, we need a continuous, nonlinear activation function. In most of our experiments we have used the *logistic* activation function in which

$$O_{pj} \quad = \quad \frac{1}{1 + e^{-(\sum_i w_{ji}O_{pi} + \theta_j)}} \tag{15}$$

where θ_j is a bias similar in function to a threshold.[2] In order to apply our learning rule, we need to know the derivative of this function with respect to its total input, net_{pj}, where $net_{pj} = \sum w_{ji}O_{pi} + \theta_j$. It is easy to show that this derivative is given by

$$\frac{dO_{pj}}{dnet_{pj}} \quad = \quad O_{pj}(1 - O_{pj}).$$

---

[2]Note that the value of the bias, θ_j, can be learned just like any other weights. We simply imagine that θ_j is the weight from a unit that is always on.

134

Thus, for the logistic activation function, the error signal, δ_{pj}, for an output unit is given by

$$\delta_{pj} \;\; = \;\; (t_{pj} - O_{pj})O_{pj}(1 - O_{pj}), \tag{16}$$

and the error for an arbitrary hidden unit u_j is given by

$$\delta_{pj} \;\; = \;\; O_{pj}(1 - O_{pj})\sum_k \delta_{pk}w_{kj}. \tag{17}$$

It should be noted that the derivative, $O_{pj}(1-O_{pj})$, reaches its maximum for $O_{pj} = 0.5$ and, since $0 \le O_{pj} \le 1$, approaches its minimum as O_{pj} approaches zero or one. Since the amount of change in a given weight is proportional to this derivative, weights will be changed most for those units that are near their midrange and, in some sense, not yet committed to being either on or off. This feature, we believe, contributes to the stability of the learning of the system.

One other feature of this activation function should be noted. The system can not actually reach its extreme values of 1 or 0 without infinitely large weights. Therefore, in a practical learning situation in which the desired outputs are binary $\{0, 1\}$, the system can never actually achieve these values. Therefore, we typically use the values of 0.1 and 0.9 as the targets, even though we will talk as if values of $\{0, 1\}$ are sought.

The learning rate. Our learning procedure requires only that the change in weight be proportional to $\partial E_p/\partial w$. True gradient descent requires that infinitesimal steps be taken. The constant of proportionality is the learning rate in our procedure. The larger this constant, the larger the changes in the weights. For practical purposes we choose a learning rate that is as large as possible without leading to oscillation. This offers the most rapid learning. One way to increase the learning rate without leading to oscillation is to modify the generalized delta rule to include a *momentum* term. This can be accomplished by the following rule:

$$\Delta w_{ji}(n+1) \;\; = \;\; \eta(\delta_{pj}O_{pi}) + \alpha\Delta w_{ji}(n) \tag{18}$$

where the subscript n indexes the presentation number, η is the learning rate, and α is a constant which determines the effect of past weight changes on the current direction of movement in weight space. This provides a kind of momentum in weight space that effectively filters out high-frequency variations of the error-surface in the weight space. This is useful in spaces containing long ravines that are characterized by sharp curvature across the ravine and a gently sloping floor. The sharp curvature tends to cause divergent oscillations across the ravine. To prevent these it is necessary to take very small steps, but this causes very slow process along the ravine. The momentum filters out the high curvature and thus allows the effective weight steps to be bigger. In most of our simulations α was about 0.9. Our experience has been that we get the same solutions by setting $\alpha = 0$ and reducing the size of η, but the system learns much faster overall with larger values of α and η.

Symmetry breaking. Our learning procedure has one more problem that can be readily overcome and this is the problem of symmetry breaking. If all weights start out with equal values and if the solution requires that unequal weights be developed, the system can never learn. This is because error is propagated back through the weights in proportion to the values of the weights. This means that all hidden units connected directly to the output inputs will get identical error signals, and, since the weight changes depend on the error signals, the weights from those units to the output units must always be the same. The system is starting out at a kind of *local maximum*, which keeps the weights equal, but it is a maximum of the error function, so once it escapes it will never return. We counteract this problem by starting the system with small random weights. Under these conditions symmetry problems of this kind do not arise.

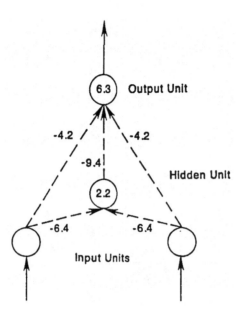

Figure 3: Observed XOR network. The connection weights are written on the arrows and the biases are written in the circles. Note a positive bias means that the unit is on unless turned off.

The XOR Problem

It is useful to begin with the exclusive-or problem since it is the classic problem requiring hidden units and since many other difficult problems involve an XOR as a subproblem. We have run the XOR problem many times and with a couple of exceptions discussed below, the system has always solved the problem. Figure 3 shows one of the solutions to the problem.

This solution was reached after 558 sweeps through the four stimulus patterns with a learning rate of $\eta = 0.5$. In this case, both the hidden unit and the output unit have *positive biases* so they are on unless turned off. The hidden unit turns on if neither input unit is on. When it is on, it turns off the output unit. The connections from input to output units arranged themselves so that they turn off the output unit whenever both inputs are on. In this case, the network has settled to a solution which is a sort of mirror image of the one illustrated in Figure 2.

We have taught the system to solve the XOR problem hundreds of times. Sometimes we have used a single hidden unit and direct connections to the output unit as illustrated here, and other times we have allowed two hidden units and set the connections from the input units to the outputs to be zero, as shown in Figure 4. In only two cases has the system encountered a *local minimum* and thus been unable to solve the problem. Both cases involved the two hidden units version of the problem and both ended up in the same local minimum. Figure 5 shows the weights for the local minimum. In this case, the system correctly responds to two of the patterns—namely, the patterns 00 and 10. In the cases of the other two patterns 11 and 01, the output unit gets a net input of zero. This leads to an output value of 0.5 for both of these patterns. This state was after 6,587 presentations of each pattern with $\eta = 0.25$.[3] Although many problems require more presentations for learning to occur, further trials on this problem merely increase the magnitude of the weights but do not lead to any improvement in performance. We do not know the frequency of such local minima, but our experience with this and other problems is that they are quite rare. We have found only one other situation in which a local minimum has occurred in many hundreds of problems of various sorts. We will

---

[3]If we set $\eta = 0.5$ or above, the system escapes this minimum. In general, however, the best way to avoid local minima is probably to use very small values of η.

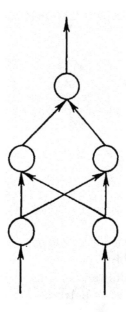

Figure 4: A simple architecture for solving XOR with two hidden units and no direct connections from input to output.

discus this case below.

The XOR problem has proved a useful test case for a number of other studies. Using the architecture illustrated in Figure 4, a student in our laboratory, Yves Chauvin, has studied the effect of varying the number of hidden units and varying the learning rate on time to solve the problem. Using as a learning criterion an error of 0.01 per pattern, Yves found that the average number of presentations to solve the problem with $\eta = 0.25$ varied from about 245 for the case with two hidden units to about 120 presentations for 32 hidden units. The results can

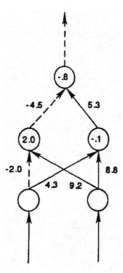

Figure 5: A network at a local minimum for the exclusive-or problem. The dashed lines indicate negative weights. Note that whenever the right-most input unit is on it turns on *both* hidden units. The weights connecting the hidden units to the output are arranged so that when both hidden units are on, the output unit gets a net input of zero. This leads to an output value of 0.5. In the other cases the network provides the correct answer.

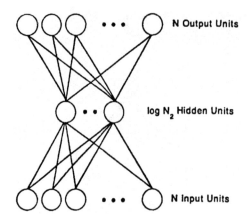

Figure 6: A network for solving the encoder problem. In this problem there are N orthogonal input patterns each paired with one of N orthogonal output patterns. There are only $\log_2 N$ hidden units. Thus, if the hidden units take on binary values, the hidden units must form a binary number to encode each of the input patterns. This is exactly what the system learns to do.

be summarized by $P = 280 - 33 \log_2 H$, where P is the required number of presentations and H is the number of hidden units employed. Thus, the time to solve XOR is reduced linearly with the logarithm of the number of hidden units. This result holds for values of H up to about 40 in the case of XOR. The general result that the time to solution is reduced by increasing the number of hidden units has been observed in virtually all of our simulations. Yves also studied the time to solution as a function of learning rate for the case of eight hidden units. He found an average of about 450 presentations with $\eta = 0.1$ to about 68 presentations with $\eta = 0.75$. He also found that learning rates larger than this led to unstable behavior. However, within this range larger learning rates speeded the learning substantially. In most of our problems we have employed learning rates of $\eta = 0.25$ or smaller and have had no difficulty.

The Encoding Problem

Ackley, Hinton, and Sejnowski (1985) have posed a problem in which a set of orthogonal input patterns are mapped to a set of orthogonal output patterns through a small set of hidden units. In such cases the internal representations of the patterns on the hidden units must be rather efficient. Suppose that we attempt to map N input patterns onto N output patterns. Suppose further that $\log_2 N$ hidden units are provided. In this case, we expect that the system will learn

Table 3:

Input Patterns		Output Patterns
10000000	$\rightarrow$	10000000
01000000	$\rightarrow$	01000000
00100000	$\rightarrow$	00100000
00010000	$\rightarrow$	00010000
00001000	$\rightarrow$	00001000
00000100	$\rightarrow$	00000100
00000010	$\rightarrow$	00000010
00000001	$\rightarrow$	00000001

Table 4:

Input Patterns		Hidden Unit Patterns		Output Patterns
10000000	$\rightarrow$	0.5 0 0	$\rightarrow$	10000000
01000000	$\rightarrow$	0 1 0	$\rightarrow$	01000000
00100000	$\rightarrow$	1 1 0	$\rightarrow$	00100000
00010000	$\rightarrow$	1 1 1	$\rightarrow$	00010000
00001000	$\rightarrow$	0 1 1	$\rightarrow$	00001000
00000100	$\rightarrow$	0.5 0 1	$\rightarrow$	00000100
00000010	$\rightarrow$	1 0 0.5	$\rightarrow$	00000010
00000001	$\rightarrow$	0 0 0.5	$\rightarrow$	00000001

to use the hidden units to form a binary code with a distinct binary pattern for each of the N input patterns. Figure 6 illustrates the basic architecture for the encoder problem. Essentially, the problem is to learn an encoding of an N bit pattern into a $\log_2 N$ bit pattern and then learn to decode this representation into the output pattern. We have presented the system with a number of these problems. Here we present a problem with eight input patterns. eight output patterns, and three hidden units. In this case the required mapping is the identity mapping illustrated in Table 3. The problem is simply to turn on the same bit in the output as in the input. Table 4 shows the mapping generated by our learning system on this example. It is of some interest that the system employed its ability to use intermediate values in solving this problem. It could, of course, have found a solution in which the hidden units took on only the values of zero and one. Often it does just that, but in this instance, and many others, there are solutions that use the intermediate values, and the learning system finds them even though it has a bias toward extreme values.

CONCLUSION

Minsky and Papert (1969) in their pessimistic discussion of perceptrons finally, near the end of their book, discuss *multilayer machines*. They state:

> The perceptron has shown itself worthy of study despite (and even because of!) its severe limitations. It has many features that attract attention: its linearity; its intriguing learning theorem; its clear paradigmatic simplicity as a kind of parallel computation. There is no reason to suppose that any of these virtues carry over to the many-layered version. Nevertheless, we consider it to be an important research problem to elucidate (or reject) our intuitive judgement that the extension is sterile. Perhaps some powerful convergence theorem will be discovered, or some profound reason for the failure to produce an interesting "learning theorem" for the multilayered machine will be found. (pp. 231-232)

Although our learning results do not *guarantee* that we can find a solution for all solvable problems, our analyses and results have shown that as a practical matter, the error propagation scheme leads to solutions in virtually every case. In short, we believe that we have answered Minsky and Papert's challenge and *have* found a learning result sufficiently powerful to demonstrate that their pessimism about learning in multilayer machines was misplaced.

One way to view the procedure we have been describing is as a parallel computer that, having been shown the appropriate input/output exemplars specifying some function, programs itself to compute that function in general. Parallel computers are notoriously difficult to program. Here we have a mechanism whereby we do not actually have to know how to write the program in order to get the system to do it. Parker (1985) has emphasized this point.

On many occasions we have been surprised to learn of new methods of computing interesting functions by observing the behavior of our learning algorithm. This also raised the question of generalization. In most of the cases presented above, we have presented the system with the entire set of exemplars. It is interesting to ask what would happen if we presented only a subset of the exemplars at training time and then watched the system generalize to remaining exemplars. In small problems such as those presented here, the system sometimes finds solutions to the problems which do not properly generalize. However, preliminary results on larger problems are very encouraging in this regard. This research is still in progress and cannot be reported here. This is currently a very active interest of ours.

REFERENCES

Ackley, D. H., Hinton, G. E., & Sejnowski, T. J. (1985). A learning algorithm for Boltzmann machines. *Cognitive Science*, 9, 147-169.

Barto, A. G. *Learning by statistical cooperation of self-interested neuron-like computing elements* (COINS Tech. Rep. 85-11). Amherst: University of Massachusetts, Department of Computer and Information Science.

Barto. A. G., & Anandan, P. (1985). Pattern recognizing stochastic learning automata. *IEEE Transactions on Systems, Man, and Cybernetics*.

Fukushima, K. (1980). Neocognitron: A self-organizing neural network model for a mechanism of pattern recognition unaffected by shift in position. *Biological Cybernetics*, 36, 193-202.

Kienker, P. K., Sejnowski, T. J., Hinton. G. E., & Schumacher, L. E. (1985). *Separating figure from ground with a network*. Unpublished manuscript.

Le Cun, Y. (1985, June). Une procedure d'apprentissage pour reseau a seuil assymetrique [A learning procedure for asymmetric threshold network]. *Proceedings of Cognitiva 85*, 599-604, Paris.

McClelland, J. L., & Rumelhart, D. E. (1981). An interactive activation model of context effects in letter perception: Part 1. An account of basic findings. *Psychological Review*, 88, 375-407.

Minsky, M. L., & Papert, S. (1969). *Perceptrons*. Cambridge, MA: MIT Press.

Parker, D. B. (1985). *Learning-logic* (TR-47). Cambridge, MA: Massachusetts Institute of Technnology, Center for Computational Research in Economics and Management Science.

Rumelhart, D. E., & McClelland, J. L. (1982). An interactive activation model of context effects in letter perception: Part 2. The contextual enhancement effect and some tests and extensions of the model. *Psychological Review*, 89, 60-94.

Widrow, G., & Hoff, M. E. (1960). Adaptive switching circuits. *Institute of Radio Engineers, Western Electric Show and Convention, Convention Record, Part 4*, 96-104.

Note

This paper was copied from a technical report (with permission from the authors) that later became a chapter in the book *Parallel Distributed Processing, Volume 1: Explorations in the Microstructure of Cognition: Foundations*. Cambridge MA: Bradford Books/MIT Press, 1986. Some sections of the original chapter have been omitted for the sake of brevity.

Chapter 6

Reinforcement Learning – 1983

Context

The problem confronting any learning system is that the consequences of each action often occur some time after the action has been executed. And in the interval between the action and its consequences, many other actions are taken, so disentangling how much past actions contribute to subsequent rewards or punishments is a difficult problem.

For example, if a rat eats a tainted food item then nausea usually follows several hours later, during which time it has probably eaten other items, so it is hard for the rat to know which item caused the nausea. Similarly, the final outcome of a chess game depends on every move within the game, so the wisdom of each move can only be evaluated after the game is over. For the apparently simple problem of balancing a pole (Figure 6.1), the consequences of each action (a nudge to the right or left) continue to evolve for many seconds and are mixed with the effects of previous and subsequent actions. Evaluating the cost or benefit of an action lies at the heart of optimal behaviour and is called the *temporal credit assignment problem*.

The ability to retrospectively assign credit to each action would be useless if an animal could not use the benefits of hindsight to foresee the likely outcome of its current actions. It follows that an animal's ability to solve the temporal credit assignment problem involves predicting future states of the world around itself. Fortunately, it turns out to be much easier to predict the future if that future is determined by, or at least affected by, the animal's own actions. By implication, the ability of an animal to take actions that maximise rewards depends heavily on its ability to predict the outcome of its actions.

The foregoing exemplifies how reinforcement learning differs from the *supervised learning* (i.e. learning the mapping from a set of input vectors to a set of output vectors) of neural networks in two important respects. First, supervised learning depends on pairs of input/output vectors, so that the neural network is informed of the correct output vector for each input it receives. Reinforcement learning is similar inasmuch as each input vector is used to generate an action (in the form of an output vector), but feedback regarding the consequences of that action is simply a scalar reward signal. Consequently, there is no explicit information regarding how to alter each element of the output vector to improve matters.

The second difference is that supervised learning involves immediate feedback. Sometimes this is also the case for reinforcement learning (*immediate reinforcement learning*), but usually the feedback received in reinforcement learning does not appear until some time after the current action (*delayed reinforcement learning*); and when it does appear, the feedback signal depends not only on the current action but also on actions taken before and after the current action. So the problem addressed by reinforcement learning is substantially harder than the supervised learning problem solved by a neural network. It is noteworthy that reinforcement learning problems are similar to problems encountered in the natural world, where the feedback for actions consists of scalar variables in the form of rewards and punishments, and these are usually delayed until some time after the actions have been executed.

In summary, with supervised learning, the response to each input vector is an output vector that receives immediate vector-valued feedback specifying the correct (target) output, and this feedback refers uniquely to the input vector just received; in contrast, each reinforcement learning output vector (action) receives scalar-valued feedback often some time after the action, and this feedback signal depends on actions taken before and after the current action.

Historically, the origins of reinforcement learning algorithms can be traced back to Shannon (1950), who described a program for playing chess (Shannon also invented *information theory*[106;125]). Even though his program did not learn, he proposed an idea that, to all intents and purposes, is reinforcement learning: "a higher level program which changes the terms and coefficients involved in the evaluation function depending on the results of games the machine has played" (Shannon's evaluation function corresponds to the action-value function used in reinforcement learning). Indeed, this idea seems to have been the inspiration for Samuel (1959), who designed a program for learning to play draughts (checkers). A later version of Samuel's program beat the checkers champion for the U.S. state of Connecticut in 1961. Since that time, reinforcement learning has been developed by many scientists, but its modern incarnation is due principally to Sutton and Barto (2018).

The Player of Games. The list of successful game-playing applications of reinforcement learning is impressive. These applications follow in the footsteps of an early success in backgammon, known as TD-Gammon[131], which was used to produce the best backgammon player in the world[130]. An intriguing aspect of TD-Gammon is that it developed a style of playing that was novel and which was subsequently widely adopted by grandmasters of the game. This foreshadowed the winning strategy of AlphaGo, which also generated novel moves that surprised and (initially) mystified human observers, but which led to successful outcomes.

The game of Go involves several simultaneous short-range and long-range battles for territory, and has about 10^{170} legal board positions – more than the number of atoms in the known universe. In 2016 the widely publicised AlphaGo beat Lee Sedol, an 18-time world champion[107]. A year later, AlphaGo beat a team of the world's top five players. Whereas AlphaGo initially learned from observing 160,000 human games, AlphaGo Zero learned through sheer trial and error before beating AlphaGo 100 games to none[108]. Both AlphaGo and AlphaGo Zero relied on a combination of reinforcement learning and deep learning (see Chapter 8).

Remarkably, without altering the pre-learning parameter values, the algorithm of AlphaGo Zero was then used to learn chess[101]; let's call the result AlphaChess Zero. The best traditional computer chess program (Stockfish) already played at super-human levels of performance. In a tournament of 100 games between Stockfish and AlphaChess Zero, 72 were a draw, and AlphaChess Zero won the remaining 28. As with AlphaGo Zero, some of the moves made by AlphaChess Zero seemed strange, but then it was realised that those strange moves were instrumental in winning the game.

Just as TD-Gammon altered the strategies used by humans to play backgammon, there is evidence from Brinkmann et al. (2023) that AlphaGo Zero is changing the strategies that humans use to play Go; and it is entirely possible that AlphaChess Zero will do the same for chess. So, in a sense, humans are starting to learn from machines that learn.

The importance of AlphaGo Zero in beating the machine (AlphaGo) that beat the human world champion cannot be overstated. We can try to rationalise the achievements of AlphaGo Zero by pointing out that it played many more games than a human could possibly play in a lifetime. But the fact remains that a computer program has learned to play a game so well that it can beat every one of the 8 billion people on the planet. Statistically speaking, that places AlphaGo Zero above the 99.9999999th percentile in terms of performance.

Prior to AlphaGo, success at classic arcade-style Atari games was achieved by a combination of deep learning and reinforcement learning. This resulted in super-human levels of performance in games such as pong, which is a computer version of table tennis[76;77]. The input used by the algorithm is the pixel

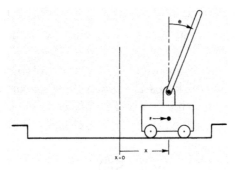

Figure 6.1. The cart–pole system has four state variables $(x, \dot{x}, \theta, \dot{\theta})$. This is Figure 1 in the paper[8].

array seen by human players, rather than the x and y positions of objects in the game. Interestingly, the same *deep Q-network* learned to play all 49 different arcade games[77].

In some respects, the early successes of reinforcement learning are even more impressive than the later successes, which involved playing games. As most humans know, learning to ride a bicycle is not easy. By simulating the dynamics of a bicycle in software, a reinforcement learning algorithm was able to keep a virtual bicycle upright for long periods of time[90]. Interestingly, learning was most rapid when a series of incremental sub-goals were used, similar to the scaffolding that adults use when teaching a child to ride a bicycle. More recently, reinforcement learning has been applied to the problems of learning to walk[33;49] and learning to fly. Using a simulation, a glider learned to gain height by circling around thermals; encouraged by this development, reinforcement learning was then used in a physical glider with considerable success[32;91;92].

Technical Summary

The Cart–Pole Problem. The cart–pole problem involves applying a series of *control actions* (impulses or nudges) to a cart so that the pole mounted on the cart remains as vertical as possible. The cart is on tracks, so it can only move *left* and *right*, as shown in Figure 6.1. Each *run* begins with the cart in the centre of the track and ends in *failure* when either: 1) the cart reaches one end of the track; or 2) the angle of the pole exceeds 45°. A run is terminated successfully if 1,000 actions are taken without violating condition 1) or 2).

The physical state of the cart–pole system is defined by four *state variables*, each of which is represented in terms of a small number of discrete intervals:

x position of the cart, discretised into three intervals:
  $[-2.4, -0.8], [-0.8, 0.8], [0.8, 2.4]$ m.
θ pole angle (vertical $= 0°$), discretised into six intervals:
  $[-12, -6], [-6, -1], [-1, 0], [0, 1], [1, 6], [6, 12]$ degrees.
$\dot{x}$ speed of the cart, discretised into three intervals:
  $[-\infty, -0.5], [-0.5, 0.5], [0.5, \infty]$ m/s.
$\dot{\theta}$ rate of change of the pole angle, discretised into three intervals:
  $[-\infty, -50], [-50, 50], [50, \infty]$ degrees/s.

Discretising each variable into a small number of intervals yields a total of $n = 3 \times 6 \times 3 \times 3 = 162$ possible combinations of states.

Michie and Chambers' Boxes System. The system described in this paper draws much inspiration from the *boxes system* proposed by Michie and Chambers (1968), which was also applied to the cart–pole problem. The boxes system divided the possible states of a cart–pole system into $n = 162$ discrete states (see above), where each state is associated with a single box. Each box has two possible *control actions*: push the cart either to the *left* or to the *right*, where all actions have a single fixed magnitude.

The control action chosen at any time depends on the outcome of previous *left/right* actions. Specifically, over many runs, each box keeps a running average of how much longer runs last after it executes an action. The basic idea is that if a box's *right* control actions (for example) almost always lead to longer runs than *left* control actions then that box should favour *right* control actions in future.

6.1. The Associative Search Element (ASE)

The associative search element (ASE) system has two different types of inputs: a scalar *reinforcement input*, r, and a *state vector*, as shown in Figure 6.2. The reinforcement input $r(t)$ is zero at every time step t, except when a run ends in failure, when it is set to $r(t) = -1$.

The state vector consists of four *state variables*, $(x, \dot{x}, \theta, \dot{\theta})$, which are updated at every time step. A *decoder* converts these state variables into a vector of 162 binary *signals* (see the description of the cart–pole system above),

$$X = (x_1, \ldots, x_n). \tag{6.1}$$

(The use of x to represent both position and ASE inputs is confusing, but that is the notation used in the paper.) For consistency with the terminology used in the paper, we will refer to the state associated with x_i as a *box*, on the understanding that this corresponds to a set of four intervals in the state of the cart–pole system.

The decoder sets every element of X to zero except for the element x_i that corresponds to the box of the current state vector, which is set to 1. The input to the ASE is the vector of n signals in X. The ASE takes a *left/right* control action that depends on the outcome of previous actions associated with x_i. For example, if the control action *right* has resulted in long runs (measured from time x_i was set to 1 by the decoder) during which the state vector had $x_i = 1$ then the next control action should probably also be *right*. Specifically, the balance between *left* and *right* actions is implicit in the value of a *synapse* or weight w_i associated with x_i. The action taken is

$$y(t) = f\left[\sum_{i=1}^{n} w_i(t)x_i(t) + \text{noise}\right], \tag{6.2}$$

where $y(t) = +1$ (*right*) if the term in square brackets is above zero and $y(t) = -1$ (*left*) otherwise.

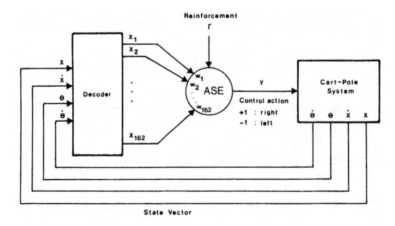

Figure 6.2. The associative search element (ASE). This is Figure 2 in the paper[8].

144

To motivate the learning rule used in the paper, consider a naive rule (not used in the paper) for updating the ith weight,

$$w_i(t+1) \quad = \quad w_i(t) + \alpha r(t) y(t) x_i(t), \tag{6.3}$$

where α is a small positive constant that determines the learning rate. The reinforcement signal $r(t)$ is zero unless the run ends at time t, in which case it is $r(t) = -1$. For example, if the action that preceded failure is $y(t) = +1$ (*right*) then (because $r(t) = -1$ at failure) w_i would be decremented, making a *right* action from x_i less probable in future. Conversely, if the action that preceded failure is $y(t) = -1$ (*left*) then w_i would be incremented, making a *right* action from x_i more probable in future.

However, the naive rule in Equation 6.3 allows only one weight to be updated, and only at the end of a run (i.e. after failure). To fix this shortcoming, each synapse is assigned its own *eligibility trace*.

The Eligibility Trace. Ideally, we would like to keep a complete record of every state that has ever been visited, the time it was visited, and the reward or punishment received at that time. However, the cost of doing so can be prohibitive. Instead, we can assign an eligibility trace to each state.

The value of the trace $e_i(t)$ of a state x_i is given a boost whenever the state x_i is encountered, but the value of $e_i(t)$ decays towards zero thereafter, as shown in Figure 6.3. Essentially, the trace value indicates how much time has elapsed between the action at x_i and the end of the run (i.e. failure); thus, the trace keeps a record of the extent to which each action was responsible for the failure. The bigger the value of e_i at failure, the more recently (or more frequently) an action was executed at x_i (and the more likely it is to have been responsible for the failure), and so the more its weight w_i should be altered.

The paper employs the eligibility trace with the following learning rule (Equation 2 in the paper):

$$w_i(t+1) \quad = \quad w_i(t) + \alpha r(t) e_i(t), \tag{6.4}$$

where (Equation 3 in the paper)

$$e_i(t+1) \quad = \quad \delta e_i(t) + (1 - \delta) y(t) x_i(t). \tag{6.5}$$

Here δ (taken to be 0.9) determines the rate at which e_i approaches zero. At the start of each run, all trace values are set to zero ($e_i = 0$).

For example, if the system encounters the state x_i at time t then $x_i(t)$ is set to 1; and if the action taken is $y(t) = +1$ (*right*) then e_i decays towards zero according to the equation

$$e_i(t+1) \quad = \quad 0.9 e_i(t) + 0.1, \tag{6.6}$$

where $e_i(t+1)$ is the trace value at time $t+1$ and $e_i(t)$ is the residual value of the trace from past encounters of the state x_i. Note that the sign of $e_i(t+1)$ is the same as the sign of the action $y(t)$ and that e_i approaches zero at an exponential rate, as shown in Figure 6.3.

Using the rule in Equation 6.4, if the end of the run occurs at time t then every state x_i encountered before time t will have its synapse weight w_i updated by an amount proportional to the value of its eligibility trace $e_i(t)$ at time t. Crucially, even though the update to each weight occurs only at the end of each run, every box x_i encountered in the current run has a non-zero eligibility trace which ensures that its weight gets updated (but often by a negligible amount; see below).

It is worth noting that the ASE did not work as well as the boxes system on which it was based. Having said that, the ASE acted as a stepping stone to a more effective system, which augmented the ASE with an *adaptive critic element* (ACE), described next.

6.2. The Adaptive Critic Element (ACE)

On its own, the ASE has to use a reward signal received at the end of each run to update the weight of each box x_i; and if the actions of a box were taken long before the run ended then the eligibility trace of that box will be close to zero (so w_i will remain almost unchanged).

To allow the ASE to receive feedback about its performance at every time step, the adaptive critic element (ACE) supplies a constant stream of *internal reinforcement* signals $\hat{r}(t)$, as shown in Figure 6.4. Of course, success relies on the assumption that the ACE's internal reinforcement signals are accurate, which is addressed next.

At each time step, the ACE makes a prediction $p(t)$ of the *return $r(t)$*, which is the cumulative total reward between time t and the end of a run. Here, the return can be interpreted as the length of time remaining before the run ends in failure. The ACE uses the temporal derivative of its own predicted returns to generate an internal reinforcement signal $\hat{r}(t)$. The objective is for the ACE to learn to make accurate predictions, so that $\hat{r}(t) = r(t)$.

Both the ASE and the ACE are affected by the reinforcement signal r received from the environment, which is non-zero only at the end of each run that ends in failure. The ACE uses the reinforcement signal r to improve the accuracy of its predictions regarding returns, which improves the accuracy of the internal reinforcement signal $\hat{r}$ that the ACE sends to the ASE. On the other hand, the ASE uses the ACE's internal reinforcement signal $\hat{r}$ to take actions that increase the returns. So the performance of the ASE depends on the performance of the ACE, and vice versa.

As in Section 6.1, the system has two different types of inputs, a scalar reinforcement input r and a state vector. In the ACE/ASE system, it is the ACE that receives the reinforcement input r, which is set to zero except at failure, when it is set to $r = -1$. Both the ASE and the ACE receive the 162-element input vector $X(t)$, supplied by the decoder, at every time step. The ACE generates an *internal reinforcement signal $\hat{r}(t)$* at every time step, which is used instead of r in Equation 6.4 for ASE to adapt its weights.

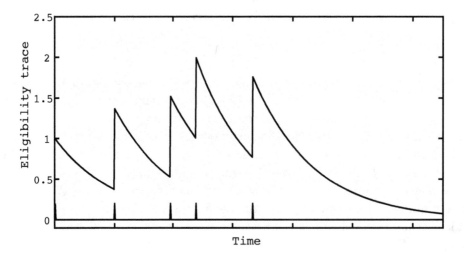

Figure 6.3. The eligibility trace $e_i(t)$ of a state x_i decays exponentially after x_i is encountered, but frequent visits to that state have a cumulative boosting effect. Each vertical bar at the bottom of the graph indicates a time t at which the state x_i was encountered. Reproduced from Stone (2019).

The principal job of the ACE is to produce an internal reinforcement signal for the ASE. To do this, the ACE learns a prediction of the cumulative reward, which is defined as (Equation 4 in the paper)

$$p(t) = \sum_{i=1}^{n} v_i(t) x_i(t), \tag{6.7}$$

where $v_i(t)$ is an adaptable weight. On p841 of the paper it is stated that $x_i(t)$ is real-valued, but it is actually binary, as it is defined in the context of the boxes system. Thus, only one box has $x_i(t) = 1$ (which corresponds to the state of the cart–pole system), and the remaining elements of X are zero. This means that there is only one non-zero product in Equation 6.7; so if the current state is $x_i(t)$ then $p(t) = v_i(t)$. In other words, v_i represents the return to be expected from the box x_i, as predicted by the ACE.

The update rule for the ACE's weights is defined as (Equation 5 in the paper)

$$v_i(t+1) = v_i(t) + \beta \big[r(t) + \gamma p(t) - p(t-1) \big] \overline{x}_i(t), \tag{6.8}$$

where β is a small positive constant learning rate. The constant $\gamma = 0.95$ is just a little less than 1, which ensures that predictions gradually approach zero if no further inputs are received. The value of $\overline{x}_i(t)$ is a running average of previous values of x_i, which decays exponentially towards zero according to (Equation 6 in the paper)

$$\overline{x}_i(t+1) = \lambda \overline{x}_i(t) + (1-\lambda) x_i(t), \tag{6.9}$$

where the *trace decay rate* constant λ (taken to be 0.8) determines the rate at which $\overline{x}_i$ approaches zero.

The internal reinforcement signal supplied by the ACE to the ASE is (Equation 7 in the paper)

$$\hat{r}(t) = r(t) + \gamma p(t) - p(t-1). \tag{6.10}$$

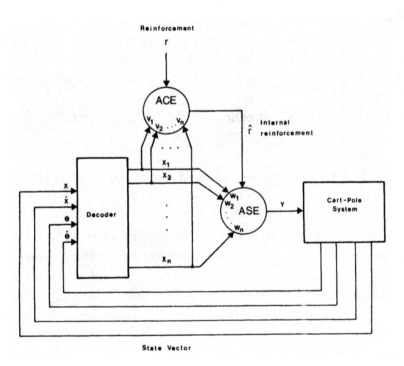

Figure 6.4. The adaptive critic element (ACE). This is Figure 3 in the paper[8].

Recall that $r(t) = 0$ at every time step except for the final step in a run, when $r(t) = -1$; and if $r(t) = 0$ then $\hat{r}(t)$ indicates whether the predicted return is increasing or decreasing. The predictions $p(t-1)$ and $p(t)$ in Equation 6.10 are associated with a transition from the box $x_i(t-1)$ to the box $x_i(t)$. This means that the difference

$$\Delta p \;=\; p(t) - p(t-1) \tag{6.11}$$

indicates the ACE's estimate of whether the predicted return increased or decreased in the transition from $x_i(t-1)$ to $x_i(t)$. In other words, did the predicted outlook just get better or worse? (We have ignored γ here, which is justified because it is close to 1.) Equation 6.10 implies that if the transition to $x_i(t)$ increased the value of p then $\hat{r}(t)$ should increase, but if the transition to $x_i(t)$ decreased p then $\hat{r}(t)$ should decrease.

At first glance, the update rule in Equation 6.8 seems hopeful at best. Indeed, Sutton and Barto described this as "learning a guess from a guess". Despite such misgivings, it can be proved[130] that, under mild conditions, the system converges to a stable solution. Finally, it is worth noting that, in modern terminology, the collective values of v_i in the ACE correspond to a *state value function*, and the method used by the ACE to predict returns is a form of *temporal difference learning*.

6.3. Results

The main result is that the ACE/ASE system performed much better than the boxes system, which performed better than the ASE system by itself. In particular, the dramatic improvement achieved by the ACE/ASE system (shown the paper's Figure 4) provided a stimulus for further research on reinforcement learning.

6.4. List of Mathematical Symbols

ACE   adaptive critic element.

ASE   associative search element.

$e_i(t)$ real-valued eligibility trace of the ith box (state) at time t.

$p(t)$ the ACE's prediction of the cumulative total reward between time t and the end of a run.

r binary reinforcement input from the environment; $r = 0$ or -1.

$r(t)$ cumulative total reward between time t and the end of a run; length of time remaining before a run ends in failure.

$\hat{r}(t)$ real-valued internal reinforcement produced by the ACE at time t.

$v_i(t)$ ACE input weight.

$w_i(t)$ ASE synapse weight.

X input vector $X = (x_1(t), \ldots, x_n(t))$, consisting of discretised states of the cart–pole system.

x position of the cart.

$\dot{x}$ speed of the cart.

$x_i(t)$ binary value of the ith box (state) of the cart–pole system at time t.

$\overline{x}_i(t)$ running average of previous values of x_i, which decays exponentially towards zero.

$y(t)$ ASE output at time t.

γ ACE discount factor, which determines the rate at which $\hat{r}$ approaches zero.

δ ASE eligibility trace parameter, which determines the rate at which e_i approaches zero.

θ pole angle (vertical $= 0°$).

$\dot{\theta}$ rate of change in the pole angle.

λ ACE trace decay rate parameter, which determines the rate at which $\overline{x}_i$ approaches zero.

6.5. Comments by the Paper's Authors: A Barto, R Sutton and C Anderson

The authors' brief comments are given here, followed by an extensive retrospective commentary, 'Looking Back on the Actor–Critic Architecture'[9](reproduced with permission).

This paper is an early result of a project that explored the idea that neurons are not simple logic-gate-like devices but are much more sophisticated mechanisms that learn from the consequences of their actions upon their later inputs. The impetus for this was the idea due to physiologist A.H. Klopf (1972,1982) that progress in artificial intelligence could be made by modeling neurons as miniature hedonists. Instead of adjusting synaptic weights to make their outputs better match given target outputs (as in supervised learning, SL), these neuron-like devices would adjust weights so that their outputs would influence their environments toward delivering their most preferred future inputs according to some preference ordering. This basic idea would later became known as reinforcement learning, RL.

We dove into this project in several directions. Wasn't this already studied long ago? Is RL really any different from SL? Did this seem like a good idea? After a lot of reading, thinking, and programming, we concluded that indeed this was something new, or, at least, that it was insufficiently studied. A point came when we wanted to demonstrate in a vivid way what we had been working on. A colleague suggested that we tackle the classic problem of balancing a pole hinged to a movable card, or the inverted pendulum problem. Systems that learned to balance a pole were described by Widrow and Smith (1964) and Michie and Chambers (1968). We thought that just one or our neuron-like elements could do it better. Although we ended up using two, different, neuron-like elements, in what became the actor–critic architecture introduced in our 1983 paper, the point was still made that maybe neurons are capable of much more sophisticated processing than commonly thought. Our 2021 'Looking Back' paper, included here, fills in many details of this story.

Research Paper:
Neuronlike Adaptive Elements That Can Solve Difficult Learning Control Problems

Reference: Barto, A., Sutton, R., and Anderson, C. (1983). Neuronlike adaptive elements that can solve difficult learning control problems. *IEEE Trans. on Systems, Man, and Cybernetics*, 13(5):834–846. https://ieeexplore.ieee.org/abstract/document/6313077 (Restricted access).
Reproduced with permission from IEEE.
Corrected versions of the cart–pole equations can be found in Florian (2007).
https://coneural.org/florian/papers/05_cart_pole.pdf

Looking Back on the Actor–Critic Architecture

Andrew G. Barto, *Life Fellow, IEEE*, Richard S. Sutton,
and Charles W. Anderson, *Senior Member, IEEE*

Abstract—This retrospective describes the overall research project that gave rise to the authors' paper "Neuronlike adaptive elements that can solve difficult learning control problems" that was published in the 1983 Neural and Sensory Information Processing special issue of the IEEE Transactions on Systems, Man, and Cybernetics. This look back explains how this project came about, presents the ideas and previous publications that influenced it, and describes our most closely related subsequent research. It concludes by pointing out some noteworthy aspects of this article that have been eclipsed by its main contributions, followed by commenting on some of the directions and cautions that should inform future research.

Index Terms—Actor–critic, hedonistic neuron, pole blancing, reinforcement learning (RL), temporal-difference learning.

I. INTRODUCTION

WHEN our paper "Neuronlike adaptive elements that can solve difficult learning control problems" was published in the 1983 Neural and Sensory Information Processing special issue of the IEEE TRANSACTIONS ON SYSTEMS, MAN, AND CYBERNETICS [1], there was no way to predict that it would have a lasting influence, or that reinforcement learning (RL), the class of machine learning methods to which it contributed, would now be one of the most active areas of artificial intelligence. In this retrospective, we describe the overall research project that gave rise to our paper, explain how this project came about, present the ideas and previous publications that influenced it, and describe our most closely related subsequent research. We do not, however, attempt to do justice to earlier related work of which we were unaware in 1983; much of that can be found in [2]. We end by pointing out some noteworthy aspects of this article that have been eclipsed by its main contributions, followed by commenting on some of the directions and cautions that should inform future research.

In the late 1970s and early 1980s, we had the opportunity to participate in a research project aimed at assessing the scientific merit of a hypothesis proposed by physiologist A. Harry Klopf, a senior scientist with the Avionics Directorate

Manuscript received November 15, 2020; accepted November 24, 2020. Date of publication December 24, 2020; date of current version January 12, 2021. This article was recommended by Associate Editor D. Liu. *(Corresponding author: Andrew G. Barto.)*

Andrew G. Barto is with the College of Information and Computer Sciences, University of Massachusetts Amherst, Amherst, MA 01003 USA (e-mail: barto@cs.umass.edu).

Richard S. Sutton is with the Department of Computing Science, University of Alberta, Edmonton, AB T6G 2R3, Canada, and also with DeepMind, Edmonton, AB, Canada.

Charles W. Anderson is with the Department of Computer Science, Colorado State University, Fort Collins, CO 80523 USA.

Digital Object Identifier 10.1109/TSMC.2020.3041775

of the Air Force Office of Scientific Research (AFOSR). Klopf was dissatisfied with the great importance attributed to equilibrium-seeking processes for explaining natural intelligence and for providing a basis for machine intelligence. These include homeostasis and error-correction learning methods for pattern classification. He argued that systems that try to maximize something (whatever that might be) are qualitatively different from equilibrium-seeking systems, and he further argued that maximizing systems hold the key to understanding important aspects of natural intelligence and for building artificial intelligences. In particular, Klopf hypothesized that neurons, the major components of our brains, are individually "hedonists" that work to maximize a neuron-local analog of pleasure while minimizing a neuron-local analog of pain [3], [4].

A project with the goal of assessing Klopf's ideas to determine if they were novel and if they were worth pursuing, was funded through an AFOSR contract to principal investigators Michael Arbib, William Kilmer, and Nico Spinelli, professors at the University of Massachusetts Amherst and founders of the Cybernetics Center for Systems Neuroscience, a far-sighted center focusing on the intersection of neuroscience and artificial intelligence. Andrew Barto, a recent Ph.D. from the University of Michigan, was hired as a post-doc in 1977, shortly joined by graduate students Richard Sutton and Charles Anderson, who later received Ph.D.s under Barto's direction after he became a UMass faculty member. It was our good fortune that the project's funding and its PIs gave us wide latitude to explore the study of learning in artificial intelligence, including its early history, its connections to experimental data and theories of animal learning from psychology, and its connections to data and theories about the neural basis of learning from neuroscience.

II. MINIMIZING OR MAXIMIZING?

Klopf argued that a system that attempts to maximize a quantity is distinctly different from one that attempts to minimize a quantity, such as a system that seeks stability by minimizing the difference between its current state and a desired state. He coined the term "heterostat" to distinguish a maximizing system from a "homeostat", Ashby's term for a system that maintains stability in a changing environment [5].

There is, of course, no mathematical difference between minimizing and maximizing (just change the sign), but one of the first things we realized was that there is a qualitative difference between being directed by a signed error vector, indicating, for example, the difference between a current and a desired state, and being directed by scalar evaluations that—in themselves—do not indicate a direction of improvement.

This turns out to be a major distinction between supervised learning, which is fundamentally an equilibrium-seeking process (zeroing out errors) and RL, in which the learner has to do more work to determine how it should change in response to its experiences. An equilibrium-seeking process can stop when equilibrium is attained (zero error), but an evaluation-driven system, not knowing what evaluation is best, has to be incessantly active, continually exploring for directions of improvement. Both error correction and RL are optimization processes, but error correction is a restricted special case where RL is more general. It seemed to us that Klopf's intuition about the relevance of this to intelligence might be on the mark.

III. Initial Progress

From our exploration of earlier research on building artificial learning systems, we came to a rather surprising conclusion. Despite the prominence of evaluation-driven learning in some of the earliest artificial learning systems, this form of learning had been largely overshadowed by error-correcting learning due to supervised learning's ability to learn to recognize patterns by being exposed to training examples. There were clear and important exceptions, but this sparsity of earlier research on what we now call RL confirmed Klopf's contention that something was missing from approaches to machine learning that were then current. This encouraged us as we launched into the subject. An account of this history is beyond our scope here but can be found in [2, Sec. 1.7] and in [6].

There was also an important form of neglect in psychology in that studies of animal learning became unfashionable, supplanted by the "cognitive revolution." In particular, learning from the rewarding and punishing consequences of behavior, studied as instrumental conditioning in psychology, suffered from neglect, despite always being seen as a key principle of learning. This principle was famously stated in Thorndike's "Law of Effect" [7], which says that if an animal's response to a situation is closely followed by the animal's satisfaction, then that response becomes more strongly connected to the situation and is, therefore, more likely to be produced when the animal faces that situation again; conversely, if a response is followed by discomfort, the connection is weakened, making the animal less likely to produce the response when that situation recurs. This "law" has endured to the present, though not without much revision and controversy. It describes the common sense process of learning by trial and error (though it is misleading to equate the word error with the error vectors of supervised learning).

Klopf's idea of hedonistic neurons was that neurons implement a neuron-local version of the law of effect. He hypothesized that the synaptic weights of neurons change with experience according to the following. When a neuron fires an action potential, all the synapses that were active in contributing to the action potential become eligible to undergo changes in their efficacies. If the action potential is followed within an appropriate time period by an increase in reward, the efficacies of all eligible synapses then increase (or decrease in the case of punishment). In this way, synapses change so as to alter the neuron's firing patterns in the service of increasing the neuron's probability of being rewarded, and decreasing its probability of being penalized, by its environment.

In Klopf's hypothesis, reward and punishment were delivered to a neuron via the same inputs that excited or inhibited its electrical activity. He objected to the idea that there is a single specialized reward signal that drives learning. Our algorithms departed from this by using a single specialized input to deliver rewards, but we did not completely discount his objection to it, as we discuss in Section- III-B below.

The actor–critic architecture presented in our 1983 paper brought together two lines of research that we had been pursuing from the beginning of the AFOSR project. One line focussed on developing a neuron-like adaptive element following Klopf's idea of a hedonistic neuron. We called this an associative search element, or ASE, later to be known as the actor component of the actor–critic architecture. The other line of research focused on issues of signal timing and prediction. This line led to the neuron-like adaptive element we called the adaptive critic element, or ACE, which became the other main component of the actor–critic architecture. This component was the source of the reward and penalty signals evaluating the actions of the ASE.

We discuss the ASE and ACE before discussing their combination in the actor–critic architecture. Both elements were "neuron like" in the same abstract way that McCulloch-Pitts' formal neurons were [8]. We decided that trying to model real neurons in any detail would distract us from the project's main computational objective.

A. Associative Search Element

The idea of storing information distributed across large areas of a physical structure had gained prominence by the late 1970s for both its computational promise and as a model of how information might be stored in brain (e.g., [9]–[11]). Called associative memories, the simplest were based on correlation matrices, and storing information consisted of presenting "keys" paired with "patterns" to store key-pattern associations. As a learning process, this was supervised learning because the desired pairings of keys and patterns were explicitly provided to the memory systems, though the systems were able to generalize beyond these training pairings as in pattern recognition uses of supervised learning.

We decided that an RL version of an associative memory would be a good way to illustrate the difference between RL and supervised learning. Instead of being given the desired patterns to be associated with the keys, the RL associative memory network had to search for the pattern that maximized an externally supplied reward signal. As this kind of learning proceeded, each key tended to cause the retrieval of better—more rewarding—choices for the pattern to be associated with it. The only part of the system having prior knowledge about what associations were best was the evaluator, or critic, which computed the reward signal.

Details of this network and some simulation results were presented by Barto *et al.* in 1981 [12]. In one example, the network was a single layer of 25 ASEs. Input keys were 8-D vectors delivered to each ASE; output patterns consisted of the binary responses of each ASE, and each ASE received the same reward signal. We called this an associative search network (ASN). The ASEs' learning algorithm generally followed Klopf's hypothesis about how the synaptic weights of neurons might be adjusted by an RL process, but we borrowed from earlier RL algorithms known as stochastic learning automata. These algorithms originated with the work of Tsetlin [13] (reviewed in [14] and [15]) and were not well known in artificial intelligence circles. We also borrowed from Tzanakou and Harth's Alopex method that these authors proposed as a stochastic model of the development of visual receptive fields [16], [17].

Stochastic learning automata and the Alopex method use randomness to search among alternative actions for those that deliver the most reward. Actions are selected according to probabilities that are altered on the basis of reward feedback so as to allocate more probability to higher performing actions. Actions have to be tried out to find out how they perform. This process is *selectional* in the same way that natural evolution is selectional in favoring higher fitness organisms. We added a random number to the activation level of each ASE so that it randomly tried out all of its actions and biased the random selection toward actions yielding more reward. Randomness was essential to provide the variety needed to drive the search, just as animal populations need variety to drive evolution.

ASEs differed from stochastic learning automata, and from the Alopex method, in an important respect. Stochastic learning automata did not normally receive input other than the reward signal, whereas all the ASEs making up the ASN received input vectors coding the associative memory keys in addition to the reward signal. As learning continued, the input keys became associated with better and better output patterns as scored by the reward signal. Where a stochastic learning automaton attempted to find a single best action, an ASE attempted to find the best action for each input key. In this respect, an ASE worked more like the law of effect in forming connections between situations, here the keys, and responses, here the associated patterns.

In more theoretical terms, a stochastic learning automaton faces a multiarmed bandit problem (e.g., [15] and [18]). An ASE, and therefore, the ASN as well, faced what we called an "associative search problem," now commonly referred to as a "contextual bandit problem." This problem involves remembering, in the form of associative links, the results of conducting multiple searches. It is, therefore, closely related what computer science calls "memoization," which is the process of saving results of a calculation in memory so that results that have been calculated previously can be retrieved from memory instead of being calculated again [19], [20]. In RL, the calculation is an ongoing search for higher rewarding actions. Consequently, at its base, RL is a kind of *contextual memorized search.*

It remained for us to decide how good and bad evaluations would be represented for delivery to the ASEs making up the ASN. A natural way to do this was to use *changes* in a scalar reward signal for adjusting each ASE's synaptic weights, instead of using reward signal itself: a reward increase made the element's response more likely in the present context; a reward signal decrease made it less likely. These reward signal changes were the reinforcement signals that directed changes in each ASE's synaptic weights; not the reward signal itself.[1]

It was necessary to include in the ASN a special reward-predictor element that learned to predict the amount of reward an ASE should expect when acting in the current context. This enabled reward signal changes to act correctly as reinforcement. In our simulations the context vectors—the keys—changed randomly at each time step. The network's output influenced the immediate context-dependent reward signal, but it had no influence on what key would be presented next (unlike the situation in our later actor–critic simulations). We had to prevent a change in reward due to a random change in the key from being attributed to the element's action. To do this, the reward change was computed by comparing the current reward with the reward expected when acting in the current context. This foreshadowed the ACE's role in the actor–critic architecture.

B. Adaptive Critic Element

We chose the term *critic* for the ACE component of the actor–critic architecture after Widrow *et al.*'s use of this term to contrast "learning with a critic" from "learning with a teacher," as supervised learning is often called [21]. A teacher provides the learner with desired or correct actions, whereas a critic merely evaluates a learner's actions. Evaluations might be the result of comparing the learner's actions with desired actions (as was the case for our ASN simulations), in which case evaluations are based on how much the learner's actions differ from the desired actions. However, evaluations do not need to be based on any knowledge of what the desired actions should be. In fact, the critic does not even need to have access to the learner's actions; it can base its evaluations on the consequences of those actions on the learner's environment. The pole-balancing control problem of our 1983 paper illustrates this because the ACE evaluates the behavior of the cart–cart system, not the ASE's actions, to which it does not have access.

We included the term *adaptive* in the ACE because it was a learning system itself, capable of learning to make more informative evaluations. In the case of the actor–critic architecture, this meant learning to evaluate the long-term performance of the actor by learning to predict how much reward would be expected to accrue over the future. The predictions themselves then became the reinforcing input to the actor.

The inspiration for the ACE came from animal learning psychology, in particular, from classical, or Pavlovian, conditioning. This form of learning enables an animal to act in

---

[1]The terms reward and reinforcement are sometimes used interchangeably, but we distinguish between them. An RL system's reward signal sets the learner's objective, which is to maximize the amount of reward received over time. Reinforcement, on the other hand, is the quantity that an RL learning rule uses to adjust the parameters determining its action probabilities.

anticipation of upcoming inputs from its environment, allowing the animal to prepare for, or to avoid, those inputs. The animal effectively learns to predict aspects of its future. The feature of classical conditioning most relevant to the actor–critic architecture is the phenomenon of higher-order conditioning (and the similar phenomenon of secondary reinforcement in instrumental conditioning). This occurs when events that predict the arrival of reward become rewarding themselves. Higher order conditioning and secondary reinforcement remove the need to wait until a final reward or penalty is received in order to learn.

A few years before our 1983 actor–critic paper, we presented the basics of the ACE algorithm as a model of classical conditioning [22], [23], which subsequently evolved into temporal difference (TD) algorithms. Sutton's 1984 Ph.D. dissertation [24] developed TD algorithms further, and Sutton's 1988 paper [25] extended the theoretical treatment, laying the groundwork for the extensive development that followed within the modern RL framework, e.g., [2].

The name TD derives from the algorithm's use of changes, or differences, in predictions over successive time steps to drive the learning process. The prediction at any given time step is updated to bring it closer to the prediction of the same quantity at the next time step. It is a self-supervised learning process that works to reduce errors between current and later predictions, taking intervening incoming data into account. Used in RL, TD algorithms learn to predict a measure of the total amount of reward expected over the future.

In employing a TD algorithm, the ACE of the actor–critic architecture addressed two requirements for successful learning. One requirement was to address the "delayed reward problem," which is when the relevant consequences of an action occur after some nontrivial time interval, making it difficult to assign credit or blame to the appropriate action, or actions, of the learner. The other requirement was to provide an appropriate reinforcement signal to the ASE. Both of these functions were clearly illustrated in the pole-balancing problem tackled in our 1983 paper.

The ACE also went part of the way toward addressing Klopf's rejection of a single unitary reward signal in his hedonistic neuron hypothesis. He argued that whatever generated this signal would have to be so intelligent itself that assuming its existence would beg the question of how intelligence arises. Klopf proposed what he called "generalized reinforcement" as a way to avoid a unitary reward signal. The TD idea is not unrelated. It uses ordinary (nonreward) input to play an important role in rewarding action. The actor is trying to maximize the excitation of the critic as well as maximize the reward, thus a kind of generalized reinforcement. But unlike Klopf's desire to eliminate a reward signal altogether, TD learning is tied ultimately to reward, which is necessary in order to create a well-defined optimization problem.

IV. POLE BALANCING

Our 1983 paper featured the problem of learning to balance a pole hinged onto a movable cart. This idea came from our discovery of the 1968 paper by Michie and Chambers entitled "BOXES: An Experiment in Adaptive Control" [26]. BOXES was a true RL system, and the paper's description of how it worked and the ideas underlying it helped shape our thinking about RL and how to explain our research. We thought that the version of the pole-balancing problem tackled by BOXES would provide a vivid illustration of the capabilities of the algorithms we had been working on, would clearly illustrate how RL differed from supervised learning, and would help to establish the utility of RL.

The BOXES pole-balancing task was adapted from the 1964 work of Widrow and Smith [27], who used supervised learning, assuming instruction from a teacher already able to balance the pole. But instead of receiving action-by-action instructions that could be copied, the sole training information available to BOXES was a failure signal when the pole fell past a certain angle or the cart hit the end of its track. This created a difficult delayed-reward problem (or delayed-penalty problem in this case) making the credit (or blame) assignment difficult.

In our earlier work with the ASN, described above, the pattern output of the network was evaluated by the critic, but the output pattern did not influence which key, or context vector, was presented next. The key presented at each step was selected uniformly at random from a finite set of keys. But in a more general setting, the RL system's actions would influence the stream of context inputs in addition to the critic's evaluations. Applying RL to a control problem like pole balancing was a natural way for the RL system's actions to influence the state of the system being controlled, with each state generating context input to the RL system. Furthermore, the critic's evaluations could be based on observing the behavior of the controlled system rather than on observing the RL system's actions themselves, thus illustrating an important property of RL. We therefore decided that a control problem would be an excellent testbed for our RL algorithms, and that Michie and Chambers' pole-balancing task would be a good place to start.

We followed the setup Michie and Chambers used in designing their BOXES system for the pole-balancing task. Like BOXES, our RL controller had no knowledge about the system being controlled, only receiving at each time step a vector describing the controlled system's state or a failure signal if the pole fell past a critical angle or the cart hit the end of the track. We used exactly their state representation, which was to divide the 4-D continuous state space into 225 "boxes" on the basis of the thresholds they selected, and to inform the controller which box the system's state was currently in. Using the picturesque "demon" terminology introduced into AI by Selfridge's 1959 Pandemonium program [28], Michie and Chambers described how BOXES worked like this:

> In order to envision how the ⋯ algorithm works it is easiest to imagine each one of the 225 boxes as being occupied by a local demon, with a global demon acting as a supervisor over all the local demons ⋯ Each local demon is armed with a left-right switch and a scoreboard. His only job is to set his switch

from time to time in the light of data which he accumulates on his scoreboard [26, p. 148].

The rule used by the local demons was quite complicated but essentially depended on records for each control decision, left or right, of the number of decisions taken before a run failed.

It was easy to translate this organization into our neural network terms, ending up with the entire system being implemented by a single neuron-like element, in particular, by a single ASE. At each time step, the ASE's input vector would be one of 225 standard-unit-basis vectors: all zeros except a one in the position corresponding to the box occupied by the current state. Then, a local demon would correspond a synapse whose influence on the element's output would correspond to its left–right switch. Its scoreboard, then, would correspond to its synaptic efficacy, or connection weight. The global demon would correspond to the activation rule of the neuron-like element that would convert the active local demon's decision, plus a random number contribution, into the element's binary output by thresholding.

A side benefit of our neuron-like implementation was that it could easily accommodate state representations more complicated than the one used by BOXES. The full function approximation power of neural networks could be enlisted, including multilayer neural networks, as subsequent advances demonstrated. See Section VII below.

Due to the delayed-reward problem presented by the pole-balancing task, which was not present in our earlier work with the ASN, we modified the ASE learning rule by adding *eligibility traces*. Recall that according to Klopf's hedonistic neuron hypothesis, all the synapses that were active in contributing to the neuron firing would become eligible to undergo changes in their efficacies, with the changes happening if reinforcement arrived during the period of eligibility. He envisioned that eligibility would be implemented by the concentration of a synaptically-local chemical that began increasing when the synapse was active in firing the neuron, reached a maximum shortly after this, and thereafter decayed to zero after a time interval long enough to register delayed reinforcement. This concentration would be a trace of past activity called an eligibility trace.

We added eligibility traces to the ASE in the simplest way we could think of. Each eligibility-triggering event added to the ongoing trace at the appropriate synapse; otherwise that trace decayed exponentially with a time constant selected as a parameter of the simulation. The result was not too different from the records kept by the local demons of BOXES. Sutton and Barto [2] extensively discuss eligibility traces in the context of ideas for synaptic tags proposed by neuroscientists.

We conjectured that our ASE algorithm, with eligibility traces that decayed sufficiently slowly, would be superior to the BOXES algorithm. Our main reasoning was that Michie and Chambers did not seriously concern themselves with the necessity for variety in the controller's actions. In other words, BOXES did only very limited exploration. It was purely deterministic except for using pseudorandom numbers to break

ties in selecting actions and for selecting the initial state for each learning trial, where a trial lasted from state reset until failure.[2]

An ASE, on the other hand, selected every action randomly, with probabilities adjusted as in stochastic learning automata. We thought that with eligibility traces lasting long enough to deal with the delayed-reward problem, a single ASE could learn faster than BOXES.

In the Summer of 1982, we saw an announcement for the "Neural and Sensory Information Processing" special issue of IEEE TRANSACTIONS ON SYSTEMS, MAN, AND CYBERNETICS in which our 1983 paper would eventually appear. We thought this would provide an excellent opportunity to publicize our research—if we could make it to the deadline for submission. So we decided to implement our system along with BOXES with the intention of comparing their performances, thinking that our stochastic approach would easily surpass the performance of BOXES.

As the special issue deadline approached, we struggled to get our system to outperform BOXES, which worked better than we had expected. Very near the deadline, we decided to insert a TD algorithm into our pole-balancer in the form of the ACE. Sutton had been developing the TD idea that was to become a major part of his 1984 Ph.D. dissertation [24]. With the ACE providing reinforcement signals to the ASE, the system was able to learn better than BOXES could. This combination of the ASE and ACE became known as the actor-critic architecture.

But at almost literally the last minute before the special issue deadline, there was another setback: we discovered a bug in our implementation of the cart-pole simulation. Our procedure for updating the pole's angle with respect to the cart accepted as input the current pole angle expressed in radians, but returned the new angle in degrees.[3] Consequently, unbeknownst to us, our system had been learning to control a very different cart-pole system than we had intended, or that BOXES had learned to control in Michie and Chambers' paper. We quickly fixed the bug, but then struggled to tune our controller and simulation to produce effective learning. Finally, after an all-night session, we achieved adequate results and submitted this article.[4]

[2]Michie and Chambers explicitly avoided probabilistic decision making, stating that "such devices cannot be optimal," regarding the two-armed bandit problem as a "famous unsolved problem of mathematics." It was not until the 1970s that a version of the problem was solved with Gittins Indices [29], and we knew that stochastic learning automata could achieve ϵ-optimality with ϵ being arbitrarily small [14].

[3]In 1999, a similar error caused the loss of NASA's 125-million dollar Mars Climate Orbiter due to the use of both English and metric units of acceleration. This put us in good company, though happy that our error was less costly.

[4]This late night struggle explains the unusual values we published for the coefficients of friction of the pole on the cart ($\mu_c = 0.0005$) and the cart on the track ($\mu_t = 0.000002$). Of course, we wanted to experiment more, but it worked with these values, and we had to stop. Our haste also explains an unfortunate error that appeared in this article as published: the sign of gravity given in the Appendix has the wrong sign (although it was correct in our simulations). We learned of this from readers' attempts to duplicate our results. With the published sign of gravity, they found that no learning at all was needed to balance the down-hanging pole. This error did not hinder further research because it was quickly noted and became widely known among RL researchers.

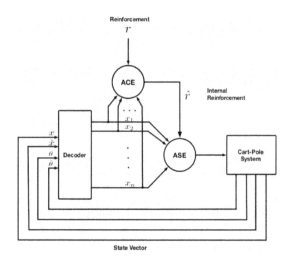

Fig. 1. Actor–critic architecture configured for the pole-balancing task as depicted in our 1983 paper [1].

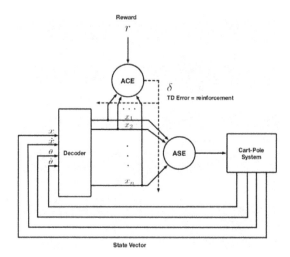

Fig. 2. Updated representation of the actor–critic architecture configured for the pole-balancing task. The input r is now labeled reward, and the dashed lines depict the TD error as the reinforcement signal for adjusting the input connection weights of both the ASE and the ACE.

In 1985, we had the opportunity to work with AI pioneer Oliver Selfridge on some additional experiments with the actor–critic architecture applied to the pole-balancing task [30]. It is more difficult to balance a short pole than it is to balance a longer pole. This suggested the following strategy for learning to balance a short pole: start with an easier-to-balance long pole and shrink its length as learning proceeds, ending up with the skill of balancing the short pole. We tried this using the actor–critic architecture of our 1983 paper. Learning to balance a 1-m pole took on average 67 failures, whereas learning to balance a 2/3-m pole took on average 119 failures. Switching to the short pole after learning to balance the long pole took just six more failures on average. This illustrated how an RL system could benefit from what psychologists call shaping [31], the training strategy that starts with an easy problem and incrementally increases its difficulty as the animal learns. Shaping is indispensable for animal training, and it can benefit learning via RL algorithms as well.

V. Some Details of the Architecture

Fig. 1 follows the figure from our 1983 paper showing the actor–critic architecture configured for the pole-balancing task. The output of the decoder is the standard unit basis vector representation of the 4-D cart-pole state space divided into $n = 225$ boxes. The ACE receives the reward signal r as input, in this case, the failure signal, and produces an "internal reinforcement" signal $\hat{r}$ which it sends to the ASE.

Fig. 2 is an updated representation of the actor–critic architecture based on our better understanding of how the architecture is related to psychology and neuroscience (see Section IX below). We distinguish between a reward signal, which sets the overall objective of the task, and a reinforcement signal, which directs the changes in the learner's parameters, here being the input connection weights of the ASE and ACE. As in Fig. 1, the ACE receives the signal r as input, but Fig. 2 labels it reward instead of reinforcement. The signal the ACE sends to the ASE is the reinforcement signal,

which here is the TD error $\delta(t) = r(t) + \gamma p(t) - p(t-1)$, where p is a prediction of the eventual reward and γ is a discount factor between 0 and 1. The TD error is also the reinforcement signal for the ACE. This is shown by the dashed lines that cross the context input lines to each element.

A unique feature of the architecture is that although the TD error $\delta(t)$ is the reinforcement signal for both the ASE and the ACE, these elements learn to perform different functions: the ASE adjusts its parameters (its synaptic weights) in order to move its action probabilities toward higher-rewarding actions, while the ACE adjusts its parameters in order to make more accurate reward predictions.[5] The TD error tells the ASE if the current prediction $p(t)$ of the amount of reward expected over the future has just increased ($\delta > 0$) or just decreased ($\delta < 0$) so that the ASE's action probabilities can be adjusted to make it more, or less, likely to execute the same action when the current context occurs again. On the other hand, the TD error tells the ACE if its prediction of future reward is too low ($\delta > 0$) or too high ($\delta < 0$) so that it can correct its prediction.

The same reinforcement signal produces these different functions because the learning rules of the ASE and ACE differ in a subtle way: they use different notions of eligibility. Eligibility for the weight of an input connection to the ACE depends solely on input via that connection. That is, eligibility traces associated with the ACE's input connections are traces of past input via those connections. In contrast, eligibility for a weight associated with an input connection to the ASE depends, in addition to input via that connection, on the ASE's

---

[5]In some presentations (e.g., [2]), the TD error at step t is $\delta(t) = r(t+1) + \gamma p(t+1) - p(t)$, that is, it depends on the reward and prediction at the next time step $t + 1$. The interpretation of this form is that it is the error in the prediction made at step t that becomes available as a signal at $t + 1$. The interpretation of the alternate form used in our 1983 paper is that $\delta(t)$ is the error in the prediction made at $t - 1$ that becomes available as a signal at time t. Both interpretations appear in the literature.

output. That is, eligibility traces for an ASE's connections are traces of past input via those connections but modulated by the element's output. According to Klopf's notion of eligibility for a neuron, a synapse that transmits an excitatory pulse to the neuron would become eligible only if that input participated in causing the neuron to fire an action potential, and if a synapse transmits an inhibitory pulse, it would become eligible only if the neuron was inhibited from firing an action potential.

We call the ACE's eligibility *noncontingent* because it is not contingent on the element's output, and we call the ASE's eligibility *contingent* because it does depend on the element's output. Output contingency makes the ASE learn about the effect of its actions on the reinforcement signal. Lacking output contingency, the ACE learns to predict reward independently of its output.

The situation is somewhat more complicated than what we have written here. For example, in [1], the ASE's output was always nonzero, being either $+1$ or -1, so that eligibility traces accumulated positive and negative contributions of past activity. Details can be found in [2]. In Section IX, we discuss these issues from the perspective of neuroscience.

VI. DECOMPOSITION INTO SUBGAMES

One of the objectives of Michie and Chambers' BOXES system was to illustrate the benefit of decomposing a large problem into many small problems, arguing (with reference to earlier work on playing Naughts and Crosses, i.e., Tic-tac-toe [32], [33]) that "it may be easier to learn to play many easy games than one difficult one." The organization of BOXES illustrated this by its division of the pole-track state space into 225 boxes, in each of which a local demon was faced with the relatively simple problem of learning how to act just for states falling in that box. BOXES illustrated that success on the large problem could be achieved in this way even if the subproblems, i.e., the problems faced by the local demons, were not independent.

The influence of this aspect of Michie and Chambers' paper is apparent in the direction we took in further developing RL. We maintained the view that favored learning in a state-dependent manner instead of treating the problem of finding an optimal policy as a monolithic, or "black box," optimization problem. In [2], Sutton and Barto referred to monolithic optimization methods as "evolutionary methods," and instead focussed on learning while interacting with an environment in order to take advantage of individual behavioral interactions, which monolithic optimization algorithms do not do.

It is fair to regard this emphasis on learning functions of states, such as value functions and policies, during interaction with the environment partly as a legacy of the influence on us of the BOXES paper. Also contributing to this emphasis was that learning via interaction fits better with our sense of how animals learn during their lifetimes. Evolutionary methods clearly have their place in machine learning where they can be the preferred methods for some problems, but to us there is

a clear difference between interaction-based RL and evolutionary, or black-box, methods. Distinguishing the complementary capabilities of each may be a prerequisite to designing algorithms that combine them as effectively as they are combined in the natural world.

VII. REINFORCEMENT LEARNING WITH MULTILAYER NEURAL NETWORKS

We adopted the BOXES state representation for our pole-balancing experiments so that our system would be as similar to BOXES as possible, differing only in the most critical features. All the while, we were well aware of the much wider set of possible state representations that neural networks could provide, including the very rich representational abilities of multilayer neural networks. In parallel with our work on developing TD algorithms, modeling classical conditioning, and illustrating RL with the pole-balancing example, we were exploring RL methods for training the hidden layers of multilayer neural networks so that state representations could be learned rather than provided from the start.

We published a series of papers showing that multilayer neural networks could learn desired nonlinear mappings if each unit in the network learned via RL and the reward signal was broadcast uniformly to each unit [34]–[37]. In this approach, inspired by the earlier work with "teams" of stochastic learning automate, e.g., [15] and [38], the gradient of the objective function was stochastically estimated rather than backpropagated as in the error backpropagation algorithm that shortly became well known through the 1986 chapter by Rumelhart *et al.* [39]. Although our RL method was much slower than the backpropagation algorithm, we argued that it was simpler and more plausible biologically [40].

Co-author Anderson experimented with the RL approach for training multilayer networks, finding that adding a trainable hidden layer improved performance in the pole-balancing task [41]. He examined the features that the hidden layer learned, noting that they captured essential aspects of the control problem. Later, in his 1986 Ph.D. dissertation [42] (also [43]), Anderson compared a number of algorithms for hidden unit learning as applied to several tasks, including pole balancing and the Tower of Hanoi task. He found that both the RL method and error backpropagation learn the solutions to the tasks much more successfully than earlier methods for training multilayer networks, and he analyzed the features developed by the hidden units in solving these tasks.

Fig. 3 shows the two-layer actor and critic networks Anderson applied to the pole-balancing task [42]. Trained by backpropagation, the two-layer system far outperformed a single-layer system, although both learned more slowly than the system with the hand-crafted boxes state representation used in our 1983 paper. Anderson attributed this to the considerable number of steps that were required for the hidden units to learn the necessary features. Anderson's work is the first instance of which we are aware in which learning algorithms for multilayer neural networks were used in RL tasks, foreshadowing current advances in deep RL.

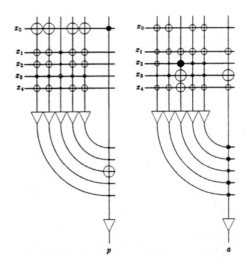

Fig. 3. Two-layer neural network actor–critic architecture applied to the pole-balancing task from Anderson's 1986 Ph.D. dissertation [42]. The left and right networks, respectively, implemented the critic and the actor. In each network, the five input lines connect to five hidden units and to the single output unit, for a total of 35 weights. The relative magnitudes and signs of the learned weights are shown by the circles on the line intersections, with the open and black circles, respectively, indicating positive and negative weights.

VIII. THEORY

RL has seen tremendous progress since our 1983 paper. New algorithms, many applications, and vastly improved theoretical understanding emerged as a result of the contributions of numerous researchers. RL has benefitted from being connected to more traditional fields, in particular, to stochastic optimal control and dynamic programming. RL is now regarded as a collection of methods for approximating solutions to Markov decision processes (MDPs), a framework into which all of our earlier work can be placed. Here, we mention a few highlights of this theoretical development that bear directly on the actor–critic architecture, omitting details that can be found in many publications, e.g., [2].

The actor–critic architecture can be understood most clearly as a policy-gradient algorithm. A policy is a function that maps environment states to control actions, often probabilistically. In a policy-gradient algorithm, the policy is represented as a parameterized function with parameters adjusted by the learning algorithm. In an actor–critic algorithm, the actor updates policy parameters by moving the policy parameter vector in the direction of an estimated gradient of a measure of long-term reward, where the gradient is estimated from sample trajectories. The ASE learning rule in our 1983 paper does not exactly perform gradient ascent, but later versions incorporating William's REINFORCE algorithm [44] do achieve stochastic gradient ascent.

The critic of the actor–critic architecture updates parameters of a parameterized state-value function, which is a function that assigns to each state an estimate of the expected long-term return (the expected cumulative reward) when a given policy is followed from that state. The critic uses a TD algorithm to adjust the value-function parameters in order to improve its prediction accuracy based on observed state transitions and rewards.

The interaction of the actor and critic is analogous to the policy-iteration algorithm of dynamic programming. Each iteration of that algorithm alternates between computing a state-value function for a current policy, and then improving the current policy according to the current state-value function. Actor–critic methods effectively perform these two phases simultaneously, interleaving single steps of state-value function estimating with single steps of policy improvement.

The theoretical properties of traditional policy iteration are well known (it converges to an optimal policy for finite MDPs under mild conditions), but the actor–critic analog is more difficult to analyze. The most comprehensive convergence results are due to Bhatnagar *et al.* [45], who prove convergence to a local maximum of the long-run average reward for several versions of the actor–critic algorithm using a two-timescale approach in which the critic learns faster than the actor.

The 1990s saw the development of RL algorithms based on estimating action-value functions instead of state-value functions, the prime example being Q-learning [46]. Action-value functions map state–action pairs to expected return. With these methods, there is no need for an explicit policy representation because actions can be selected simply by consulting the estimated values of the actions for the current state. Action-value functions have been called "action-dependent adaptive critics" [47].

Action-value algorithms came to be preferred over actor–critic algorithms because of their relative ease of implementation, but understanding their convergence properties is challenging, especially when they use function-approximation methods for learning action-value functions. As a result, interest in policy-gradient algorithms, including actor–critic algorithms, has lately increased. Their reliance on explicit parameterized policy representations offers a number of advantages, among which are the following: 1) they provide useful ways to deal with continuous action spaces; 2) they make it possible to select actions with arbitrary probabilities, and yet can converge to deterministic policies; and 3) they offer a good way to introduce prior knowledge into the learning process. Additionally, sometimes a high-performing policy is much simpler than an action-value function [48]. By accounting for this simplicity in selecting a policy parameterization, learning can be faster and lead to a better policy than possible with action-value methods.

IX. ACTOR–CRITIC IN THE BRAIN

Mounting evidence from neuroscience suggests that the nervous systems of humans and many other animals implement algorithms that correspond in striking ways to RL algorithms. The most remarkable point of contact involves dopamine, a chemical fundamentally involved in reward processing in the brains of mammals. Experiments have shown that neurons (at least many of them) that produce dopamine as a neurotransmitter respond to rewarding events with substantial bursts of activity only if the animal does not expect those events [49]. This finding suggests that many dopamine-producing neurons are signaling reward prediction errors instead of reward itself.

Experiments have also shown that as an animal learns to predict a rewarding event on the basis of preceding sensory cues, the bursting activity of dopamine-producing neurons shifts to earlier predictive cues while decreasing to later predictive cues. This parallels the backing-up effect of the TD error as the ACE learns to predict reward. It is now widely accepted that bursts of dopamine neuron activity convey reward prediction errors to brain structures where learning and decision making take place, and evidence supports the idea that the prediction errors might be TD errors [50].

Experimental results like these have led to the hypothesis that the brain might implement something like our actor–critic architecture. TD errors conveyed by the activity of dopamine neurons are reinforcement signals that train both the critic's predictions and encourage or discourage the actor's choice of actions [49], [51].

Whether a brain region performs an actor-like or a critic-like function depends on how synaptic efficacies change in each region in response to receipt of dopamine reinforcement. Synapses in a critic-like region would implement noncontingent eligibility traces. In neural terms, noncontingent eligibility means that the eligibility of a synapse is solely a function of presynaptic activity, that is, of activity that reaches the synapse as input from other neurons. Synapses in an actor-like region, on the other hand, would implement contingent eligibility, which is a function of the activity both the pre- and postsynaptic neurons. Neuroscientists would say that critic synapses have a two-factor learning rule (presynaptic activity + dopamine), whereas actor synapses have a three-factor learning rule (presynaptic activity + postsynaptic activity + dopamine). Moreover, if the neuron is to implement a kind of law of effect as Klopf conjectured, the presynaptic activity must have taken part in generating the postsynaptic activity in order for the synapse to become eligible for modification.

Even if neurons behave nothing like the ASE or the ACE of the actor–critic architecture, our 1983 paper suggested that neurons may be capable of very sophisticated processing, vastly more complex than the logic-gate analogy of old would suggest. That nearly the entire BOXES system of Michie and Chambers could be implemented by a *single* neuron-like element is a touchstone for thinking about neural networks as being more analogous to networks of computers than to logic circuits.

Furthermore, understanding how adaptive elements such as the ASE operate requires thinking about them as embedded in closed-loop interactions with their environments. They are metaphorically "swimming" in a medium composed of the rest of the neural net plus the organism's (or robot's) external environment. Their adaptive changes are sensitive to the effects that their actions have on input signals they later receive.

The brain's reward system is undoubtedly much more complicated than current RL algorithms, and the story is still unfolding as more is being learned about the brains's reward system, but the actor–critic architecture, along with other RL algorithms and theory, is proving to be enormously useful in making sense of experimental data, in suggesting new kinds of experiments, and in pointing to factors that may be critical to manipulate and to measure. See [2] for more about RL and neuroscience.

X. Future

Some of the most impressive achievements in AI have been produced by programs that include RL. Notable examples are DeepMind's Go-playing programs [52], [53]. While these programs are vastly more complex than an actor–critic architecture, and the problem they faced is vastly more difficult than pole balancing, they nevertheless carry forward some features of our RL pole balancer, such as learning from interaction with a dynamic environment, caching trial-and-error search results, and learning value predictions to address long-term goals. The future will see the methods used in these game-playing programs, and other successful learning programs, adapted and extended to address a widening range of challenging problems, including pressing real-world problems of scientific and social importance.

RL has the potential to improve the quality, efficiency, and cost effectiveness of processes on which we depend in education, healthcare, transportation, and energy management, among others, but challenges have to be addressed to realize this potential. Many design decisions are involved in applying RL. The architecture has to be designed by selecting appropriate learning algorithms, state and action representations, training procedures, hyperparameter settings, and other design details. An important goal for future research is to make RL algorithms more robust and easier to apply so that new applications can be developed by experts in the application domains instead of by teams of experts in RL and other machine learning methods.

There are ample opportunities to improve and generalize RL algorithms and architectures. Examples include: architectures for learning hierarchical polices to improve efficiency and the ability to transfer learning to new problems; efficient methods for learning with incomplete state information; expanding the role of prediction to enable agents to predict and control many signals from their environments, not just long-term reward; further development of multiagent RL; further development of model-based RL to integrate planning and higher level reasoning; and architectures for open-ended life-long learning. Progress has been made along these and other directions, but much more is possible. Additional improvements are discussed in [2] and [54].

As RL moves out into the real world, it is critical to make sure that what is learned conforms to the intentions of the application's designer, and that the learning agent does no harm to itself or to its environment, including any people in it, both during and after learning. This requires adopting risk mitigation and management strategies that exist for other risk-prone technologies and designing new methods especially targeting risks posed by machines that learn through interacting with real-world environments. Unless RL is restricted to always operate in benign environments, like game playing where one can tolerate the worst that can happen, ensuring the safety of RL applications is a critical challenge that needs careful attention.

The design of an RL system's reward function is of special importance for RL safety. This is the function that assigns reward and penalty magnitudes to states, actions, state–action pairs, and perhaps other aspects of the system. Because RL fundamentally involves optimization, with the reward function being the objective function, it shares with other optimization processes the problem that it can produce unexpected, and sometimes unwanted, possibly catastrophic, results. This possibility has long been recognized. For example, Norbert Wiener, the founder of cybernetics, warned of this problem more than half a century ago by relating the supernatural story of "The Monkey's Paw:" "⋯ it grants what you ask for, not what you should have asked for or what you intend" [55, p. 59]. The problem is featured as "perverse instantiation" in Bostrom's broadside about the dangers of AI [56]. Methods for designing reward functions are needed that go beyond the hand tuning that is common practice today, and sound methodologies are needed to assess the safety of what a reward function enables a system to learn.

Turning to the connections between RL and neuroscience, the future will see continued fruitful interaction between neuroscience and RL. Advances in RL will suggest new ways to think about the brain's decision and reward systems, and advances in neuroscience will inform the further development of RL. As neuroscience uncovers more about how reward processing works in the brain, we may see experimental support for Klopf's hypothesis that neurons—at least some of them—individually implement a kind of law of effect.

XI. CONCLUSION

Contributing to this 50th Anniversary Issue of IEEE TRANSACTIONS ON SYSTEMS, MAN, AND CYBERNETICS has given us the opportunity to revisit some of our earliest efforts in what has become the flourishing RL subarea of machine learning. The 1983 "Neural and Sensory Information Processing" special issue of the journal provided the ideal venue for our work at a time when biologically inspired machine learning was not as routine as it is today. We have been gratified and surprised by the influence our paper has had. It is among the most frequently cited of the publications by any of us, its three authors, and pole balancing has served as a testbed for many different learning architectures.

Exploring Klopf's hedonistic neuron hypothesis led us through some of the early history of AI, to psychology's theories of learning, and eventually to appreciation of stochastic optimal control and dynamic programming. The striking parallels between TD algorithms and the brain's dopamine system revealed strong connections between RL algorithms and reward processing in the brain. Although Klopf's idea of the hedonistic neuron remains a hypothesis, time will tell if it finds compelling neuroscientific support.

There is now a large international community of researchers improving RL algorithms and architectures, combining RL with other technologies, creating new RL applications, and exploring new directions inspired by RL. The actor–critic

architecture now is in the company of many other architectures that embody core RL ideas while exploiting the immensely improved computational resources available today. We fully expect that RL can help improve the quality, fairness, and sustainability of life on our planet, provided its risks can be successfully managed.

REFERENCES

[1] A. G. Barto, R. S. Sutton, and C. W. Anderson, "Neuronlike elements that can solve difficult learning control problems," *IEEE Trans. Syst., Man, Cybern.*, vol. SMC-13, no. 5, pp. 835–846, Oct. 1983.

[2] R. S. Sutton and A. G. Barto, *Reinforcement Learning: An Introduction*, 2nd ed. Cambridge, MA, USA: MIT Press, 2018.

[3] A. H. Klopf, "Brain function and adaptive systems—A heterostatic theory," Air Force Cambridge Res. Lab., Bedford, MA, USA, Rep. AFCRL-72-0164, 1972.

[4] A. H. Klopf, *The Hedonistic Neuron: A Theory of Memory, Learning, and Intelligence*. Washington, DC, USA: Hemisphere, 1982.

[5] W. R. Ashby, *Design for a Brain: The Origin of Adaptive Behavior*. London, U.K.: Assoc. Book Publ., 1960.

[6] A. G. Barto and R. S. Sutton, "Goal seeking components for adaptive intelligence: An initial assessment," Air Force Wright Aeronaut. Lab./Avionics Lab., Wright-Patterson AFB, OH, USA, Rep. AFWAL-TR-81-1070, 1981.

[7] E. L. Thorndike, *Animal Intelligence*. Darien, CT, USA: Hafner, 1911.

[8] W. S. McCulloch and W. Pitts, "A logical calculus of the ideas immanent in nervous activity," *Bull Math. Biophys.*, vol. 5, no. 4, pp. 115–133, 1943.

[9] S.-I. Amari, "Neural theory of association and concept-formation," *Biol. Cybern.*, vol. 26, no. 3, pp. 175–185, 1977.

[10] J. A. Anderson, J. W. Silverstein, S. A. Ritz, and R. S. Jones, "Distinctive features, categorical perception, and probability learning: Some applications of a neural model," *Psychol. Rev.*, vol. 84, pp. 413–451, 1977.

[11] T. Kohonen, *Associative Memory: A System Theoretic Approach*. Berlin, Germany: Springer-Verlag, 1977.

[12] A. G. Barto, R. S. Sutton, and P. S. Brouwer, "Associative search network: A reinforcement learning associative memory," *Biol. Cybern.*, vol. 40, pp. 201–211, May 1981.

[13] M. L. Tsetlin, *Automaton Theory and Modeling of Biological Systems*. New York, NY, USA: Academic, 1973.

[14] K. S. Narendra and M. A. L. Thathachar, "Learning automata—A survey," *IEEE Trans. Syst., Man, Cybern.*, vol. SMC-4, no. 4, pp. 323–334, Jul. 1974.

[15] K. Narendra and M. A. L. Thathachar, *Learning Automata: An Introduction*. Englewood Cliffs, NJ, USA: Prentice-Hall, 1989.

[16] E. Harth and E. Tzanakou, "Alopex: A stochastic method for determining visual receptive fields," *Vis. Res.*, vol. 14, pp. 1475–1482, Dec. 1974.

[17] E. Tzanakou, R. Michalak, and E. Harth, "The alopex process: Visual receptive fields by response feedback," *Biol. Cybern.*, vol. 35, no. 3, pp. 161–174, 1979.

[18] D. A. Berry and B. Fristedt, *Bandit Problems*. London, U.K.: Chapman-Hall, 1985.

[19] D. Michie, *Memo Functions: A Language Feature With "Rote-Learning" Properties* (Research Memorandum MIP-R-29), Dept. Mach. Intell. Percept., Univ. Edinburgh, Edinburgh, U.K., 1967.

[20] R. J. Popplestone, *"Memo" Functions and the Pop-2 Language* (Research Memorandum MIP-R-30), Dept. Mach. Intell. Percept., Univ. Edinburgh, Edinburgh, U.K., 1967.

[21] B. Widrow, N. K. Gupta, and S. Maitra, "Punish/reward: Learning with a critic in adaptive threshold systems," *IEEE Trans. Syst., Man, Cybern.*, vol. SMC-3, no. 5, pp. 455–465, Sep. 1973.

[22] R. S. Sutton and A. G. Barto, "Toward a modern theory of adaptive networks: Expectation and prediction," *Psychol. Rev.*, vol. 88, pp. 135–170, Mar. 1981.

[23] A. G. Barto and R. S. Sutton, "Simulation of anticipatory responses in classical conditioning by a neuron-like adaptive element," *Behav. Brain Res.*, vol. 4, pp. 221–235, Mar. 1982.

[24] R. S. Sutton, "Temporal credit assignment in reinforcement learning," Ph.D. dissertation, Dept. Comput. Inf. Sci., Univ. Massachusetts, Amherst, MA, USA, 1984.

[25] R. S. Sutton, "Learning to predict by the methods of temporal differences," *Mach. Learn.*, vol. 3, pp. 9–44, Aug. 1988.

[26] D. Michie and R. A. Chambers, "BOXES: An experiment in adaptive control," in *Machine Intelligence 2*, E. Dale and D. Michie, Eds. Edinburgh, U.K.: Oliver Boyd, 1968, pp. 137–152.

[27] B. Widrow and F. W. Smith, "Pattern-recognizing control systems," in *Computer and Information Sciences*, J. T. Tou and R. H. Wilcox, Eds. Washington, DC, USA: Spartan, 1964, pp. 288–317.

[28] O. J. Selfridge, "Pandemonium: A paradigm for learning," in *Proc. Symp. Mech. Thought Process.*, 1959, pp. 511–529.

[29] J. C. Gittins and D. M. Jones, "A dynamic allocation index for the discounted multiarmed bandit problem," *Biometrika*, vol. 66, no. 3, pp. 561–565, 1979.

[30] O. J. Selfridge, R. S. Sutton, and A. G. Barto, "Training and tracking in robotics," in *Proc. 9th Int. Joint Conf. Artif. Intell.*, 1985, pp. 670–672.

[31] B. F. Skinner, "Reinforcement today," *Amer. Psychol.*, vol. 13, no. 3, p. 94, 1958.

[32] D. Michie, "Trial and error," in *Science Survey, Part 2*, S. A. Barnett and A. McLaren, Eds. Harmondsworth, U.K.: Penguin, 1961, pp. 129–145.

[33] D. Michie, "Experiments on the mechanization of game-learning part I. Characterization of the model and its parameters," *Comput. J.*, vol. 6, no. 3, pp. 232–263, 1963.

[34] A. G. Barto, "Game-theoretic cooperativity in networks of self-interested units," in *Neural Networks for Computing*, J. S. Denker, Ed. New York, NY, USA: Amer. Inst. Phys., 1986, pp. 41–46.

[35] C. W. Anderson, "Strategy learning with multilayer connectionist representations," in *Proc. 4th Int. Workshop Mach. Learn.*, 1987, pp. 103–114.

[36] A. G. Barto, "Learning by statistical cooperation of self-interested neuron-like computing elements," *Hum. Neurobiol.*, vol. 4, no. 4, pp. 229–256, 1985.

[37] A. G. Barto, P. Anandan, and C. W. Anderson, "Cooperativity in networks of pattern recognizing stochastic learning automata," in *Adaptive and Learning Systems: Theory and Applications*, K. S. Narendra, Ed. New York, NY, USA: Plenum, 1986.

[38] K. S. Narendra and R. M. Wheeler, "An *n*-player sequential stochastic game with identical payoffs," *IEEE Trans. Syst., Man, Cybern.*, vol. SMC-13, no. 6, pp. 1154–1158, Nov./Dec. 1983.

[39] D. E. Rumelhart, G. E. Hinton, and R. J. Williams, "Learning internal representations by error propagation," in *Parallel Distributed Processing: Explorations in the Microstructure of Cognition, Vol.1: Foundations*, D. E. Rumelhart and J. L. McClelland, Eds. Cambridge, MA, USA: MIT Press, 1986.

[40] A. G. Barto and M. I. Jordan, "Gradient following without back-propagation in layered networks," in *Proc. IEEE 1st Annu. Conf. Neural Netw.*, San Diego, CA, USA, 1987, pp. II629–II636.

[41] C. W. Anderson, "Feature generation and selection by a layered network of reinforcement learning elements: Some initial experiments," Dept. Comput, Inf. Sci., Univ. Massachusetts, Amherst, MA, USA, Rep. COINS 82–12, 1982.

[42] C. W. Anderson, "Learning and problem solving with multilayer connectionist systems," Ph.D. dissertation, Dept. Comput. Inf. Sci., Univ. Massachusetts, Amherst, MA, USA, 1986.

[43] C. W. Anderson, "Learning to control an inverted pendulum using neural networks," *IEEE Control Syst. Mag.*, vol. 9, no. 3, pp. 31–37, Apr. 1989.

[44] R. J. Williams, "Simple statistical gradient-following algorithms for connectionist reinforcement learning," *Mach. Learn.*, vol. 8, pp. 229–256, May 1992.

[45] S. Bhatnagar, R. S. Sutton, M. Ghavamzadeh, and M. Lee, "Natural actor–critic algorithms," *Automatica*, vol. 45, no. 11, pp. 2471–2482, 2009.

[46] C. J. C. H. Watkins and P. Dayan, "Q-learning," *Mach. Learn.*, vol. 8, pp. 279–292, May 1992.

[47] D. Liu, X. Xiong, and Y. Zhang, "Action-dependent adaptive critic designs," in *Proc. Int. Joint Conf. Neural Netw. (Cat. No. 01CH37222)*, vol. 2. Washington, DC, USA, 2001, pp. 990–995.

[48] C. W. Anderson, "Approximating a policy can be easier than approximating a value function," Dept. Comput. Sci., Colorado State Univ., Fort Collins, CO, USA, Rep. CS-00-101, 2000.

[49] W. Schultz, "Predictive reward signal of dopamine neurons," *J. Neurophysiol.*, vol. 80, pp. 1–27, Jul. 1998.

[50] W. Schultz, P. Dayan, and P. R. Montague, "A neural substrate of prediction and reward," *Science*, vol. 275, pp. 1593–1598, Mar. 1997.

[51] A. G. Barto, "Adaptive critics and the basal ganglia," in *Models of Information Processing in the Basal Ganglia*, J. C. Houk, J. L. Davis, and D. G. Beiser, Eds. Cambridge, MA, USA: MIT Press, 1995, pp. 215–232.

[52] D. Silver *et al.*, "Mastering the game of Go with deep neural networks and tree search," *Nature*, vol. 529, no. 7587, pp. 484–489, 2016.

[53] D. Silver *et al.*, "Mastering the game of Go without human knowledge," *Nature*, vol. 550, no. 7676, pp. 354–359, 2017.

[54] M. Wiering and M. V. Otterlo, *Reinforcement Learning: State-of-the-Art.* Berlin, Germany: Springer-Verlag, 2012.

[55] N. Wiener, *God & Golem, Inc.* Cambridge, MA, USA: MIT Press, 1964.

[56] N. Bostrom, *Superintelligence: Paths, Dangers, Strategies.* Oxford, U.K.: Oxford Univ. Press, 2014.

Andrew G. Barto (Life Fellow, IEEE) received the B.S. (with distinction) degree in mathematics and the Ph.D. degree in computer science from the University of Michigan, Ann Arbor, MI, USA, in 1970 and in 1975, respectively.

He Co-Directed the Autonomous Learning Laboratory, University of Massachusetts Amherst, Amherst, MA, USA, where he is a Professor Emeritus of Computer Science having retired in 2012. He is currently an Associate Member of the Neuroscience and Behavior Program, University of Massachusetts Amherst. He has published over one hundred papers or chapters in journals, books, and conference and workshop proceedings. He has coauthored textbook *Reinforcement Learning: An Introduction*, (MIT Press, 1998, 2018).

Dr. Barto received the 2004 IEEE Neural Network Society Pioneer Award, the IJCAI-17 Award for Research Excellence, and the University of Massachusetts Neurosciences Lifetime Achievement Award in 2019. He serves as an Associate Editor of Neural Computation, as a Member of the Advisory Board of the *Journal of Machine Learning Research*, and as a Member of the Editorial Board of *Adaptive Behavior*. He is a Fellow of the American Association for the Advancement of Science.

Richard S. Sutton received the B.A. degree in psychology from Stanford University, Stanford, CA, USA, in 1978, and the Ph.D. degree in computer science from the University of Massachusetts Amherst, Amherst, MA, USA, in 1984.

He is a Distinguished Research Scientist with DeepMind, Edmonton, AB, Canada, and a Professor with the Department of Computing Science, University of Alberta, Edmonton. Prior to joining DeepMind in 2017, and the University of Alberta in 2003, he worked in industry at AT&T and GTE Labs, and in academia at the University of Massachusetts Amherst. He has coauthored textbook *Reinforcement Learning: An Introduction* from (MIT Press). His research interests center on the learning problems facing a decision-maker interacting with its environment, which he sees as central to intelligence. He has additional interests in animal learning psychology, in connectionist networks, and generally in systems that continually improve their representations and models of the world. His scientific publications have been cited more than 70 000 times. He is also a libertarian, a chess player, and a cancer survivor.

Prof. Sutton is also a Fellow of the Royal Society of Canada, the Association for the Advancement of Artificial Intelligence, the Alberta Machine Intelligence Institute, and CIFAR.

Charles W. Anderson (Senior Member, IEEE) received the B.S. degree in computer science from the University of Nebraska-Lincoln, Lincoln, NE, USA, in 1978, and the Ph.D. degree in computer science from the University of Massachusetts Amherst, Amherst, MA, USA, in 1986.

From 1986 to 1990, he was a Senior Member of Technical Staff, GTE Laboratories, Waltham, MA, USA. He is currently a Professor with the Department of Computer Science, Colorado State University, Fort Collins, CO, USA, with joint appointments with the School of Biomedical Engineering, the Molecular, Cellular and Integrative Neuroscience Program, the Graduate Degree Program in Ecology, and the Systems Engineering Program. He is the founder of Pattern Exploration, a small company that develops explainable AI solutions to prediction, classification, and control problems. His research interests are in practical algorithms for training neural networks and applying them to real-world problems in climate, environment, health, and energy.

Neuronlike Adaptive Elements That Can Solve Difficult Learning Control Problems

ANDREW G. BARTO, MEMBER, IEEE, RICHARD S. SUTTON, AND CHARLES W. ANDERSON

Abstract—It is shown how a system consisting of two neuronlike adaptive elements can solve a difficult leaning control problem. The task is to balance a pole that is hinged to a movable cart by applying forces to the cart's base. It is assumed that the equations of motion of the cart–pole system are not known and that the only feedback evaluating performance is a failure signal that occurs when the pole falls past a certain angle from the vertical, or the cart reaches an end of a track. This evaluative feedback is of much lower quality than is required by standard adaptive control techniques. It is argued that the learning problems faced by adaptive elements that are components of adaptive networks are at least as difficult as this version of the pole-balancing problem. The learning system consists of a single *associative search element* (ASE) and a single *adaptive critic element* (ACE). In the course of learning to balance the pole, the ASE constructs associations between input and output by searching under the influence of reinforcement feedback, and the ACE constructs a more informative evaluation function than reinforcement feedback alone can provide. The differences between this approach and other attempts to solve problems using neuronlike elements are discussed, as is the relation of this work to classical and instrumental conditioning in animal learning studies and its possible implications for research in the neurosciences.

Manuscript received August 1, 1982; revised April 20, 1983. This work was supported by AFOSR and the Air Force Wright Aeronautical Laboratory under Contract F33615-80-C-1088.

The authors are with the Department of Computer and Information Science, University of Massachusetts, Amherst, MA 01003.

163

MATHEMATICALLY formulated networks of neuronlike elements have been studied both as models of specific neural circuits and as abstract, though biologically inspired, computational architectures. As models of specific neural circuits, network models can provide theories to explain anatomical and physiological data. As computational architectures, they represent attempts to explore possible substrates for intelligent behavior, both natural and artificial. Networks of this second category are relevant to brain and behavioral science to the extent that their behavior can be related to phenomena of animal behavior for which no plausible mechanisms are known, thereby suggesting novel lines of empirical research. They are relevant to artificial intelligence to the extent that they exhibit forms of problem solving, knowledge acquisition, or data storage that are difficult to achieve by more conventional means.

In this article we illustrate an abstract neural network approach that we believe can have relevance for both neuroscience and computer science. Advances in our appreciation of the complexity of biological cells make it clear that the 35-year old metaphor that places the neuron at the level of the computer logic gate is inadequate. Neurons and synapses have information processing capabilities that make use of both short- and long-term information storage, locally implemented by complex biochemical mechanisms. Biochemical networks within cells are known to perform functions that had previously been attributed to networks of interacting cells. These facts call for new neural metaphors. Moreover, advances in computer science suggest the possibility of achieving sophisticated problem-solving capacity through networks of interacting components that are themselves powerful problem-solving systems (e.g., [1] and [2]). In our approach, network components are neuronlike in their basic structure and behavior and communicate by means of excitatory and inhibitory signals rather than by symbolic messages, but they are much more complex than neuronlike adaptive elements studied in the past. Rather than asking how very primitive components can be interconnected in order to solve problems, we are pursuing questions about how components that are themselves capable of solving relatively difficult problems can interact in order to solve problems that are even more difficult.

This article is devoted to the justification of the design of two types of neuronlike adaptive elements and an illustration of the problem-solving capacities of a system consisting of a single element of each type. We call one element an *associative search element* (ASE) and the other an *adaptive critic element* (ACE). As a vehicle for introducing our adaptive elements, we describe an earlier adaptive problem-solving system, called "boxes," developed by Michie and Chambers [3], [4]. We show that a learning strategy similar to theirs can be implemented by a *single* ASE, and we show how its learning performance can be improved by the addition of a *single* ACE. To illustrate the

problem-solving capabilities of these elements, we use the pole-balancing control problem posed by Michie and Chambers to illustrate their boxes algorithm, and we compare the performance of their system with that of our own. We conclude with a brief discussion of behavioral interpretations of our adaptive elements and their possible implications for neuroscience. A strong analogy exists between the behavior of the ACE and animal behavior in classical conditioning experiments, and parallels can be seen between the behavior of the ASE/ACE system and animal behavior in instrumental learning experiments. The adaptive elements we describe are refinements of those we have discussed previously [5]–[10] and were suggested by the work of Klopf [11], [12]. Our approach also has similarities with the work of Widrow and colleagues [13], [14] on what they called "bootstrap adaptation."

The significance of endowing single adaptive elements with this level of problem-solving capability is twofold. First, we wish to suggest neural metaphors, constrained by the computational demands of problem-solving, that postulate functions for the complex cellular mechanisms that are rapidly being elucidated as the study of the cellular basis of learning progresses. Second, we wish to suggest that if adaptive elements are to learn effectively *as network components*, then they must possess adaptive capabilities at least as robust as those of the elements discussed here. As we argue in the following, the learning problem faced by an adaptive element that is deeply embedded in the interior of a network is characterized by some of the same types of complexities that are present in the pole-balancing task considered here.

Thus, although the algorithms that we implement by means of single adaptive elements can obviously be implemented by networks of many simpler elements, we are attempting to delineate those properties required of components if they are to learn how to function as interconnected, cooperating components of networks. The extensive history of attempts to construct powerful adaptive networks and the generally acknowledged failure of these attempts suggest that network components as simple as those usually considered are not adequate. This lesson from previous theoretical studies, together with our contention that the view of neural function that constrained these studies was too limited, leads us to study elements as complex as the ASE and ACE. Despite our ultimate interest in networks, we do not present results in this paper that show that the elements discussed here are able to learn as components of powerful adaptive networks. However, previous simulation experiments with networks of similar elements have provided preliminary support for our approach to adaptive networks [5], [6], [8], and the research discussed here represents an initial attempt to move toward more difficult learning problems.

Although we intend to raise questions about the level in the functional hierarchy of the nervous system at which neurons can be said to act, we are not claiming that there is necessarily a strict correspondence between single neurons and ACE's and ASE's. Some of the features of these elements clearly are not neuronlike but can be implemented in standard ways by elements more faithful to neural limitations. For example, the ASE can "fire" with both negative and positive output values, but it can be implemented by a pair of reciprocally inhibiting elements, each capable only of positive "spikes." Consequently, by the term "neuronlike element" we do not mean a literal neuron model, and we purposefully exclude well-known neuron properties which would have no clear functional role in the present problem.

Our interest in the pole-balancing problem arises from its convenience as a test bed for exploring a variety of algorithms that may enable elements to learn effectively when embedded in networks. We are not interested in pole balancing *per se*, and our formulation of the problem, following that of Michie and Chambers [3], [4], makes it much more difficult than it would need to be if one were simply interested in controlling this type of dynamical system. We assume that the controller's design must be based upon very little knowledge of the controlled system's dynamics and that the evaluative feedback provided to the controller is of much lower quality than is required by standard adaptive control methods. These constraints produce a difficult learning control problem and reflect some of the conditions that we believe characterize the tasks faced by network components. While a variety of well-developed adaptive control methods can be (and have been) successfully applied to pole balancing, we know of none that are directly applicable to the problem subject to the constraints we impose. Additionally, the algorithm we describe can be applied to nonnumerical problems as well as to problems requiring the control of dynamical systems.

II. Learning within Networks

Many of the previous studies of adaptive networks of neuronlike elements focused on adaptive elements that are capable of solving certain types of pattern classification problems. Elements such as the ADALINE (adaptive linear element [16]) and those employed in the Perceptron [15] perform supervised learning pattern classification (see, for example, [17]). These elements form linear discrimination rules by adjusting a set of "synaptic" weights in an attempt to match their response to each training input pattern with a desired response, or correct classification, that is provided by a "teacher." The resulting discrimination rule can be used to classify new pattern instances (perhaps incorrectly), thereby providing a form of generalization. The algorithms implemented by these adaptive elements are closely related to iterative regression methods used in adaptive control for the identification of unknown system parameters [17].

Unfortunately, a network composed of these types of adaptive elements can only learn if its environment contains a teacher that can supply *each* component adaptive element with its individual desired response for each pattern in a training sequence. This is the Achilles' heel of supervised learning pattern classifiers as network compo-

nents. In many problem-solving tasks, the network's environment may be able to provide assessments of certain *consequences* of the *collective activity* of all of the network components but the environment cannot know the desired responses of individual elements or even evaluate the behavior of individual elements. To use terms encountered in the artificial intelligence literature (e.g., [18]), the network's internal mechanism is not very "transparent" to the "critic."

Other approaches to the problem of learning within adaptive networks rely on adaptive elements that require neither teachers nor critics. These elements employ some form of unsupervised learning, or clustering, algorithm, often based on Hebb's [19] hypothesis that repeated pairing of pre- and postsynaptic activity strengthens synaptic efficacy. While clustering is likely to play an important role in sophisticated problem-solving systems, it does not by itself provide the necessary means for a system to improve performance in tasks determined by factors external to the system, such as, for example, the task of controlling an environment having initially unknown dynamics. For these types of tasks, a learning system must not just cluster information but must form those clusters that are useful in terms of the system's interaction with its environment. Thus it seems necessary to consider networks that learn under the influence of some sort of evaluative feedback, but this feedback cannot be so informative as to provide individualized instruction to each adaptive element.

These considerations have led us to study adaptive elements that are capable of learning to improve performance with respect to an evaluation function that assesses the consequences, which may be quite indirect, of element actions but does not directly specify these actions. Further, these elements are capable of improving performance under conditions of considerable uncertainty. Since evaluative feedback, or reinforcement feedback, will generally assess the performance of the entire network rather than the performance of individual elements, a high degree of uncertainty is necessarily present in the optimization problem faced by any individual component. Additional uncertainty arises from any delay that might exist between the time of an element's action and the time it receives the resulting reinforcement. The reinforcement feedback received by a network component at any time will generally depend upon factors other than its own action taken some fixed time earlier; it will additionally depend upon the actions of a large number of components taken at a variety of earlier times.

The ASE implements one part of our approach to these problems. Since we assume its environment is unable to provide desired responses, the ASE must *discover* what responses lead to improvements in performance. It employs a trial-and-error, or generate-and-test, search process. In the presence of input signals, it generates actions by a random process. Based on feedback that evaluates the problem-solving consequences of the actions, the ASE "tunes in" input signals to bias the action generation process, conditionally on the input, so that it will more likely generate the actions leading to improved performance. Different actions can be optimal when taken in the presence of different input signals. Actions that lead to improved performance when taken in the presence of certain input signals become associated with those signals in a developing input–output mapping. This type of stochastic search allows the ASE to improve performance under conditions of uncertainty. We have called this general process *associative search* [8] to emphasize both its association formation and generate-and-test search aspects.

In providing elements with these capabilities, we have been guided by the hypothesis of Klopf [11], [12] that neurons implement a strategy for attempting to maximize the frequency of occurrence of one type of input signal and minimize the frequency of occurrence of another type. According to this hypothesis, in other words, neurons can be conditioned in an operant or instrumental manner, where certain types of inputs act as rewarding stimuli and others act as punishing stimuli. A neuron learns how to attain certain types of inputs and avoid others by adjusting the transmission efficacy of its synapses according to the consequences of its discharges as fed back through pathways both internal to the nervous system and external to the animal. The ASE departs in several ways from Klopf's hypothesis, but his underlying idea remains the same.

III. Error Correction Versus Reinforcement Learning

Considerable misunderstanding is evident in the literature about how this type of "reinforcement learning" differs from supervised learning pattern classification as performed, for example, by Perceptrons and ADALINE's. It is important to emphasize these differences before we describe our adaptive elements. Supervised learning pattern classification elements are sometimes formulated in such a manner that the training process occurs as follows. A training pattern is presented to the element which responds as directed by its current set of weights; based on knowledge of the correct response, the element's environment feeds back an error signal giving the difference between the actual and correct resonses; the element uses this error signal to update its weight values. This sequence is repeated for all of the training patterns until the error signals become zero. These error signals are response-contingent feedback to the adaptive element, but it is misleading to view this process as a general form of reinforcement learning.

One important difference between the error-correction process just described and reinforcement learning as implemented by the ASE is that the latter does not rely exclusively on its weight values to determine its actions. Instead, it generates actions by a random process that is merely *biased* by the combination of its weight values and the input patterns. Actions are thus not appropriately viewed strictly as *responses* to input patterns. The random component of the generation process introduces the variety that is necessary to serve as the basis for subsequent selection by

evaluative feedback. The ASE therefore searches in its action space in a manner that supervised learning pattern classification machines do not.

Additionally, significant differences exist between general performance evaluation signals and the signed error signals required by supervised learning pattern classification elements. To supply a signed error signal, the environment must know both what the actual action was and what it should have been.[1] Evaluation of performance, on the other hand, may be based on a relative assessment of certain consequences of the element's actions rather than on knowledge of both the correct and actual actions. Widrow *et al.* [13] used the phrase "learning with a critic" to distinguish this type of process from learning with a teacher, as supervised learning pattern classification is sometimes called.

Very few studies have been made of neuronlike elements capable of learning under reinforcement feedback that is less informative than are signed error signals (Farley and Clark [20]; Minsky [21]; and Widrow *et al.* [13]). Indeed, considerable confusion arises from an unfortunate inconsistency in the usage of the term "error." What psychologists mean by trial-and-error learning is not the same as the error-correction process used by supervised learning pattern classification machines. Like the process employed by our ASE, trial-and-error learning is a "selectional" rather than an "instructional" process (cf. the usage of these terms by Edelman [22], although the selectional mechanism of the ASE is quite different from the one he proposes). Much more could be said about these issues, but we shall let the following example further clarify them. It will be apparent that elements such as Perceptrons and ADALINE's cannot by themselves solve the control problem we will consider.

IV. The Credit Assignment Problem

One can view the uncertainty discussed in the foregoing as a result of a fundamental problem that faces any learning system, whether it is natural or artificial, that has been called the "credit-assignment" problem by artificial intelligence researchers [18], [23]. This is the problem of determining what parts of a complex interacting or interlocking set of mechanisms, decisions, or actions deserve credit (blame) for improvements (decrements) in the overall performance of the system. The credit-assignment problem is especially acute when evaluative feedback to the learning system occurs infrequently, for example, upon the completion of a long series of decisions or actions.

Given the widely acknowledged importance of the credit-assignment problem for adaptive problem-solving systems, it is surprising that techniques for its solution have not been more intensely studied. The most successful,

---

[1]It is thus possible to formulate this training paradigm as one in which the learning machine's environment provides training patterns together with their desired responses (as we have done in Section II), and the system itself determines its error. This formulation does not involve feedback that passes through the machine's environment and more clearly reveals the limited nature of this type of process.

and perhaps the most extensible, solution to date was used in the checkers-playing program written by Samuel [24] more than two decade ago. A few isolated studies using similar techniques have been undertaken (Doran [25]; Holland [26]; Minsky [21], [23]; and Witten [27]), but the current approaches to the credit-assignment problem in artificial intelligence largely rely on providing the critic with domain-specific knowledge [18], [27]. Samuel's method, on the other hand, is one by which the system improves its own internal critic by a learning process.

The ACE implements a strategy most closely related to the methods of Samuel [24] and Witten [27] for reducing the severity of the credit-assignment problem. It adaptively develops an evaluation function that is more informative than the one directly available from the learning system's environment. This reduces the uncertainty under which the ASE must learn. The ACE was developed primarily by Sutton as a refinement of the adaptive element model of classical conditioning introduced by Sutton and Barto [9].

V. A LEARNING CONTROL PROBLEM: POLE BALANCING

Fig. 1 shows a schematic representation of a cart to which a rigid pole is hinged. The cart is free to move within the bounds of a one-dimensional track. The pole is free to move only in the vertical plane of the cart and track. The controller can apply an impulsive "left" or "right" force F of fixed magnitude to the cart at discrete time intervals. The cart–pole system was simulated by digital computer using a very detailed model that includes all of the nonlinearities and reactive forces of the physical system (the Appendix provides details of the cart–pole model and simulations). The cart–pole model has four state variables:

x position of the cart on the track,
θ angle of the pole with the vertical,
$\dot{x}$ cart velocity, and
$\dot{\theta}$ rate of change of the angle.

Parameters specify the pole length and mass, cart mass, coefficients of friction between the cart and the track and at the hinge between the pole and the cart, the impulsive control force magnitude, the force due to gravity, and the simulation time step size.

The control problem we pose is identical to the one studied by Michie and Chambers. We assume that the equations of motion of the cart–pole system are not known and that there is no preexisting controller that can be imitated. At each time step, the controller receives a vector giving the cart–pole system's state at that instant. If the pole falls or the cart hits the track boundary, the controller receives a failure signal, the cart–pole system (but not the controller's memory) is reset to its initial state, and another learning trial begins. The controller must attempt to generate controlling forces in order to avoid the failure signal for as long as possible. No evaluative feedback other than the failure signal is available.

Learning to avoid the failure signal under these constraints is a very different problem than learning to balance

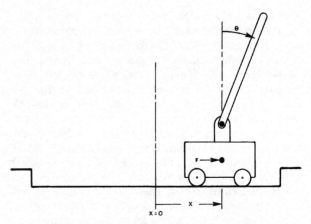

Fig. 1. Cart–pole system to be controlled. Solution of system's equations of motion approximated numerically (see Appendix).

the pole under the conditions usually assumed by control theorists. Since the failure signal will occur only after a long sequence of individual control decisions, a difficult credit-assignment problem arises in the attempt to determine which decisions were responsible for the failure. Neither a continuously available error signal nor a continuously available performance evaluation signal exists, as is the case in more conventional formulations of pole balancing. For example, Widrow and Smith [14] used a linear regression method, implemented by an ADALINE, to approximate the bang-bang control law required for balancing the pole. In order to use this method, however, they had to supply the controller with a signed error signal at each time step whose determination required external knowledge of the correct control decision for that time step. The present formulation of the problem, on the other hand, requires the learning system to discover for itself which control decisions are correct, and in so doing, solve a difficult credit-assignment problem that is completely absent in the usual versions of this problem.

VI. THE BOXES SYSTEM

By first describing Michie and Chambers' [3], [4] boxes system, we can provide much of the justification for the design of our adaptive elements. The strategy of these authors was to decompose the pole-balancing problem into a number of independent subproblems and to use an identical generate-and-test rule for learning to solve each subproblem. They divided the four-dimensional cart–pole state space into disjoint regions (or boxes) by quantizing the four state variables. They distinguished three grades of cart position, six of the pole angle, three of cart velocity, and three of pole angular velocity [4]. We use a similar partition of the state space based on the following quantization thresholds:

1) x: ± 0.8, ± 2.4 m,
2) θ: 0, ± 1, ± 6, $\pm 12°$,
3) $\dot{x}$: ± 0.5, $\pm \infty$ m/s,
4) $\dot{\theta}$: ± 50, $\pm \infty°$/s.

This yields $3 \times 3 \times 6 \times 3 = 162$ regions corresponding to all of the combinations of the intervals. The physical units of these thresholds differ from those used in [3] and [4]. We chose these values and units to produce what seemed like a physically realistic control problem, given our parameterization of the cart–pole simulation (Michie and Chambers did not publish the parameters of their cart–pole simulation. See the Appendix for our parameter values). At present we assume, as Michie and Chambers did, that this quantization is provided from the start (see Section X).

Each box is imagined to contain a *local demon* whose job is to choose a control action (left or right) whenever the system state enters its box. The local demon must learn to choose the action that will tend to be correlated with long system lifetime, that is, a long time until the occurrence of the failure signal. A *global demon* inspects the incoming state vector at each time step and alerts the local demon whose box contains that system state. When a failure signal is received, the global demon distributes it to all local demons. Each local demon maintains estimates of the expected lifetimes of the system following a left decision and following a right decision. A local demon's estimate of the expected lifetime for left is a weighted average of actual system lifetimes over all past occasions that the system state entered the demon's box and the decision left was made. The expected lifetime for the decision right is determined in the same way for occasions in which a right decision was made.

More specifically, upon being signaled by the global demon that the system state has entered its box, a local demon does the following.

1) It chooses the control action left or right according to which has the longest lifetime estimate. The control system emits the control action as soon as the decision is made.

2) It remembers which action was just taken and begins to count time steps.

3) When a failure signal is received, it uses its current count to update the left or right lifetime estimate, depending on which action it chose when its box was entered.

Michie and Chambers' actual algorithm is somewhat more complicated than this, but this description is sufficient for our present purposes. Details are provided in [3] where it is shown that the system is capable of learning to balance the pole for extended periods of time (in one reported run, the pole was balanced for a time approximately corresponding to one hour of real time). Notice that since the effect of a demon's decision will depend on the decisions made by other demons whose boxes are visited during a trial (where a trial is the time period from reset to failure), the environment of a local demon, consisting of the other demons as well as the cart–pole system, does not consistently evaluate the demon's actions.

VII. THE ASSOCIATIVE SEARCH ELEMENT (ASE)

Obviously, many possibilities exist for implementing a system like boxes using neuronlike elements. We know, for example, that any algorithm can be implemented by a network of McCulloch–Pitts abstract neurons acting as logic gates and delay units. Such an implementation would illustrate the neural metaphor resulting from the very earliest contact between neuroscience and digital technology [29]. More recent neural metaphors suggest that each local demon might be implemented by a network of adaptive neurons that would be set into reverberatory activity under conditions corresponding to the demon's box being entered by the state vector. Upon receipt of the failure signal, the magnitude of this reverberatory activity would somehow alter synapses used for triggering control actions. The global demon might be implemented by a neural network responsible for quantizing the system state vectors, conjunctively combining the results, and activating appropriate local demon networks (a neural decoder—see Section X). Finally, an element or network of elements would be required for channeling the action of each local demon network to a common efferent pathway.

In the neuronlike implementation we are pursuing, however, a local demon corresponds to the mechanism of a single synapse (to use the language of neural metaphor), and the output pathway of the postsynaptic element (the ASE) provides the common efferent pathway for control signals. At each synapse of the ASE are both a long-term memory trace that determines control actions and a short-term memory trace that is required to update the long-term trace, a role similar to that of a local demon's counter in the boxes algorithm. To accomplish the global demon's job of activating the appropriate local demon, we assume the existence of a decoder that has four real-valued input pathways (for the system state vector) and 162 binary valued output pathways corresponding to the boxes of Michie and Chambers' system (Fig. 2). The decoder transforms each state vector into a 162-component binary vector whose components are all zeros except for a single one in the position corresponding to the box containing the state vector. This vector is provided as input to the ASE and effectively selects the synapse corresponding to the appropriate box. For the other job of the global demon, that of distributing a failure signal to all of the local demons, we just let the adaptive element receive the failure signal via its reinforcement pathway and distribute the information to all of of its afferent synapses. In this way the entire boxes algorithm can be implemented by a single neuronlike ASE and an appropriate decoder.

In more detail, an ASE is defined as follows. The element has a reinforcement input pathway, n pathways for nonreinforcement input, and a single output pathway (Fig. 2). Let $x_i(t), 1 \le i \le n$, denote the real-valued signal on the ith nonreinforcement input pathway at time t, and let $y(t)$ denote the output at time t. Associated with each nonreinforcement input pathway i is a real-valued weight with value at time t denoted by $w_i(t)$.

The element's output $y(t)$ is determined from the input vector $X(t) = (x_1(t), \cdots, x_n(t))$ as follows:

$$y(t) = f\left[\sum_{i=1}^{n} w_i(t) x_i(t) + \text{noise}(t) \right] \quad (1)$$

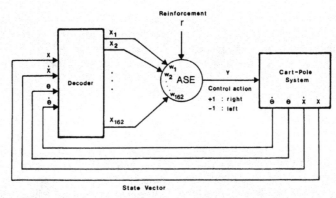

Fig. 2. ASE controller for cart–pole system. ASE's input is determined from current cart–pole state vector by decoder that produces output vector consisting of zeros with single one indicating which of 162 boxes contains state vector. ASE's output determines force applied to cart. Reinforcement is constant throughout trial and becomes -1 to signal failure.

where noise (t) is a real random variable with probability density function d and f is either a threshold, sigmoid, or identity function. For the pole-balancing illustration, d is the mean zero Gaussian distribution with variance σ^2, and f is the following threshold function:

$$f(x) = \begin{cases} +1, & \text{if } x \geq 0 \quad \text{(control action right)} \\ -1, & \text{if } x < 0 \quad \text{(control action left).} \end{cases}$$

This follows the usual linear threshold convention common in adaptive network studies, but our approach does not depend strongly on the specifics of the input/output function of the element.

According to (1), actions are emitted even in the absence of nonzero input signals. The element's output is determined by chance, with a probability biased by the weighted sum of the input signals. If that sum is zero, the left and right control actions are equally probable. Assuming the decoder input shown in Fig. 2, a positive weight w_i, for example, would make the decision right more probable than left when box i is entered by the system state vector. The value of a weight, therefore, plays a role corresponding to the difference between the expected lifetimes for the left and right actions stored by a local demon in the boxes system. However, unlike the boxes system, the weight only determines the probability of an action rather than the action itself. The learning process updates the action probabilities. Also note that an input vector need not be of the restricted form produced by the decoder in order for (1) and the equations that follow to the meaningful.

The weights $w_i, 1 \leq i \leq n$, change over (discrete) time as follows:

$$w_i(t + 1) = w_i(t) + \alpha r(t)e_i(t) \qquad (2)$$

where

α positive constant determining the rate of change of w_i,

$r(t)$ real-valued *reinforcement* at time t, and

$e_i(t)$ *eligibility* at time t of input pathway i.

The basic idea expressed by (2) is that whenever certain conditions (to be discussed later) hold for input pathway i, then that pathway becomes eligible to have its weight

modified, and it remains eligible for some period of time after the conditions cease to hold. How w_i changes depends on the reinforcement received during periods of eligibility. If the reinforcement indicates improved performance, then the weights of the eligible pathways are changed so as to make the element more likely to do whatever it did that made those pathways eligible. If reinforcement indicates decreased performance, then the weights of the eligible pathways are changed to make the element more likely to do something else. The term "eligibility" and this weight update scheme are derived from the theory of Klopf [11], [12] and have precursors in the work of Farley and Clark [20], Minsky [21], and others. This general approach to reinforcement learning is related to the theory of stochastic learning automata [30], [31], which has its roots in the work of Bush and Mosteller [32] and Tsetlin [33].

Reinforcement: Positive r indicates the occurrence of a rewarding event and negative r indicates the occurrence of a punishing event.[2] It can be regarded as a measure of the *change* in the value of a performance criterion as commonly used in control theory. For the pole-balancing problem, r remains zero throughout a trial and becomes -1 when failure occurs.

Eligibility: Klopf [11] proposed that a pathway should reach maximum eligibility a short time after the occurrence of a pairing of a nonzero input signal on that pathway with the "firing" of the element. Eligibility should decay thereafter toward zero. Thus, when the consequences of the element's firing are fed back to the element, credit or blame can be assigned to the weights that will alter the firing probability when a similar input pattern occurs in the future. More generally, the eligibility of a pathway reflects the extent to which input activity on that pathway was paired in the past with element output activity. The eligibility of pathway i at time t is therefore a *trace* of the product $y(\tau)x_i(\tau)$ for times τ preceding t. If either or both of the quantities $y(\tau)$ and $x_i(\tau)$ are negative (as they can be for the ASE defined earlier), then credit is assigned

---

[2] A negative value of r is not the same as a psychologists's "negative reinforcement." In psychology, negative reinforcement is reinforcement due to the cessation of an aversive stimulus.

appropriately via (2) if eligibility is a trace of the signed product $y(\tau)x_i(\tau)$.

For computational simplicity, we generate exponentially decaying eligibility traces e_i using the following linear difference equation:

$$e_i(t + 1) = \delta e_i(t) + (1 - \delta) y(t) x_i(t), \qquad (3)$$

where $\delta, 0 \leq \delta < 1$, determines the trace decay rate. Note that each synapse has its own local eligibility trace.

Eligibility plays a role analogous to the part of the boxes local-demon algorithm that, when the demon's box is entered and an action has been chosen, remembers what action was chosen and begins to count. The factor $x_i(t)$ in (3) triggers the eligibility trace, a kind of count, or contributes to an ongoing trace, whenever box i is entered ($x_i(t) = 1$). Instead of explicitly remembering what action was chosen, our system contributes a different amount to the eligibility trace depending on what action was chosen (via the term $y(t)$ in (3)). Thus the trace contains information not only about how long ago a box was entered but also about what decision was made when it was entered.

Unlike the count initiated by a local demon in the boxes system, however, the eligibility trace effectively counts down rather than up (more precisely, its magnitude decays toward zero). Recall that reinforcement r remains zero until a failure occurs, at which time it becomes -1. Thus whatever control decision was made when a box was visited will always be made *less* likely when the failure occurs, but the longer the time interval between the decision and the occurrence of the failure signal, the less this decrease in probability will be. From one perspective, this process seems appropriate. Since the failure signal always eventually occurs, the action that was taken may deserve some of the blame for the failure. However, this view misses the point that even though both actions inevitably lead to failure, one action is probably *better* than the other. The learning process defined by (1)–(3) needs to be more subtle to ensure convergence to the actions that yield the least punishment in cases in which only punishment is available. In the present article, we build this subtlety into the ACE rather than into the ASE. Among its other functions, the ACE constructs predictions of reinforcement so that if punishment is less than its expected level, it acts as reward. For the pole-balancing task, the ASE as defined here must operate in conjunction with the ACE.

Although the boxes system and the version of the pole-balancing problem described earlier serve well to make an ASE's design understandable, the ASE does not represent an attempt to duplicate the boxes algorithm in neuronlike form. We are interested in tasks more general than the pole-balancing problem and in learning systems that are more general than the boxes system. An ASE is less restricted than the boxes system in several ways. First, the boxes system is based on the subdivision of the problem space into a finite number of nonoverlapping regions, and no generalization is attempted between regions. It develops a control rule that is effectively specified by means of a lookup table. Although a form of generalization can be easily added to the boxes algorithm by using an averaging process over neighboring boxes (see Section X) it is not immediately obvious how to extend the algorithm to take advantage of the other forms of generalization that would be possible if the controlled system's states could be represented by arbitrary vectors rather than only by the standard unit basis vectors which are produced by a suitable decoder. The ASE can accept arbitrary input vectors and, although we do not illustrate it in this article, can be regarded as a step toward extending the type of generalization produced by error-correction supervised learning pattern classification methods to the less restricted reinforcement learning paradigm (see Section III).

The boxes system is also restricted in that its design was based on the *a priori* knowledge that the time until failure was to serve as the evaluation criterion and that the learning process would be divided into distinct trials that would always end with a failure signal. This knowledge permitted Michie and Chambers to reduce the uncertainty in the problem by restricting each local demon to choosing the same action each time its box was entered during any given trial. The ASE, on the other hand, is capable of working to achieve rewarding events and to avoid punishing events which might occur at any time. It is not exclusively failure-driven, and its operation is specified without reference to the notion of a trial.

VIII. THE ADAPTIVE CRITIC ELEMENT (ACE)

Fig. 3 shows an ASE together with an ACE configured for the pole-balancing task. The ACE receives the externally supplied reinforcement signal which it uses to determine how to compute, on the basis of the current cart–pole state vector, an improved reinforcement signal that it sends to the ASE. Expressed in terms of the boxes system, the job of the ACE is to store in each box a prediction or expectation of the reinforcement that can eventually be obtained from the environment by choosing an action for that box. The ACE uses this prediction to determine a reinforcement signal that it delivers to the ASE whenever the box is entered by the cart–pole state, thus permitting learning to occur throughout the pole-balancing trials rather than solely upon failure. This greatly decreases the uncertainty faced by the ASE. The central idea behind the ACE algorithm is that predictions are formed that predict not just reinforcement but also future predictions of reinforcement.

Like the ASE, the ACE has a reinforcement input pathway, n pathways for nonreinforcement input, and a single output pathway (Fig. 3). Let $r(t)$ denote the real-valued reinforcement at time t; let $x_i(t), 1 \leq i \leq n$, denote the real-valued signal on the ith nonreinforcement input pathway at time t; and let $\hat{r}(t)$ denote the real-valued output signal at time t. Each nonreinforcement input pathway i has a weight with real value $v_i(t)$ at time t. The output $\hat{r}$ is the improved reinforcement signal that is used by the ASE in place of r in (2).

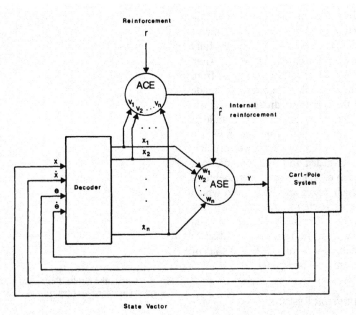

Fig. 3.   ASE and ACE configured for pole-balancing task. ACE receives same nonreinforcing input as ASE and uses it to compute an improved or internal reinforcement signal to be used by ASE.

In order to produce $\hat{r}(t)$, the ACE must determine a prediction $p(t)$ of eventual reinforcement that is a function of the input vector $X(t)$ (which in the boxes paradigm, simply selects a box). We let

$$p(t) = \sum_{i=1}^{n} v_i(t) x_i(t) \qquad (4)$$

and seek a means of updating the weights v_i so that $p(t)$ converges to an accurate prediction. The updating rule we use is

$$v_i(t + 1) = v_i(t) + \beta [r(t) + \gamma p(t) - p(t - 1)] \bar{x}_i(t), \qquad (5)$$

where β is a positive constant determining the rate of change of v_i; $\gamma, 0 < \gamma \leqslant 1$, is a constant to be explained below; $r(t)$ is the reinforcement signal supplied by the environment at time t; and $\bar{x}_i(t)$ is the value at time t of a *trace* of the input variable x_i.

It is beyond the scope of the present paper to explain the derivation of this learning rule fully (see [7] and [9]). Very briefly, the trace $\bar{x}_i$ acts much like the eligibility trace e_i defined by (3). Here, however, an input pathway gains positive eligibility whenever a nonzero signal is present on that pathway, irrespective of what the element's action is. We compute $\bar{x}_i$ using the following linear difference equation (cf. (3)):

$$\bar{x}_i(t + 1) = \lambda \bar{x}_i(t) + (1 - \lambda) x_i(t), \qquad (6)$$

where $\lambda, 0 \leqslant \lambda < 1$ determines the trace decay rate.

According to (6), an eligible pathways's weight changes whenever the actual reinforcement $r(t)$ plus the current prediction $p(t)$ differs from the value $p(t - 1)$ that was predicted for this sum. Closely related to the ADALINE learning rule and related regression techniques, this rule provides a means of finding weight values such that $p(t - 1)$ approximates $r(t) + \gamma p(t)$, or, equivalently, such that $p(t)$ approximates $r(t + 1) + \gamma p(t + 1)$. By attempting to

predict its own prediction, the learning rule produces predictions that tend to be the earliest possible indictions of eventual reinforcement. The constant γ, related to Witten's [27] "discount factor," provides for eventual extinction of predictions in the absence of external reinforcement. If $\gamma = 1$, predictions will be self-sustaining in the absence of external reinforcement; whereas if $0 < \gamma < 1$, predictions will decay in the absence of external reinforcement. In our simulations, $\gamma = 0.95$.

The ACE's output, the improved or internal reinforcement signal, is computed from these predictions as follows:

$$\hat{r}(t) = r(t) + \gamma p(t) - p(t - 1). \qquad (7)$$

This is the same expression appearing in (5). The reader should note that with $\hat{r}$ substituted for r in (2), the weight updating rules for the ASE and ACE ((2) and (5), respectively) differ only in their forms of eligibility traces. The ASE's traces are conditional on its output, whereas the ACE's are not.

Although this process works for arbitrary input vectors, it is easiest to justify (7) by again specializing to the boxes input representation. According to (7), as the cart–pole state moves between boxes without failure occurring (i.e., $r(t) = 0$), the reinforcement $\hat{r}(t)$ sent to the ASE is the difference between the prediction of reinforcement of the current box (discounted by γ) and the prediction of reinforcement of the previous box. Increases in reinforcement prediction therefore become rewarding events (assuming $\gamma = 1$), and decreases become penalizing events.

When failure occurs, the situation is slightly different. Given the way the control problem is represented, when failure occurs the cart–pole state is not in any box. Thus all $x_i(t)$ are equal to zero at failure, and according to (4), so is $p(t)$. Upon failure, then, the reinforcement sent to the ASE is the externally supplied reinforcement $r(t) = -1$, minus the previous prediction $p(t - 1)$. Consequently, an unpredicted failure results in $\hat{r}(t)$ being negative. This both punishes the actions made preceding the failure and, via (5), increments the predictions of failure (i.e., decrements the p's) of the boxes entered before the failure. A fully predicted failure generates no punishment. However, when a box with such a high prediction of failure is entered from a box with a lower prediction of failure, the recently made actions are punished and the recent predictions of failure are incremented, just as they were initially upon failure. Similarly, if the cart–pole state moves from a box with a higher prediction of failure to a box with a lower prediction, the recent actions are rewarded and recent predictions of failure are decremented (i.e., the p's are incremented). The system thus learns which boxes are "safe" and which are "dangerous." It punishes itself for moving from any box to a more dangerous box and rewards itself for moving from any box to a safer box. In the following we discuss the relation between the behavior of the ACE and that of animals in classical conditioning experiments.

IX. SIMULATION RESULTS

We implemented the boxes system as described in [3] and [4] as well as our systems shown in Fig. 2 (ASE alone) and Fig. 3 (ASE with ACE). We wanted to determine what kinds of neuronlike elements could attain or exceed the performance of the boxes system. Our results suggest that a system using an ASE with internal reinforcement supplied by an ACE is easily able to outperform the boxes system. We must emphasize at the outset, however, that it is not our intention to criticize Michie and Chambers' program: the boxes system they described was in an initial state of development and clearly could be extended to include a mechanism analogous to our ACE. We make comparisons with the performance of the boxes system because it provides a convenient reference point.

We simulated a series of runs of each learning system attempting to control the same cart–pole simulation (see the Appendix). Each run consisted of a sequence of trials, where each trial began with the cart–pole state $x = 0$, $\dot{x} = 0$, $\theta = 0$, $\dot{\theta} = 0$, and ended with a failure signal indicating that θ left the interval $[-12°, 12°]$ or x left the interval $[-2.4 \text{ m}, 2.4 \text{ m}]$. We also set all the trace variables e_i to zero at the start of each trial. The learning systems were "naive" at the start of each run (i.e., all the weights w_i and v_i were set to zero). At the start of each boxes run, we supplied a different seed value to the pseudorandom number generator that we used to initialize the state of the learning system and to break ties in comparing expected lifetimes in order to choose control actions. We did not reset the cart–pole state to a randomly chosen state at the start of each trial as was done in the experiments reported in [3] and [4]. At the start of each run of an ASE system, we supplied a different seed to the pseudorandom number generator that we used to generate the noise used in (1). We approximated this Gaussian random variable by the usual procedure of summing uniformly distributed random variables (we used an eightfold sum). Since the ASE runs began with weight vectors equal to zero, initial actions for each box were equiprobable, and initial ACE predictions were zero. Except for the random number generator seeds,

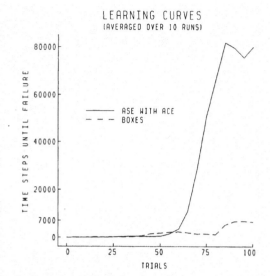

Fig. 4. Simulation results. Performance of boxes system and ASE/ACE system averaged over ten runs. See text for complete explanation.

identical parameter values were used for all runs. Runs consisted of 100 trials unless the run's duration exceeded 500 000 time steps (approximately 2.8 h of simulated real time), in which case the run was terminated. For our implementation of the boxes system, we used the parameter values published in [3]. We experimented with other parameter values without obtaining consistently better performance. We did not attempt to optimize the performance of the systems using the ASE. We picked values that seemed reasonable based on our previous experience with similar adaptive elements.

Figs. 4 and 5 show the results of our simulations of boxes and the ASE/ACE system. The graphs of Fig. 4 are averages of performance over the ten runs that produced the individual graphs shown in Fig. 5. In both figures, a single point is plotted for each bin of five trials giving the number of time steps until failure averaged over those five trials. Almost all runs of the ASE/ACE system, and one run of the boxes system, were terminated after 500 000 time steps before all 100 trials took place (those whose graphs terminate short of 100 trials in Fig. 5). We stopped the simulation before failure on the last trials of these runs. To produce the averages for all 100 trials shown in Fig. 4, we needed to make special provision for the interrupted runs. If the duration of the trial that was underway when the run was interrupted was less than the duration of the immediately preceding (and therefore complete) trial, then we assigned to fictitious remaining trials the duration of that preceding trial. Otherwise, we assigned to fictitious remaining trials the duration of the last trial when it was interrupted. We did this to prevent any short interrupted trials from producing deceptively low averages.

The ASE/ACE system achieved much longer runs than did the boxes system. Fig. 5 shows that the ACE/ASE system tended to solve the problem before it had experienced 100 failures, whereas the boxes system tended not to. Obviously, we cannot make definitive statements about the relative performance of these systems, or about the general utility of the ASE/ACE system solely on the basis of these

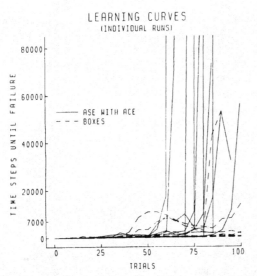

LEARNING CURVES
(INDIVIDUAL RUNS)

Fig. 5. Simulation results. Performance of boxes system and ASE/ACE system in individual runs that were averaged to produce Fig. 4. See text for complete explanation.

experiments. However, these results encourage us to continue developing the principles upon which the ASE/ACE system is based.

The parameter values used in producing these results were $\alpha = 1000$, $\beta = 0.5$, $\delta = 0.9$, $\gamma = 0.95$, $\lambda = 0.8$, and $\sigma = 0.01$. Except for one set of extreme values, these were the only values we tried. The large value of α was chosen so that large changes in the weights w_i occurred upon reinforcement. This caused the probability of a rewarded action to become nearly one and the probability of a penalized action to become nearly zero. We did this in an attempt to implement in our system the feature of the boxes system that causes each local demon to choose the same action each time its box is entered in any given trial. This greatly reduces the uncertainty in the problem but would be inappropriate, we think, for problems in which other reinforcing events could occur during trials. The parameters δ and λ determine the durations of the eligibility traces. Their values, 0.9 and 0.8, respectively, cause long, slowly decaying traces to form, as seemed appropriate given the nature of the problem.

The good performance of the ASE/ACE system was almost entirely due to the ACE's supplying reinforcement throughout trials. For the boxes system and for an ASE without an ACE, learning occurs only upon failure, an event that becomes less frequent as learning proceeds. With the ACE in place, an ASE can receive feedback on every time step. The learning produced by this feedback causes the system to attempt to enter particular parts of the state space and to avoid others. We simulated the control problem using an ASE without an ACE, using the same parameter settings that worked well for the ASE/ACE experiments. The ASE was not able to attain the level of performance shown by the boxes system. These shortcomings of the ASE are due to difficulties in the convergence process in tasks involving only penalizing feedback, as

discussed in Section VII. The use of reinforcement computed by the ACE markedly changes this property of the pole-balancing problem. At present, we have little experience with ASE-like elements operating without ACE supplied reinforcement in the pole-balancing problem.

X. The Decoder

We have assumed the existence of a decoder that effectively divides the cart–pole state space into a number of disjoint regions by transforming each state vector into a vector having 162 components, all but one of which is zero. We call this a decoder after a similar device used in computer memory circuits to transform each memory address into a signal on the wire connected to the physical location having that address. With this decoder providing its input, the ASE essentially fills in a lookup table of control actions. Similarly, the ACE fills in a table of reinforcement predictions. Each item of information is stored by the setting of the value of a single synaptic weight at a given location.

As a consequence of this localized storage scheme, no generalization occurs beyond the confines of a given box. Given the relative smoothness of the cart–pole dynamics, learning would be faster if information stored in a box could be extrapolated to neighboring boxes (using the Euclidean metric). This can be accomplished by using a kind of decoder that produces activity on overlapping sets of output pathways. It is interesting to note that in several theories of sensorimotor learning, it is postulated that the granular layer of the cerebellum implements just this kind of decoder and that Purkinje cells are adaptive elements [34], [35].[3]

Localized extrapolation is not the only type of generalization that can be useful. There has lately been increasing interest in "associative memory networks" that use distributed representations in which dispersed rather than localized patterns of activity encode information [36], [37]. Rather than implementing table-lookup storage, associative memories use weighted summations to compute output vectors from input vectors. This style of information storage provides generalization among patterns according to where they lie with respect to a set of linear discriminant functions. Since the ASE and ACE use weighted summations that are defined for arbitrary input vectors, they implement linear discriminant functions and are capable of forming information storage networks having all of the properties that have generated interest in associative memory networks. Unlike the associative memory networks discussed in the literature, however, networks of ASE-like components are capable of discovering via reinforcement learning what information is useful to store. These aspects

---

[3]In these theories, the adaptive elements perform supervised learning pattern classification, with climbing fiber input providing the desired responses, and not the type of reinforcement learning with which the present article is concerned. If the adaptive capacity of a Purkinje cell were limited to that postulated in these theories, then a Purkinje cell would not be able to solve the type of problem illustrated by the pole-balancing task described in this article.

of ASE's are emphasized by Barto *et al.* [8] where *associative search* networks are discussed.

Whether environmental states are represented using localized or distributed patterns, the problem remains of how to choose the specifics of the representations in order to facilitate learning. In this article we followed Michie and Chambers in choosing a state–space partition based on special knowledge of the control task. As they point out, it is easy to choose a partition that makes the task impossible [4]. For the next stage of development of the boxes system, Michie and Chambers planned to give the system the ability to change the boundaries of the boxes by the processes of "splitting" and "lumping" [4]. We are not aware of any results they published on these processes, but we were motivated in part by their comments to experiment with layered networks of ASE-like adaptive elements in order to examine the feasibility of implementing a kind of adaptive decoder. Some preliminary results, reported in [5], were encouraging, and we are continuing our investigation in this direction.

XI. Animal Learning

Minsky has pointed out [23] that methods for reducing the severity of the credit-assignment problem like the one used in Samuel's checkers player are suggestive of secondary or conditioned reinforcement phenomena in animal learning studies. A stimulus acquires reinforcing qualities (i.e., becomes a secondary reinforcer) if it predicts either primary reinforcement (e.g., food or shock) or some other secondary reinforcer. It is generally held that higher order classical conditioning, whereby previously conditioned conditioned stimuli (CS's) can act as unconditioned stimuli (US's) for earlier potential CS's, is the basis for the development of secondary reinforcement [38].

The ACE is a refinement of the model of classical conditioning that was presented in [9]. That model's behavior is consistent with the Rescorla–Wagner model of classical conditioning [39]. While not without certain problems, the Rescorla–Wagner model has been the most influential model of classical conditioning for the last ten years [40]. One interpretation of the basic premise of the Rescorla–Wagner model is that the degree to which an event is "unexpected" or "surprising" determines the degree to which it enters into associations with earlier events. Stimuli lose their reinforcing qualities to the extent that they are expected on the basis of the occurrence of earlier stimuli. The model upon which the ACE is based extends the basic mechanism of the Rescorla–Wagner model to provide for some of the features of higher order conditioning, the influence of relative event timing within trials, and the occurrence of conditioned responses (CR's) that anticipate the US. In these terms, the failure signal r corresponds to the US, the signals x_i from the decoder correspond to potential CS's, and the prediction p corresponds to a component of the CR. We have not yet thoroughly investigated the extent to which the ACE/ASE system is a valid model of animal behavior in instrumental conditioning experiments.

XII. Conclusion

It should be clear that our approach differs from that of the pioneering adaptive neural-network theorists of the 1950's and 1960's. We have built into single neuronlike adaptive elements a problem-solving capacity that in many respects exceeds that achieved in the past by entire simulated neural networks. The metaphor for neural function suggested by this approach provides, at least to us, the first convincing inkling of how nervous tissue could possibly be capable of its exquisite feats of problem solving and control.

We argued that components of powerful adaptive networks must be at least as sophisticated as the components described in this article. If this were true for biological networks as well as for artificial networks, then it would suggest that parallels might exist between neurons and the adaptive elements described here. It would suggest, for example, that 1) there are single neuron analogs of instrumental conditioning and chemically specialized reinforcing neurons that may themselves be adaptive (see [41]); 2) the random component of an instrumental neuron's behavior is necessary for generating variety to serve as the basis for subsequent selection; and 3) mechanisms exist for maintaining relatively long-lasting synaptically local traces of activity that modulate changes in synaptic efficacy. Although some of these implications are supported in varying degrees by existing data, there are no data that provide direct support for the existence of the specific mechanisms used in our adaptive elements. By showing how neuronlike elements can solve genuinely difficult problems that are solved routinely by many animals, we hope to stimulate interest in the relevant experimental research.

Appendix
Details of the Cart–Pole Simulation

The cart–pole system is modeled by the following nonlinear differential equations (see [42]):

$$\ddot{\theta}_t = \frac{g \sin \theta_t + \cos \theta_t \left[\dfrac{-F_t - ml\dot{\theta}_t^2 \sin \theta_t + \mu_c \operatorname{sgn}(\dot{x}_t)}{m_c + m} \right] - \dfrac{\mu_p \dot{\theta}_t}{ml}}{l \left[\dfrac{4}{3} - \dfrac{m \cos^2 \theta_t}{m_c + m} \right]}$$

$$\ddot{x}_t = \frac{F_t + ml \left[\dot{\theta}_t^2 \sin \theta_t - \ddot{\theta}_t \cos \theta_t \right] - \mu_c \operatorname{sgn}(\dot{x}_t)}{m_c + m}$$

where

g = -9.8 m/s^2, acceleration due to gravity,

m_c = 1.0 kg, mass of cart

m = 0.1 kg, mass of pole,

l = 0.5 m, half-pole length,

μ_c = 0.0005, coefficient of friction of cart on track,

μ_p = 0.000002, coefficient of friction of pole on cart,

F_t = ± 10.0 newtons, force applied to cart's center of mass at time t.

We initially used the Adams–Moulton predictor–corrector method to approximate numerically the solution of these equations, but the results reported in this article were produced using Euler's method with a time step of 0.02 s for the sake of computational speed. Comparisons of solutions generated by the Adams–Moulton methods and the less accurate Euler method did not reveal discrepencies that we deemed significant for the purposes of this article.

ACKNOWLEDGMENT

We thank A. H. Klopf for bringing to us a set of ideas filled with possibilities. We are grateful also to D. N. Spinelli, M. A. Arbib, and S. Epstein for their valuable comments and criticisms; to D. Lawton for first making us aware of Michie and Chambers' boxes system; to D. Politis and W. Licata for pointing out an important error in our original cart–pole simulations; and to S. Parker for essential help in preparing the manuscript.

REFERENCES

[1] B. Chandrasekaran, "Natural and social system metaphors for distributed problem solving: Introduction to the issue," *IEEE Trans. Syst., Man., Cybern.*, vol. SMC-11, pp. 1–5, 1981.

[2] V. R. Lesser and D. D. Corkill, "Functionally-accurate, cooperative distributed systems," *IEEE Trans. Syst., Man, Cybern.*, vol. SMC-11, pp. 81–96, 1981.

[3] D. Michie and R. A. Chambers, "BOXES: An experiment in adaptive control," in *Machine Intelligence 2*, E. Dale and D. Michie, Eds. Edinburgh: Oliver and Boyd, 1968, pp. 137–152.

[4] D. Michie and R. A. Chambers, "'Boxes' as a model of pattern-formation," in *Towards a Theoretical Biology*, vol. 1, *Prolegomena*, C. H. Waddington, Ed. Edinburgh: Edinburgh Univ. Press, 1968, pp. 206–215.

[5] A. G. Barto, C. W. Anderson, and R. S. Sutton, "Synthesis of nonlinear control surfaces by a layered associative search network," *Biol. Cybern.*, vol. 43, pp. 175–185, 1982.

[6] A. G. Barto and R. S. Sutton, "Landmark learning: An illustration of associative search," *Biol. Cybern.*, vol. 42, pp. 1–8, 1981.

[7] ____, "Simulation of anticipatory responses in classical conditioning by a neuron-like adaptive element," *Behavioral Brain Res.*, vol. 4, pp. 221–235, 1982.

[8] A. G. Barto, R. S. Sutton, and P. S. Brouwer, "Associative search network: A reinforcement learning associative memory," *Biol. Cybern.*, vol. 40, pp. 201–211, 1981.

[9] R. S. Sutton and A. G. Barto, "Toward a modern theory of adaptive networks: Expectation and prediction," *Psychol. Rev.*, vol. 88, pp. 135–171, 1981.

[10] ____, "An adaptive network that constructs and uses an internal model of its world," *Cognition and Brain Theory*, vol. 4, pp. 213–246, 1981.

[11] A. H. Klopf, "Brain function and adaptive systems—A heterostatic theory," Air Force Cambridge Res. Lab. Res. Rep., AFCRL-72-0164, Bedford, MA, 1972. (A summary appears in *Proc. Int. Conf. Syst., Man, Cybern.*, 1974).

[12] A. H. Klopf, *The Hedonistic Neuron: A Theory of Memory, Learning, and Intelligence*. Washington, DC: Hemisphere, 1982.

[13] B. Widrow, N. K. Gupta, and S. Maitra, "Punish/reward: learning with a critic in adaptive threshold systems," *IEEE Trans. Syst., Man, Cybern.*, vol. SMC-3, pp. 455–465, 1973.

[14] B. Widrow and F. W. Smith, "Pattern-recognizing control systems," in *Computer and Information Sciences*, J. T. Tow and R. H. Wilcox, Eds. Clever Hume Press, 1964, pp. 288–317.

[15] F. Rosenblatt, *Principles of Neurodynamics*. New York: Spartan, 1962.

[16] B. Widrow and M. E. Hoff, "Adaptive switching circuits," in *1960 WESCON Conv. Record*, part IV, 1960, pp. 96–104.

[17] R. O. Duda and P. E. Hart, *Pattern Classification and Scene Analysis*. New York: Wiley, 1973.

[18] P. R. Cohen and E. A. Feigenbaum, *The Handbook of Artificial Intelligence*, vol. 3. Los Altos, CA: Kauffman, 1982.

[19] D. O. Hebb, *Organization of Behavior*. New York: Wiley, 1949.

[20] B. G. Farley and W. A. Clark, "Simulation of self-organizing systems by digital computer," *IRE Trans. Inform. Theory*, vol. PGIT-4, pp. 76–84, 1954

[21] M. L. Minsky, "Theory of neural-analog reinforcement systems and its application to the brain-model problem," Ph.D. dissertation, Princeton Univ., Princeton, NJ, 1954.

[22] G. M. Edelman, "Group selection and phasic reentrant signaling: A theory of higher brain function," in *The Mindful Brain: Cortical Organization and the Group-Selective Theory of Higher Brain Function*, G. M. Edelman and V. B. Mountcastle, Eds. Cambridge, MA: MIT Press, 1978.

[23] M. L. Minsky, "Steps toward artificial intelligence," *Proc. IRE*, vol. 49, pp. 8–30, 1961.

[24] A. L. Samuel, "Some studies in machine learning using the game of checkers," *IBM J. Res. Develop.*, vol. 3, pp. 210–229, 1959.

[25] J. Doran, "An approach to automatic problem solving," in *Machine Intelligence*, vol. 1, E. L. Collins and D. Michie, Eds. Edinburgh: Oliver and Boyd, 1967, pp. 105–123.

[26] J. H. Holland, "Adaptive algorithms for discovering and using general patterns in growing knowledge-bases," *Int. J. Policy Anal. Inform. Syst.*, vol. 4, pp. 217–240, 1980.

[27] I. H. Witten, "An adaptive optimal controller for discrete-time Markov environments," *Inform. Contr.*, vol. 34, pp. 286–295, 1977.

[28] T. D. Dietterich and B. G. Buchanan, "The role of the critic in learning systems," Stanford Univ. Tech. Rep., STAN-CS-81-891, 1981.

[29] W. S. McCulloch and W. H. Pitts, "A logical calculus of the ideas immanent in nervous activity," *Bull. Math. Biophys.*, vol. 5, pp. 115–133, 1943.

[30] K. S. Narendra and M. A. L. Thatachar, "Learning automata—A survey," *IEEE Trans. Syst., Man, Cybern.*, vol. SMC-4, pp. 323–334, 1974.

[31] S. Lakshmivarahan, *Learning Algorithms Theory and Applications*. New York: Springer-Verlag, 1981.

[32] R. R. Bush and F. Mosteller, *Stochastic Models for Learning*. New York: Wiley, 1958.

[33] M. L. Tsetlin, *Automaton Theory and Modelling of Biological Systems*. New York: Academic, 1973.

[34] J. A. Albus, *Brains, Behavior, and Robotics*. Peterborough, NH: BYTE Books, 1981.

[35] D. Marr, "A theory of cerebellar cortex," *J. Physiol.*, vol. 202, pp. 437–470, 1969.

[36] G. E. Hinton and J. A. Anderson, Eds., *Parallel Models of Associative Memory*. Hillsdale, NJ: Erlbaum, 1981.

[37] T. Kohonen, *Associative Memory: A System Theoretic Approach*. Berlin, Germany: Springer, 1977.

[38] R. A. Rescorla, *Pavlovian Second-Order Conditioning: Studies in Associative Learning*. Hillsdale, NJ: Erlbaum, 1980.

[39] R. A. Rescorla and A. R. Wagner, "A theory of Pavlovian conditioning: Variations in the effectiveness of reinforcement and nonreinforcement," in *Classical Conditioning II: Current Research and Theory*, A. H. Black and W. R. Prokasy, Eds. New York: Appleton-Century-Crofts, 1972.

[40] A. Dickinson, *Contemporary Animal Learning Theory*. Cambridge: Cambridge Univ. Press, 1980.

[41] L. Stein and J. D. Belluzzi, "Beyond the reflex arc: A neuronal model of operant conditioning," in *Changing Concepts of the Nervous System*. New York: Academic, 1982.

[42] R. H. Cannon, *Dynamics of Physical Systems*. New York: McGraw-Hill, 1967.

Chapter 7

Convolutional Neural Networks – 1989

Context

Ever since computers were invented, scientists have tried to write programs that allow a machine to *see*. However, the word 'see' encompasses a multitude of skills, only some of which have been acquired by computers. Despite decades of effort by the vision research community using conventional engineering methods, most of the recent progress in building a seeing computer stems from advances in artificial neural networks. This is particularly interesting from a neurophysiological perspective, because the only type of neural network that can definitely see is a biological neural network (i.e. a brain). As noted in the paper[63] discussed in this chapter:

> *Some kernels synthesized by the network can be interpreted as feature detectors remarkably similar to those found to exist in biological vision systems (Hubel and Wiesel 1962).*

Accordingly, it should be unsurprising to learn that a major advance in getting artificial neural networks to see was achieved by copying key architectural features of the brain's visual system[118]. Probably the most important of these features is *convolution*, as used in *convolutional neural networks* (CNNs) by LeCun et al. (1989). As noted by Sejnowski (2008):

> *Yann [LeCun] didn't slavishly try to duplicate the cortex. He tried many different variations, but the ones he converged onto were the ones that nature converged onto. This is an important observation. The convergence of nature and AI has a lot to teach us and there's much farther to go.*

Today, CNNs are ubiquitous in computer vision applications, such as facial recognition, object detection, medical image analysis and autonomous driving.

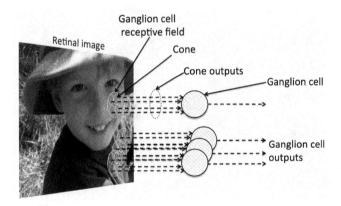

Figure 7.1. Retinal cones (photoreceptors) within a small region send their outputs to several ganglion cells. The region from which a ganglion cell collates photoreceptor outputs is its receptive field, and the receptive fields of adjacent ganglion cells overlap. Here, each ganglion cell receptive field receives the outputs of four photoreceptors. Reproduced from Stone (2019).

Technical Summary

In order to understand convolutional neural networks, it is first necessary to understand convolution. Because convolutional neural networks were inspired by the neurophysiology of vision in humans, we begin with a brief account of image processing in the retina.

The human retina contains 126 million photoreceptors, which can be thought of as pixels in a digital image. The outputs of the photoreceptors connect (via several layers of neurons) to *ganglion cells* within the retina. Each retinal ganglion cell collates the outputs from photoreceptors within a small area of the retina, which is the ganglion cell's *receptive field* (Figures 7.1 and 7.2). Some photoreceptors have stronger connections than others, and the consequent differential weighting of photoreceptor outputs determines the structure of the ganglion cell's receptive field; this structure determines which visual features (e.g. contrast edges) evoke the strongest response from the ganglion cell. The values of these differential weightings define a *filter* or *kernel*.

Interpreted in terms of neural networks, the photoreceptors correspond to units of the input layer, called *visible units*, and the filters correspond to units in the next layer of the network, called *hidden units*. The total input to a hidden unit is the sum of products of the outputs of visible units that connect to the hidden unit and the weights of those connections. For example, the individual weights in the 3×3 two-dimensional filter shown in Figure 7.2 can be represented as a matrix,

$$\mathbf{w} = \begin{pmatrix} w(-1,-1) & w(-1,0) & w(-1,+1) \\ w(0,-1) & w(0,0) & w(0,+1) \\ w(+1,-1) & w(+1,0) & w(+1,+1) \end{pmatrix}, \tag{7.1}$$

where $w(\alpha, \beta)$ is the value of the weight at position (α, β) within the matrix of weights.

To see how convolution works, consider a filter $\mathbf{w}$ centred at a point (x, y) in the image X, where x and y define positions along the horizontal and vertical axes of the image, respectively. This filter gathers its inputs from the values of pixels in a small region of the image (its receptive field), and the effective connection strength between the filter and each pixel defines the filter kernel, as shown in Figure 7.2. If a hidden unit has a receptive field centred at (x, y) in the image then the input to that

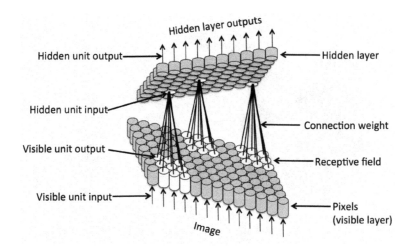

Figure 7.2. Each hidden unit (filter) receives inputs from a small region of the visible layer, which defines its receptive field. Reproduced from Stone (2019).

unit is a weighted average of pixel values within the unit's receptive field,

$$\mathbf{v}(x,y) \;=\; \sum_{\alpha=-k}^{k} \sum_{\beta=-k}^{k} \mathbf{w}(\alpha,\beta)\, X(x+\alpha, y+\beta), \tag{7.2}$$

where k defines the size of the receptive field (e.g. $k = 1$ for the filter in Equation 7.1). Note that $\mathbf{v}(x,y)$ is the total input to just one hidden unit, and the sum of products in Equation 7.2 must be computed for every hidden unit, as shown in Figure 7.2.

Suppose that the image X consists of a square array of $N \times N$ pixels. If we place the centre of the same filter $\mathbf{w}$ over each pixel then (using Equation 7.2) the resultant $N \times N$ 'image' of filter inputs $\mathbf{v}$ can be expressed using the *convolution operator* $*$ as

$$\mathbf{v} \;=\; \mathbf{w} * X. \tag{7.3}$$

The filter outputs define a new image, sometimes called a *convolution image*. If the output of the filter is equal to its input then the convolution image is $\mathbf{v}$, which corresponds to a *feature map* (see below). Examples of convolution images are shown in Figures 7.3 and 7.4.

7.1. The Convolutional Neural Network

A landmark paper in the history of neural networks was published by LeCun et al.[63] in 1989, which introduced the network shown in Figure 7.5. The main innovation employed in this paper is that the filters (sets of weights) of hidden units were *learned* from the data, rather than being hand-

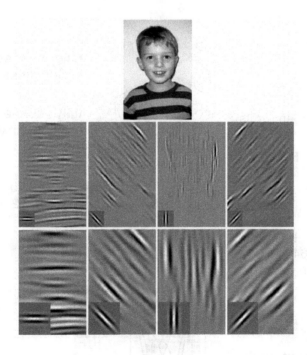

Figure 7.3. Examples of convolution of an image of a boy. Top: Image to be convolved with filters. Bottom: Eight convolution images. The two-dimensional matrix of weights shown in the bottom left corner of each convolution image acts like a filter. Each of the eight convolution images is obtained by using the corresponding filter to convolve the image at the top. These (Gabor) filters are similar to filters found in the human visual system and similar to the filters learned by neural networks. Each convolution image corresponds to the activity within one feature map of a convolutional network. Reproduced from Stone (2019).

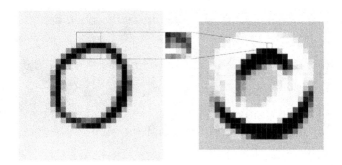

Figure 7.4. Left: Image of a zero, showing the 5×5 receptive field of a hidden unit. Middle: Weights within the 5×5 receptive field of a hidden unit. Right: Convolution image (hidden unit outputs). The lines show how the input to a hidden unit is obtained by multiplying each weight by the corresponding pixel within the unit's receptive field. All hidden units in one feature map have the same weights (middle), and their outputs can be represented as an image (right). The number of pixels and hidden units is the same here, but there would usually be fewer hidden units than pixels, as in Figure 7.5. Reproduced with permission from LeCun et al. (1989).

crafted. When combined with *weight sharing* and *pooling* (see below), this allowed the network to attain impressive performance in classifying handwritten digits.

The network has a total of five layers: an input layer, three hidden layers, and an output layer. All units except those in the input layer have hyperbolic tangent (tanh) activation functions (the tanh function has a similar shape to the sigmoidal function shown in Figure 5.1, except that it has an output between -1 and $+1$). The input layer is a 16×16 array of units with linear activation functions. This layer is effectively the retina of the network, which registers the image to be recognised.

The first hidden layer consists of 12 *feature maps*, where each feature map is an 8×8 array of hidden units. All units in a single feature map have the same 5×5 array of weights, which defines the

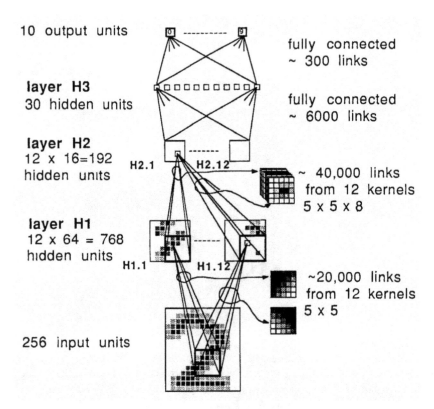

Figure 7.5. Architecture of the neural network, from LeCun et al. (1989).

receptive field of each hidden unit in that feature map. This ensures that the same visual feature can be detected in different parts of the image. Crucially, these shared weights are learned, rather than being hand-crafted. Because all of the hidden units in each feature map have the same weights, this is known as *weight sharing*.

Adjacent hidden units have overlapping receptive fields in the image, so each feature map effectively tiles the input image. Specifically, adjacent hidden units in a given feature map have receptive field centres separated by two pixels in the input image. As noted above, the outputs of units in a given feature map constitute a convolution image – the result of convolving the unit weights (which are identical for every hidden unit in a single feature map) with the input image and then passing the total input to each unit through an activation function. Because there are more input units (pixels) than hidden units in a given feature map, each of the 12 feature maps effectively *sub-samples* the image, producing an 8×8 feature map from a 16×16 image. Even though all hidden units in a given feature map have the same weights, different units have unique biases, which are learned during training. For a hidden unit with a receptive field centred at a given position (a, b) within the image, this bias reflects the frequency with which the feature detected by that unit is likely to be present at (a, b). Units in successively higher-order layers also have unique biases, where the 'image' for units in each layer consists of the outputs of units in the preceding layer.

There are two principal reasons for using feature maps. First, sharing the same weights between all units in each feature map reduces the number of free parameters (i.e. weights). Second, because all units in a feature map have the same weights, each feature map is able to detect a single feature in any part of the image. Notice that using 12 feature maps implicitly assumes that 12 types of localised features are sufficient to recognise digits.

The second hidden layer follows a similar design principle to the first hidden layer, with 12 feature maps, each of which is an array of 4×4 units. This layer performs the same type of operation on the output of the first hidden layer as the first hidden layer performs on the input image. Each hidden unit has a 5×5 receptive field at the same location as for feature maps in the first hidden layer. However, the first hidden layer contains 12 feature maps, but each unit in the second hidden layer receives inputs from a subset of eight of those 12 maps. The reason for having multiple layers of feature detectors is (for example) that the second hidden layer should find combinations of the low-order features discovered by units in the first hidden layer. This cumulative 'features of features' strategy is analogous to the hierarchical processing thought to occur in the human visual system.

The third hidden layer consists of 30 units, each of which is fully connected to every unit in the second hidden layer. Finally, the output layer consists of 10 units, each of which is fully connected to every unit in the third hidden layer. In total, there are 1,256 units and 9,760 weights. The network

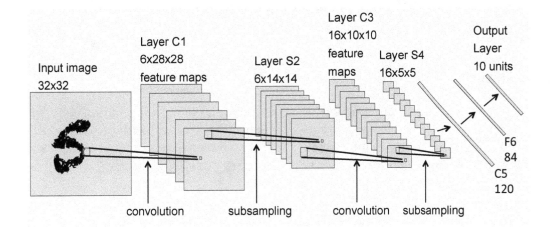

Figure 7.6. Schematic diagram of LeNet5. Reproduced from Stone (2019).

was trained using backprop on 7,291 images in a training set, and was tested on 2,007 images in a test set. After 23 training iterations (each iteration involving 7,291 images, which took three days of CPU time), the error on the training set was 0.14%, compared with 5% on the test set. This video from 1989 shows Yann LeCun demonstrating his convolutional network for recognising handwritten digits: https://www.youtube.com/watch?v=FwFduRA_L6Q.

7.2. LeNet5: Convolutional Neural Networks in 1998

Almost a decade after this paper was published, a network called LeNet5 was introduced[64] in 1998. The main improvement over the 1989 network was the introduction of *average pooling*. Pooling is a mechanism used to provide invariance with respect to a chosen image transformation, such the orientation of a feature. Average pooling can be understood more easily if we first consider a related method called *max pooling*.

Consider four units that are at the same position in four different feature maps of the first hidden layer. Suppose all of these units respond to the image of the digit 5, but at different orientations. If a 'higher-order' unit in the second hidden layer can adopt the maximum state of those four units then this higher-order unit will behave as if it is an orientation-invariant detector of the digit 5; this is max pooling. In contrast, the average pooling used in LeNet5 consists of a unit taking the average output of four units in the first hidden layer. This average is multiplied by a learned weight, and the result is transformed by the unit's sigmoidal activation function. Notice that if one of the four units dominates the input to the higher-order unit then the end result is similar to max pooling.

LeNet5 consists of an input image of 32×32 pixels, plus seven layers, as shown in Figure 7.6. The entire network contains 340,908 connections, but because some of these are shared, there are a total of 60,000 adjustable weights. The test error on the standard data set of MNIST digits (see Glossary) was 0.95%. Demonstrations of LeNet5 can be viewed at http://yann.lecun.com/exdb/lenet/index.html.

7.3. Results of the LeCun et al. (1989) paper

The data comprised 9,298 numerals of handwritten zip codes; 7,291 were used to train the network, and 2,007 were used for testing. Each numeral was rescaled to a 16×16 pixel image, and the grey-levels were rescaled to the range $[-1, 1]$.

The main result is shown in Figure 2 in the paper[†], which indicates how the error rate on the training data and the test data fall during training. The final training error rate is 0.14% on the training data (10 misclassified images) and 5% on the test data (102 misclassified images). From the perspective of the modern era, this represents a tiny training set, and it would be surprising if any system could perform much better on such a small data set. For example, the LeNet5 network described above was trained using the MNIST data set of 60,000 images.

Research Paper:
Backpropagation Applied to Handwritten Zip Code Recognition

Reference: LeCun, Y., Boser, B., Denker, J. S., Henderson, R. E., Hubbard, W., and Jackel, L. D. (1989). Backpropagation applied to handwritten ZIP code recognition. *Neural Comp.*, 1:541–551. https://direct.mit.edu/neco/article-abstract/1/4/541/5515/Backpropagation-Applied-to-Handwritten-Zip-Code (Restricted access).
Reproduced with permission from MIT Press.

[†]The captions of Figures 2 and 3 are the wrong way round.

Backpropagation Applied to Handwritten Zip Code Recognition

Y. LeCun
B. Boser
J. S. Denker
D. Henderson
R. E. Howard
W. Hubbard
L. D. Jackel

AT&T Bell Laboratories, Holmdel, NJ 07733 USA

The ability of learning networks to generalize can be greatly enhanced
by providing constraints from the task domain. This paper demonstrates how such constraints can be integrated into a backpropagation
network through the architecture of the network. This approach has
been successfully applied to the recognition of handwritten zip code
digits provided by the U.S. Postal Service. A single network learns the
entire recognition operation, going from the normalized image of the
character to the final classification.

1 Introduction

Previous work performed on recognizing simple digit images (LeCun 1989)
has shown that good generalization on complex tasks can be obtained
by designing a network architecture that contains a certain amount of
a priori knowledge about the task. The basic design principle is to
reduce the number of free parameters in the network as much as possible
without overly reducing its computational power. Application of this
principle increases the probability of correct generalization because it
results in a specialized network architecture that has a reduced entropy
(Denker et al. 1987; Patarnello and Carnevali 1987; Tishby et al. 1989; Le
Cun 1989) and a reduced Vapnik-Chervonenkis dimensionality (Baum
and Haussler 1989).

In this paper we apply the backpropagation algorithm (Rumelhart et
al. 1986) to a real-world problem in recognizing handwritten digits taken
from the U.S. Mail. Unlike previous work on the subject (Denker et al.
1989), the learning network is directly fed with images, rather than feature vectors, thus demonstrating the ability of backpropagation networks
to deal with large amounts of low-level information.

Backpropagation Applied to Handwritten Zip Code Recognition

Y. LeCun
B. Boser
J. S. Denker
D. Henderson
R. E. Howard
W. Hubbard
L. D. Jackel
AT&T Bell Laboratories, Holmdel, NJ 07733 USA

The ability of learning networks to generalize can be greatly enhanced by providing constraints from the task domain. This paper demonstrates how such constraints can be integrated into a backpropagation network through the architecture of the network. This approach has been successfully applied to the recognition of handwritten zip code digits provided by the U.S. Postal Service. A single network learns the entire recognition operation, going from the normalized image of the character to the final classification.

1 Introduction

Previous work performed on recognizing simple digit images (LeCun 1989) showed that good generalization on complex tasks can be obtained by designing a network architecture that contains a certain amount of a priori knowledge about the task. The basic design principle is to reduce the number of free parameters in the network as much as possible without overly reducing its computational power. Application of this principle increases the probability of correct generalization because it results in a specialized network architecture that has a reduced entropy (Denker *et al.* 1987; Patarnello and Carnevali 1987; Tishby *et al.* 1989; Le-Cun 1989), and a reduced Vapnik–Chervonenkis dimensionality (Baum and Haussler 1989).

In this paper, we apply the backpropagation algorithm (Rumelhart *et al.* 1986) to a real-world problem in recognizing handwritten digits taken from the U.S. Mail. Unlike previous results reported by our group on this problem (Denker *et al.* 1989), the learning network is directly fed with images, rather than feature vectors, thus demonstrating the ability of backpropagation networks to deal with large amounts of low-level information.

Neural Computation **1,** 541–551 (1989) © 1989 Massachusetts Institute of Technology

2.1 Data Base. The data base used to train and test the network consists of 9298 segmented numerals digitized from handwritten zip codes that appeared on U.S. mail passing through the Buffalo, NY post office. Examples of such images are shown in Figure 1. The digits were written by many different people, using a great variety of sizes, writing styles, and instruments, with widely varying amounts of care; 7291 examples are used for training the network and 2007 are used for testing the generalization performance. One important feature of this data base is that both the training set and the testing set contain numerous examples that are ambiguous, unclassifiable, or even misclassified.

2.2 Preprocessing. Locating the zip code on the envelope and separating each digit from its neighbors, a very hard task in itself, was performed by Postal Service contractors (Wang and Srihari 1988). At this point, the size of a digit image varies but is typically around 40 by 60 pixels. A linear transformation is then applied to make the image fit in a 16 by 16 pixel image. This transformation preserves the aspect ratio of the character, and is performed after extraneous marks in the image have been removed. Because of the linear transformation, the resulting image is not binary but has multiple gray levels, since a variable number of pixels in the original image can fall into a given pixel in the target image. The gray levels of each image are scaled and translated to fall within the range −1 to 1.

3 Network Design _____

3.1 Input and Output. The remainder of the recognition is entirely performed by a multilayer network. All of the connections in the network are adaptive, although heavily constrained, and are trained using backpropagation. This is in contrast with earlier work (Denker *et al.* 1989) where the first few layers of connections were hand-chosen constants implemented on a neural-network chip. The input of the network is a 16 by 16 normalized image. The output is composed of 10 units (one per class) and uses place coding.

3.2 Feature Maps and Weight Sharing. Classical work in visual pattern recognition has demonstrated the advantage of extracting local features and combining them to form higher order features. Such knowledge can be easily built into the network by forcing the hidden units to combine only local sources of information. Distinctive features of an object can appear at various locations on the input image. Therefore it seems judicious to have a set of feature detectors that can detect a particular

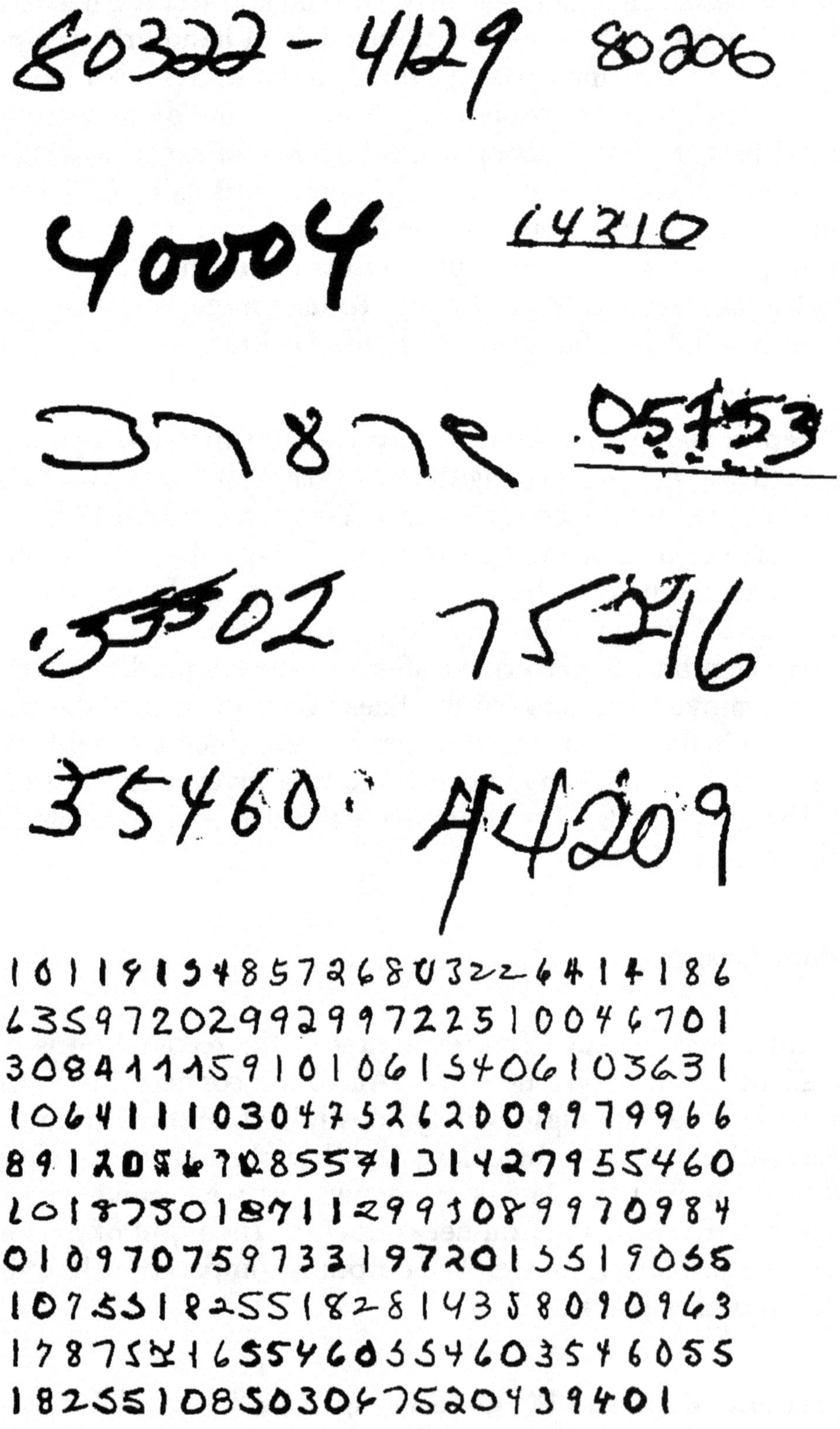

Figure 1: Examples of original zip codes (top) and normalized digits from the testing set (bottom).

instance of a feature anywhere on the input plane. Since the *precise* location of a feature is not relevant to the classification, we can afford to lose some position information in the process. Nevertheless, *approximate* position information must be preserved, to allow the next levels to detect higher order, more complex features (Fukushima 1980; Mozer 1987).

The detection of a particular feature at any location on the input can be easily done using the "weight sharing" technique. Weight sharing was described in Rumelhart *et al.* (1986) for the so-called T-C problem and consists in having several connections (links) controlled by a single parameter (weight). It can be interpreted as imposing equality constraints among the connection strengths. This technique can be implemented with very little computational overhead.

Weight sharing not only greatly reduces the number of free parameters in the network but also can express information about the geometry and topology of the task. In our case, the first hidden layer is composed of several planes that we call *feature maps*. All units in a plane share the same set of weights, thereby detecting the same feature at different locations. Since the exact position of the feature is not important, the feature maps need not have as many units as the input.

3.3 Network Architecture.

The network is represented in Figure 2. Its architecture is a direct extension of the one proposed in LeCun (1989). The network has three hidden layers named H1, H2, and H3, respectively. Connections entering H1 and H2 are local and are heavily constrained.

H1 is composed of 12 groups of 64 units arranged as 12 independent 8 by 8 feature maps. These 12 feature maps will be designated by H1.1, H1.2, ..., H1.12. Each unit in a feature map takes input on a 5 by 5 neighborhood on the input plane. For units in layer H1 that are one unit apart, their receptive fields (in the input layer) are two pixels apart. Thus, the input image is *undersampled* and some position information is eliminated. A similar two-to-one undersampling occurs going from layer H1 to H2. The motivation is that high resolution may be needed to detect the presence of a feature, while its exact position need not be determined with equally high precision.

It is also known that the kinds of features that are important at one place in the image are likely to be important in other places. Therefore, corresponding connections on each unit in a given feature map are constrained to have the same weights. In other words, each of the 64 units in H1.1 uses the same set of 25 weights. Each unit performs the same operation on corresponding parts of the image. The function performed by a feature map can thus be interpreted as a nonlinear subsampled convolution with a 5 by 5 kernel.

Of course, units in another map (say H1.4) share *another* set of 25 weights. Units do not share their biases (thresholds). Each unit thus has 25 input lines plus a bias. Connections extending past the boundaries of the input plane take their input from a virtual background plane whose

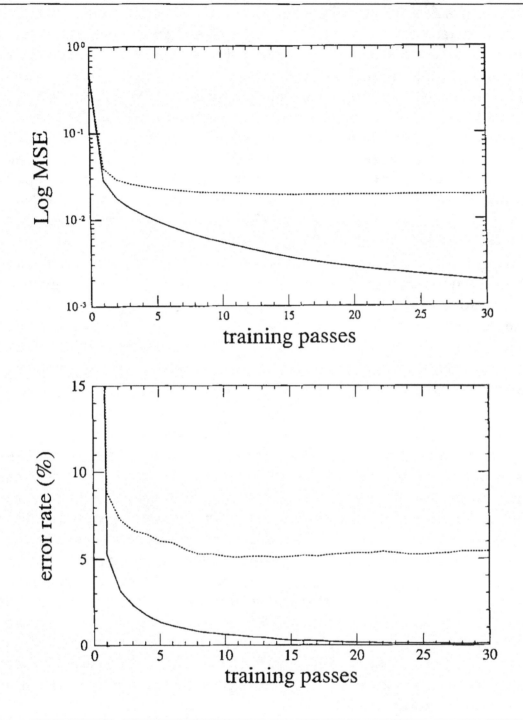

Figure 2: Network architecture.

state is equal to a constant, predetermined background level, in our case −1. Thus, layer H1 comprises 768 units (8 by 8 times 12), 19,968 connections (768 times 26), but only 1068 free parameters (768 biases plus 25 times 12 feature kernels) since many connections share the same weight.

Layer H2 is also composed of 12 features maps. Each feature map contains 16 units arranged in a 4 by 4 plane. As before, these feature maps will be designated as H2.1, H2.2, ..., H2.12. The connection scheme between H1 and H2 is quite similar to the one between the input and H1, but slightly more complicated because H1 has multiple two-dimensional maps. Each unit in H2 combines local information coming from 8 of the 12 different feature maps in H1. Its receptive field is composed of eight 5 by 5 neighborhoods centered around units that are at identical positions within each of the eight maps. Thus, a unit in H2 has 200 inputs, 200 weights, and a bias. Once again, all units in a given map are constrained to have identical weight vectors. The eight maps in H1 on which a map in H2 takes its inputs are chosen according a scheme that will not be described here. Connections falling off the boundaries are treated like as in H1. To summarize, layer H2 contains 192 units (12 times 4 by 4) and there is a total of 38, 592 connections between layers H1 and H2 (192 units times 201 input lines). All these connections are controlled by only 2592 free parameters (12 feature maps times 200 weights plus 192 biases).

Layer H3 has 30 units, and is fully connected to H2. The number of connections between H2 and H3 is thus 5790 (30 times 192 plus 30 biases). The output layer has 10 units and is also fully connected to H3, adding another 310 weights. In summary, the network has 1256 units, 64, 660 connections, and 9760 independent parameters.

4 Experimental Environment

All simulations were performed using the backpropagation simulator SN (Bottou and LeCun 1988) running on a SUN-4/260.

The nonlinear function used at each node was a scaled hyperbolic tangent. Symmetric functions of that kind are believed to yield faster convergence, although the learning can be extremely slow if some weights are too small (LeCun 1987). The target values for the output units were chosen within the quasilinear range of the sigmoid. This prevents the weights from growing indefinitely and prevents the output units from operating in the flat spot of the sigmoid. The output cost function was the mean squared error.

Before training, the weights were initialized with random values using a uniform distribution between $-2.4/F_i$ and $2.4/F_i$ where F_i is the number of inputs (fan-in) of the unit to which the connection belongs. This technique tends to keep the total inputs within the operating range of the sigmoid.

During each learning experiment, the patterns were repeatedly presented in a constant order. The weights were updated according to the so-called stochastic gradient or "on-line" procedure (updating after each presentation of a single pattern) as opposed to the "true" gradient procedure (averaging over the whole training set before updating the weights). From empirical study (supported by theoretical arguments), the stochastic gradient was found to converge much faster than the true gradient, especially on large, redundant data bases. It also finds solutions that are more robust.

All experiments were done using a special version of Newton's algorithm that uses a positive, diagonal approximation of the Hessian matrix (LeCun 1987; Becker and LeCun 1988). This algorithm is not believed to bring a tremendous increase in learning speed but it converges reliably without requiring extensive adjustments of the parameters.

5 Results

After each pass through the training set, the performance was measured both on the training and on the test set. The network was trained for 23 passes through the training set (167, 693 pattern presentations).

After these 23 passes, the MSE averaged over the patterns and over the output units was 2.5×10^{-3} on the training set and 1.8×10^{-2} on the test set. The percentage of misclassified patterns was 0.14% on the training set (10 mistakes) and 5.0% on the test set (102 mistakes). As can be seen in Figure 3, the convergence is extremely quick, and shows that backpropagation *can* be used on fairly large tasks with reasonable training times. This is due in part to the high redundancy of real data.

In a realistic application, the user usually is interested in the number of rejections necessary to reach a given level of accuracy rather than in the raw error rate. We measured the percentage of test patterns that must be rejected in order to get 1% error rate on the *remaining* test patterns. Our main rejection criterion was that the difference between the activity levels of the two most active units should exceed a given threshold.

The percentage of rejections was then 12.1% for 1% classification error on the remaining (nonrejected) test patterns. It should be emphasized that the rejection thresholds were obtained using performance measures on the *test set*.

Some kernels synthesized by the network can be interpreted as feature detectors remarkably similar to those found to exist in biological vision systems (Hubel and Wiesel 1962) and/or designed into previous artificial character recognizers, such as spatial derivative estimators or off-center/on-surround type feature detectors.

Most misclassifications are due to erroneous segmentation of the image into individual characters. Segmentation is a very difficult problem, especially when the characters overlap extensively. Other mistakes are due to ambiguous patterns, low-resolution effects, or writing styles not present in the training set.

Other networks with fewer feature maps were tried, but produced worse results. Various fully connected, unconstrained networks were also tried, but generalization performances were quite bad. For example, a fully connected network with one hidden layer of 40 units (10, 690 connections total) gave the following results: 1.6% misclassification on the

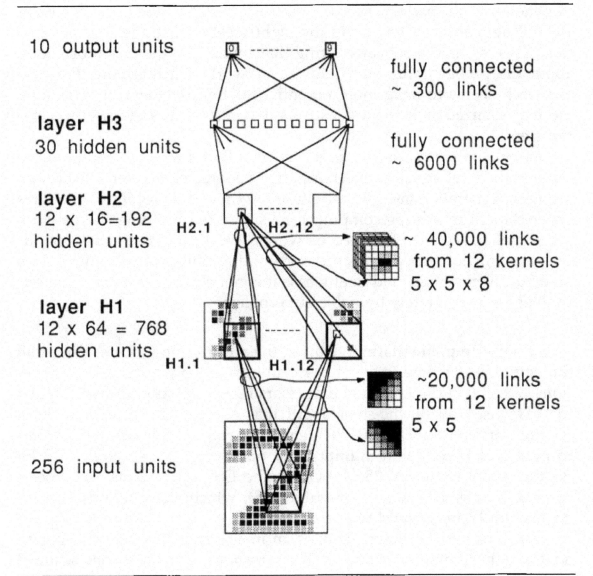

10 output units   fully connected ~ 300 links

layer H3
30 hidden units   fully connected ~ 6000 links

layer H2
12 x 16=192 hidden units

H2.1   H2.12   ~ 40,000 links from 12 kernels 5 x 5 x 8

layer H1
12 x 64 = 768 hidden units

H1.1   H1.12   ~20,000 links from 12 kernels 5 x 5

256 input units

Figure 3: Log mean squared error (MSE) (top) and raw error rate (bottom) versus number of training passes.

training set, 8.1% misclassifications on the test set, and 19.4% rejections for 1% error rate on the remaining test patterns. A full comparative study will be described in another paper.

5.1 Comparison with Other Work. The first several stages of processing in our previous system (described in Denker *et al.* 1989) involved convolutions in which the coefficients had been laboriously hand designed. In the present system, the first two layers of the network are constrained to be convolutional, but the system automatically learns the coefficients that make up the kernels. This "constrained backpropagation" is the key to success of the present system: it not only builds in shift-invariance, but vastly reduces the entropy, the Vapnik–Chervonenkis dimensionality, and the number of free parameters, thereby proportionately reducing the amount of training data required to achieve a given level of generalization performance (Denker *et al.* 1987; Baum and Haussler 1989). The present system performs slightly better than the previous system. This is remarkable considering that much less specific information about the problem was built into the network. Furthermore, the new approach seems to have more potential for improvement by designing more specialized architectures with more connections and fewer free parameters. [1]

Waibel (1989) describes a large network (but still small compared to ours) with about 18,000 connections and 1800 free parameters, trained on a speech recognition task. Because training time was prohibitive (18 days on an Alliant mini-supercomputer), he suggested building the network from smaller, separately trained networks. We did not need such a modular construction procedure since our training times were "only" 3 days on a Sun workstation, and in any case it is not clear how to partition our problem into separately trainable subproblems.

5.2 DSP Implementation. During the recognition process, almost all the computation time is spent performing multiply accumulate operations, a task that digital signal processors (DSP) are specifically designed for. We used an off-the-shelf board that contains 256 kbytes of local memory and an AT&T DSP-32C general purpose DSP with a peak performance of 12.5 million multiply add operations per second on 32 bit floating point numbers (25 MFLOPS). The DSP operates as a coprocessor; the host is a personal computer (PC), which also contains a video acquisition board connected to a camera.

The personal computer digitizes an image and binarizes it using an adaptive thresholding technique. The thresholded image is then scanned and each connected component (or segment) is isolated. Components that are too small or too large are discarded; remaining components are sent to the DSP for normalization and recognition. The PC gives a variable sized pixel map representation of a single digit to the DSP, which performs the normalization and the classification.

The overall throughput of the digit recognizer including image acquisition is 10 to 12 classifications per second and is limited mainly by the normalization step. On normalized digits, the DSP performs more than 30 classifications per second.

6 Conclusion

We have successfully applied backpropagation learning to a large, real-world task. Our results appear to be at the state of the art in digit recognition. Our network was trained on a low-level representation of

---

[1] A network similar to the one described here with 100,000 connections and 2600 free parameters recently achieved 9% rejection for 1% error rate. That is about 30% better than the best of the hand-coded-kernel networks.

data that had minimal preprocessing (as opposed to elaborate feature extraction). The network had many connections but relatively few free parameters. The network architecture and the constraints on the weights were designed to incorporate geometric knowledge about the task into the system. Because of the redundant nature of the data and because of the constraints imposed on the network, the learning time was relatively short considering the size of the training set. Scaling properties were far better than one would expect just from extrapolating results of backpropagation on smaller, artificial problems.

The final network of connections and weights obtained by backpropagation learning was readily implementable on commercial digital signal processing hardware. Throughput rates, from camera to classified image, of more than 10 digits per second were obtained.

This work points out the necessity of having flexible "network design" software tools that ease the design of complex, specialized network architectures.

Acknowledgments

We thank the U.S. Postal Service and its contractors for providing us with the data base. The Neural Network simulator SN is the result of a collaboration between Léon-Yves Bottou and Yann LeCun.

References

Baum, E. B., and Haussler, D. 1989. What size net gives valid generaliztion? *Neural Comp.* **1**, 151–160.

Becker, S., and LeCun, Y. 1988. *Improving the Convergence of Back-Propagation Learning With Second-Order Methods*. Tech. Rep. CRG-TR-88-5, University of Toronto Connectionist Research Group.

Bottou, L.-Y., and LeCun, Y. 1988. Sn: A simulator for connectionist models. In *Proceedings of NeuroNimes 88*, Nimes, France.

Denker, J., Schwartz, D., Wittner, B., Solla, S. A., Howard, R., Jackel, L., and Hopfield, J. 1987. Large automatic learning, rule extraction and generalization. *Complex Syst.* **1**, 877–922.

Denker, J. S., Gardner, W. R., Graf, H. P., Henderson, D., Howard, R. E., Hubbard, W., Jackel, L. D., Baird, H. S., and Guyon, I. 1989. Neural network recognizer for hand-written zip code digits. In D. Touretzky, ed., *Advances in Neural Information Processing Systems*, pp. 323–331. Morgan Kaufmann, San Mateo, CA.

Fukushima, K. 1980. Neocognitron: A self-organizing neural network model for a mechanism of pattern recognition unaffected by shift in position. *Biol. Cybernet.* **36**, 193–202.

Hubel, D. H., and Wiesel, T. N. 1962. Receptive fields, binocular interaction and functional architecture in the cat's visual cortex. *J. of Physiol.* **160**, 106–154.

LeCun, Y. 1987. Modèles connexionnistes de l'apprentissage. Ph.D. thesis, Université Pierre et Marie Curie, Paris, France.

LeCun, Y. 1989. Generalization and network design strategies. In *Connectionism in Perspective*, R. Pfeifer, Z. Schreter, F. Fogelman, and L. Steels, eds. North-Holland, Amsterdam.

Mozer, M. C. 1987. Early parallel processing in reading: A connectionist approach. In *Attention and Performance, XII: The Psychology of Reading*, M. Coltheart, ed., Vol. XII, pp. 83–104. Erlbaum, Hillsdale, NY.

Patarnello, S., and Carnevali, P. 1987. Learning networks of neurons with boolean logic. *Europhys. Lett.* **4**(4), 503–508.

Rumelhart, D. E., Hinton, G. E., and Williams, R. J. 1986. Learning internal representations by error propagation. In *Parallel Distributed Processing: Explorations in the Microstructure of Cognition*, D. E. Rumelhart and J. L. McClelland, eds., Vol. I, pp. 318–362. Bradford Books, Cambridge, MA.

Tishby, N., Levin, E., and Solla, S. A. 1989. Consistent inference of probabilities in layered networks: Predictions and generalization. In *Proceedings of the International Joint Conference on Neural Networks*, Washington DC.

Waibel, A. 1989. Consonant recognition by modular construction of large phonemic time-delay neural networks. In *Advances in Neural Information Processing Systems*, D. Touretzky, ed., pp. 215–223. Morgan Kaufmann, San Mateo, CA.

Wang, C. H., and Srihari, S. N. 1988. A framework for object recognition in a visually complex environment and its application to locating address blocks on mail pieces. *Int. J. Computer Vision* **2**, 125.

Received 7 July 1989; accepted 12 September 1989.

Chapter 8

Deep Convolutional Neural Networks – 2012

Context

By any standards, progress on solving image recognition problems over the past few decades has been extraordinary. This surge in progress began with LeCun et al.'s five-layer convolutional network[63] discussed in Chapter 7, which demonstrated the potential of neural networks to classify images of handwritten digits. In 2012, a seven-layer deep convolutional neural network, now known as *AlexNet*, was published by Alex Krizhevsky, Ilya Sutskever and Geoffrey Hinton. AlexNet was the first convolutional network to win the *ImageNet challenge*. This paper represents a breakthrough in image recognition and precipitated the rapid adoption of deep learning by the computer vision community.

Technical Summary

8.1. AlexNet Architecture

The network had seven hidden layers, plus *max pooling* layers (see Section 7.2), where the first few layers were convolutional (see Figure 8.1a). The images used to evaluate performance were from the standard ImageNet data set. The number of images in this set varies from year to year, but is never less than a million. In 2012, the training set consisted of 1.2 million 256×256 training images from 1,000 different classes, and a separate test set contained 100,000 images.

8.2. Training

In order to increase the effective size of the training set, the network was trained using backprop on random 224×224 patches from the images. The authors claimed that a key contribution to AlexNet's

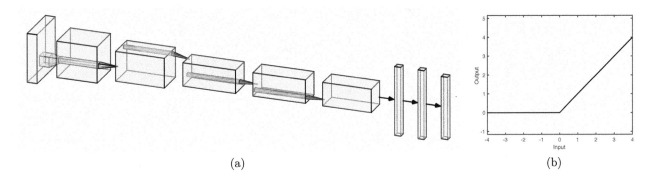

(a)   (b)

Figure 8.1. (a) Schematic diagram of AlexNet. Reproduced with permission from http://alexlenail.me/NN-SVG/AlexNet.html. (b) The rectified linear unit (ReLU) activation function.

success is the use of *rectified linear units* (ReLU, Figure 8.1b) in every layer. In order to reduce *over-fitting*, a technique called *dropout* was used. This means that half of the hidden units in a layer are randomly removed during each update of the weights. This dropout technique is remarkably effective, but the reasons for its effectiveness remain unclear[37].

8.3. Results

Classification performance is often reported as a percentage based on a 'top-n' threshold. After training, the probability that a given input belongs to each of 1,000 classes is computed. If inputs are classified as belonging to the top five classes 80% of the time, this is reported as a top-5 performance of 80%. A top-1 performance specifies how often the network classification is the correct class label. On the test data, AlexNet achieved a top-5 error of 17.0% and a top-1 error of 37.5%.

Research Paper:
ImageNet Classification With Deep Convolutional Neural Networks

Reference: Krizhevsky, A., Sutskever, I., and Hinton, G. E. (2012). ImageNet classification with deep convolutional neural networks. In *Advances in Neural Information Processing Systems 25 (NIPS 2012)*, pages 1097–1105.
https://papers.nips.cc/paper/2012/hash/c399862d3b9d6b76c8436e924a68c45b-Abstract.html.
Reproduced with permission.

ImageNet Classification with Deep Convolutional Neural Networks

Alex Krizhevsky
University of Toronto
kriz@cs.utoronto.ca

Ilya Sutskever
University of Toronto
ilya@cs.utoronto.ca

Geoffrey E. Hinton
University of Toronto
hinton@cs.utoronto.ca

Abstract

We trained a large, deep convolutional neural network to classify the 1.2 million high-resolution images in the ImageNet LSVRC-2010 contest into the 1000 different classes. On the test data, we achieved top-1 and top-5 error rates of 37.5% and 17.0% which is considerably better than the previous state-of-the-art. The neural network, which has 60 million parameters and 650,000 neurons, consists of five convolutional layers, some of which are followed by max-pooling layers, and three fully-connected layers with a final 1000-way softmax. To make training faster, we used non-saturating neurons and a very efficient GPU implementation of the convolution operation. To reduce overfitting in the fully-connected layers we employed a recently-developed regularization method called "dropout" that proved to be very effective. We also entered a variant of this model in the ILSVRC-2012 competition and achieved a winning top-5 test error rate of 15.3%, compared to 26.2% achieved by the second-best entry.

1 Introduction

Current approaches to object recognition make essential use of machine learning methods. To improve their performance, we can collect larger datasets, learn more powerful models, and use better techniques for preventing overfitting. Until recently, datasets of labeled images were relatively small — on the order of tens of thousands of images (e.g., NORB [16], Caltech-101/256 [8, 9], and CIFAR-10/100 [12]). Simple recognition tasks can be solved quite well with datasets of this size, especially if they are augmented with label-preserving transformations. For example, the current-best error rate on the MNIST digit-recognition task (<0.3%) approaches human performance [4]. But objects in realistic settings exhibit considerable variability, so to learn to recognize them it is necessary to use much larger training sets. And indeed, the shortcomings of small image datasets have been widely recognized (e.g., Pinto et al. [21]), but it has only recently become possible to collect labeled datasets with millions of images. The new larger datasets include LabelMe [23], which consists of hundreds of thousands of fully-segmented images, and ImageNet [6], which consists of over 15 million labeled high-resolution images in over 22,000 categories.

To learn about thousands of objects from millions of images, we need a model with a large learning capacity. However, the immense complexity of the object recognition task means that this problem cannot be specified even by a dataset as large as ImageNet, so our model should also have lots of prior knowledge to compensate for all the data we don't have. Convolutional neural networks (CNNs) constitute one such class of models [16, 11, 13, 18, 15, 22, 26]. Their capacity can be controlled by varying their depth and breadth, and they also make strong and mostly correct assumptions about the nature of images (namely, stationarity of statistics and locality of pixel dependencies). Thus, compared to standard feedforward neural networks with similarly-sized layers, CNNs have much fewer connections and parameters and so they are easier to train, while their theoretically-best performance is likely to be only slightly worse.

197

Despite the attractive qualities of CNNs, and despite the relative efficiency of their local architecture, they have still been prohibitively expensive to apply in large scale to high-resolution images. Luckily, current GPUs, paired with a highly-optimized implementation of 2D convolution, are powerful enough to facilitate the training of interestingly-large CNNs, and recent datasets such as ImageNet contain enough labeled examples to train such models without severe overfitting.

The specific contributions of this paper are as follows: we trained one of the largest convolutional neural networks to date on the subsets of ImageNet used in the ILSVRC-2010 and ILSVRC-2012 competitions [2] and achieved by far the best results ever reported on these datasets. We wrote a highly-optimized GPU implementation of 2D convolution and all the other operations inherent in training convolutional neural networks, which we make available publicly[1]. Our network contains a number of new and unusual features which improve its performance and reduce its training time, which are detailed in Section 3. The size of our network made overfitting a significant problem, even with 1.2 million labeled training examples, so we used several effective techniques for preventing overfitting, which are described in Section 4. Our final network contains five convolutional and three fully-connected layers, and this depth seems to be important: we found that removing any convolutional layer (each of which contains no more than 1% of the model's parameters) resulted in inferior performance.

In the end, the network's size is limited mainly by the amount of memory available on current GPUs and by the amount of training time that we are willing to tolerate. Our network takes between five and six days to train on two GTX 580 3GB GPUs. All of our experiments suggest that our results can be improved simply by waiting for faster GPUs and bigger datasets to become available.

2 The Dataset

ImageNet is a dataset of over 15 million labeled high-resolution images belonging to roughly 22,000 categories. The images were collected from the web and labeled by human labelers using Amazon's Mechanical Turk crowd-sourcing tool. Starting in 2010, as part of the Pascal Visual Object Challenge, an annual competition called the ImageNet Large-Scale Visual Recognition Challenge (ILSVRC) has been held. ILSVRC uses a subset of ImageNet with roughly 1000 images in each of 1000 categories. In all, there are roughly 1.2 million training images, 50,000 validation images, and 150,000 testing images.

ILSVRC-2010 is the only version of ILSVRC for which the test set labels are available, so this is the version on which we performed most of our experiments. Since we also entered our model in the ILSVRC-2012 competition, in Section 6 we report our results on this version of the dataset as well, for which test set labels are unavailable. On ImageNet, it is customary to report two error rates: top-1 and top-5, where the top-5 error rate is the fraction of test images for which the correct label is not among the five labels considered most probable by the model.

ImageNet consists of variable-resolution images, while our system requires a constant input dimensionality. Therefore, we down-sampled the images to a fixed resolution of 256×256. Given a rectangular image, we first rescaled the image such that the shorter side was of length 256, and then cropped out the central 256×256 patch from the resulting image. We did not pre-process the images in any other way, except for subtracting the mean activity over the training set from each pixel. So we trained our network on the (centered) raw RGB values of the pixels.

3 The Architecture

The architecture of our network is summarized in Figure 2. It contains eight learned layers — five convolutional and three fully-connected. Below, we describe some of the novel or unusual features of our network's architecture. Sections 3.1-3.4 are sorted according to our estimation of their importance, with the most important first.

---

[1] http://code.google.com/p/cuda-convnet/

3.1 ReLU Nonlinearity

The standard way to model a neuron's output f as a function of its input x is with $f(x) = \tanh(x)$ or $f(x) = (1 + e^{-x})^{-1}$. In terms of training time with gradient descent, these saturating nonlinearities are much slower than the non-saturating nonlinearity $f(x) = \max(0, x)$. Following Nair and Hinton [20], we refer to neurons with this nonlinearity as Rectified Linear Units (ReLUs). Deep convolutional neural networks with ReLUs train several times faster than their equivalents with tanh units. This is demonstrated in Figure 1, which shows the number of iterations required to reach 25% training error on the CIFAR-10 dataset for a particular four-layer convolutional network. This plot shows that we would not have been able to experiment with such large neural networks for this work if we had used traditional saturating neuron models.

We are not the first to consider alternatives to traditional neuron models in CNNs. For example, Jarrett et al. [11] claim that the nonlinearity $f(x) = |\tanh(x)|$ works particularly well with their type of contrast normalization followed by local average pooling on the Caltech-101 dataset. However, on this dataset the primary concern is preventing overfitting, so the effect they are observing is different from the accelerated ability to fit the training set which we report when using ReLUs. Faster learning has a great influence on the performance of large models trained on large datasets.

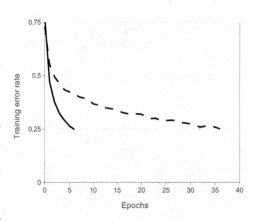

Figure 1: A four-layer convolutional neural network with ReLUs (**solid line**) reaches a 25% training error rate on CIFAR-10 six times faster than an equivalent network with tanh neurons (**dashed line**). The learning rates for each network were chosen independently to make training as fast as possible. No regularization of any kind was employed. The magnitude of the effect demonstrated here varies with network architecture, but networks with ReLUs consistently learn several times faster than equivalents with saturating neurons.

3.2 Training on Multiple GPUs

A single GTX 580 GPU has only 3GB of memory, which limits the maximum size of the networks that can be trained on it. It turns out that 1.2 million training examples are enough to train networks which are too big to fit on one GPU. Therefore we spread the net across two GPUs. Current GPUs are particularly well-suited to cross-GPU parallelization, as they are able to read from and write to one another's memory directly, without going through host machine memory. The parallelization scheme that we employ essentially puts half of the kernels (or neurons) on each GPU, with one additional trick: the GPUs communicate only in certain layers. This means that, for example, the kernels of layer 3 take input from all kernel maps in layer 2. However, kernels in layer 4 take input only from those kernel maps in layer 3 which reside on the same GPU. Choosing the pattern of connectivity is a problem for cross-validation, but this allows us to precisely tune the amount of communication until it is an acceptable fraction of the amount of computation.

The resultant architecture is somewhat similar to that of the "columnar" CNN employed by Cireşan et al. [5], except that our columns are not independent (see Figure 2). This scheme reduces our top-1 and top-5 error rates by 1.7% and 1.2%, respectively, as compared with a net with half as many kernels in each convolutional layer trained on one GPU. The two-GPU net takes slightly less time to train than the one-GPU net[2].

---

[2]The one-GPU net actually has the same number of kernels as the two-GPU net in the final convolutional layer. This is because most of the net's parameters are in the first fully-connected layer, which takes the last convolutional layer as input. So to make the two nets have approximately the same number of parameters, we did not halve the size of the final convolutional layer (nor the fully-conneced layers which follow). Therefore this comparison is biased in favor of the one-GPU net, since it is bigger than "half the size" of the two-GPU net.

3.3 Local Response Normalization

ReLUs have the desirable property that they do not require input normalization to prevent them from saturating. If at least some training examples produce a positive input to a ReLU, learning will happen in that neuron. However, we still find that the following local normalization scheme aids generalization. Denoting by $a^i_{x,y}$ the activity of a neuron computed by applying kernel i at position (x, y) and then applying the ReLU nonlinearity, the response-normalized activity $b^i_{x,y}$ is given by the expression

$$b^i_{x,y} = a^i_{x,y} / \left(k + \alpha \sum_{j=\max(0,i-n/2)}^{\min(N-1,i+n/2)} (a^j_{x,y})^2 \right)^\beta$$

where the sum runs over n "adjacent" kernel maps at the same spatial position, and N is the total number of kernels in the layer. The ordering of the kernel maps is of course arbitrary and determined before training begins. This sort of response normalization implements a form of lateral inhibition inspired by the type found in real neurons, creating competition for big activities amongst neuron outputs computed using different kernels. The constants k, n, α, and β are hyper-parameters whose values are determined using a validation set; we used $k = 2$, $n = 5$, $\alpha = 10^{-4}$, and $\beta = 0.75$. We applied this normalization after applying the ReLU nonlinearity in certain layers (see Section 3.5).

This scheme bears some resemblance to the local contrast normalization scheme of Jarrett et al. [11], but ours would be more correctly termed "brightness normalization", since we do not subtract the mean activity. Response normalization reduces our top-1 and top-5 error rates by 1.4% and 1.2%, respectively. We also verified the effectiveness of this scheme on the CIFAR-10 dataset: a four-layer CNN achieved a 13% test error rate without normalization and 11% with normalization[3].

3.4 Overlapping Pooling

Pooling layers in CNNs summarize the outputs of neighboring groups of neurons in the same kernel map. Traditionally, the neighborhoods summarized by adjacent pooling units do not overlap (e.g., [17, 11, 4]). To be more precise, a pooling layer can be thought of as consisting of a grid of pooling units spaced s pixels apart, each summarizing a neighborhood of size $z \times z$ centered at the location of the pooling unit. If we set $s = z$, we obtain traditional local pooling as commonly employed in CNNs. If we set $s < z$, we obtain overlapping pooling. This is what we use throughout our network, with $s = 2$ and $z = 3$. This scheme reduces the top-1 and top-5 error rates by 0.4% and 0.3%, respectively, as compared with the non-overlapping scheme $s = 2, z = 2$, which produces output of equivalent dimensions. We generally observe during training that models with overlapping pooling find it slightly more difficult to overfit.

3.5 Overall Architecture

Now we are ready to describe the overall architecture of our CNN. As depicted in Figure 2, the net contains eight layers with weights; the first five are convolutional and the remaining three are fully-connected. The output of the last fully-connected layer is fed to a 1000-way softmax which produces a distribution over the 1000 class labels. Our network maximizes the multinomial logistic regression objective, which is equivalent to maximizing the average across training cases of the log-probability of the correct label under the prediction distribution.

The kernels of the second, fourth, and fifth convolutional layers are connected only to those kernel maps in the previous layer which reside on the same GPU (see Figure 2). The kernels of the third convolutional layer are connected to all kernel maps in the second layer. The neurons in the fully-connected layers are connected to all neurons in the previous layer. Response-normalization layers follow the first and second convolutional layers. Max-pooling layers, of the kind described in Section 3.4, follow both response-normalization layers as well as the fifth convolutional layer. The ReLU non-linearity is applied to the output of every convolutional and fully-connected layer.

The first convolutional layer filters the $224 \times 224 \times 3$ input image with 96 kernels of size $11 \times 11 \times 3$ with a stride of 4 pixels (this is the distance between the receptive field centers of neighboring

---

[3]We cannot describe this network in detail due to space constraints, but it is specified precisely by the code and parameter files provided here: http://code.google.com/p/cuda-convnet/.

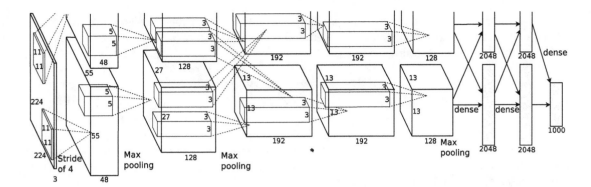

Figure 2: An illustration of the architecture of our CNN, explicitly showing the delineation of responsibilities between the two GPUs. One GPU runs the layer-parts at the top of the figure while the other runs the layer-parts at the bottom. The GPUs communicate only at certain layers. The network's input is 150,528-dimensional, and the number of neurons in the network's remaining layers is given by 253,440–186,624–64,896–64,896–43,264–4096–4096–1000.

neurons in a kernel map). The second convolutional layer takes as input the (response-normalized and pooled) output of the first convolutional layer and filters it with 256 kernels of size $5 \times 5 \times 48$. The third, fourth, and fifth convolutional layers are connected to one another without any intervening pooling or normalization layers. The third convolutional layer has 384 kernels of size $3 \times 3 \times 256$ connected to the (normalized, pooled) outputs of the second convolutional layer. The fourth convolutional layer has 384 kernels of size $3 \times 3 \times 192$, and the fifth convolutional layer has 256 kernels of size $3 \times 3 \times 192$. The fully-connected layers have 4096 neurons each.

4   Reducing Overfitting

Our neural network architecture has 60 million parameters. Although the 1000 classes of ILSVRC make each training example impose 10 bits of constraint on the mapping from image to label, this turns out to be insufficient to learn so many parameters without considerable overfitting. Below, we describe the two primary ways in which we combat overfitting.

4.1   Data Augmentation

The easiest and most common method to reduce overfitting on image data is to artificially enlarge the dataset using label-preserving transformations (e.g., [25, 4, 5]). We employ two distinct forms of data augmentation, both of which allow transformed images to be produced from the original images with very little computation, so the transformed images do not need to be stored on disk. In our implementation, the transformed images are generated in Python code on the CPU while the GPU is training on the previous batch of images. So these data augmentation schemes are, in effect, computationally free.

The first form of data augmentation consists of generating image translations and horizontal reflections. We do this by extracting random 224×224 patches (and their horizontal reflections) from the 256×256 images and training our network on these extracted patches[4]. This increases the size of our training set by a factor of 2048, though the resulting training examples are, of course, highly interdependent. Without this scheme, our network suffers from substantial overfitting, which would have forced us to use much smaller networks. At test time, the network makes a prediction by extracting five 224×224 patches (the four corner patches and the center patch) as well as their horizontal reflections (hence ten patches in all), and averaging the predictions made by the network's softmax layer on the ten patches.

The second form of data augmentation consists of altering the intensities of the RGB channels in training images. Specifically, we perform PCA on the set of RGB pixel values throughout the ImageNet training set. To each training image, we add multiples of the found principal components,

---

[4]This is the reason why the input images in Figure 2 are $224 \times 224 \times 3$-dimensional.

with magnitudes proportional to the corresponding eigenvalues times a random variable drawn from a Gaussian with mean zero and standard deviation 0.1. Therefore to each RGB image pixel $I_{xy} = [I_{xy}^R, I_{xy}^G, I_{xy}^B]^T$ we add the following quantity:

$$[\mathbf{p}_1, \mathbf{p}_2, \mathbf{p}_3][\alpha_1\lambda_1, \alpha_2\lambda_2, \alpha_3\lambda_3]^T$$

where $\mathbf{p}_i$ and λ_i are ith eigenvector and eigenvalue of the 3×3 covariance matrix of RGB pixel values, respectively, and α_i is the aforementioned random variable. Each α_i is drawn only once for all the pixels of a particular training image until that image is used for training again, at which point it is re-drawn. This scheme approximately captures an important property of natural images, namely, that object identity is invariant to changes in the intensity and color of the illumination. This scheme reduces the top-1 error rate by over 1%.

4.2 Dropout

Combining the predictions of many different models is a very successful way to reduce test errors [1, 3], but it appears to be too expensive for big neural networks that already take several days to train. There is, however, a very efficient version of model combination that only costs about a factor of two during training. The recently-introduced technique, called "dropout" [10], consists of setting to zero the output of each hidden neuron with probability 0.5. The neurons which are "dropped out" in this way do not contribute to the forward pass and do not participate in back-propagation. So every time an input is presented, the neural network samples a different architecture, but all these architectures share weights. This technique reduces complex co-adaptations of neurons, since a neuron cannot rely on the presence of particular other neurons. It is, therefore, forced to learn more robust features that are useful in conjunction with many different random subsets of the other neurons. At test time, we use all the neurons but multiply their outputs by 0.5, which is a reasonable approximation to taking the geometric mean of the predictive distributions produced by the exponentially-many dropout networks.

We use dropout in the first two fully-connected layers of Figure 2. Without dropout, our network exhibits substantial overfitting. Dropout roughly doubles the number of iterations required to converge.

5 Details of learning

We trained our models using stochastic gradient descent with a batch size of 128 examples, momentum of 0.9, and weight decay of 0.0005. We found that this small amount of weight decay was important for the model to learn. In other words, weight decay here is not merely a regularizer: it reduces the model's training error. The update rule for weight w was

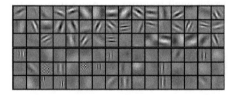

Figure 3: 96 convolutional kernels of size $11 \times 11 \times 3$ learned by the first convolutional layer on the $224 \times 224 \times 3$ input images. The top 48 kernels were learned on GPU 1 while the bottom 48 kernels were learned on GPU 2. See Section 6.1 for details.

$$v_{i+1} := 0.9 \cdot v_i - 0.0005 \cdot \epsilon \cdot w_i - \epsilon \cdot \left\langle \frac{\partial L}{\partial w}\Big|_{w_i} \right\rangle_{D_i}$$

$$w_{i+1} := w_i + v_{i+1}$$

where i is the iteration index, v is the momentum variable, ϵ is the learning rate, and $\left\langle \frac{\partial L}{\partial w}\big|_{w_i} \right\rangle_{D_i}$ is the average over the ith batch D_i of the derivative of the objective with respect to w, evaluated at w_i.

We initialized the weights in each layer from a zero-mean Gaussian distribution with standard deviation 0.01. We initialized the neuron biases in the second, fourth, and fifth convolutional layers, as well as in the fully-connected hidden layers, with the constant 1. This initialization accelerates the early stages of learning by providing the ReLUs with positive inputs. We initialized the neuron biases in the remaining layers with the constant 0.

We used an equal learning rate for all layers, which we adjusted manually throughout training. The heuristic which we followed was to divide the learning rate by 10 when the validation error rate stopped improving with the current learning rate. The learning rate was initialized at 0.01 and reduced three times prior to termination. We trained the network for roughly 90 cycles through the training set of 1.2 million images, which took five to six days on two NVIDIA GTX 580 3GB GPUs.

6 Results

Our results on ILSVRC-2010 are summarized in Table 1. Our network achieves top-1 and top-5 test set error rates of **37.5%** and **17.0%**[5]. The best performance achieved during the ILSVRC-2010 competition was 47.1% and 28.2% with an approach that averages the predictions produced from six sparse-coding models trained on different features [2], and since then the best published results are 45.7% and 25.7% with an approach that averages the predictions of two classifiers trained on Fisher Vectors (FVs) computed from two types of densely-sampled features [24].

We also entered our model in the ILSVRC-2012 competition and report our results in Table 2. Since the ILSVRC-2012 test set labels are not publicly available, we cannot report test error rates for all the models that we tried. In the remainder of this paragraph, we use validation and test error rates interchangeably because in our experience they do not differ by more than 0.1% (see Table 2). The CNN described in this paper achieves a top-5 error rate of 18.2%. Averaging the predictions

Model	Top-1	Top-5
Sparse coding [2]	*47.1%*	*28.2%*
SIFT + FVs [24]	*45.7%*	*25.7%*
CNN	**37.5%**	**17.0%**

Table 1: Comparison of results on ILSVRC-2010 test set. In *italics* are best results achieved by others.

of five similar CNNs gives an error rate of 16.4%. Training one CNN, with an extra sixth convolutional layer over the last pooling layer, to classify the entire ImageNet Fall 2011 release (15M images, 22K categories), and then "fine-tuning" it on ILSVRC-2012 gives an error rate of 16.6%. Averaging the predictions of two CNNs that were pre-trained on the entire Fall 2011 release with the aforementioned five CNNs gives an error rate of **15.3%**. The second-best contest entry achieved an error rate of 26.2% with an approach that averages the predictions of several classifiers trained on FVs computed from different types of densely-sampled features [7].

Finally, we also report our error rates on the Fall 2009 version of ImageNet with 10,184 categories and 8.9 million images. On this dataset we follow the convention in the literature of using half of the images for training and half for testing. Since there is no established test set, our split necessarily differs from the splits used by previous authors, but this does not affect the results appreciably. Our top-1 and top-5 error rates on this dataset are **67.4%** and

Model	Top-1 (val)	Top-5 (val)	Top-5 (test)
SIFT + FVs [7]	—	—	*26.2%*
1 CNN	40.7%	18.2%	—
5 CNNs	38.1%	16.4%	**16.4%**
1 CNN*	39.0%	16.6%	—
7 CNNs*	36.7%	15.4%	**15.3%**

Table 2: Comparison of error rates on ILSVRC-2012 validation and test sets. In *italics* are best results achieved by others. Models with an asterisk* were "pre-trained" to classify the entire ImageNet 2011 Fall release. See Section 6 for details.

40.9%, attained by the net described above but with an additional, sixth convolutional layer over the last pooling layer. The best published results on this dataset are 78.1% and 60.9% [19].

6.1 Qualitative Evaluations

Figure 3 shows the convolutional kernels learned by the network's two data-connected layers. The network has learned a variety of frequency- and orientation-selective kernels, as well as various colored blobs. Notice the specialization exhibited by the two GPUs, a result of the restricted connectivity described in Section 3.5. The kernels on GPU 1 are largely color-agnostic, while the kernels on on GPU 2 are largely color-specific. This kind of specialization occurs during every run and is independent of any particular random weight initialization (modulo a renumbering of the GPUs).

---

[5]The error rates without averaging predictions over ten patches as described in Section 4.1 are 39.0% and 18.3%.

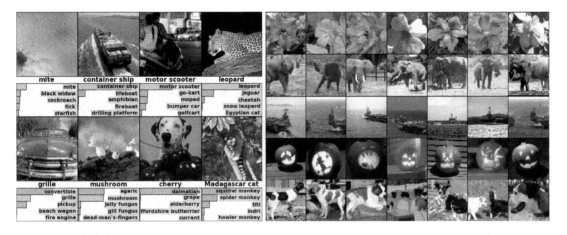

Figure 4: **(Left)** Eight ILSVRC-2010 test images and the five labels considered most probable by our model. The correct label is written under each image, and the probability assigned to the correct label is also shown with a red bar (if it happens to be in the top 5). **(Right)** Five ILSVRC-2010 test images in the first column. The remaining columns show the six training images that produce feature vectors in the last hidden layer with the smallest Euclidean distance from the feature vector for the test image.

In the left panel of Figure 4 we qualitatively assess what the network has learned by computing its top-5 predictions on eight test images. Notice that even off-center objects, such as the mite in the top-left, can be recognized by the net. Most of the top-5 labels appear reasonable. For example, only other types of cat are considered plausible labels for the leopard. In some cases (grille, cherry) there is genuine ambiguity about the intended focus of the photograph.

Another way to probe the network's visual knowledge is to consider the feature activations induced by an image at the last, 4096-dimensional hidden layer. If two images produce feature activation vectors with a small Euclidean separation, we can say that the higher levels of the neural network consider them to be similar. Figure 4 shows five images from the test set and the six images from the training set that are most similar to each of them according to this measure. Notice that at the pixel level, the retrieved training images are generally not close in L2 to the query images in the first column. For example, the retrieved dogs and elephants appear in a variety of poses. We present the results for many more test images in the supplementary material.

Computing similarity by using Euclidean distance between two 4096-dimensional, real-valued vectors is inefficient, but it could be made efficient by training an auto-encoder to compress these vectors to short binary codes. This should produce a much better image retrieval method than applying auto-encoders to the raw pixels [14], which does not make use of image labels and hence has a tendency to retrieve images with similar patterns of edges, whether or not they are semantically similar.

7 Discussion

Our results show that a large, deep convolutional neural network is capable of achieving record-breaking results on a highly challenging dataset using purely supervised learning. It is notable that our network's performance degrades if a single convolutional layer is removed. For example, removing any of the middle layers results in a loss of about 2% for the top-1 performance of the network. So the depth really is important for achieving our results.

To simplify our experiments, we did not use any unsupervised pre-training even though we expect that it will help, especially if we obtain enough computational power to significantly increase the size of the network without obtaining a corresponding increase in the amount of labeled data. Thus far, our results have improved as we have made our network larger and trained it longer but we still have many orders of magnitude to go in order to match the infero-temporal pathway of the human visual system. Ultimately we would like to use very large and deep convolutional nets on video sequences where the temporal structure provides very helpful information that is missing or far less obvious in static images.

References

[1] R.M. Bell and Y. Koren. Lessons from the netflix prize challenge. *ACM SIGKDD Explorations Newsletter*, 9(2):75–79, 2007.

[2] A. Berg, J. Deng, and L. Fei-Fei. Large scale visual recognition challenge 2010. www.image-net.org/challenges. 2010.

[3] L. Breiman. Random forests. *Machine learning*, 45(1):5–32, 2001.

[4] D. Cireşan, U. Meier, and J. Schmidhuber. Multi-column deep neural networks for image classification. *Arxiv preprint arXiv:1202.2745*, 2012.

[5] D.C. Cireşan, U. Meier, J. Masci, L.M. Gambardella, and J. Schmidhuber. High-performance neural networks for visual object classification. *Arxiv preprint arXiv:1102.0183*, 2011.

[6] J. Deng, W. Dong, R. Socher, L.-J. Li, K. Li, and L. Fei-Fei. ImageNet: A Large-Scale Hierarchical Image Database. In *CVPR09*, 2009.

[7] J. Deng, A. Berg, S. Satheesh, H. Su, A. Khosla, and L. Fei-Fei. *ILSVRC-2012*, 2012. URL http://www.image-net.org/challenges/LSVRC/2012/.

[8] L. Fei-Fei, R. Fergus, and P. Perona. Learning generative visual models from few training examples: An incremental bayesian approach tested on 101 object categories. *Computer Vision and Image Understanding*, 106(1):59–70, 2007.

[9] G. Griffin, A. Holub, and P. Perona. Caltech-256 object category dataset. Technical Report 7694, California Institute of Technology, 2007. URL http://authors.library.caltech.edu/7694.

[10] G.E. Hinton, N. Srivastava, A. Krizhevsky, I. Sutskever, and R.R. Salakhutdinov. Improving neural networks by preventing co-adaptation of feature detectors. *arXiv preprint arXiv:1207.0580*, 2012.

[11] K. Jarrett, K. Kavukcuoglu, M. A. Ranzato, and Y. LeCun. What is the best multi-stage architecture for object recognition? In *International Conference on Computer Vision*, pages 2146–2153. IEEE, 2009.

[12] A. Krizhevsky. Learning multiple layers of features from tiny images. Master's thesis, Department of Computer Science, University of Toronto, 2009.

[13] A. Krizhevsky. Convolutional deep belief networks on cifar-10. *Unpublished manuscript*, 2010.

[14] A. Krizhevsky and G.E. Hinton. Using very deep autoencoders for content-based image retrieval. In *ESANN*, 2011.

[15] Y. Le Cun, B. Boser, J.S. Denker, D. Henderson, R.E. Howard, W. Hubbard, L.D. Jackel, et al. Handwritten digit recognition with a back-propagation network. In *Advances in neural information processing systems*, 1990.

[16] Y. LeCun, F.J. Huang, and L. Bottou. Learning methods for generic object recognition with invariance to pose and lighting. In *Computer Vision and Pattern Recognition, 2004. CVPR 2004. Proceedings of the 2004 IEEE Computer Society Conference on*, volume 2, pages II–97. IEEE, 2004.

[17] Y. LeCun, K. Kavukcuoglu, and C. Farabet. Convolutional networks and applications in vision. In *Circuits and Systems (ISCAS), Proceedings of 2010 IEEE International Symposium on*, pages 253–256. IEEE, 2010.

[18] H. Lee, R. Grosse, R. Ranganath, and A.Y. Ng. Convolutional deep belief networks for scalable unsupervised learning of hierarchical representations. In *Proceedings of the 26th Annual International Conference on Machine Learning*, pages 609–616. ACM, 2009.

[19] T. Mensink, J. Verbeek, F. Perronnin, and G. Csurka. Metric Learning for Large Scale Image Classification: Generalizing to New Classes at Near-Zero Cost. In *ECCV - European Conference on Computer Vision*, Florence, Italy, October 2012.

[20] V. Nair and G. E. Hinton. Rectified linear units improve restricted boltzmann machines. In *Proc. 27th International Conference on Machine Learning*, 2010.

[21] N. Pinto, D.D. Cox, and J.J. DiCarlo. Why is real-world visual object recognition hard? *PLoS computational biology*, 4(1):e27, 2008.

[22] N. Pinto, D. Doukhan, J.J. DiCarlo, and D.D. Cox. A high-throughput screening approach to discovering good forms of biologically inspired visual representation. *PLoS computational biology*, 5(11):e1000579, 2009.

[23] B.C. Russell, A. Torralba, K.P. Murphy, and W.T. Freeman. Labelme: a database and web-based tool for image annotation. *International journal of computer vision*, 77(1):157–173, 2008.

[24] J. Sánchez and F. Perronnin. High-dimensional signature compression for large-scale image classification. In *Computer Vision and Pattern Recognition (CVPR), 2011 IEEE Conference on*, pages 1665–1672. IEEE, 2011.

[25] P.Y. Simard, D. Steinkraus, and J.C. Platt. Best practices for convolutional neural networks applied to visual document analysis. In *Proceedings of the Seventh International Conference on Document Analysis and Recognition*, volume 2, pages 958–962, 2003.

[26] S.C. Turaga, J.F. Murray, V. Jain, F. Roth, M. Helmstaedter, K. Briggman, W. Denk, and H.S. Seung. Convolutional networks can learn to generate affinity graphs for image segmentation. *Neural Computation*, 22(2):511–538, 2010.

Chapter 9

Variational Autoencoders – 2013

Context

The main objective of a variational autoencoder is to squeeze high-dimensional data (e.g. images) through a low-dimensional information bottleneck. By definition, the bottleneck is narrow, which forces it to extract as much information as possible about the data (Kingma and Welling, 2014; Rezende et al., 2014; Alemi et al., 2017). A variational autoencoder consists of two modules, an encoder and a decoder, as shown in Figure 9.1. The encoder is a neural network that transforms each input image into a vector of unit states in its output layer, which represents the values of the *latent variables* in the input image. The decoder uses the states of units in the encoder output layer as its input, and attempts to generate an output that is a reconstruction the encoder's input image. Variational autoencoders do not seem to suffer from the problem of over-fitting, probably because the training method includes a regularisation term.

Variational autoencoders represent a synthesis of ideas immanent in generative models such as Boltzmann machine autoencoders (Section 4.7), RBM deep autoencoder networks (Hinton and Salakhutdinov, 2006), denoising autoencoders (Vincent et al., 2010) and the *information bottleneck* (Tishby et al., 2000), which have their roots in the idea of *analysis by synthesis* (Selfridge, 1958). Unlike Boltzmann machines, which rely only on computationally intensive *Gibbs sampling*, variational autoencoders employ fast variational methods. For readers unfamiliar with these terms, Gibbs sampling provides unbiased estimates of the network weights, but those estimates have high variance, whereas variational methods provide biased estimates that have low variance. In essence, this is the nature of the trade-off between sampling and variational methods. The method described here is based on Kingma and Welling (2014), and a more detailed account of different methods for learning in variational autoencoders can be found in Kim et al. (2018).

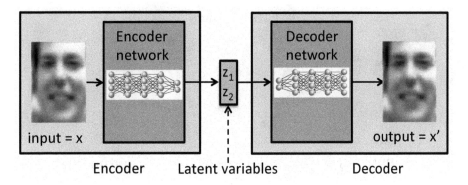

Figure 9.1. Schematic of a variational autoencoder, which comprises an encoder and a decoder. The encoder maps each input image X to several output unit states (z_1 and z_2), which are assumed to represent the latent variables. The outputs of the encoder act as inputs to the decoder, which produces an approximation X' of the input image. All images of faces in this chapter are reproduced from Kingma (2017) with permission.

Variational autoencoders represent a step-change in neural network algorithms for three reasons. First, they provide a relatively efficient method for estimating latent variables. Second, because each input is mapped to itself, variational autoencoders do not require training data that have been labelled. Third, variational autoencoders provide a coherent theoretical framework that includes a principled method for separating the reconstruction error from the regularisation term.

Technical Summary

9.1. Overview

The main problem solved by variational autoencoders is learning which physical latent variables describe the images in a training set. In practice, these physical latent variables are usually only implicit in the states of the encoder output units, but we will treat these output unit states as if they were physical latent variables for now.

The method used to learn is intimately related to a measure of how well latent variables have been extracted. Specifically, if a variational autoencoder has extracted the latent variables of a particular image, then we would expect that the image could be reconstructed based only on the values of these latent variables. In other words, if an image is mapped to its latent variables, then the values of these variables should be sufficient to produce an accurate rendition of the image.

Before we explore the details of the variational autoencoder, let's consider what we would expect under ideal conditions. In a perfect world, we want the variational autoencoder to extract the underlying latent variables implicit in an image. For example, given the input image of a child wearing a hat, the encoder output units should have states that explicitly represent the colour, texture and position of the hat in the image. Because the number of latent variables is small in relation to the number of pixels in a typical image, we expect that only a relatively small number of units will be required to represent the latent variables. In practice, this number can be as high as 100, but in order to visualise the latent variables, we can force the network to use just two latent variables, which are represented as two units in the encoder output layer.

Consider Figure 9.2, depicting an example used to train a variational autoencoder on images in which facial expression and head orientation varied from image to image. The number of hidden units in the input layer of the decoder was artificially restricted to two, which effectively forced the network to extract just two latent variables. The reason for restricting the number of decoder input units to two is that it enables us to visualise the nature of the latent variables being used by the variational autoencoder to reproduce the encoder input images at the decoder output layer. Because there are

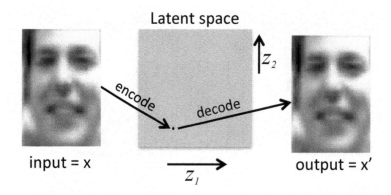

Figure 9.2. An input image X corresponds to one location $\mathbf{z} = (z_1, z_2)$ in a latent space of two variables z_1 and z_2. In this example, z_1 indicates the left–right head orientation, whereas z_2 indicates facial expression, as shown in Figure 9.3. The values of these two variables can then be used to recover an approximation X' to X.

just two decoder input units, the states of these units, $\mathbf{z} = (z_1, z_2)$, can be represented as points in a two-dimensional coordinate system. For example, if we fix the values of the decoder input units to $\mathbf{z} = (0.2, 0.1)$ then we can observe the image X' associated with this point as the output of the decoder when it is given the input $\mathbf{z}$. If our coordinate system has a range between 0 and 1 along each axis, then we can place X' at the location $\mathbf{z}$ used to generate X'. Of course, this exercise can be repeated for uniformly spaced intervals along the z_1 and z_2 axes, which defines a regular grid of m locations $\mathbf{z}_1, \ldots, \mathbf{z}_m$, such that each location corresponds to a unique decoder output. When presented with a vector $\mathbf{z}_i$, the decoder output image is placed at the location $\mathbf{z}_i = (z_{1i}, z_{2i})$, as in Figure 9.3. The result is a visual representation of the *embedding* of the input image into the space of two latent variables.

9.2. Latent Variables and Manifolds

The value of a two-element vector $\mathbf{z} = (z_1, z_2)$ can be plotted as a point on a plane (Figure 9.2). Similarly, an image of 10×10 pixels defines a vector $X = (x_1, \ldots, x_{100})$, which specifies a point in 100-dimensional space, denoted by $\mathbb{R}^{100}$. If we consider a large number of images of faces, say, we would not expect the corresponding points to be scattered at random in $\mathbb{R}^{100}$. In fact, the majority of these images will occupy a relatively small volume in $\mathbb{R}^{100}$ (see Section 9.4).

Suppose we restrict the images by including only photographs of a single face, but taken at different angles and with expressions that vary between sad and happy (e.g. Figure 9.3). In this case, we have effectively defined a set of images that can be described with the values of just two variables, $z_1 = $ orientation and $z_2 = $ expression. If we consider the images formed by varying head orientation, they define a smooth trajectory through $\mathbb{R}^{100}$. Because the relationship between head orientation and image pixel values is nonlinear, this smooth trajectory is a curved line in $\mathbb{R}^{100}$. Similarly, if we consider the images formed by varying facial expression, they define a different smooth curve through $\mathbb{R}^{100}$. Taken together, varying head orientation and facial expression defines a two-dimensional curved surface, or *manifold*, in $\mathbb{R}^{100}$, as shown in Figure 9.4. Note that this is not a vague analogy, but rather a precise account of how a population of images of 10×10 pixels represented in terms of two variables defines a two-dimensional manifold in $\mathbb{R}^{100}$.

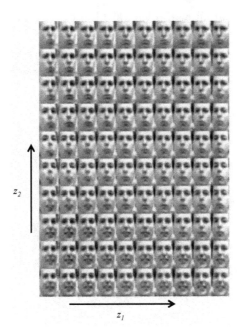

Figure 9.3. Embeddings of faces in two dimensions generated by the decoder. Head orientation varies from left to right, whereas facial expression varies along the vertical direction. From Kingma (2017) with permission.

9.3. Key Quantities

It is common practice in the literature to refer to the states of decoder input units as latent variables. For consistency, we follow that practice here, but we make a distinction between the *latent variables* $\mathbf{z}$ represented by decoder input states and *physical latent variables* such as colour and speed. The status of each variable is specified in parentheses (e.g. known, unknown, learned).

X_t Encoder input vector (e.g. image). The set of T training vectors is represented as $\{X\} = \{X_1, \ldots, X_T\}$ or simply X here. (Known.)

$p(X_t)$ The *marginal likelihood* or *model evidence* of the image X_t. When considered over all images, $p(X)$ is the probability distribution of the data to be learned. (Unknown; learned during training.)

J Number of latent variables, which are represented by $2J$ units: J means and J standard deviations (see below). (Known; specified before training.)

$\boldsymbol{\mu}_t$ Vector of J encoder output unit states (means) from $2J$ encoder output units. (Unknown; learned during training.)

Σ_t Covariance matrix of the J-dimensional (posterior) Gaussian distribution with mean and standard deviation specified by the $2J$ encoder output states. Σ_t is diagonal, which encourages different units to represent independent physical quantities and allows Σ_t to be expressed as a vector $\boldsymbol{\sigma}_t^2 = (\sigma_{1t}^2, \ldots, \sigma_{Jt}^2)$. (Unknown; learned during training.)

$\mathbf{z}_{it}$ The ith sample from the Gaussian distribution defined by states $\boldsymbol{\mu}_t$ and $\boldsymbol{\sigma}_t$ of the $2J$ encoder output units, $\mathbf{z}_{it} \sim \mathcal{N}(\boldsymbol{\mu}_t, \boldsymbol{\sigma}_t^2)$, referred to as latent variables. (Unknown; learned during training.)

$p(\mathbf{z})$ The prior distribution of the latent variables, implemented as a J-dimensional Gaussian distribution with zero mean and unit variance (i.e. $\mathcal{N}(\mathbf{0}, \Sigma^{\text{prior}})$). A distribution with $J = 2$ is shown in Figure 9.5. (Known; specified before training.)

X_t' Decoder output image. (Unknown; learned during training.)

$q(\mathbf{z}|X_t)$ Variational distribution of the latent variables used as decoder input vectors, where $\mathbf{z} \sim \mathcal{N}(\boldsymbol{\mu}_t, \boldsymbol{\sigma}_t^2)$. (Unknown; parametric form, i.e. Gaussian, specified before training; $\boldsymbol{\mu}_t$ and $\boldsymbol{\sigma}_t$ values learned during training.)

$p(\mathbf{z}|X_t)$ Posterior distribution of latent variables. (Unknown; estimated during training as $q(\mathbf{z}|X_t) = \mathcal{N}(\boldsymbol{\mu}_t, \boldsymbol{\sigma}_t^2)$, where $\boldsymbol{\mu}_t$ and $\boldsymbol{\sigma}_t$ are encoder output states.)

$p(X_t|\mathbf{z})$ Conditional distribution of decoder output vectors, where the probability density of X_t given $\mathbf{z}$ is determined by the decoder. (Unknown; learned during training.)

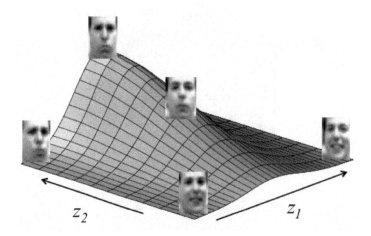

Figure 9.4. Schematic example of a two-dimensional curved manifold. Each point $\mathbf{z} = (z_1, z_2)$ on the manifold corresponds to the values of two latent variables, $z_1 = $ head orientation and $z_2 = $ facial expression.

9.4. How Variational Autoencoders Work

As stated earlier, a variational autoencoder consists of two modules, an encoder and a decoder. The data are a set of T images $\{X\} = \{X_1, \ldots, X_T\}$. The encoder maps each input X_t to a layer of $2J$ encoder output units, which represent the values of a small number J of latent variables.

The reason each latent variable is represented by two encoder output units is that each pair of encoder output units represents the mean μ and standard deviation σ of a Gaussian distribution, as shown in Figure 9.6. Thus, each encoder output state consists of two J-element vectors $\boldsymbol{\mu} = (\mu_1, \ldots, \mu_J)$ and $\boldsymbol{\sigma} = (\sigma_1, \ldots, \sigma_J)$. For now, we restrict our attention to the encoder output units that represent the means $\boldsymbol{\mu}$.

In practice, we rarely get to see the J physical latent variables represented by the encoder output states, because the state of each encoder output unit is a mixture of several physical latent variables, unless special measures are taken during training (e.g. Figure 9.3). For example, the state of an encoder output unit might represent a combination of contrast and the curvature of lines in the input image. However, the important point is that values of key physical latent variables are encoded in the collective states of the encoder output units. Ideally, after training, the decoder maps each sampled value of the encoder output state $\mathbf{z} \sim \mathcal{N}(\boldsymbol{\mu}, \boldsymbol{\sigma}^2)$ to a decoder output X_t', such that $X_t' \approx X_t$. Thus, the combined effect of the encoder and decoder is to map each encoder input X_t to a similar decoder output X_t'.

Mapping Latent Variables to Images. As a toy example, consider a set of images where each image $X_t = (x_{1t}, x_{2t})$ consists of two pixels, as in Figure 2.4. This is an unusual example because the number ($J = 2$) of latent variables equals the number of elements in each input vector (image). Because each image has two pixels, it can be plotted as a point on the plane, as in Figure 9.7.

For simplicity we consider just two classes, C_A (e.g. images of A) and C_B (e.g. images of B), of inputs with distributions $p(X_A)$ and $p(X_B)$, respectively. Now, if an instance X_1 is chosen from C_A, it yields an encoder output consisting of means $\boldsymbol{\mu}_1 = (\mu_1, \mu_2)$ and standard deviations $\boldsymbol{\sigma}_1 = (\sigma_1, \sigma_2)$, which are the elements of the diagonal covariance matrix Σ_1. These outputs define the Gaussian distribution $q(\mathbf{z}|X_1) = \mathcal{N}(\boldsymbol{\mu}_1, \Sigma_1)$ shown in the box on the left of Figure 9.7. Given a sample $\mathbf{z}_t$ drawn from $q(\mathbf{z}|X_1)$, the decoder maps it to a point X_1'. When considered over all points in $q(\mathbf{z}|X_1)$, the decoder maps this entire distribution to the distribution $p(X'|\mathbf{z}(\boldsymbol{\mu}_1))$, which is labelled as $p(X'|\boldsymbol{\mu}_1)$ in the box on the right of Figure 9.7. After training, the encoder should map each input vector X from class C_A to a point X' close to $\boldsymbol{\mu}_1$, so that $p(X'|\boldsymbol{\mu}_1)$ is roughly the same as the distribution $p(X_A)$ of vectors in class C_A. Similarly, we should have $p(X'|\boldsymbol{\mu}_2) \approx p(X_B)$ for vectors in class C_B.

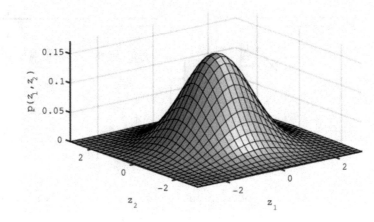

Figure 9.5. Multivariate Gaussian function with $J = 2$. The height at $\mathbf{z} = (z_1, z_2)$ is $p(\mathbf{z})$, obtained by replacing $(X - X')$ with $\mathbf{z}$ in Equation 9.27. Reproduced from Stone (2019).

Naming Distributions. Ideally, the inputs to the decoder would have a *posterior distribution* $p(\mathbf{z}|X_t)$ such that the $\mathbf{z}$ values reflect the underlying latent variables of the input image X_t. The process of using the input image X_t to estimate $p(\mathbf{z}|X_t)$ is called *inference*. The posterior distribution is often written as $p(\mathbf{z}|X)$, where X implicitly refers to the set of all possible values of the inputs X_t. The reason we write X_t here is to emphasize which quantities are given (i.e. the single image X_t) and which are inferred (i.e. the distribution $p(\mathbf{z}|X_t)$).

Using *Bayes' theorem* (Appendix D), the posterior distribution can be written as

$$p(\mathbf{z}|X_t) \quad = \quad \frac{p(X_t|\mathbf{z})\,p(\mathbf{z})}{p(X_t)}, \tag{9.1}$$

where $p(X_t|\mathbf{z})$ is the *likelihood*, $p(\mathbf{z})$ is the *prior distribution*, and $p(X_t)$ is the *marginal likelihood* or *model evidence*. The difficulty in evaluating the posterior distribution lies with the marginal likelihood.

Why Evaluating the Marginal Likelihood $p(X_t)$ **is Hard.** In principle, we could estimate $p(X_t)$ using a uniform sampling of N points $\{\mathbf{z}_1, \ldots, \mathbf{z}_N\}$ on a grid of $\mathbf{z}$ values, so that

$$p(X_t) \quad \approx \quad \sum_{i=1}^{N} p(X_t|\mathbf{z}_i)p(\mathbf{z}_i). \tag{9.2}$$

However, even though we know how to evaluate every term in this sum, the sum itself is intractable. For example, if each of J variables $\mathbf{z} = (z_1, \ldots, z_J)$ is sampled at n uniformly spaced points then the number of samples will be $N = n^J$, which becomes impractical as J increases. Worse, almost all values $\mathbf{z}_i$ have a density $p(\mathbf{z}_i) \approx 0$, as explained next.

We assume that each vector $\mathbf{z}$ is drawn from a multivariate Gaussian distribution, $\mathbf{z} \sim \mathcal{N}(\boldsymbol{\mu}, \boldsymbol{\sigma}^2)$, where $\boldsymbol{\mu} = (\mu_1, \ldots, \mu_J)$ and $\boldsymbol{\sigma} = (\sigma_1, \ldots, \sigma_J)$. If all the σ_j values are the same, this defines a spherical distribution. For large values of J, it can be shown[14] that almost all Gaussian vectors $\mathbf{z}$ are located within the *typical set*[71], an extremely thin spherical shell at distance $r \approx \sigma\sqrt{J}$ from $\boldsymbol{\mu}$. Therefore, the volume of the typical set is less than the volume v of a J-dimensional hypersphere with radius r. For large values of J, it can be shown that $v \propto 1/\sqrt{J}$, so that $v \to 0$ as $J \to \infty$. Therefore, the space in which $\mathbf{z}$ resides is almost entirely empty, except near the surface of a vanishingly small hypersphere. If the covariance matrix Σ is not diagonal, then this hypersphere becomes a hyperellipsoid, but the

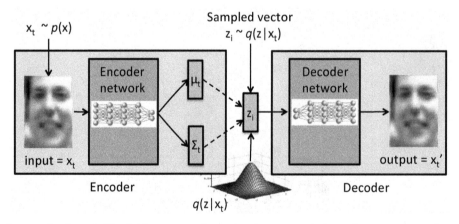

Figure 9.6. Each encoder input X_t yields an encoder output represented as J means $\boldsymbol{\mu}_t = (\mu_{1t}, \ldots, \mu_{Jt})$ and J standard deviations $\boldsymbol{\sigma}_t = (\sigma_{1t}, \ldots, \sigma_{Jt})$ of a Gaussian distribution $\mathcal{N}(\boldsymbol{\mu}_t, \boldsymbol{\sigma}_t^2)$, where $\boldsymbol{\sigma}_t^2$ denotes the vector of diagonal elements of a $J \times J$ diagonal covariance matrix Σ_t. The decoder input is a sample $\mathbf{z}_i$ from $\mathcal{N}(\boldsymbol{\mu}_t, \boldsymbol{\sigma}_t^2)$, which yields a decoder output X_t'. Reproduced from Stone (2019).

same reasoning regarding the typical set applies. Thus, if we try to estimate $p(X_t)$ (Equation 9.2) by randomly sampling $\mathbf{z}$, we are doomed to failure because $p(\mathbf{z}) \approx 0$ for almost all values of $\mathbf{z}$. More generally, random sampling is ineffective because the volume occupied by a probability distribution shrinks rapidly as the number J of dimensions increases, a phenomenon called *concentration of measure*.

Competing Objectives. The main goal of the variational autoencoder is to train a network to minimise the difference between the (known) estimate of the posterior $q(\mathbf{z}|X_t)$ and the true (but unknown) posterior $p(\mathbf{z}|X_t)$. This is achieved by (simultaneously) training two interdependent networks: 1) an encoder network, which minimises the difference between $q(\mathbf{z}|X_t)$ and an assumed (known) prior distribution $p(\mathbf{z})$; and 2) a decoder network, which uses $\mathbf{z}$ to minimise the difference between its output X_t' and the original input X_t.

We wish to find parameter values $\mathbf{w}$ (i.e. weights) such that (roughly speaking) the probability distribution of decoder outputs $p(X|\mathbf{z}_t)$ is maximised, subject to the constraint that $q(\mathbf{z}|X_t) \approx p(\mathbf{z}|X_t)$. The autoencoder is summarised as

$$X_t \xrightarrow[\text{encoder}]{q(\mathbf{z}|X_t)} \mathbf{z}_t \xrightarrow[\text{decoder}]{p(X|\mathbf{z}_t)} X_t'. \tag{9.3}$$

Thus, the autoencoder has two competing objectives. First, the decoder, given an input $\mathbf{z}_t$ from the encoder, tries to generate an output X_t' similar to the original input X_t. Second, the encoder tries to generate outputs which yield decoder inputs $\mathbf{z}_t$ such that the encoder output distribution $q(\mathbf{z}|X_t)$ is similar to the (unknown) posterior distribution $p(\mathbf{z}|X_t)$.

To understand the source of these competing objectives, we express the model evidence as a marginal distribution. Specifically, the model evidence $p(X_t)$ can be expressed as the marginal distribution of the joint distribution $p(X, \mathbf{z})$ at $X = X_t$. Given that $p(X_t, \mathbf{z}) = p(X_t|\mathbf{z})p(\mathbf{z})$,

$$p(X_t) = \int_{\mathbf{z}} p(X_t|\mathbf{z})p(\mathbf{z}) \, d\mathbf{z}, \tag{9.4}$$

which is the continuous version of Equation 9.2. To learn the correct weights, we need to know how to evaluate $p(X_t)$.

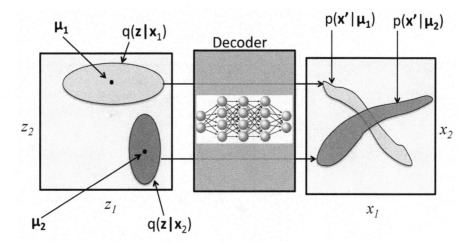

Figure 9.7. Decoding two classes C_A and C_B, with distributions $p(X_A)$ and $p(X_B)$ (not shown). If a typical item X_1 in C_A is presented to the encoder then its output $(\boldsymbol{\mu}_1, \Sigma_1)$ defines the Gaussian distribution $q(\mathbf{z}|X_1) = \mathcal{N}(\boldsymbol{\mu}_1, \Sigma_1)$. After training, the decoder maps $q(\mathbf{z}|X_1)$ to a distribution $p(X'|\boldsymbol{\mu}_1)$, which should be similar to $p(X_A)$; similarly, we would like $p(X'|\boldsymbol{\mu}_2) \approx p(X_B)$. Reproduced from Stone (2019).

9.5. The Evidence Lower Bound

Consider a variational autoencoder with weights that have been initialised with random values. When presented with a training vector X_t, the encoder produces outputs that yield decoder inputs $\mathbf{z}$ with a distribution $q(\mathbf{z}|X_t)$. However, as explained above, the resultant decoder outputs have vanishingly small probabilities $p(X_t|\mathbf{z})$. Consequently, we need to adjust the weights so that the decoder yields outputs $X_t' \approx X_t$. To this end, imagine an *ideal encoder* that implements $p(\mathbf{z}|X_t)$ (the true posterior) rather than the distribution $q(\mathbf{z}|X_t)$ of decoder inputs provided by the encoder. If we knew $p(\mathbf{z}|X_t)$ then the dissimilarity between $q(\mathbf{z}|X_t)$ and $p(\mathbf{z}|X_t)$ could be measured by the *Kullback–Leibler divergence*,

$$D\big(q(\mathbf{z}|X_t)\|p(\mathbf{z}|X_t)\big) \quad = \quad \int_z q(\mathbf{z}|X_t) \log \frac{q(\mathbf{z}|X_t)}{p(\mathbf{z}|X_t)} \, d\mathbf{z}, \tag{9.5}$$

where $D \geq 0$, with $D = 0$ if and only if $q(\mathbf{z}|X_t) = p(\mathbf{z}|X_t)$. Given a sample $\{\mathbf{z}_i\}$ of $\mathbf{z}$ values, the Kullback–Leibler divergence is estimated as

$$D\big(q(\mathbf{z}|X_t)\|p(\mathbf{z}|X_t)\big) \quad = \quad \sum_i q(\mathbf{z}_i|X_t) \log \frac{q(\mathbf{z}_i|X_t)}{p(\mathbf{z}_i|X_t)}. \tag{9.6}$$

Next, we will use the Kullback–Leibler divergence to derive two expressions for a key quantity called the *evidence lower bound* (ELBO). The form of the first expression (Equation 9.17) makes it obvious that we wish to minimise it, though it is not obvious how to do so. The second expression (Equation 9.26) can be minimised, but it is less obvious that we would want to do so – apart from the fact that it is also an expression for the ELBO.

The First Expression for the ELBO. Because

$$p(\mathbf{z}, X_t) \quad = \quad p(X_t|\mathbf{z})p(\mathbf{z}), \tag{9.7}$$

Equation 9.1 can be written as $p(\mathbf{z}|X_t) = p(\mathbf{z}, X_t)/p(X_t)$, and taking logarithms gives

$$\log p(\mathbf{z}|X_t) \quad = \quad \log p(\mathbf{z}, X_t) - \log p(X_t). \tag{9.8}$$

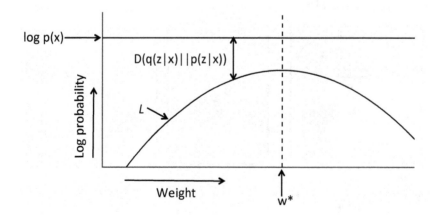

Figure 9.8. The lower bound L of $\log p(X)$ can be maximised by varying the encoder weights $\mathbf{w}$ to decrease the Kullback–Leibler divergence $D(q(\mathbf{z}|X)\|p(\mathbf{z}|X))$ between the variational distribution $q(\mathbf{z}|X)$ and the posterior distribution $p(\mathbf{z}|X)$. The graph shows a cross-section through $D(q(\mathbf{z}|X)\|p(\mathbf{z}|X))$ for a single weight. The optimal weight is labelled w^*. Reproduced from Stone (2019).

If we invert the log ratio in Equation 9.6 then we get

$$D\big(q(\mathbf{z}|X_t)\|p(\mathbf{z}|X_t)\big) \;=\; -\sum_i q(\mathbf{z}_i|X_t) \log \frac{p(\mathbf{z}_i|X_t)}{q(\mathbf{z}_i|X_t)}. \tag{9.9}$$

Substituting Equation 9.8 into Equation 9.9 and recombining the logarithms yields

$$
\begin{aligned}
D\big(q(\mathbf{z}|X_t)\|p(\mathbf{z}|X_t)\big) \;&=\; -\sum_i q(\mathbf{z}_i|X_t)\big[\log p(\mathbf{z}_i|X_t) - \log q(\mathbf{z}_i|X_t)\big] \\
&=\; -\sum_i q(\mathbf{z}_i|X_t)\big[\log p(\mathbf{z}_i, X_t) - \log p(X_t) - \log q(\mathbf{z}_i|X_t)\big] \\
&=\; -\sum_i q(\mathbf{z}_i|X_t)\left[\log \frac{p(\mathbf{z}_i, X_t)}{q(\mathbf{z}_i|X_t)} - \log p(X_t)\right] \\
&=\; -\sum_i q(\mathbf{z}_i|X_t) \log \frac{p(\mathbf{z}_i, X_t)}{q(\mathbf{z}_i|X_t)} + \sum_i q(\mathbf{z}_i|X_t) \log p(X_t).
\end{aligned} \tag{9.10}
$$

Since $\sum_i q(\mathbf{z}_i|X_t) = 1$, the second term on the right-hand side of Equation 9.10 is

$$\sum_i q(\mathbf{z}_i|X_t) \log p(X_t) \;=\; \log p(X_t) \sum_i q(\mathbf{z}_i|X_t) \;=\; \log p(X_t). \tag{9.11}$$

For convenience, we define the first summation on the right in Equation 9.10 as

$$L_t \;=\; \sum_i q(\mathbf{z}_i|X_t) \log \frac{p(\mathbf{z}_i, X_t)}{q(\mathbf{z}_i|X_t)}. \tag{9.12}$$

Therefore Equation 9.10 can be written as

$$D\big(q(\mathbf{z}|X_t)\|p(\mathbf{z}|X_t)\big) \;=\; -L_t + \log p(X_t), \tag{9.13}$$

and rearranging yields

$$L_t \;=\; \log p(X_t) - D\big(q(\mathbf{z}|X_t)\|p(\mathbf{z}|X_t)\big). \tag{9.14}$$

Because $D\big(q(\mathbf{z}|X_t)\|p(\mathbf{z}|X_t)\big) \geq 0$, it follows that L_t must be less than (or equal to) the log evidence $\log p(X_t)$, so L_t defines a *lower bound* on $\log p(X_t)$; this is why L_t is called the *evidence lower bound* (ELBO) or *variational lower bound*. The tightness of this bound depends on how well the variational distribution $q(\mathbf{z}|X_t)$ approximates the posterior distribution $p(\mathbf{z}|X_t)$. Clearly, if $q(\mathbf{z}|X_t) \approx p(\mathbf{z}|X_t)$ then D is close to zero and so the bound is tight, $L_t \approx \log p(X_t)$. In this case, maximising L_t is almost as effective as maximising $\log p(X_t)$, as shown in Figure 9.8.

When considered over the whole set of input vectors $\{X\}$,

$$L \;=\; \sum_t L_t \tag{9.15}$$

$$=\; \log p(\{X\}) - D\big(q(\mathbf{z}|\{X\})\|p(\mathbf{z}|\{X\})\big). \tag{9.16}$$

For brevity, we omit the set brackets { } and write our first expression for the ELBO as

$$L \;=\; \log p(X) - D\big(q(\mathbf{z}|X)\|p(\mathbf{z}|X)\big). \tag{9.17}$$

The Second Expression for the ELBO. Substituting Equation 9.7 into Equation 9.12 gives

$$L_t = \sum_i q(\mathbf{z}_i|X_t) \log \frac{p(X_t|\mathbf{z}_i)p(\mathbf{z}_i)}{q(\mathbf{z}_i|X_t)} \tag{9.18}$$

$$= \sum_i q(\mathbf{z}_i|X_t) \left[\log p(X_t|\mathbf{z}_i) + \log \frac{p(\mathbf{z}_i)}{q(\mathbf{z}_i|X_t)} \right], \tag{9.19}$$

and expanding the brackets yields

$$L_t = \sum_i q(\mathbf{z}_i|X_t) \log p(X_t|\mathbf{z}_i) + \sum_i q(\mathbf{z}_i|X_t) \log \frac{p(\mathbf{z}_i)}{q(\mathbf{z}_i|X_t)}. \tag{9.20}$$

This equation is used to define two key quantities. First, we define

$$L_t^{\text{dec}} = \sum_i q(\mathbf{z}_i|X_t) \log p(X_t|\mathbf{z}_i) \tag{9.21}$$

$$= \mathbb{E}_{\mathbf{z} \sim q(\mathbf{z}|X_t)} \big[\log p(X_t|\mathbf{z}) \big], \tag{9.22}$$

where $\mathbf{z} \sim q(\mathbf{z}|X_t)$ means that $\mathbf{z}$ values are samples from the distribution $q(\mathbf{z}|X_t)$. Thus, Equation 9.22 is the expected log likelihood given that $\mathbf{z}_i \sim q(\mathbf{z}|X_t)$. Note that the mapping from the decoder input $\mathbf{z}_i$ to the decoder output X_t' is deterministic, so the weights $\mathbf{w}$ that maximise the likelihood $p(X_t|\mathbf{z}(\mathbf{w}))$ also maximise the likelihood $p(X_t|X_t'(\mathbf{w}))$, where we have made the dependence on $\mathbf{w}$ explicit here. Using the second term in Equation 9.20 we define

$$L_t^{\text{enc}} = \sum_i q(\mathbf{z}_i|X_t) \log \frac{p(\mathbf{z}_i)}{q(\mathbf{z}_i|X_t)} \tag{9.23}$$

$$= -D\big(q(\mathbf{z}|X_t)\|p(\mathbf{z})\big), \tag{9.24}$$

which is minus the Kullback–Leibler divergence between the (known) variational distribution $q(\mathbf{z}|X_t)$ and the prior distribution $p(\mathbf{z})$. Therefore, Equation 9.20 can be written as

$$L_t = L_t^{\text{dec}} + L_t^{\text{enc}} \tag{9.25}$$

$$= \mathbb{E}_{\mathbf{z} \sim q(\mathbf{z}|X_t)} \big[\log p(X_t|\mathbf{z}) \big] - \beta D\big(q(\mathbf{z}|X_t)\|p(\mathbf{z})\big), \tag{9.26}$$

where we have introduced the parameter β to vary the trade-off between decoder reconstruction and encoder inference [19;38] ($\beta = 1$ in Equation 9.20). When summed over T training vectors (Equation 9.15), we obtain our second expression for the ELBO.

It is important to note that the unknown posterior of the ideal encoder $p(\mathbf{z}|X)$ in Equation 9.17 has been replaced by the known prior $p(\mathbf{z})$ in Equation 9.26. Thus, the lower bound in Equation 9.17 can be evaluated, and maximised, using the equivalent Equation 9.26.

The quantity $\log p(X_t|\mathbf{z})$ in Equation 9.26 is the log probability that the network generates the image X_t given a vector $\mathbf{z}$ derived stochastically from the encoder output. Because X_t is fixed, $p(X_t|\mathbf{z})$ varies as a deterministic function of $\mathbf{z}$, and is called a *likelihood function*. Further, because $\mathbf{z}$ depends on the weights $\mathbf{w}$, the likelihood function is proportional to the likelihood of the weights. Thus, by adjusting the weights to maximise $\log p(X_t|\mathbf{z})$, we are really performing *maximum likelihood estimation* of the weights (i.e. of $\log p(X_t|\mathbf{w})$). However, we actually maximise a lower bound L, with L^{enc} acting as a constraint, so variational autoencoders perform a type of constrained or *penalised maximum likelihood estimation* of the weights.

9.6. Maximising the Lower Bound

Knowing that L is a lower bound matters because, given that we wish to learn weights $\mathbf{w}$ that maximise the log likelihood $\log p(X|\mathbf{z})$, we can instead maximise $\log p(X|\mathbf{z})$ by gradient ascent on L (Equation 9.26). Maximising L might seem like a poor substitute for maximising $\log p(X|\mathbf{z})$ itself. However, using Equation 9.26 to maximise L implicitly forces the variational (encoder output) distribution $q(\mathbf{z}|X)$ to be similar to the unknown posterior $p(\mathbf{z}|X)$, so that $D(q(\mathbf{z}|X)\|p(\mathbf{z}|X)) \approx 0$ in Equation 9.17. Thus, as learning progresses, maximising the lower bound L can be almost as good as maximising $\log p(X|\mathbf{z})$ itself.

The reconstruction accuracy is measured by the probability $p(X_t|X_t')$ that the encoder input X_t is reconstructed at the decoder output, given the observed decoder output X_t'. In general, given an n-dimensional vector X_t', the probability distribution of an n-dimensional Gaussian is

$$p(X_t|X_t') = \frac{\exp\left[-(X_t - X_t')^\top \Sigma_X^{-1}(X_t - X_t')/2\right]}{\left[(2\pi)^n \det(\Sigma_X)\right]^{1/2}}, \tag{9.27}$$

where Σ_X is an $n \times n$ covariance matrix and $\det(\Sigma_X)$ is its determinant. A multivariate Gaussian distribution with $n = 2$ is shown in Figure 9.5, and its cross-section is shown in Figure 9.9 (see Appendix C). Because X_t' is a deterministic function of $\mathbf{z}$, this implies $p(X_t|X_t') = p(X_t|\mathbf{z})$, so that

$$\log p(X_t|\mathbf{z}) = \log p(X_t|X_t'). \tag{9.28}$$

Substituting Equation 9.27 into Equation 9.28 yields

$$\log p(X_t|\mathbf{z}) = c - \frac{1}{2}(X_t - X_t')^\top \Sigma_X^{-1}(X_t - X_t'), \tag{9.29}$$

where $c = -\log\left[(2\pi)^n \det(\Sigma_X)\right]^{1/2}$. If we assume that Σ_X is proportional to the identity matrix I (so that it has identical diagonal elements σ_X^2, i.e. $\Sigma_X = \sigma_X^2 I$) then $\Sigma_X^{-1} = I/\sigma_X^2$, and therefore

$$\log p(X_t|\mathbf{z}) = c - \frac{1}{2\sigma_X^2}|X_t - X_t'|^2, \tag{9.30}$$

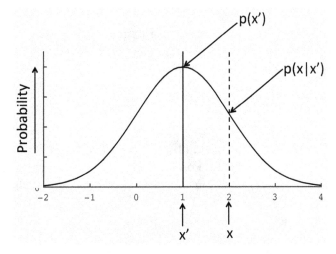

Figure 9.9. Cross-section of a multivariate Gaussian function (Equation 9.27, Figure 9.5). If the decoder output is $X' = 1$ then the probability $p(X|X')$ that the output is the same as the encoder input $X = 2$ is a Gaussian function centred at X', so it is a Gaussian function of $|X - X'|$. See Appendix C. Reproduced from Stone (2019).

which measures the similarity between the encoder input X_t and the decoder output X_t'. Substituting Equation 9.30 into Equation 9.22, and assuming $\sigma_X^2 = 1$, yields

$$
\begin{aligned}
L_t^{\text{dec}} &= \mathbb{E}_{\mathbf{z} \sim q(\mathbf{z}|X_t)} \left[c - \frac{1}{2} |X_t - X_t'|^2 \right] \\
&= c - \frac{1}{2} \mathbb{E}_{\mathbf{z} \sim q(\mathbf{z}|X_t)} \left[|X_t - X_t'|^2 \right].
\end{aligned}
\tag{9.31}
$$

Given that the decoder input is $\mathbf{z} \sim q(\mathbf{z}|X_t)$ and that each sampled value $\mathbf{z}_{it}$ produces a corresponding decoder output X_{it}', we can estimate the expectation in Equation 9.31 using N_z samples of $\mathbf{z}$, so that (for each input X_t)

$$
L_t^{\text{dec}} = c - \frac{1}{2N_z} \sum_{i=1}^{N_z} |X_t - X_{it}'|^2.
\tag{9.32}
$$

To use backprop with the encoder term in Equation 9.24, we need the *reparametrisation trick*. This allows us to use a sample $\eta \sim \mathcal{N}(0,1)$ to obtain a sample $z \sim \mathcal{N}(\mu_{jt}, \sigma_{jt}^2)$, where (μ_{jt}, σ_{jt}) is the state of the jth pair of encoder output units in response to the encoder input X_t. Given a sample η, the corresponding sample z that would be obtained from $\mathcal{N}(\mu_{jt}, \sigma_{jt}^2)$ is $z = \mu_{jt} + \eta \, \sigma_{jt}$. When considered over all J pairs of encoder output units, this yields $\mathbf{z} \sim \mathcal{N}(\boldsymbol{\mu}_t, \Sigma_t)$. Assuming that the $J \times J$ covariance matrix Σ_t is diagonal, the encoder regularisation term evaluates to[56]

$$
L_t^{\text{enc}} = \frac{1}{2} \sum_{j=1}^{J} (1 + \log \sigma_{jt}^2 - \sigma_{jt}^2 - \mu_{jt}^2),
\tag{9.33}
$$

which attains its maximal value when $\boldsymbol{\mu}_t = \mathbf{0}$ and $\boldsymbol{\sigma}_t^2 = \mathbf{1}$, that is, when the estimate $q(\mathbf{z}|X_t) = \mathcal{N}(\boldsymbol{\mu}_t, \Sigma_t)$ of the posterior distribution $p(\mathbf{z}|X_t)$ equals the prior distribution $p(\mathbf{z}) = \mathcal{N}(\mathbf{0}, \mathbf{1})$. Therefore, L_t^{enc} measures the extent to which the estimated posterior distribution $q(\mathbf{z}|X_t)$ deviates from the prior distribution $p(\mathbf{z})$, where $q(\mathbf{z}|X_t)$ is effectively implemented as the encoder.

Substituting Equations 9.32 and 9.33 into Equation 9.25 yields

$$
L_t = c - \frac{1}{2N_z} \sum_{i=1}^{N_z} |X_t - X_{it}'|^2 + \frac{1}{2} \sum_{j=1}^{J} (1 + \log \sigma_{jt}^2 - \sigma_{jt}^2 - \mu_{jt}^2).
\tag{9.34}
$$

In practice, gradient ascent with backprop is used to maximise $L = \sum_t L_t$ based on batches of inputs, as in Burda et al. (2015). The gradient of L with respect to the network weights is calculated using an automatic gradient tool such as autograd[141].

9.7. Results

Variational autoencoders and related information bottleneck methods seem promising, but their ability to solve realistic tasks is still being explored. For example, after training on images of digits between 0 and 9, a small modification allows the hidden unit states to be used for classification, which yielded an impressive test error of 1.13% on the MNIST digit set (Alemi et al., 2017).

Using test images of objects with different sizes, positions and orientations, Higgins et al. (2016) found that setting the regularisation parameter to $\beta = 4$ (Equation 9.26) was reasonably successful at forcing the hidden units to represent different physical parameters. Surprisingly, even though networks with $\beta = 1$ and $\beta = 4$ were equally good at reconstructing simple geometric objects (that were part of the training data), those with $\beta = 4$ performed substantially better on objects that were not part of the training data. This suggests that (as we might expect) forcing networks to represent independent physical parameters (e.g. position, orientation) in different hidden units yields networks that generalise well beyond their training data.

It is worth noting that results comparable to the embedding of faces in Figure 9.3 were obtained by combining variational autoencoders with generative adversarial networks (Larsen et al., 2015; see Chapter 10). However, training variational autoencoders tends to be more straightforward than training generative adversarial networks.

Variational autoencoders can generate molecules with pre-specified desirable physiological properties. Specifically, training a conditional variational autoencoder on molecules labelled with known properties allows novel molecules to be generated by treating a subset of the decoder input units as a 'one-hot vector' of desirable properties[31;66;84]. Variational autoencoders have also been used for speech recognition[47] and to recover images being viewed from electroencephalograph (EEG) data[52]. For a review, see Maaløe et al. (2019).

9.8. List of Mathematical Symbols

See Section 9.3.

A Note on the Reproduction of Colour Figures. Because the paper is reproduced here without colour, the plotted lines in Figures 2 and 3 are hard to distinguish. For each graph in Figure 2, the lower pair of curves represent 'Wake-Sleep (test)' (dashed) and 'Wake-Sleep (train)' (solid), whereas the upper pair of curves represent 'AEVB (test)' (dashed) and 'AEVB (train)' (solid). In Figure 3, the curves in the left graph represent (from bottom to top): 'Wake-Sleep (test)', 'AEVB (test)', 'MCEM (test)', 'Wake-Sleep (train)', 'MCEM (train)' and 'AEVB (train)'. In Figure 3, the curves in the right graph represent (from bottom to top): 'MCEM (test)', 'MCEM (train)', 'Wake-Sleep (test)', 'AEVB (test)', 'Wake-Sleep (train)' and 'AEVB (train)'.

The original paper can be viewed with colour figures at `https://arxiv.org/pdf/1312.6114v11.pdf`.

Research Paper: Auto-Encoding Variational Bayes

Reference: Kingma, D. P. and Welling, M. (2013). Auto-encoding variational Bayes. *arXiv e-prints*. Reproduced with permission.

Auto-Encoding Variational Bayes

Diederik P. Kingma
Machine Learning Group
Universiteit van Amsterdam
dpkingma@gmail.com

Max Welling
Machine Learning Group
Universiteit van Amsterdam
welling.max@gmail.com

Abstract

How can we perform efficient inference and learning in directed probabilistic models, in the presence of continuous latent variables with intractable posterior distributions, and large datasets? We introduce a stochastic variational inference and learning algorithm that scales to large datasets and, under some mild differentiability conditions, even works in the intractable case. Our contributions are two-fold. First, we show that a reparameterization of the variational lower bound yields a lower bound estimator that can be straightforwardly optimized using standard stochastic gradient methods. Second, we show that for i.i.d. datasets with continuous latent variables per datapoint, posterior inference can be made especially efficient by fitting an approximate inference model (also called a recognition model) to the intractable posterior using the proposed lower bound estimator. Theoretical advantages are reflected in experimental results.

1 Introduction

How can we perform efficient approximate inference and learning with directed probabilistic models whose continuous latent variables and/or parameters have intractable posterior distributions? The variational Bayesian (VB) approach involves the optimization of an approximation to the intractable posterior. Unfortunately, the common mean-field approach requires analytical solutions of expectations w.r.t. the approximate posterior, which are also intractable in the general case. We show how a reparameterization of the variational lower bound yields a simple differentiable unbiased estimator of the lower bound; this SGVB (Stochastic Gradient Variational Bayes) estimator can be used for efficient approximate posterior inference in almost any model with continuous latent variables and/or parameters, and is straightforward to optimize using standard stochastic gradient ascent techniques.

For the case of an i.i.d. dataset and continuous latent variables per datapoint, we propose the Auto-Encoding VB (AEVB) algorithm. In the AEVB algorithm we make inference and learning especially efficient by using the SGVB estimator to optimize a recognition model that allows us to perform very efficient approximate posterior inference using simple ancestral sampling, which in turn allows us to efficiently learn the model parameters, without the need of expensive iterative inference schemes (such as MCMC) per datapoint. The learned approximate posterior inference model can also be used for a host of tasks such as recognition, denoising, representation and visualization purposes. When a neural network is used for the recognition model, we arrive at the *variational auto-encoder*.

2 Method

The strategy in this section can be used to derive a lower bound estimator (a stochastic objective function) for a variety of directed graphical models with continuous latent variables. We will restrict ourselves here to the common case where we have an i.i.d. dataset with latent variables per datapoint, and where we like to perform maximum likelihood (ML) or maximum a posteriori (MAP) inference on the (global) parameters, and variational inference on the latent variables. It is, for example,

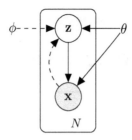

Figure 1: The type of directed graphical model under consideration. Solid lines denote the generative model $p_\theta(\mathbf{z})p_\theta(\mathbf{x}|\mathbf{z})$, dashed lines denote the variational approximation $q_\phi(\mathbf{z}|\mathbf{x})$ to the intractable posterior $p_\theta(\mathbf{z}|\mathbf{x})$. The variational parameters ϕ are learned jointly with the generative model parameters θ.

straightforward to extend this scenario to the case where we also perform variational inference on the global parameters; that algorithm is put in the appendix, but experiments with that case are left to future work. Note that our method can be applied to online, non-stationary settings, e.g. streaming data, but here we assume a fixed dataset for simplicity.

2.1 Problem scenario

Let us consider some dataset $\mathbf{X} = \{\mathbf{x}^{(i)}\}_{i=1}^N$ consisting of N i.i.d. samples of some continuous or discrete variable $\mathbf{x}$. We assume that the data are generated by some random process, involving an unobserved continuous random variable $\mathbf{z}$. The process consists of two steps: (1) a value $\mathbf{z}^{(i)}$ is generated from some prior distribution $p_{\theta^*}(\mathbf{z})$; (2) a value $\mathbf{x}^{(i)}$ is generated from some conditional distribution $p_{\theta^*}(\mathbf{x}|\mathbf{z})$. We assume that the prior $p_{\theta^*}(\mathbf{z})$ and likelihood $p_{\theta^*}(\mathbf{x}|\mathbf{z})$ come from parametric families of distributions $p_\theta(\mathbf{z})$ and $p_\theta(\mathbf{x}|\mathbf{z})$, and that their PDFs are differentiable almost everywhere w.r.t. both θ and $\mathbf{z}$. Unfortunately, a lot of this process is hidden from our view: the true parameters θ^* as well as the values of the latent variables $\mathbf{z}^{(i)}$ are unknown to us.

Very importantly, we *do not* make the common simplifying assumptions about the marginal or posterior probabilities. Conversely, we are here interested in a general algorithm that even works efficiently in the case of:

1. *Intractability*: the case where the integral of the marginal likelihood $p_\theta(\mathbf{x}) = \int p_\theta(\mathbf{z})p_\theta(\mathbf{x}|\mathbf{z})\,d\mathbf{z}$ is intractable (so we cannot evaluate or differentiate the marginal likelihood), where the true posterior density $p_\theta(\mathbf{z}|\mathbf{x}) = p_\theta(\mathbf{x}|\mathbf{z})p_\theta(\mathbf{z})/p_\theta(\mathbf{x})$ is intractable (so the EM algorithm cannot be used), and where the required integrals for any reasonable mean-field VB algorithm are also intractable. These intractabilities are quite common and appear in cases of moderately complicated likelihood functions $p_\theta(\mathbf{x}|\mathbf{z})$, e.g. a neural network with a nonlinear hidden layer.

2. *A large dataset*: we have so much data that batch optimization is too costly; we would like to make parameter updates using small minibatches or even single datapoints. Sampling-based solutions, e.g. Monte Carlo EM, would in general be too slow, since it involves a typically expensive sampling loop per datapoint.

We are interested in, and propose a solution to, three related problems in the above scenario:

1. Efficient approximate ML or MAP estimation for the parameters θ. The parameters can be of interest themselves, e.g. if we are analyzing some natural process. They also allow us to mimic the hidden random process and generate artificial data that resembles the real data.

2. Efficient approximate posterior inference of the latent variable $\mathbf{z}$ given an observed value $\mathbf{x}$ for a choice of parameters θ. This is useful for coding or data representation tasks.

3. Efficient approximate marginal inference of the variable $\mathbf{x}$. This allows us to perform all kinds of inference tasks where a prior over $\mathbf{x}$ is required. Common applications in computer vision include image denoising, inpainting and super-resolution.

For the purpose of solving the above problems, let us introduce a recognition model $q_\phi(\mathbf{z}|\mathbf{x})$: an approximation to the intractable true posterior $p_\theta(\mathbf{z}|\mathbf{x})$. Note that in contrast with the approximate posterior in mean-field variational inference, it is not necessarily factorial and its parameters ϕ are not computed from some closed-form expectation. Instead, we'll introduce a method for learning the recognition model parameters ϕ jointly with the generative model parameters θ.

From a coding theory perspective, the unobserved variables $\mathbf{z}$ have an interpretation as a latent representation or *code*. In this paper we will therefore also refer to the recognition model $q_\phi(\mathbf{z}|\mathbf{x})$ as a probabilistic *encoder*, since given a datapoint $\mathbf{x}$ it produces a distribution (e.g. a Gaussian) over the possible values of the code $\mathbf{z}$ from which the datapoint $\mathbf{x}$ could have been generated. In a similar vein we will refer to $p_\theta(\mathbf{x}|\mathbf{z})$ as a probabilistic *decoder*, since given a code $\mathbf{z}$ it produces a distribution over the possible corresponding values of $\mathbf{x}$.

2.2 The variational bound

The marginal likelihood is composed of a sum over the marginal likelihoods of individual datapoints $\log p_\theta(\mathbf{x}^{(1)}, \cdots, \mathbf{x}^{(N)}) = \sum_{i=1}^{N} \log p_\theta(\mathbf{x}^{(i)})$, which can each be rewritten as:

$$\log p_\theta(\mathbf{x}^{(i)}) = D_{KL}(q_\phi(\mathbf{z}|\mathbf{x}^{(i)})||p_\theta(\mathbf{z}|\mathbf{x}^{(i)})) + \mathcal{L}(\theta, \phi; \mathbf{x}^{(i)}) \tag{1}$$

The first RHS term is the KL divergence of the approximate from the true posterior. Since this KL-divergence is non-negative, the second RHS term $\mathcal{L}(\theta, \phi; \mathbf{x}^{(i)})$ is called the (variational) *lower bound* on the marginal likelihood of datapoint i, and can be written as:

$$\log p_\theta(\mathbf{x}^{(i)}) \geq \mathcal{L}(\theta, \phi; \mathbf{x}^{(i)}) = \mathbb{E}_{q_\phi(\mathbf{z}|\mathbf{x})} \left[-\log q_\phi(\mathbf{z}|\mathbf{x}) + \log p_\theta(\mathbf{x}, \mathbf{z}) \right] \tag{2}$$

which can also be written as:

$$\mathcal{L}(\theta, \phi; \mathbf{x}^{(i)}) = -D_{KL}(q_\phi(\mathbf{z}|\mathbf{x}^{(i)})||p_\theta(\mathbf{z})) + \mathbb{E}_{q_\phi(\mathbf{z}|\mathbf{x}^{(i)})} \left[\log p_\theta(\mathbf{x}^{(i)}|\mathbf{z}) \right] \tag{3}$$

We want to differentiate and optimize the lower bound $\mathcal{L}(\theta, \phi; \mathbf{x}^{(i)})$ w.r.t. both the variational parameters ϕ and generative parameters θ. However, the gradient of the lower bound w.r.t. ϕ is a bit problematic. The usual (naïve) Monte Carlo gradient estimator for this type of problem is: $\nabla_\phi \mathbb{E}_{q_\phi(\mathbf{z})} \left[f(\mathbf{z}) \right] = \mathbb{E}_{q_\phi(\mathbf{z})} \left[f(\mathbf{z}) \nabla_{q_\phi(\mathbf{z})} \log q_\phi(\mathbf{z}) \right] \simeq \frac{1}{L} \sum_{l=1}^{L} f(\mathbf{z}) \nabla_{q_\phi(\mathbf{z}^{(l)})} \log q_\phi(\mathbf{z}^{(l)})$ where $\mathbf{z}^{(l)} \sim q_\phi(\mathbf{z}|\mathbf{x}^{(i)})$. This gradient estimator exhibits exhibits very high variance (see e.g. [BJP12]) and is impractical for our purposes.

2.3 The SGVB estimator and AEVB algorithm

In this section we introduce a practical estimator of the lower bound and its derivatives w.r.t. the parameters. We assume an approximate posterior in the form $q_\phi(\mathbf{z}|\mathbf{x})$, but please note that the technique can be applied to the case $q_\phi(\mathbf{z})$, i.e. where we do not condition on $\mathbf{x}$, as well. The fully variational Bayesian method for inferring a posterior over the parameters is given in the appendix.

Under certain mild conditions outlined in section 2.4 for a chosen approximate posterior $q_\phi(\mathbf{z}|\mathbf{x})$ we can reparameterize the random variable $\widetilde{\mathbf{z}} \sim q_\phi(\mathbf{z}|\mathbf{x})$ using a differentiable transformation $g_\phi(\epsilon, \mathbf{x})$ of an (auxiliary) noise variable ϵ:

$$\widetilde{\mathbf{z}} = g_\phi(\epsilon, \mathbf{x}) \quad \text{with} \quad \epsilon \sim p(\epsilon) \tag{4}$$

See section 2.4 for general strategies for chosing such an approriate distribution $p(\epsilon)$ and function $g_\phi(\epsilon, \mathbf{x})$. We can now form Monte Carlo estimates of expectations of some function $f(\mathbf{z})$ w.r.t. $q_\phi(\mathbf{z}|\mathbf{x})$ as follows:

$$\mathbb{E}_{q_\phi(\mathbf{z}|\mathbf{x}^{(i)})} \left[f(\mathbf{z}) \right] = \mathbb{E}_{p(\epsilon)} \left[f(g_\phi(\epsilon, \mathbf{x}^{(i)})) \right] \simeq \frac{1}{L} \sum_{l=1}^{L} f(g_\phi(\epsilon^{(l)}, \mathbf{x}^{(i)})) \quad \text{where} \quad \epsilon^{(l)} \sim p(\epsilon) \tag{5}$$

We apply this technique to the variational lower bound (eq. (2)), yielding our generic Stochastic Gradient Variational Bayes (SGVB) estimator $\widetilde{\mathcal{L}}^A(\theta, \phi; \mathbf{x}^{(i)}) \simeq \mathcal{L}(\theta, \phi; \mathbf{x}^{(i)})$:

$$\widetilde{\mathcal{L}}^A(\theta, \phi; \mathbf{x}^{(i)}) = \frac{1}{L} \sum_{l=1}^{L} \log p_\theta(\mathbf{x}^{(i)}, \mathbf{z}^{(i,l)}) - \log q_\phi(\mathbf{z}^{(i,l)}|\mathbf{x}^{(i)})$$

$$\text{where} \quad \mathbf{z}^{(i,l)} = g_\phi(\epsilon^{(i,l)}, \mathbf{x}^{(i)}) \quad \text{and} \quad \epsilon^{(l)} \sim p(\epsilon) \tag{6}$$

Algorithm 1 Minibatch version of the Auto-Encoding VB (AEVB) algorithm. Either of the two SGVB estimators in section 2.3 can be used. We use settings $M = 100$ and $L = 1$ in experiments.

$\boldsymbol{\theta}, \boldsymbol{\phi} \leftarrow$ Initialize parameters
repeat
    $\mathbf{X}^M \leftarrow$ Random minibatch of M datapoints (drawn from full dataset)
    $\boldsymbol{\epsilon} \leftarrow$ Random samples from noise distribution $p(\boldsymbol{\epsilon})$
    $\mathbf{g} \leftarrow \nabla_{\boldsymbol{\theta},\boldsymbol{\phi}} \widetilde{\mathcal{L}}^M(\boldsymbol{\theta}, \boldsymbol{\phi}; \mathbf{X}^M, \boldsymbol{\epsilon})$ (Gradients of minibatch estimator (8))
    $\boldsymbol{\theta}, \boldsymbol{\phi} \leftarrow$ Update parameters using gradients $\mathbf{g}$ (e.g. SGD or Adagrad [DHS10])
until convergence of parameters $(\boldsymbol{\theta}, \boldsymbol{\phi})$
return $\boldsymbol{\theta}, \boldsymbol{\phi}$

Often, the KL-divergence $D_{KL}(q_{\boldsymbol{\phi}}(\mathbf{z}|\mathbf{x}^{(i)})\|p_{\boldsymbol{\theta}}(\mathbf{z}))$ of eq. (3) can be integrated analytically (see appendix B), such that only the expected reconstruction error $\mathbb{E}_{q_{\boldsymbol{\phi}}(\mathbf{z}|\mathbf{x}^{(i)})}\left[\log p_{\boldsymbol{\theta}}(\mathbf{x}^{(i)}|\mathbf{z})\right]$ requires estimation by sampling. The KL-divergence term can then be interpreted as regularizing $\boldsymbol{\phi}$, encouraging the approximate posterior to be close to the prior $p_{\boldsymbol{\theta}}(\mathbf{z})$. This yields a second version of the SGVB estimator $\widetilde{\mathcal{L}}^B(\boldsymbol{\theta}, \boldsymbol{\phi}; \mathbf{x}^{(i)}) \simeq \mathcal{L}(\boldsymbol{\theta}, \boldsymbol{\phi}; \mathbf{x}^{(i)})$, corresponding to eq. (3), which typically has less variance than the generic estimator:

$$\widetilde{\mathcal{L}}^B(\boldsymbol{\theta}, \boldsymbol{\phi}; \mathbf{x}^{(i)}) = -D_{KL}(q_{\boldsymbol{\phi}}(\mathbf{z}|\mathbf{x}^{(i)})\|p_{\boldsymbol{\theta}}(\mathbf{z})) + \frac{1}{L}\sum_{l=1}^{L}(\log p_{\boldsymbol{\theta}}(\mathbf{x}^{(i)}|\mathbf{z}^{(i,l)}))$$

$$\text{where} \quad \mathbf{z}^{(i,l)} = g_{\boldsymbol{\phi}}(\boldsymbol{\epsilon}^{(i,l)}, \mathbf{x}^{(i)}) \quad \text{and} \quad \boldsymbol{\epsilon}^{(l)} \sim p(\boldsymbol{\epsilon}) \tag{7}$$

Given multiple datapoints from a dataset $\mathbf{X}$ with N datapoints, we can construct an estimator of the marginal likelihood lower bound of the full dataset, based on minibatches:

$$\mathcal{L}(\boldsymbol{\theta}, \boldsymbol{\phi}; \mathbf{X}) \simeq \widetilde{\mathcal{L}}^M(\boldsymbol{\theta}, \boldsymbol{\phi}; \mathbf{X}^M) = \frac{N}{M}\sum_{i=1}^{M}\widetilde{\mathcal{L}}(\boldsymbol{\theta}, \boldsymbol{\phi}; \mathbf{x}^{(i)}) \tag{8}$$

where the minibatch $\mathbf{X}^M = \{\mathbf{x}^{(i)}\}_{i=1}^{M}$ is a randomly drawn sample of M datapoints from the full dataset $\mathbf{X}$ with N datapoints. In our experiments we found that the number of samples L per datapoint can be set to 1 as long as the minibatch size M was large enough, e.g. $M = 100$. Derivatives $\nabla_{\boldsymbol{\theta},\boldsymbol{\phi}} \widetilde{\mathcal{L}}(\boldsymbol{\theta}; \mathbf{X}^M)$ can be taken, and the resulting gradients can be used in conjunction with stochastic optimization methods such as SGD or Adagrad [DHS10]. See algorithm 1 for a basic approach to compute the stochastic gradients.

A connection with auto-encoders becomes clear when looking at the objective function given at eq. (7). The first term is (the KL divergence of the approximate posterior from the prior) acts as a regularizer, while the second term is a an expected negative reconstruction error. The function $g_{\boldsymbol{\phi}}(.)$ is chosen such that it maps a datapoint $\mathbf{x}^{(i)}$ and a random noise vector $\boldsymbol{\epsilon}^{(l)}$ to a sample from the approximate posterior for that datapoint: $\mathbf{z}^{(i,l)} = g_{\boldsymbol{\phi}}(\boldsymbol{\epsilon}^{(l)}, \mathbf{x}^{(i)})$ where $\mathbf{z}^{(i,l)} \sim q_{\boldsymbol{\phi}}(\mathbf{z}|\mathbf{x}^{(i)})$. Subsequently, the sample $\mathbf{z}^{(i,l)}$ is then input to function $\log p_{\boldsymbol{\theta}}(\mathbf{x}^{(i)}|\mathbf{z}^{(i,l)})$, which equals the probability density (or mass) of datapoint $\mathbf{x}^{(i)}$ under the generative model, given $\mathbf{z}^{(i,l)}$. This term is a negative *reconstruction error* in auto-encoder parlance.

2.4 The reparameterization trick

In order to solve our problem we invoked an alternative method for generating samples from $q_{\boldsymbol{\phi}}(\mathbf{z}|\mathbf{x})$. The essential parameterization trick is quite simple. Let $\mathbf{z}$ be a continuous random variable, and $\mathbf{z} \sim q_{\boldsymbol{\phi}}(\mathbf{z}|\mathbf{x})$ be some conditional distribution. It is then often possible to express the random variable $\mathbf{z}$ as a deterministic variable $\mathbf{z} = g_{\boldsymbol{\phi}}(\boldsymbol{\epsilon}, \mathbf{x})$, where $\boldsymbol{\epsilon}$ is an auxiliary variable with independent marginal $p(\boldsymbol{\epsilon})$, and $g_{\boldsymbol{\phi}}(.)$ is some vector-valued function parameterized by $\boldsymbol{\phi}$.

This reparameterization is useful for our case since it can be used to rewrite an expectation w.r.t $q_{\boldsymbol{\phi}}(\mathbf{z}|\mathbf{x})$ such that the Monte Carlo estimate of the expectation is differentiable w.r.t. $\boldsymbol{\phi}$. A proof is as follows. Given the deterministic mapping $\mathbf{z} = g_{\boldsymbol{\phi}}(\boldsymbol{\epsilon}, \mathbf{x})$ we know that $q_{\boldsymbol{\phi}}(\mathbf{z}|\mathbf{x})\prod_i dz_i = p(\boldsymbol{\epsilon})\prod_i d\epsilon_i$. Therefore[1], $\int q_{\boldsymbol{\phi}}(\mathbf{z}|\mathbf{x})f(\mathbf{z})\, d\mathbf{z} = \int p(\boldsymbol{\epsilon})f(\mathbf{z})\, d\boldsymbol{\epsilon} = \int p(\boldsymbol{\epsilon})f(g_{\boldsymbol{\phi}}(\boldsymbol{\epsilon}, \mathbf{x}))\, d\boldsymbol{\epsilon}$. It follows

---

[1] Note that for infinitesimals we use the notational convention $d\mathbf{z} = \prod_i dz_i$

that a differentiable estimator can be constructed: $\int q_\phi(\mathbf{z}|\mathbf{x}) f(\mathbf{z})\, d\mathbf{z} \simeq \frac{1}{L} \sum_{l=1}^{L} f(g_\phi(\mathbf{x}, \boldsymbol{\epsilon}^{(l)}))$ where $\boldsymbol{\epsilon}^{(l)} \sim p(\boldsymbol{\epsilon})$. In section 2.3 we applied this trick to obtain a differentiable estimator of the variational lower bound.

Take, for example, the univariate Gaussian case: let $z \sim p(z|x) = \mathcal{N}(\mu, \sigma^2)$. In this case, a valid reparameterization is $z = \mu + \sigma\epsilon$, where ϵ is an auxiliary noise variable $\epsilon \sim \mathcal{N}(0,1)$. Therefore, $\mathbb{E}_{\mathcal{N}(z;\mu,\sigma^2)}[f(z)] = \mathbb{E}_{\mathcal{N}(\epsilon;0,1)}[f(\mu + \sigma\epsilon)] \simeq \frac{1}{L} \sum_{l=1}^{L} f(\mu + \sigma\epsilon^{(l)})$ where $\epsilon^{(l)} \sim \mathcal{N}(0,1)$.

For which $q_\phi(\mathbf{z}|\mathbf{x})$ can we choose such a differentiable transformation $g_\phi(.)$ and auxiliary variable $\epsilon \sim p(\epsilon)$? Three basic approaches are:

1. Tractable inverse CDF. In this case, let $\epsilon \sim \mathcal{U}(\mathbf{0}, \mathbf{I})$, and let $g_\phi(\boldsymbol{\epsilon}, \mathbf{x})$ be the inverse CDF of $q_\phi(\mathbf{z}|\mathbf{x})$. Examples: Exponential, Cauchy, Logistic, Rayleigh, Pareto, Weibull, Reciprocal, Gompertz, Gumbel and Erlang distributions.

2. Analogous to the Gaussian example, for any "location-scale" family of distributions we can choose the standard distribution (with location = 0, scale = 1) as the auxiliary variable ϵ, and let $g(.) = \text{location} + \text{scale} \cdot \epsilon$. Examples: Laplace, Elliptical, Student's t, Logistic, Uniform, Triangular and Gaussian distributions.

3. Composition: It is often possible to express random variables as different transformations of auxiliary variables. Examples: Log-Normal (exponentiation of normally distributed variable), Gamma (a sum over exponentially distributed variables), Dirichlet (weighted sum of Gamma variates), Beta, Chi-Squared, and F distributions.

When all three approaches fail, good approximations to the inverse CDF exist requiring computations with time complexity comparable to the PDF (see e.g. [Dev86] for some methods).

3 Example: Variational Auto-Encoder

In this section we'll give an example where we use a neural network for the probabilistic encoder $q_\phi(\mathbf{z}|\mathbf{x})$ (the approximation to the posterior of the generative model $p_\theta(\mathbf{x}, \mathbf{z})$) and where the parameters ϕ and θ are optimized jointly with the AEVB algorithm.

Let the prior over the latent variables be the centered isotropic multivariate Gaussian $p_\theta(\mathbf{z}) = \mathcal{N}(\mathbf{z}; \mathbf{0}, \mathbf{I})$. Note that in this case, the prior lacks parameters. We let $p_\theta(\mathbf{x}|\mathbf{z})$ be a multivariate Gaussian (in case of real-valued data) or Bernoulli (in case of binary data) whose distribution parameters are computed from $\mathbf{z}$ with a MLP (a fully-connected neural network with a single hidden layer, see appendix C). Note the true posterior $p_\theta(\mathbf{z}|\mathbf{x})$ is in this case intractable. While there is much freedom in the form $q_\phi(\mathbf{z}|\mathbf{x})$, we'll assume the true (but intractable) posterior takes on a approximate Gaussian form with an approximately diagonal covariance. In this case, we can let the variational approximate posterior be a multivariate Gaussian with a diagonal covariance structure[2]:

$$\log q_\phi(\mathbf{z}|\mathbf{x}^{(i)}) = \log \mathcal{N}(\mathbf{z}; \boldsymbol{\mu}^{(i)}, \boldsymbol{\sigma}^{2(i)}\mathbf{I}) \tag{9}$$

where the mean and s.d. of the approximate posterior, $\boldsymbol{\mu}^{(i)}$ and $\boldsymbol{\sigma}^{(i)}$, are outputs of the encoding MLP, i.e. nonlinear functions of datapoint $\mathbf{x}^{(i)}$ and the variational parameters ϕ (see appendix C).

As explained in section 2.4, we sample from the posterior $\mathbf{z}^{(i,l)} \sim q_\phi(\mathbf{z}|\mathbf{x}^{(i)})$ using $\mathbf{z}^{(i,l)} = g_\phi(\mathbf{x}^{(i)}, \boldsymbol{\epsilon}^{(l)}) = \boldsymbol{\mu}^{(i)} + \boldsymbol{\sigma}^{(i)} \odot \boldsymbol{\epsilon}^{(l)}$ where $\boldsymbol{\epsilon}^{(l)} \sim \mathcal{N}(\mathbf{0}, \mathbf{I})$. With $\odot$ we signify an element-wise product. In this model both $p_\theta(\mathbf{z})$ (the prior) and $q_\phi(\mathbf{z}|\mathbf{x})$ are Gaussian; in this case, we can use the estimator of eq. (7) where the KL divergence can be computed and differentiated without estimation (see appendix B). The resulting estimator for this model and datapoint $\mathbf{x}^{(i)}$ is:

$$\mathcal{L}(\boldsymbol{\theta}, \boldsymbol{\phi}; \mathbf{x}^{(i)}) \simeq \frac{1}{2} \sum_{j=1}^{J} \left(1 + \log((\sigma_j^{(i)})^2) - (\mu_j^{(i)})^2 - (\sigma_j^{(i)})^2\right) + \frac{1}{L} \sum_{l=1}^{L} \log p_\theta(\mathbf{x}^{(i)}|\mathbf{z}^{(i,l)})$$

$$\text{where} \quad \mathbf{z}^{(i,l)} = \boldsymbol{\mu}^{(i)} + \boldsymbol{\sigma}^{(i)} \odot \boldsymbol{\epsilon}^{(l)} \quad \text{and} \quad \boldsymbol{\epsilon}^{(l)} \sim \mathcal{N}(\mathbf{0}, \mathbf{I}) \tag{10}$$

As explained above and in appendix C, the decoding term $\log p_\theta(\mathbf{x}^{(i)}|\mathbf{z}^{(i,l)})$ is a Bernoulli or Gaussian MLP, depending on the type of data we are modelling.

---

[2]Note that this is just a (simplifying) choice, and not a limitation of our method.

4 Related work

The wake-sleep algorithm [HDFN95] is, to the best of our knowledge, the only other on-line learning method in the literature that is applicable to the same general class of continuous latent variable models. Like our method, the wake-sleep algorithm employs a recognition model that approximates the true posterior. A drawback of the wake-sleep algorithm is that it requires a concurrent optimization of two objective functions, which together do not correspond to optimization of (a bound of) the marginal likelihood. An advantage of wake-sleep is that it also applies to models with discrete latent variables. Wake-Sleep has the same computational complexity as AEVB per datapoint.

Stochastic variational inference [HBWP13] has recently received increasing interest. Recently, [BJP12] introduced a control variate schemes to reduce the high variance of the naïve gradient estimator discussed in section 2.1, and applied to exponential family approximations of the posterior. In [RGB13] some general methods, i.e. a control variate scheme, were introduced for reducing the variance of the original gradient estimator. In [SK13], a similar reparameterization as in this paper was used in an efficient version of a stochastic variational inference algorithm for learning the natural parameters of exponential-family approximating distributions.

The AEVB algorithm exposes a connection between directed probabilistic models (trained with a variational objective) and auto-encoders. A connection between *linear* auto-encoders and a certain class of generative linear-Gaussian models has long been known. In [Row98] it was shown that PCA corresponds to the maximum-likelihood (ML) solution of a special case of the linear-Gaussian model with a prior $p(\mathbf{z}) = \mathcal{N}(0, \mathbf{I})$ and a conditional distribution $p(\mathbf{x}|\mathbf{z}) = \mathcal{N}(\mathbf{x}; \mathbf{Wz}, \epsilon\mathbf{I})$, specifically the case with infinitesimally small ϵ.

In relevant recent work on autoencoders [VLL$^+$10] it was shown that the training criterion of unregularized autoencoders corresponds to maximization of a lower bound (see the infomax principle [Lin89]) of the mutual information between input X and latent representation Z. Maximizing (w.r.t. parameters) of the mutual information is equivalent to maximizing the conditional entropy, which is lower bounded by the expected loglikelihood of the data under the autoencoding model [VLL$^+$10], i.e. the negative reconstruction error. However, it is well known that this reconstruction criterion is in itself not sufficient for learning useful representations [BCV13]. Regularization techniques have been proposed to make autoencoders learn useful representations, such as denoising, contractive and sparse autoencoder variants [BCV13]. The SGVB objective contains a regularization term dictated by the variational bound (e.g. eq. (10)), lacking the usual nuisance regularization hyperparameter required to learn useful representations. Related are also encoder-decoder architectures such as the predictive sparse decomposition (PSD) [KRL08], from which we drew some inspiration. Also relevant are the recently introduced Generative Stochastic Networks [BTL13] where noisy auto-encoders learn the transition operator of a Markov chain that samples from the data distribution. In [SL10] a recognition model was employed for efficient learning with Deep Boltzmann Machines. These methods are targeted at either unnormalized models (i.e. undirected models like Boltzmann machines) or limited to sparse coding models, in contrast to our proposed algorithm for learning a general class of directed probabilistic models.

The recently proposed DARN method [GMW13], also learns a directed probabilistic model using an auto-encoding structure, however their method applies to binary latent variables. Even more recently, [RMW14] also make the connection between auto-encoders, directed proabilistic models and stochastic variational inference using the reparameterization trick we describe in this paper. Their work was developed independently of ours and provides an additional perspective on AEVB.

5 Experiments

We trained generative models of images from the MNIST and Frey Face datasets[3] and compared learning algorithms in terms of the variational lower bound, and the estimated marginal likelihood.

The generative model (encoder) and variational approximation (decoder) from section 3 were used, where the described encoder and decoder have an equal number of hidden units. Since the Frey Face data are continuous, we used a decoder with Gaussian outputs, identical to the encoder, except that the means were constrained to the interval $(0, 1)$ using a sigmoidal activation function at the

---

[3]Available at http://www.cs.nyu.edu/~roweis/data.html

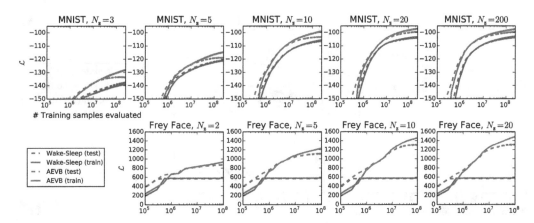

Figure 2: Comparison of our AEVB method to the wake-sleep algorithm, in terms of optimizing the lower bound, for different dimensionality of latent space ($N_{\mathbf{z}}$). Our method converged considerably faster and reached a better solution in all experiments. Interestingly enough, more latent variables does not result in more overfitting, which is explained by the regularizing effect of the lower bound. Vertical axis: the estimated average variational lower bound per datapoint. The estimator variance was small (< 1) and omitted. Horizontal axis: amount of training points evaluated. Computation took around 20-40 minutes per million training samples with a Intel Xeon CPU running at an effective 40 GFLOPS.

decoder output. Note that with *hidden units* we refer to the hidden layer of the neural networks of the encoder and decoder.

Parameters are updated using stochastic gradient ascent where gradients are computed by differentiating the lower bound estimator $\nabla_{\boldsymbol{\theta},\boldsymbol{\phi}}\mathcal{L}(\boldsymbol{\theta},\boldsymbol{\phi};\mathbf{X})$ (see algorithm 1), plus a small weight decay term corresponding to a prior $p(\boldsymbol{\theta}) = \mathcal{N}(0,\mathbf{I})$. Optimization of this objective is equivalent to approximate MAP estimation, where the likelihood gradient is approximated by the gradient of the lower bound.

We compared performance of AEVB to the wake-sleep algorithm [HDFN95]. We employed the same encoder (also called recognition model) for the wake-sleep algorithm and the variational auto-encoder. All parameters, both variational and generative, were initialized by random sampling from $\mathcal{N}(0, 0.01)$, and were jointly stochastically optimized using the MAP criterion. Stepsizes were adapted with Adagrad [DHS10]; the Adagrad global stepsize parameters were chosen from {0.01, 0.02, 0.1} based on performance on the training set in the first few iterations. Minibatches of size $M = 100$ were used, with $L = 1$ samples per datapoint.

Likelihood lower bound We trained generative models (decoders) and corresponding encoders (a.k.a. recognition models) having 500 hidden units in case of MNIST, and 200 hidden units in case of the Frey Face dataset (to prevent overfitting, since it is a considerably smaller dataset). The chosen number of hidden units is based on prior literature on auto-encoders, and the relative performance of different algorithms was not very sensitive to these choices. Figure 2 shows the results when comparing the lower bounds. Interestingly, superfluous latent variables did not result in overfitting, which is explained by the regularizing nature of the variational bound.

Marginal likelihood For very low-dimensional latent space it is possible to estimate the marginal likelihood of the learned generative models using an MCMC estimator. More information about the marginal likelihood estimator is available in the appendix. For the encoder and decoder we again used neural networks, this time with 100 hidden units, and 3 latent variables; for higher dimensional latent space the estimates became unreliable. Again, the MNIST dataset was used. The AEVB and Wake-Sleep methods were compared to Monte Carlo EM (MCEM) with a Hybrid Monte Carlo (HMC) [DKPR87] sampler; details are in the appendix. We compared the convergence speed for the three algorithms, for a small and large training set size. Results are in figure 3.

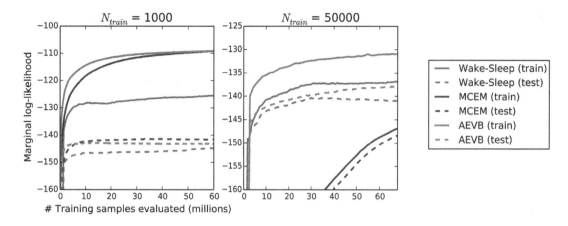

Figure 3: Comparison of AEVB to the wake-sleep algorithm and Monte Carlo EM, in terms of the estimated marginal likelihood, for a different number of training points. Monte Carlo EM is not an on-line algorithm, and (unlike AEVB and the wake-sleep method) can't be applied efficiently for the full MNIST dataset.

Visualisation of high-dimensional data If we choose a low-dimensional latent space (e.g. 2D), we can use the learned encoders (recognition model) to project high-dimensional data to a low-dimensional manifold. See appendix A for visualisations of the 2D latent manifolds for the MNIST and Frey Face datasets.

6 Conclusion

We have introduced a novel estimator of the variational lower bound, Stochastic Gradient VB (SGVB), for efficient approximate inference with continuous latent variables. The proposed estimator can be straightforwardly differentiated and optimized using standard stochastic gradient methods. For the case of i.i.d. datasets and continuous latent variables per datapoint we introduce an efficient algorithm for efficient inference and learning, Auto-Encoding VB (AEVB), that learns an approximate inference model using the SGVB estimator. The theoretical advantages are reflected in experimental results.

7 Future work

Since the SGVB estimator and the AEVB algorithm can be applied to almost any inference and learning problem with continuous latent variables, there are plenty of future directions: (i) learning hierarchical generative architectures with deep neural networks (e.g. convolutional networks) used for the encoders and decoders, trained jointly with AEVB; (ii) time-series models (i.e. dynamic Bayesian networks); (iii) application of SGVB to the global parameters; (iv) supervised models with latent variables, useful for learning complicated noise distributions.

References

[BCV13]   Yoshua Bengio, Aaron Courville, and Pascal Vincent. Representation learning: A review and new perspectives. 2013.

[BJP12]   David M Blei, Michael I Jordan, and John W Paisley. Variational Bayesian inference with Stochastic Search. In *Proceedings of the 29th International Conference on Machine Learning (ICML-12)*, pages 1367–1374, 2012.

[BTL13]   Yoshua Bengio and Éric Thibodeau-Laufer. Deep generative stochastic networks trainable by backprop. *arXiv preprint arXiv:1306.1091*, 2013.

[Dev86]   Luc Devroye. Sample-based non-uniform random variate generation. In *Proceedings of the 18th conference on Winter simulation*, pages 260–265. ACM, 1986.

[DHS10]   John Duchi, Elad Hazan, and Yoram Singer. Adaptive subgradient methods for online learning and stochastic optimization. *Journal of Machine Learning Research*, 12:2121–2159, 2010.

[DKPR87] Simon Duane, Anthony D Kennedy, Brian J Pendleton, and Duncan Roweth. Hybrid monte carlo. *Physics letters B*, 195(2):216–222, 1987.

[GMW13]   Karol Gregor, Andriy Mnih, and Daan Wierstra. Deep autoregressive networks. *arXiv preprint arXiv:1310.8499*, 2013.

[HBWP13] Matthew D Hoffman, David M Blei, Chong Wang, and John Paisley. Stochastic variational inference. *The Journal of Machine Learning Research*, 14(1):1303–1347, 2013.

[HDFN95] Geoffrey E Hinton, Peter Dayan, Brendan J Frey, and Radford M Neal. The" wake-sleep" algorithm for unsupervised neural networks. *SCIENCE*, pages 1158–1158, 1995.

[KRL08]   Koray Kavukcuoglu, Marc'Aurelio Ranzato, and Yann LeCun. Fast inference in sparse coding algorithms with applications to object recognition. Technical Report CBLL-TR-2008-12-01, Computational and Biological Learning Lab, Courant Institute, NYU, 2008.

[Lin89]   Ralph Linsker. *An application of the principle of maximum information preservation to linear systems*. Morgan Kaufmann Publishers Inc., 1989.

[RGB13]   Rajesh Ranganath, Sean Gerrish, and David M Blei. Black Box Variational Inference. *arXiv preprint arXiv:1401.0118*, 2013.

[RMW14]   Danilo Jimenez Rezende, Shakir Mohamed, and Daan Wierstra. Stochastic backpropagation and variational inference in deep latent gaussian models. *arXiv preprint arXiv:1401.4082*, 2014.

[Row98]   Sam Roweis. EM algorithms for PCA and SPCA. *Advances in neural information processing systems*, pages 626–632, 1998.

[SK13]   Tim Salimans and David A Knowles. Fixed-form variational posterior approximation through stochastic linear regression. *Bayesian Analysis*, 8(4), 2013.

[SL10]   Ruslan Salakhutdinov and Hugo Larochelle. Efficient learning of deep boltzmann machines. In *International Conference on Artificial Intelligence and Statistics*, pages 693–700, 2010.

[VLL+10] Pascal Vincent, Hugo Larochelle, Isabelle Lajoie, Yoshua Bengio, and Pierre-Antoine Manzagol. Stacked denoising autoencoders: Learning useful representations in a deep network with a local denoising criterion. *The Journal of Machine Learning Research*, 9999:3371–3408, 2010.

A   Visualisations

See figures 4 and 5 for visualisations of latent space and corresponding observed space of models learned with SGVB.

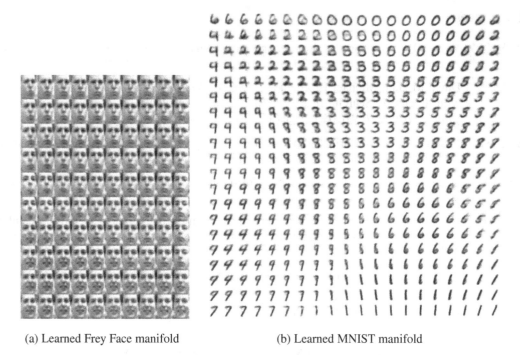

(a) Learned Frey Face manifold   (b) Learned MNIST manifold

Figure 4: Visualisations of learned data manifold for generative models with two-dimensional latent space, learned with AEVB. Since the prior of the latent space is Gaussian, linearly spaced coordinates on the unit square were transformed through the inverse CDF of the Gaussian to produce values of the latent variables $\mathbf{z}$. For each of these values $\mathbf{z}$, we plotted the corresponding generative $p_\theta(\mathbf{x}|\mathbf{z})$ with the learned parameters θ.

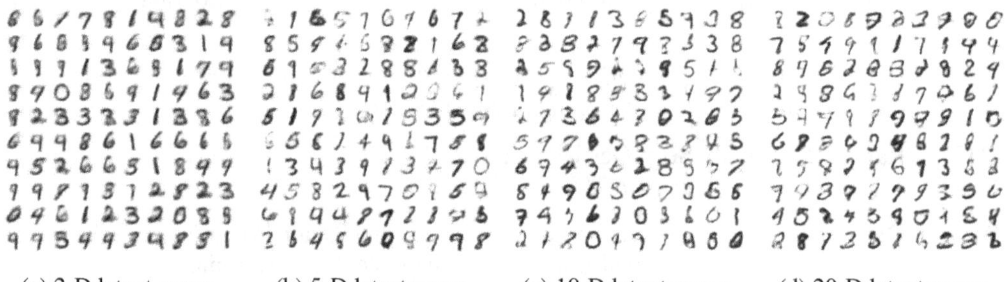

(a) 2-D latent space   (b) 5-D latent space   (c) 10-D latent space   (d) 20-D latent space

Figure 5: Random samples from learned generative models of MNIST for different dimensionalities of latent space.

B   Solution of $-D_{KL}(q_\phi(\mathbf{z})||p_\theta(\mathbf{z}))$, Gaussian case

The variational lower bound (the objective to be maximized) contains a KL term that can often be integrated analytically. Here we give the solution when both the prior $p_\theta(\mathbf{z}) = \mathcal{N}(0, \mathbf{I})$ and the posterior approximation $q_\phi(\mathbf{z}|\mathbf{x}^{(i)})$ are Gaussian. Let J be the dimensionality of $\mathbf{z}$. Let $\boldsymbol{\mu}$ and $\boldsymbol{\sigma}$ denote the variational mean and s.d. evaluated at datapoint i, and let μ_j and σ_j simply denote the j-th element of these vectors. Then:

$$\int q_\theta(\mathbf{z}) \log p(\mathbf{z}) \, d\mathbf{z} = \int \mathcal{N}(\mathbf{z}; \boldsymbol{\mu}, \boldsymbol{\sigma}^2) \log \mathcal{N}(\mathbf{z}; \mathbf{0}, \mathbf{I}) \, d\mathbf{z}$$

$$= -\frac{J}{2} \log(2\pi) - \frac{1}{2} \sum_{j=1}^{J} (\mu_j^2 + \sigma_j^2)$$

And:

$$\int q_{\boldsymbol{\theta}}(\mathbf{z}) \log q_{\boldsymbol{\theta}}(\mathbf{z}) \, d\mathbf{z} = \int \mathcal{N}(\mathbf{z}; \boldsymbol{\mu}, \boldsymbol{\sigma}^2) \log \mathcal{N}(\mathbf{z}; \boldsymbol{\mu}, \boldsymbol{\sigma}^2) \, d\mathbf{z}$$

$$= -\frac{J}{2} \log(2\pi) - \frac{1}{2} \sum_{j=1}^{J} (1 + \log \sigma_j^2)$$

Therefore:

$$-D_{KL}((q_{\phi}(\mathbf{z})||p_{\boldsymbol{\theta}}(\mathbf{z})) = \int q_{\boldsymbol{\theta}}(\mathbf{z}) \left(\log p_{\boldsymbol{\theta}}(\mathbf{z}) - \log q_{\boldsymbol{\theta}}(\mathbf{z}) \right) \, d\mathbf{z}$$

$$= \frac{1}{2} \sum_{j=1}^{J} \left(1 + \log((\sigma_j)^2) - (\mu_j)^2 - (\sigma_j)^2 \right)$$

When using a recognition model $q_{\phi}(\mathbf{z}|\mathbf{x})$ then $\boldsymbol{\mu}$ and s.d. $\boldsymbol{\sigma}$ are simply functions of $\mathbf{x}$ and the variational parameters ϕ, as exemplified in the text.

C   MLP's as probabilistic encoders and decoders

In variational auto-encoders, neural networks are used as probabilistic encoders and decoders. There are many possible choices of encoders and decoders, depending on the type of data and model. In our example we used relatively simple neural networks, namely multi-layered perceptrons (MLPs). For the encoder we used a MLP with Gaussian output, while for the decoder we used MLPs with either Gaussian or Bernoulli outputs, depending on the type of data.

C.1   Bernoulli MLP as decoder

In this case let $p_{\boldsymbol{\theta}}(\mathbf{x}|\mathbf{z})$ be a multivariate Bernoulli whose probabilities are computed from $\mathbf{z}$ with a fully-connected neural network with a single hidden layer:

$$\log p(\mathbf{x}|\mathbf{z}) = \sum_{i=1}^{D} x_i \log y_i + (1 - x_i) \cdot \log(1 - y_i)$$

$$\text{where} \quad \mathbf{y} = f_{\sigma}(\mathbf{W}_2 \tanh(\mathbf{W}_1 \mathbf{z} + \mathbf{b}_1) + \mathbf{b}_2) \tag{11}$$

where $f_{\sigma}(.)$ is the elementwise sigmoid activation function, and where $\boldsymbol{\theta} = \{\mathbf{W}_1, \mathbf{W}_2, \mathbf{b}_1, \mathbf{b}_2\}$ are the weights and biases of the MLP.

C.2   Gaussian MLP as encoder or decoder

In this case let encoder or decoder be a multivariate Gaussian with a diagonal covariance structure:

$$\log p(\mathbf{x}|\mathbf{z}) = \log \mathcal{N}(\mathbf{x}; \boldsymbol{\mu}, \boldsymbol{\sigma}^2 \mathbf{I})$$
$$\text{where} \quad \boldsymbol{\mu} = \mathbf{W}_4 \mathbf{h} + \mathbf{b}_4$$
$$\log \boldsymbol{\sigma}^2 = \mathbf{W}_5 \mathbf{h} + \mathbf{b}_5$$
$$\mathbf{h} = \tanh(\mathbf{W}_3 \mathbf{z} + \mathbf{b}_3) \tag{12}$$

where $\{\mathbf{W}_3, \mathbf{W}_4, \mathbf{W}_5, \mathbf{b}_3, \mathbf{b}_4, \mathbf{b}_5\}$ are the weights and biases of the MLP and part of $\boldsymbol{\theta}$ when used as decoder. Note that when this network is used as an encoder $q_{\phi}(\mathbf{z}|\mathbf{x})$, then $\mathbf{z}$ and $\mathbf{x}$ are swapped, and the weights and biases are variational parameters ϕ.

D   Marginal likelihood estimator

We derived the following marginal likelihood estimator that produces good estimates of the marginal likelihood as long as the dimensionality of the sampled space is low (less then 5 dimensions), and sufficient samples are taken. Let $p_{\boldsymbol{\theta}}(\mathbf{x}, \mathbf{z}) = p_{\boldsymbol{\theta}}(\mathbf{z}) p_{\boldsymbol{\theta}}(\mathbf{x}|\mathbf{z})$ be the generative model we are sampling from, and for a given datapoint $\mathbf{x}^{(i)}$ we would like to estimate the marginal likelihood $p_{\boldsymbol{\theta}}(\mathbf{x}^{(i)})$.

The estimation process consists of three stages:

1. Sample L values $\{z^{(l)}\}$ from the posterior using gradient-based MCMC, e.g. Hybrid Monte Carlo, using $\nabla_z \log p_\theta(z|x) = \nabla_z \log p_\theta(z) + \nabla_z \log p_\theta(x|z)$.

2. Fit a density estimator $q(z)$ to these samples $\{z^{(l)}\}$.

3. Again, sample L new values from the posterior. Plug these samples, as well as the fitted $q(z)$, into the following estimator:

$$p_\theta(x^{(i)}) \simeq \left(\frac{1}{L} \sum_{l=1}^{L} \frac{q(z^{(l)})}{p_\theta(z)p_\theta(x^{(i)}|z^{(l)})} \right)^{-1} \quad \text{where} \quad z^{(l)} \sim p_\theta(z|x^{(i)})$$

Derivation of the estimator:

$$\frac{1}{p_\theta(x^{(i)})} = \frac{\int q(z)\, dz}{p_\theta(x^{(i)})} = \frac{\int q(z) \frac{p_\theta(x^{(i)},z)}{p_\theta(x^{(i)},z)}\, dz}{p_\theta(x^{(i)})}$$

$$= \int \frac{p_\theta(x^{(i)},z)}{p_\theta(x^{(i)})} \frac{q(z)}{p_\theta(x^{(i)},z)}\, dz$$

$$= \int p_\theta(z|x^{(i)}) \frac{q(z)}{p_\theta(x^{(i)},z)}\, dz$$

$$\simeq \frac{1}{L} \sum_{l=1}^{L} \frac{q(z^{(l)})}{p_\theta(z)p_\theta(x^{(i)}|z^{(l)})} \quad \text{where} \quad z^{(l)} \sim p_\theta(z|x^{(i)})$$

E Monte Carlo EM

The Monte Carlo EM algorithm does not employ an encoder, instead it samples from the posterior of the latent variables using gradients of the posterior computed with $\nabla_z \log p_\theta(z|x) = \nabla_z \log p_\theta(z) + \nabla_z \log p_\theta(x|z)$. The Monte Carlo EM procedure consists of 10 HMC leapfrog steps with an automatically tuned stepsize such that the acceptance rate was 90%, followed by 5 weight updates steps using the acquired sample. For all algorithms the parameters were updated using the Adagrad stepsizes (with accompanying annealing schedule).

The marginal likelihood was estimated with the first 1000 datapoints from the train and test sets, for each datapoint sampling 50 values from the posterior of the latent variables using Hybrid Monte Carlo with 4 leapfrog steps.

F Full VB

As written in the paper, it is possible to perform variational inference on both the parameters θ and the latent variables z, as opposed to just the latent variables as we did in the paper. Here, we'll derive our estimator for that case.

Let $p_\alpha(\theta)$ be some hyperprior for the parameters introduced above, parameterized by α. The marginal likelihood can be written as:

$$\log p_\alpha(X) = D_{KL}(q_\phi(\theta)\|p_\alpha(\theta|X)) + \mathcal{L}(\phi; X) \tag{13}$$

where the first RHS term denotes a KL divergence of the approximate from the true posterior, and where $\mathcal{L}(\phi; X)$ denotes the variational lower bound to the marginal likelihood:

$$\mathcal{L}(\phi; X) = \int q_\phi(\theta) \left(\log p_\theta(X) + \log p_\alpha(\theta) - \log q_\phi(\theta) \right) d\theta \tag{14}$$

Note that this is a lower bound since the KL divergence is non-negative; the bound equals the true marginal when the approximate and true posteriors match exactly. The term $\log p_\theta(X)$ is composed of a sum over the marginal likelihoods of individual datapoints $\log p_\theta(X) = \sum_{i=1}^{N} \log p_\theta(x^{(i)})$, which can each be rewritten as:

$$\log p_\theta(x^{(i)}) = D_{KL}(q_\phi(z|x^{(i)})\|p_\theta(z|x^{(i)})) + \mathcal{L}(\theta, \phi; x^{(i)}) \tag{15}$$

where again the first RHS term is the KL divergence of the approximate from the true posterior, and $\mathcal{L}(\boldsymbol{\theta}, \boldsymbol{\phi}; \mathbf{x})$ is the variational lower bound of the marginal likelihood of datapoint i:

$$\mathcal{L}(\boldsymbol{\theta}, \boldsymbol{\phi}; \mathbf{x}^{(i)}) = \int q_{\boldsymbol{\phi}}(\mathbf{z}|\mathbf{x}) \left(\log p_{\boldsymbol{\theta}}(\mathbf{x}^{(i)}|\mathbf{z}) + \log p_{\boldsymbol{\theta}}(\mathbf{z}) - \log q_{\boldsymbol{\phi}}(\mathbf{z}|\mathbf{x}) \right) d\mathbf{z} \qquad (16)$$

The expectations on the RHS of eqs (14) and (16) can obviously be written as a sum of three separate expectations, of which the second and third component can sometimes be analytically solved, e.g. when both $p_{\boldsymbol{\theta}}(\mathbf{x})$ and $q_{\boldsymbol{\phi}}(\mathbf{z}|\mathbf{x})$ are Gaussian. For generality we will here assume that each of these expectations is intractable.

Under certain mild conditions outlined in section (see paper) for chosen approximate posteriors $q_{\boldsymbol{\phi}}(\boldsymbol{\theta})$ and $q_{\boldsymbol{\phi}}(\mathbf{z}|\mathbf{x})$ we can reparameterize conditional samples $\widetilde{\mathbf{z}} \sim q_{\boldsymbol{\phi}}(\mathbf{z}|\mathbf{x})$ as

$$\widetilde{\mathbf{z}} = g_{\boldsymbol{\phi}}(\boldsymbol{\epsilon}, \mathbf{x}) \quad \text{with} \quad \boldsymbol{\epsilon} \sim p(\boldsymbol{\epsilon}) \qquad (17)$$

where we choose a prior $p(\boldsymbol{\epsilon})$ and a function $g_{\boldsymbol{\phi}}(\boldsymbol{\epsilon}, \mathbf{x})$ such that the following holds:

$$\begin{aligned} \mathcal{L}(\boldsymbol{\theta}, \boldsymbol{\phi}; \mathbf{x}^{(i)}) &= \int q_{\boldsymbol{\phi}}(\mathbf{z}|\mathbf{x}) \left(\log p_{\boldsymbol{\theta}}(\mathbf{x}^{(i)}|\mathbf{z}) + \log p_{\boldsymbol{\theta}}(\mathbf{z}) - \log q_{\boldsymbol{\phi}}(\mathbf{z}|\mathbf{x}) \right) d\mathbf{z} \\ &= \int p(\boldsymbol{\epsilon}) \left(\log p_{\boldsymbol{\theta}}(\mathbf{x}^{(i)}|\mathbf{z}) + \log p_{\boldsymbol{\theta}}(\mathbf{z}) - \log q_{\boldsymbol{\phi}}(\mathbf{z}|\mathbf{x}) \right) \Bigg|_{\mathbf{z} = g_{\boldsymbol{\phi}}(\boldsymbol{\epsilon}, \mathbf{x}^{(i)})} d\boldsymbol{\epsilon} \end{aligned} \qquad (18)$$

The same can be done for the approximate posterior $q_{\boldsymbol{\phi}}(\boldsymbol{\theta})$:

$$\widetilde{\boldsymbol{\theta}} = h_{\boldsymbol{\phi}}(\boldsymbol{\zeta}) \quad \text{with} \quad \boldsymbol{\zeta} \sim p(\boldsymbol{\zeta}) \qquad (19)$$

where we, similarly as above, choose a prior $p(\boldsymbol{\zeta})$ and a function $h_{\boldsymbol{\phi}}(\boldsymbol{\zeta})$ such that the following holds:

$$\begin{aligned} \mathcal{L}(\boldsymbol{\phi}; \mathbf{X}) &= \int q_{\boldsymbol{\phi}}(\boldsymbol{\theta}) \left(\log p_{\boldsymbol{\theta}}(\mathbf{X}) + \log p_{\boldsymbol{\alpha}}(\boldsymbol{\theta}) - \log q_{\boldsymbol{\phi}}(\boldsymbol{\theta}) \right) d\boldsymbol{\theta} \\ &= \int p(\boldsymbol{\zeta}) \left(\log p_{\boldsymbol{\theta}}(\mathbf{X}) + \log p_{\boldsymbol{\alpha}}(\boldsymbol{\theta}) - \log q_{\boldsymbol{\phi}}(\boldsymbol{\theta}) \right) \Bigg|_{\boldsymbol{\theta} = h_{\boldsymbol{\phi}}(\boldsymbol{\zeta})} d\boldsymbol{\zeta} \end{aligned} \qquad (20)$$

For notational conciseness we introduce a shorthand notation $f_{\boldsymbol{\phi}}(\mathbf{x}, \mathbf{z}, \boldsymbol{\theta})$:

$$f_{\boldsymbol{\phi}}(\mathbf{x}, \mathbf{z}, \boldsymbol{\theta}) = N \cdot (\log p_{\boldsymbol{\theta}}(\mathbf{x}|\mathbf{z}) + \log p_{\boldsymbol{\theta}}(\mathbf{z}) - \log q_{\boldsymbol{\phi}}(\mathbf{z}|\mathbf{x})) + \log p_{\boldsymbol{\alpha}}(\boldsymbol{\theta}) - \log q_{\boldsymbol{\phi}}(\boldsymbol{\theta}) \qquad (21)$$

Using equations (20) and (18), the Monte Carlo estimate of the variational lower bound, given datapoint $\mathbf{x}^{(i)}$, is:

$$\mathcal{L}(\boldsymbol{\phi}; \mathbf{X}) \simeq \frac{1}{L} \sum_{l=1}^{L} f_{\boldsymbol{\phi}}(\mathbf{x}^{(l)}, g_{\boldsymbol{\phi}}(\boldsymbol{\epsilon}^{(l)}, \mathbf{x}^{(l)}), h_{\boldsymbol{\phi}}(\boldsymbol{\zeta}^{(l)})) \qquad (22)$$

where $\boldsymbol{\epsilon}^{(l)} \sim p(\boldsymbol{\epsilon})$ and $\boldsymbol{\zeta}^{(l)} \sim p(\boldsymbol{\zeta})$. The estimator only depends on samples from $p(\boldsymbol{\epsilon})$ and $p(\boldsymbol{\zeta})$ which are obviously not influenced by $\boldsymbol{\phi}$, therefore the estimator can be differentiated w.r.t. $\boldsymbol{\phi}$. The resulting stochastic gradients can be used in conjunction with stochastic optimization methods such as SGD or Adagrad [DHS10]. See algorithm 1 for a basic approach to computing stochastic gradients.

F.1 Example

Let the prior over the parameters and latent variables be the centered isotropic Gaussian $p_{\boldsymbol{\alpha}}(\boldsymbol{\theta}) = \mathcal{N}(\mathbf{z}; \mathbf{0}, \mathbf{I})$ and $p_{\boldsymbol{\theta}}(\mathbf{z}) = \mathcal{N}(\mathbf{z}; \mathbf{0}, \mathbf{I})$. Note that in this case, the prior lacks parameters. Let's also assume that the true posteriors are approximatily Gaussian with an approximately diagonal covariance. In this case, we can let the variational approximate posteriors be multivariate Gaussians with a diagonal covariance structure:

$$\begin{aligned} \log q_{\boldsymbol{\phi}}(\boldsymbol{\theta}) &= \log \mathcal{N}(\boldsymbol{\theta}; \boldsymbol{\mu}_{\boldsymbol{\theta}}, \sigma_{\boldsymbol{\theta}}^2 \mathbf{I}) \\ \log q_{\boldsymbol{\phi}}(\mathbf{z}|\mathbf{x}) &= \log \mathcal{N}(\mathbf{z}; \boldsymbol{\mu}_{\mathbf{z}}, \sigma_{\mathbf{z}}^2 \mathbf{I}) \end{aligned} \qquad (23)$$

Algorithm 2 Pseudocode for computing a stochastic gradient using our estimator. See text for meaning of the functions f_ϕ, g_ϕ and h_ϕ.

Require: ϕ (Current value of variational parameters)
   $\mathbf{g} \leftarrow 0$
   for l is 1 to L **do**
      $\mathbf{x} \leftarrow$ Random draw from dataset $\mathbf{X}$
      $\epsilon \leftarrow$ Random draw from prior $p(\epsilon)$
      $\zeta \leftarrow$ Random draw from prior $p(\zeta)$
      $\mathbf{g} \leftarrow \mathbf{g} + \frac{1}{L}\nabla_\phi f_\phi(\mathbf{x}, g_\phi(\epsilon, \mathbf{x}), h_\phi(\zeta))$
   end for
   return $\mathbf{g}$

---

where $\boldsymbol{\mu}_\mathbf{z}$ and $\boldsymbol{\sigma}_\mathbf{z}$ are yet unspecified functions of $\mathbf{x}$. Since they are Gaussian, we can parameterize the variational approximate posteriors:

$$q_\phi(\boldsymbol{\theta}) \quad \text{as} \quad \widetilde{\boldsymbol{\theta}} = \boldsymbol{\mu}_{\boldsymbol{\theta}} + \boldsymbol{\sigma}_{\boldsymbol{\theta}} \odot \boldsymbol{\zeta} \qquad \text{where} \quad \boldsymbol{\zeta} \sim \mathcal{N}(\mathbf{0}, \mathbf{I})$$

$$q_\phi(\mathbf{z}|\mathbf{x}) \quad \text{as} \quad \widetilde{\mathbf{z}} = \boldsymbol{\mu}_\mathbf{z} + \boldsymbol{\sigma}_\mathbf{z} \odot \boldsymbol{\epsilon} \qquad \text{where} \quad \boldsymbol{\epsilon} \sim \mathcal{N}(\mathbf{0}, \mathbf{I})$$

With $\odot$ we signify an element-wise product. These can be plugged into the lower bound defined above (eqs (21) and (22)).

In this case it is possible to construct an alternative estimator with a lower variance, since in this model $p_\alpha(\boldsymbol{\theta})$, $p_{\boldsymbol{\theta}}(\mathbf{z})$, $q_\phi(\boldsymbol{\theta})$ and $q_\phi(\mathbf{z}|\mathbf{x})$ are Gaussian, and therefore four terms of f_ϕ can be solved analytically. The resulting estimator is:

$$\mathcal{L}(\phi; \mathbf{X}) \simeq \frac{1}{L}\sum_{l=1}^{L} N \cdot \left(\frac{1}{2}\sum_{j=1}^{J}\left(1 + \log((\sigma_{\mathbf{z},j}^{(l)})^2) - (\mu_{\mathbf{z},j}^{(l)})^2 - (\sigma_{\mathbf{z},j}^{(l)})^2 \right) + \log p_{\boldsymbol{\theta}}(\mathbf{x}^{(i)}\mathbf{z}^{(i)}) \right)$$

$$+ \frac{1}{2}\sum_{j=1}^{J}\left(1 + \log((\sigma_{\boldsymbol{\theta},j}^{(l)})^2) - (\mu_{\boldsymbol{\theta},j}^{(l)})^2 - (\sigma_{\boldsymbol{\theta},j}^{(l)})^2 \right) \tag{24}$$

$\mu_j^{(i)}$ and $\sigma_j^{(i)}$ simply denote the j-th element of vectors $\boldsymbol{\mu}^{(i)}$ and $\boldsymbol{\sigma}^{(i)}$.

Chapter 10

Generative Adversarial Networks – 2014

Context

A generative adversarial network (GAN) can create new and realistic images, as shown in Figure 10.1. It consists of two separate networks, a *generator network* (G) and a *discriminator network* (D), as shown in Figure 10.2 (Goodfellow et al., 2014). The discriminator network is presented with images which are either a) created by the generator network or b) copied from a training set of images.

These two networks are pitted against each other. The generator keeps trying to improve its forgeries to fool the discriminator, while the discriminator keeps getting better at spotting the fakes. In a sense, these two networks are engaged in an arms race, in which the generator network gradually learns to produce realistic synthetic images, while the discriminator network simultaneously learns to differentiate between the generator network's images and images from the training set.

Technical Summary

Goodfellow et al. (2014) conceptualise the problem faced by a GAN by considering images in the training set as being samples from an underlying distribution $p_x(X)$ of images $X_1, X_2, \ldots$ taken by a camera (i.e. images of scenes). Similarly, each image produced by the generator network is considered to be a sample from the generator's distribution $p_G(G(\mathbf{z}))$ of images $G(\mathbf{z}_1), G(\mathbf{z}_2), \ldots$. Here, for the generator network G, an input image $\mathbf{z}$ is chosen from a prior probability distribution $p_\mathbf{z}(\mathbf{z})$ of noise images.

Figure 10.1. Synthetic celebrities. These pictures were produced by a generative adversarial network that was trained on 30,000 images of celebrities. Reproduced with permission from Karras et al. (2017).

10.1. The Generative Adversarial Net Architecture

The Generator Network. The distribution $p_G(G(\mathbf{z}))$ is the generator network's estimate of $p_x(X)$; of course, initially $p_G(G(\mathbf{z}))$ is nothing like $p_x(X)$. Ideally, after training, $p_G(G(\mathbf{z})) \approx p_x(X)$, so the generator network would produce images that are indistinguishable from images in the training set. Consequently, when confronted with an equal proportion of images from the generator network and from the training set, the discriminator network would have an accuracy of 50%.

The Discriminator Network. Given an image X, the discriminator network's output $D(X)$ is an estimate of the probability that X was chosen from the distribution of training images $p_x(X)$ (rather than from the generator network's distribution $p_G(X)$). Goodfellow et al. proved that the optimal discriminator network's output is the probability that X was chosen from $p_x(X)$:

$$D^*(X) \;\; = \;\; \frac{p_x(X)}{p_x(X) + p_G(X)}. \tag{10.1}$$

10.2. Training Generative Adversarial Nets

Within each training iteration, first the weights in the discriminator network are adjusted repeatedly, and then the weights in the generator network are adjusted once.

Training the Discriminator Network. A sample $\{X\}$ of m images is chosen, and a sample $\{\mathbf{z}\}$ of m noise images is chosen. The weights are adjusted so as to increase the probability that the discriminator network will increase its output (i.e. make $D(X)$ closer to 1) for members of $\{X\}$ and decrease its output (i.e. make $D(X)$ closer to 0) for members of $\{\mathbf{z}\}$. These weight adjustments to the discriminator network are made using backprop by gradient ascent on the function

$$E_D \;\; = \;\; \frac{1}{m} \sum_{i=1}^{m} \Big[\log D(X_i) + \log\big(1 - D(G(\mathbf{z}_i))\big) \Big]. \tag{10.2}$$

Notice that maximising E_D maximises $D(X_i)$, which is the discriminator network's estimate of the probability that the image X_i belongs to the training set. Also, maximising $\log(1 - D(G(\mathbf{z}_i)))$

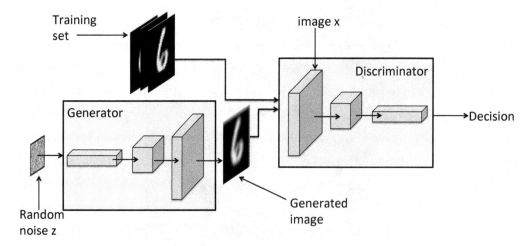

Figure 10.2. A generative adversarial network comprises two backprop networks. The discriminator network receives an image from the training set or the generator network. The output $D(X)$ of the discriminator network is the probability that the image X is from the training set. Reproduced from Stone (2019).

minimises $D(G(\mathbf{z}_i))$, which is the discriminator network's estimate of the probability that the image $G(\mathbf{z}_i)$ produced by the generator network belongs to the training set.

Training the Generator Network. A sample $\{\mathbf{z}\}$ of m images is chosen, and weight adjustments are made by gradient descent on the function

$$E_G \;\; = \;\; \frac{1}{m}\sum_{i=1}^{m}\big[\log\big(1 - D(G(\mathbf{z}_i))\big)\big]. \tag{10.3}$$

Minimising E_G is equivalent to minimising $\log(1 - D(G(\mathbf{z}_i)))$ or maximising $D(G(\mathbf{z}_i))$, which is the discriminator network's estimate of the probability that the image $G(\mathbf{z}_i)$ produced by the generator network belongs to the training set. In other words, the generator network weights are adjusted to increase the probability that the discriminator network will be fooled into classifying synthetic images $G(\mathbf{z}_i)$ as belonging to the training set $\{X\}$.

As indicated above, maximising E_D while simultaneously minimising E_G defines an arms race in which the discriminator network becomes increasingly expert at differentiating between synthetic images and training set images, and the generator network becomes increasingly proficient at producing synthetic images that the discriminator network will classify as belonging to the training set. This adversarial research framework has been extended to autoencoders (Radford et al., 2015) and to the production of high-resolution images of synthetic celebrities, as shown in Figure 10.1.

Research Paper: Generative Adversarial Nets

Reference: Goodfellow, I. J., Pouget-Abadie, J., Mirza, M., Xu, B., Warde-Farley, D., Ozair, S., Courville, A., and Bengio, Y. (2014). Generative adversarial nets. *arXiv e-prints*.
https://arxiv.org/abs/1406.2661.

Generative Adversarial Nets

Ian J. Goodfellow, Jean Pouget-Abadie,* Mehdi Mirza, Bing Xu, David Warde-Farley,
Sherjil Ozair,† Aaron Courville, Yoshua Bengio‡
Département d'informatique et de recherche opérationnelle
Université de Montréal
Montréal, QC H3C 3J7

Abstract

We propose a new framework for estimating generative models via an adversarial process, in which we simultaneously train two models: a generative model G that captures the data distribution, and a discriminative model D that estimates the probability that a sample came from the training data rather than G. The training procedure for G is to maximize the probability of D making a mistake. This framework corresponds to a minimax two-player game. In the space of arbitrary functions G and D, a unique solution exists, with G recovering the training data distribution and D equal to $\frac{1}{2}$ everywhere. In the case where G and D are defined by multilayer perceptrons, the entire system can be trained with backpropagation. There is no need for any Markov chains or unrolled approximate inference networks during either training or generation of samples. Experiments demonstrate the potential of the framework through qualitative and quantitative evaluation of the generated samples.

1   Introduction

The promise of deep learning is to discover rich, hierarchical models [2] that represent probability distributions over the kinds of data encountered in artificial intelligence applications, such as natural images, audio waveforms containing speech, and symbols in natural language corpora. So far, the most striking successes in deep learning have involved discriminative models, usually those that map a high-dimensional, rich sensory input to a class label [14, 22]. These striking successes have primarily been based on the backpropagation and dropout algorithms, using piecewise linear units [19, 9, 10] which have a particularly well-behaved gradient . Deep *generative* models have had less of an impact, due to the difficulty of approximating many intractable probabilistic computations that arise in maximum likelihood estimation and related strategies, and due to difficulty of leveraging the benefits of piecewise linear units in the generative context. We propose a new generative model estimation procedure that sidesteps these difficulties. [1]

In the proposed *adversarial nets* framework, the generative model is pitted against an adversary: a discriminative model that learns to determine whether a sample is from the model distribution or the data distribution. The generative model can be thought of as analogous to a team of counterfeiters, trying to produce fake currency and use it without detection, while the discriminative model is analogous to the police, trying to detect the counterfeit currency. Competition in this game drives both teams to improve their methods until the counterfeits are indistiguishable from the genuine articles.

---

*Jean Pouget-Abadie is visiting Université de Montréal from Ecole Polytechnique.

†Sherjil Ozair is visiting Université de Montréal from Indian Institute of Technology Delhi

‡Yoshua Bengio is a CIFAR Senior Fellow.

[1]All code and hyperparameters available at http://www.github.com/goodfeli/adversarial

This framework can yield specific training algorithms for many kinds of model and optimization algorithm. In this article, we explore the special case when the generative model generates samples by passing random noise through a multilayer perceptron, and the discriminative model is also a multilayer perceptron. We refer to this special case as *adversarial nets*. In this case, we can train both models using only the highly successful backpropagation and dropout algorithms [17] and sample from the generative model using only forward propagation. No approximate inference or Markov chains are necessary.

2 Related work

An alternative to directed graphical models with latent variables are undirected graphical models with latent variables, such as restricted Boltzmann machines (RBMs) [27, 16], deep Boltzmann machines (DBMs) [26] and their numerous variants. The interactions within such models are represented as the product of unnormalized potential functions, normalized by a global summation/integration over all states of the random variables. This quantity (the *partition function*) and its gradient are intractable for all but the most trivial instances, although they can be estimated by Markov chain Monte Carlo (MCMC) methods. Mixing poses a significant problem for learning algorithms that rely on MCMC [3, 5].

Deep belief networks (DBNs) [16] are hybrid models containing a single undirected layer and several directed layers. While a fast approximate layer-wise training criterion exists, DBNs incur the computational difficulties associated with both undirected and directed models.

Alternative criteria that do not approximate or bound the log-likelihood have also been proposed, such as score matching [18] and noise-contrastive estimation (NCE) [13]. Both of these require the learned probability density to be analytically specified up to a normalization constant. Note that in many interesting generative models with several layers of latent variables (such as DBNs and DBMs), it is not even possible to derive a tractable unnormalized probability density. Some models such as denoising auto-encoders [30] and contractive autoencoders have learning rules very similar to score matching applied to RBMs. In NCE, as in this work, a discriminative training criterion is employed to fit a generative model. However, rather than fitting a separate discriminative model, the generative model itself is used to discriminate generated data from samples a fixed noise distribution. Because NCE uses a fixed noise distribution, learning slows dramatically after the model has learned even an approximately correct distribution over a small subset of the observed variables.

Finally, some techniques do not involve defining a probability distribution explicitly, but rather train a generative machine to draw samples from the desired distribution. This approach has the advantage that such machines can be designed to be trained by back-propagation. Prominent recent work in this area includes the generative stochastic network (GSN) framework [5], which extends generalized denoising auto-encoders [4]: both can be seen as defining a parameterized Markov chain, i.e., one learns the parameters of a machine that performs one step of a generative Markov chain. Compared to GSNs, the adversarial nets framework does not require a Markov chain for sampling. Because adversarial nets do not require feedback loops during generation, they are better able to leverage piecewise linear units [19, 9, 10], which improve the performance of backpropagation but have problems with unbounded activation when used in a feedback loop. More recent examples of training a generative machine by back-propagating into it include recent work on auto-encoding variational Bayes [20] and stochastic backpropagation [24].

3 Adversarial nets

The adversarial modeling framework is most straightforward to apply when the models are both multilayer perceptrons. To learn the generator's distribution p_g over data x, we define a prior on input noise variables $p_z(z)$, then represent a mapping to data space as $G(z; \theta_g)$, where G is a differentiable function represented by a multilayer perceptron with parameters θ_g. We also define a second multilayer perceptron $D(x; \theta_d)$ that outputs a single scalar. $D(x)$ represents the probability that x came from the data rather than p_g. We train D to maximize the probability of assigning the correct label to both training examples and samples from G. We simultaneously train G to minimize $\log(1 - D(G(z)))$:

In other words, D and G play the following two-player minimax game with value function $V(G, D)$:

$$\min_G \max_D V(D, G) = \mathbb{E}_{\boldsymbol{x} \sim p_{\text{data}}(\boldsymbol{x})}[\log D(\boldsymbol{x})] + \mathbb{E}_{\boldsymbol{z} \sim p_{\boldsymbol{z}}(\boldsymbol{z})}[\log(1 - D(G(\boldsymbol{z})))]. \qquad (1)$$

In the next section, we present a theoretical analysis of adversarial nets, essentially showing that the training criterion allows one to recover the data generating distribution as G and D are given enough capacity, i.e., in the non-parametric limit. See Figure 1 for a less formal, more pedagogical explanation of the approach. In practice, we must implement the game using an iterative, numerical approach. Optimizing D to completion in the inner loop of training is computationally prohibitive, and on finite datasets would result in overfitting. Instead, we alternate between k steps of optimizing D and one step of optimizing G. This results in D being maintained near its optimal solution, so long as G changes slowly enough. This strategy is analogous to the way that SML/PCD [31, 29] training maintains samples from a Markov chain from one learning step to the next in order to avoid burning in a Markov chain as part of the inner loop of learning. The procedure is formally presented in Algorithm 1.

In practice, equation 1 may not provide sufficient gradient for G to learn well. Early in learning, when G is poor, D can reject samples with high confidence because they are clearly different from the training data. In this case, $\log(1 - D(G(\boldsymbol{z})))$ saturates. Rather than training G to minimize $\log(1 - D(G(\boldsymbol{z})))$ we can train G to maximize $\log D(G(\boldsymbol{z}))$. This objective function results in the same fixed point of the dynamics of G and D but provides much stronger gradients early in learning.

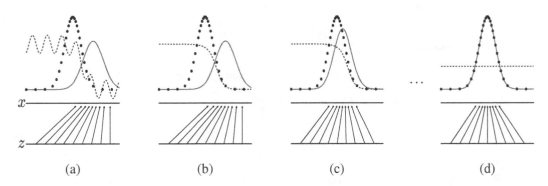

(a)   (b)   (c)   (d)

Figure 1: Generative adversarial nets are trained by simultaneously updating the **d**iscriminative distribution (D, blue, dashed line) so that it discriminates between samples from the data generating distribution (black, dotted line) $p_{\boldsymbol{x}}$ from those of the **g**enerative distribution p_g (G) (green, solid line). The lower horizontal line is the domain from which $\boldsymbol{z}$ is sampled, in this case uniformly. The horizontal line above is part of the domain of $\boldsymbol{x}$. The upward arrows show how the mapping $\boldsymbol{x} = G(\boldsymbol{z})$ imposes the non-uniform distribution p_g on transformed samples. G contracts in regions of high density and expands in regions of low density of p_g. (a) Consider an adversarial pair near convergence: p_g is similar to p_{data} and D is a partially accurate classifier. (b) In the inner loop of the algorithm D is trained to discriminate samples from data, converging to $D^*(\boldsymbol{x}) = \frac{p_{\text{data}}(\boldsymbol{x})}{p_{\text{data}}(\boldsymbol{x}) + p_g(\boldsymbol{x})}$. (c) After an update to G, gradient of D has guided $G(\boldsymbol{z})$ to flow to regions that are more likely to be classified as data. (d) After several steps of training, if G and D have enough capacity, they will reach a point at which both cannot improve because $p_g = p_{\text{data}}$. The discriminator is unable to differentiate between the two distributions, i.e. $D(\boldsymbol{x}) = \frac{1}{2}$.

4   Theoretical Results

The generator G implicitly defines a probability distribution p_g as the distribution of the samples $G(\boldsymbol{z})$ obtained when $\boldsymbol{z} \sim p_{\boldsymbol{z}}$. Therefore, we would like Algorithm 1 to converge to a good estimator of p_{data}, if given enough capacity and training time. The results of this section are done in a non-parametric setting, e.g. we represent a model with infinite capacity by studying convergence in the space of probability density functions.

We will show in section 4.1 that this minimax game has a global optimum for $p_g = p_{\text{data}}$. We will then show in section 4.2 that Algorithm 1 optimizes Eq 1, thus obtaining the desired result.

Algorithm 1 Minibatch stochastic gradient descent training of generative adversarial nets. The number of steps to apply to the discriminator, k, is a hyperparameter. We used $k = 1$, the least expensive option, in our experiments.

---

for number of training iterations **do**

    for k steps **do**

        • Sample minibatch of m noise samples $\{z^{(1)}, \ldots, z^{(m)}\}$ from noise prior $p_g(z)$.

        • Sample minibatch of m examples $\{x^{(1)}, \ldots, x^{(m)}\}$ from data generating distribution $p_{\text{data}}(x)$.

        • Update the discriminator by ascending its stochastic gradient:

$$\nabla_{\theta_d} \frac{1}{m} \sum_{i=1}^{m} \left[\log D\left(x^{(i)}\right) + \log\left(1 - D\left(G\left(z^{(i)}\right)\right)\right) \right].$$

    end for

    • Sample minibatch of m noise samples $\{z^{(1)}, \ldots, z^{(m)}\}$ from noise prior $p_g(z)$.

    • Update the generator by descending its stochastic gradient:

$$\nabla_{\theta_g} \frac{1}{m} \sum_{i=1}^{m} \log\left(1 - D\left(G\left(z^{(i)}\right)\right)\right).$$

end for

The gradient-based updates can use any standard gradient-based learning rule. We used momentum in our experiments.

4.1 Global Optimality of $p_g = p_{\text{data}}$

We first consider the optimal discriminator D for any given generator G.

Proposition 1. *For G fixed, the optimal discriminator D is*

$$D_G^*(x) = \frac{p_{data}(x)}{p_{data}(x) + p_g(x)} \tag{2}$$

Proof. The training criterion for the discriminator D, given any generator G, is to maximize the quantity $V(G, D)$

$$
\begin{aligned}
V(G, D) &= \int_x p_{\text{data}}(x) \log(D(x)) dx + \int_z p_z(z) \log(1 - D(g(z))) dz \\
&= \int_x p_{\text{data}}(x) \log(D(x)) + p_g(x) \log(1 - D(x)) dx
\end{aligned}
\tag{3}
$$

For any $(a, b) \in \mathbb{R}^2 \setminus \{0, 0\}$, the function $y \to a \log(y) + b \log(1 - y)$ achieves its maximum in $[0, 1]$ at $\frac{a}{a+b}$. The discriminator does not need to be defined outside of $Supp(p_{\text{data}}) \cup Supp(p_g)$, concluding the proof. $\square$

Note that the training objective for D can be interpreted as maximizing the log-likelihood for estimating the conditional probability $P(Y = y|x)$, where Y indicates whether x comes from p_{data} (with $y = 1$) or from p_g (with $y = 0$). The minimax game in Eq. 1 can now be reformulated as:

$$
\begin{aligned}
C(G) &= \max_D V(G, D) \\
&= \mathbb{E}_{x \sim p_{\text{data}}}[\log D_G^*(x)] + \mathbb{E}_{z \sim p_z}[\log(1 - D_G^*(G(z)))] \\
&= \mathbb{E}_{x \sim p_{\text{data}}}[\log D_G^*(x)] + \mathbb{E}_{x \sim p_g}[\log(1 - D_G^*(x))] \\
&= \mathbb{E}_{x \sim p_{\text{data}}}\left[\log \frac{p_{\text{data}}(x)}{P_{\text{data}}(x) + p_g(x)}\right] + \mathbb{E}_{x \sim p_g}\left[\log \frac{p_g(x)}{p_{\text{data}}(x) + p_g(x)}\right]
\end{aligned}
\tag{4}
$$

Theorem 1. *The global minimum of the virtual training criterion $C(G)$ is achieved if and only if $p_g = p_{data}$. At that point, $C(G)$ achieves the value $-\log 4$.*

Proof. For $p_g = p_{data}$, $D_G^*(\boldsymbol{x}) = \frac{1}{2}$, (consider Eq. 2). Hence, by inspecting Eq. 4 at $D_G^*(\boldsymbol{x}) = \frac{1}{2}$, we find $C(G) = \log \frac{1}{2} + \log \frac{1}{2} = -\log 4$. To see that this is the best possible value of $C(G)$, reached only for $p_g = p_{data}$, observe that

$$\mathbb{E}_{\boldsymbol{x} \sim p_{data}}\left[-\log 2\right] + \mathbb{E}_{\boldsymbol{x} \sim p_g}\left[-\log 2\right] = -\log 4$$

and that by subtracting this expression from $C(G) = V(D_G^*, G)$, we obtain:

$$C(G) = -\log(4) + KL\left(p_{data} \left\| \frac{p_{data} + p_g}{2}\right.\right) + KL\left(p_g \left\| \frac{p_{data} + p_g}{2}\right.\right) \tag{5}$$

where KL is the Kullback–Leibler divergence. We recognize in the previous expression the Jensen–Shannon divergence between the model's distribution and the data generating process:

$$C(G) = -\log(4) + 2 \cdot JSD\left(p_{data} \| p_g\right) \tag{6}$$

Since the Jensen–Shannon divergence between two distributions is always non-negative and zero only when they are equal, we have shown that $C^* = -\log(4)$ is the global minimum of $C(G)$ and that the only solution is $p_g = p_{data}$, i.e., the generative model perfectly replicating the data generating process. $\qquad\square$

4.2 Convergence of Algorithm 1

Proposition 2. *If G and D have enough capacity, and at each step of Algorithm 1, the discriminator is allowed to reach its optimum given G, and p_g is updated so as to improve the criterion*

$$\mathbb{E}_{\boldsymbol{x} \sim p_{data}}[\log D_G^*(\boldsymbol{x})] + \mathbb{E}_{\boldsymbol{x} \sim p_g}[\log(1 - D_G^*(\boldsymbol{x}))]$$

then p_g converges to p_{data}

Proof. Consider $V(G, D) = U(p_g, D)$ as a function of p_g as done in the above criterion. Note that $U(p_g, D)$ is convex in p_g. The subderivatives of a supremum of convex functions include the derivative of the function at the point where the maximum is attained. In other words, if $f(x) = \sup_{\alpha \in \mathcal{A}} f_\alpha(x)$ and $f_\alpha(x)$ is convex in x for every α, then $\partial f_\beta(x) \in \partial f$ if $\beta = \arg \sup_{\alpha \in \mathcal{A}} f_\alpha(x)$. This is equivalent to computing a gradient descent update for p_g at the optimal D given the corresponding G. $\sup_D U(p_g, D)$ is convex in p_g with a unique global optima as proven in Thm 1, therefore with sufficiently small updates of p_g, p_g converges to p_x, concluding the proof. $\qquad\square$

In practice, adversarial nets represent a limited family of p_g distributions via the function $G(\boldsymbol{z}; \theta_g)$, and we optimize θ_g rather than p_g itself. Using a multilayer perceptron to define G introduces multiple critical points in parameter space. However, the excellent performance of multilayer perceptrons in practice suggests that they are a reasonable model to use despite their lack of theoretical guarantees.

5 Experiments

We trained adversarial nets an a range of datasets including MNIST[23], the Toronto Face Database (TFD) [28], and CIFAR-10 [21]. The generator nets used a mixture of rectifier linear activations [19, 9] and sigmoid activations, while the discriminator net used maxout [10] activations. Dropout [17] was applied in training the discriminator net. While our theoretical framework permits the use of dropout and other noise at intermediate layers of the generator, we used noise as the input to only the bottommost layer of the generator network.

We estimate probability of the test set data under p_g by fitting a Gaussian Parzen window to the samples generated with G and reporting the log-likelihood under this distribution. The σ parameter

Model	MNIST	TFD
DBN [3]	138 ± 2	1909 ± 66
Stacked CAE [3]	121 ± 1.6	$\mathbf{2110 \pm 50}$
Deep GSN [6]	214 ± 1.1	1890 ± 29
Adversarial nets	$\mathbf{225 \pm 2}$	$\mathbf{2057 \pm 26}$

Table 1: Parzen window-based log-likelihood estimates. The reported numbers on MNIST are the mean log-likelihood of samples on test set, with the standard error of the mean computed across examples. On TFD, we computed the standard error across folds of the dataset, with a different σ chosen using the validation set of each fold. On TFD, σ was cross validated on each fold and mean log-likelihood on each fold were computed. For MNIST we compare against other models of the real-valued (rather than binary) version of dataset.

of the Gaussians was obtained by cross validation on the validation set. This procedure was introduced in Breuleux *et al.* [8] and used for various generative models for which the exact likelihood is not tractable [25, 3, 5]. Results are reported in Table 1. This method of estimating the likelihood has somewhat high variance and does not perform well in high dimensional spaces but it is the best method available to our knowledge. Advances in generative models that can sample but not estimate likelihood directly motivate further research into how to evaluate such models.

In Figures 2 and 3 we show samples drawn from the generator net after training. While we make no claim that these samples are better than samples generated by existing methods, we believe that these samples are at least competitive with the better generative models in the literature and highlight the potential of the adversarial framework.

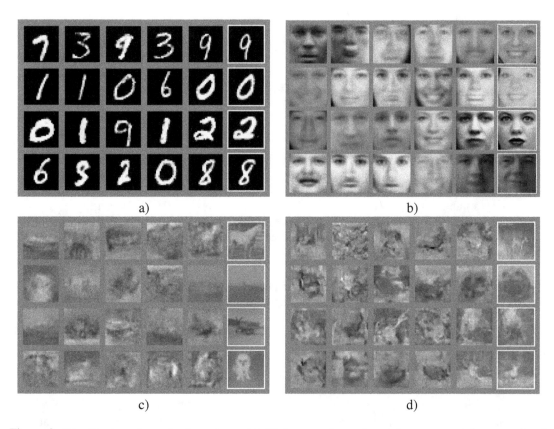

a)   b)

c)   d)

Figure 2: Visualization of samples from the model. Rightmost column shows the nearest training example of the neighboring sample, in order to demonstrate that the model has not memorized the training set. Samples are fair random draws, not cherry-picked. Unlike most other visualizations of deep generative models, these images show actual samples from the model distributions, not conditional means given samples of hidden units. Moreover, these samples are uncorrelated because the sampling process does not depend on Markov chain mixing. a) MNIST b) TFD c) CIFAR-10 (fully connected model) d) CIFAR-10 (convolutional discriminator and "deconvolutional" generator)

Figure 3: Digits obtained by linearly interpolating between coordinates in z space of the full model.

	Deep directed graphical models	Deep undirected graphical models	Generative autoencoders	Adversarial models
Training	Inference needed during training.	Inference needed during training. MCMC needed to approximate partition function gradient.	Enforced tradeoff between mixing and power of reconstruction generation	Synchronizing the discriminator with the generator. Helvetica.
Inference	Learned approximate inference	Variational inference	MCMC-based inference	Learned approximate inference
Sampling	No difficulties	Requires Markov chain	Requires Markov chain	No difficulties
Evaluating $p(x)$	Intractable, may be approximated with AIS	Intractable, may be approximated with AIS	Not explicitly represented, may be approximated with Parzen density estimation	Not explicitly represented, may be approximated with Parzen density estimation
Model design	Nearly all models incur extreme difficulty	Careful design needed to ensure multiple properties	Any differentiable function is theoretically permitted	Any differentiable function is theoretically permitted

Table 2: Challenges in generative modeling: a summary of the difficulties encountered by different approaches to deep generative modeling for each of the major operations involving a model.

6   Advantages and disadvantages

This new framework comes with advantages and disadvantages relative to previous modeling frameworks. The disadvantages are primarily that there is no explicit representation of $p_g(\boldsymbol{x})$, and that D must be synchronized well with G during training (in particular, G must not be trained too much without updating D, in order to avoid "the Helvetica scenario" in which G collapses too many values of $\mathbf{z}$ to the same value of $\mathbf{x}$ to have enough diversity to model p_{data}), much as the negative chains of a Boltzmann machine must be kept up to date between learning steps. The advantages are that Markov chains are never needed, only backprop is used to obtain gradients, no inference is needed during learning, and a wide variety of functions can be incorporated into the model. Table 2 summarizes the comparison of generative adversarial nets with other generative modeling approaches.

The aforementioned advantages are primarily computational. Adversarial models may also gain some statistical advantage from the generator network not being updated directly with data examples, but only with gradients flowing through the discriminator. This means that components of the input are not copied directly into the generator's parameters. Another advantage of adversarial networks is that they can represent very sharp, even degenerate distributions, while methods based on Markov chains require that the distribution be somewhat blurry in order for the chains to be able to mix between modes.

7   Conclusions and future work

This framework admits many straightforward extensions:

1. A *conditional generative* model $p(\boldsymbol{x} \mid \boldsymbol{c})$ can be obtained by adding $\boldsymbol{c}$ as input to both G and D.
2. *Learned approximate inference* can be performed by training an auxiliary network to predict $\boldsymbol{z}$ given $\boldsymbol{x}$. This is similar to the inference net trained by the wake-sleep algorithm [15] but with the advantage that the inference net may be trained for a fixed generator net after the generator net has finished training.

3. One can approximately model all conditionals $p(x_S \mid x_{\not S})$ where S is a subset of the indices of x by training a family of conditional models that share parameters. Essentially, one can use adversarial nets to implement a stochastic extension of the deterministic MP-DBM [11].

4. *Semi-supervised learning*: features from the discriminator or inference net could improve performance of classifiers when limited labeled data is available.

5. *Efficiency improvements:* training could be accelerated greatly by divising better methods for coordinating G and D or determining better distributions to sample $\mathbf{z}$ from during training.

This paper has demonstrated the viability of the adversarial modeling framework, suggesting that these research directions could prove useful.

Acknowledgments

We would like to acknowledge Patrice Marcotte, Olivier Delalleau, Kyunghyun Cho, Guillaume Alain and Jason Yosinski for helpful discussions. Yann Dauphin shared his Parzen window evaluation code with us. We would like to thank the developers of Pylearn2 [12] and Theano [7, 1], particularly Frédéric Bastien who rushed a Theano feature specifically to benefit this project. Arnaud Bergeron provided much-needed support with LATEX typesetting. We would also like to thank CIFAR, and Canada Research Chairs for funding, and Compute Canada, and Calcul Québec for providing computational resources. Ian Goodfellow is supported by the 2013 Google Fellowship in Deep Learning. Finally, we would like to thank Les Trois Brasseurs for stimulating our creativity.

References

[1] Bastien, F., Lamblin, P., Pascanu, R., Bergstra, J., Goodfellow, I. J., Bergeron, A., Bouchard, N., and Bengio, Y. (2012). Theano: new features and speed improvements. Deep Learning and Unsupervised Feature Learning NIPS 2012 Workshop.

[2] Bengio, Y. (2009). *Learning deep architectures for AI*. Now Publishers.

[3] Bengio, Y., Mesnil, G., Dauphin, Y., and Rifai, S. (2013a). Better mixing via deep representations. In *ICML'13*.

[4] Bengio, Y., Yao, L., Alain, G., and Vincent, P. (2013b). Generalized denoising auto-encoders as generative models. In *NIPS26*. Nips Foundation.

[5] Bengio, Y., Thibodeau-Laufer, E., and Yosinski, J. (2014a). Deep generative stochastic networks trainable by backprop. In *ICML'14*.

[6] Bengio, Y., Thibodeau-Laufer, E., Alain, G., and Yosinski, J. (2014b). Deep generative stochastic networks trainable by backprop. In *Proceedings of the 30th International Conference on Machine Learning (ICML'14)*.

[7] Bergstra, J., Breuleux, O., Bastien, F., Lamblin, P., Pascanu, R., Desjardins, G., Turian, J., Warde-Farley, D., and Bengio, Y. (2010). Theano: a CPU and GPU math expression compiler. In *Proceedings of the Python for Scientific Computing Conference (SciPy)*. Oral Presentation.

[8] Breuleux, O., Bengio, Y., and Vincent, P. (2011). Quickly generating representative samples from an RBM-derived process. *Neural Computation*, **23**(8), 2053–2073.

[9] Glorot, X., Bordes, A., and Bengio, Y. (2011). Deep sparse rectifier neural networks. In *AISTATS'2011*.

[10] Goodfellow, I. J., Warde-Farley, D., Mirza, M., Courville, A., and Bengio, Y. (2013a). Maxout networks. In *ICML'2013*.

[11] Goodfellow, I. J., Mirza, M., Courville, A., and Bengio, Y. (2013b). Multi-prediction deep Boltzmann machines. In *NIPS'2013*.

[12] Goodfellow, I. J., Warde-Farley, D., Lamblin, P., Dumoulin, V., Mirza, M., Pascanu, R., Bergstra, J., Bastien, F., and Bengio, Y. (2013c). Pylearn2: a machine learning research library. *arXiv preprint arXiv:1308.4214*.

[13] Gutmann, M. and Hyvarinen, A. (2010). Noise-contrastive estimation: A new estimation principle for unnormalized statistical models. In *AISTATS'2010*.

[14] Hinton, G., Deng, L., Dahl, G. E., Mohamed, A., Jaitly, N., Senior, A., Vanhoucke, V., Nguyen, P., Sainath, T., and Kingsbury, B. (2012a). Deep neural networks for acoustic modeling in speech recognition. *IEEE Signal Processing Magazine*, **29**(6), 82–97.

[15] Hinton, G. E., Dayan, P., Frey, B. J., and Neal, R. M. (1995). The wake-sleep algorithm for unsupervised neural networks. *Science*, **268**, 1558–1161.

[16] Hinton, G. E., Osindero, S., and Teh, Y. (2006). A fast learning algorithm for deep belief nets. *Neural Computation*, **18**, 1527–1554.

[17] Hinton, G. E., Srivastava, N., Krizhevsky, A., Sutskever, I., and Salakhutdinov, R. (2012b). Improving neural networks by preventing co-adaptation of feature detectors. Technical report, arXiv:1207.0580.

[18] Hyvärinen, A. (2005). Estimation of non-normalized statistical models using score matching. *J. Machine Learning Res.*, **6**.

[19] Jarrett, K., Kavukcuoglu, K., Ranzato, M., and LeCun, Y. (2009). What is the best multi-stage architecture for object recognition? In *Proc. International Conference on Computer Vision (ICCV'09)*, pages 2146–2153. IEEE.

[20] Kingma, D. P. and Welling, M. (2014). Auto-encoding variational bayes. In *Proceedings of the International Conference on Learning Representations (ICLR)*.

[21] Krizhevsky, A. and Hinton, G. (2009). Learning multiple layers of features from tiny images. Technical report, University of Toronto.

[22] Krizhevsky, A., Sutskever, I., and Hinton, G. (2012). ImageNet classification with deep convolutional neural networks. In *NIPS'2012*.

[23] LeCun, Y., Bottou, L., Bengio, Y., and Haffner, P. (1998). Gradient-based learning applied to document recognition. *Proceedings of the IEEE*, **86**(11), 2278–2324.

[24] Rezende, D. J., Mohamed, S., and Wierstra, D. (2014). Stochastic backpropagation and approximate inference in deep generative models. Technical report, arXiv:1401.4082.

[25] Rifai, S., Bengio, Y., Dauphin, Y., and Vincent, P. (2012). A generative process for sampling contractive auto-encoders. In *ICML'12*.

[26] Salakhutdinov, R. and Hinton, G. E. (2009). Deep Boltzmann machines. In *AISTATS'2009*, pages 448–455.

[27] Smolensky, P. (1986). Information processing in dynamical systems: Foundations of harmony theory. In D. E. Rumelhart and J. L. McClelland, editors, *Parallel Distributed Processing*, volume 1, chapter 6, pages 194–281. MIT Press, Cambridge.

[28] Susskind, J., Anderson, A., and Hinton, G. E. (2010). The Toronto face dataset. Technical Report UTML TR 2010-001, U. Toronto.

[29] Tieleman, T. (2008). Training restricted Boltzmann machines using approximations to the likelihood gradient. In W. W. Cohen, A. McCallum, and S. T. Roweis, editors, *ICML 2008*, pages 1064–1071. ACM.

[30] Vincent, P., Larochelle, H., Bengio, Y., and Manzagol, P.-A. (2008). Extracting and composing robust features with denoising autoencoders. In *ICML 2008*.

[31] Younes, L. (1999). On the convergence of Markovian stochastic algorithms with rapidly decreasing ergodicity rates. *Stochastics and Stochastic Reports*, **65**(3), 177–228.

Chapter 11

Diffusion Models – 2015

Context

Diffusion models were introduced in 2015 by Soh-Dickstein, Weiss, Maheswaranathan and Ganguli, but they only became widely used after being modified by Ho, Jain and Abbeel, who demonstrated in 2020 that these models could be used to generate faces, as shown in Figure 11.1.

In essence, diffusion models work by reversing the process of progressively adding noise to an image. Specifically, adding noise in small steps to an image eventually produces an image that consists of pure noise. Conversely, subtracting noise from an image of pure noise recovers the original image.

While this is obviously true of a single image, the general principle can also be applied to a whole population of training images (e.g. of faces). The result is a system that takes a pure noise image as input and produces an entirely novel image as output. Crucially, in the process of learning to reverse the diffusion process, the system learns the statistical structure of the population of training images. Consequently, each image generated by the reverse process looks as if it were taken from the training images, but (and this is an important point) without being the same as any one of those images. In practice, this means that a system trained on faces can produce realistic images of faces, none of which exist in the training set.

For readers familiar with variational autoencoders (VAEs; Chapter 9), it is worth noting a few key differences between diffusion models and VAEs. First, whereas encoding in VAEs is a single step that is learned, encoding in diffusion models is a multi-step diffusion process in which an image is transformed into pure noise. Second, the latent space in VAEs consists of a relatively small number of dimensions, whereas the latent space of diffusion models has the same number of dimensions as the input. Third, decoding in VAEs is learned as a single-step process, whereas decoding in diffusion models is learned as a multi-step reverse diffusion process.

Figure 11.1. Synethetic faces generated by the denoising diffusion probabilistic model (DDPM) introduced by Ho, Jain and Abbeel (2020).

Technical Summary

Notation. An image with N_r rows and N_c columns of pixels has a total of $N = N_r \times N_c$ pixels, which can be represented by an N-element vector $\mathbf{x}$. The paper considers both binomial and Gaussian noise, but for simplicity we will assume that the noise is Gaussian. Accordingly, the noise value (e.g. grey-level) of a single pixel is assumed to be a sample η from a Gaussian distribution with mean μ and variance σ^2,

$$\eta \;\sim\; \mathcal{N}(\mu, \sigma^2). \tag{11.1}$$

When noise is applied to the N pixels in an image, the result is a noise vector $\boldsymbol{\eta}$ drawn from a multivariate Gaussian distribution with a vector mean of $\mathbf{0}$ and an $N \times N$ covariance matrix $\boldsymbol{\Sigma}$. To simplify the analysis, the covariance matrix is represented as a scaled identity matrix $\boldsymbol{\Sigma} = \beta \mathbf{I}$ with variance β, so that

$$\boldsymbol{\eta} \;\sim\; \mathcal{N}(\mathbf{0}, \boldsymbol{\Sigma}). \tag{11.2}$$

This notation allows the addition of noise $\boldsymbol{\eta}$ to an image $\mathbf{x}$ to be represented succinctly as $\mathbf{x} + \boldsymbol{\eta}$. The notation below follows Sohl-Dickstein et al. (2015), except that instead of superscripts in parentheses we use subscripts.

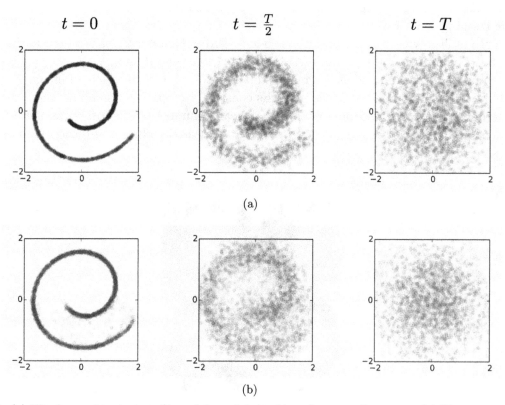

Figure 11.2. (a) The forward trajectory (from left to right) adds noise to an image $\mathbf{x}_0$. (b) The reverse trajectory (from right to left) subtracts noise from an image. From Sohl-Dickstein et al. (2015).

11.1. The Forward Trajectory: Encoding as Diffusion

The process of iteratively adding noise $\boldsymbol{\eta}$ to an image $\mathbf{x}_0$ is called the *forward trajectory*, which consists of T steps,

$$\mathbf{x}_1 = \sqrt{1 - \beta_1}\,\mathbf{x}_0 + \sqrt{\beta_1}\,\boldsymbol{\eta}_0, \tag{11.3}$$

$$\vdots \qquad\qquad \vdots$$

$$\mathbf{x}_T = \sqrt{1 - \beta_T}\,\mathbf{x}_{T-1} + \sqrt{\beta_T}\,\boldsymbol{\eta}_{T-1}, \tag{11.4}$$

where $\boldsymbol{\eta}_t$ is a random sample of noise added at the tth step and the final iteration yields an image of almost pure noise. The parameters $\beta_t \in [0, 1]$ are adjusted according to a *noise schedule*, which determines how quickly the image $\mathbf{x}_0$ is transformed into pure noise. Note that in the context of diffusion noise models, the latent space is considered to be $\{\mathbf{x}_t : t > 0\}$ (i.e. noisy versions of $\mathbf{x}_0$).

The noise has a Gaussian distribution, so the forward trajectory can be written as

$$q(\mathbf{x}_t | \mathbf{x}_{t-1}) = \mathcal{N}\!\left(\sqrt{1 - \beta_t}\,\mathbf{x}_{t-1}, \beta_t \boldsymbol{\Sigma}\right), \tag{11.5}$$

where $q(\mathbf{x}_t | \mathbf{x}_{t-1})$ is the probability of obtaining $\mathbf{x}_t$ by adding noise to $\mathbf{x}_{t-1}$. This defines a *Markov chain* because the probability of each $\mathbf{x}_t$ can be determined based only on the immediately preceding image $\mathbf{x}_{t-1}$.

Using Bayes' theorem (Appendix D), the joint probability distribution after the first forward step is

$$q(\mathbf{x}_0, \mathbf{x}_1) = q(\mathbf{x}_1 | \mathbf{x}_0)\, q(\mathbf{x}_0). \tag{11.6}$$

And, because Equation 11.6 also holds for any consecutive pair of images,

$$q(\mathbf{x}_{t-1}, \mathbf{x}_t) = q(\mathbf{x}_t | \mathbf{x}_{t-1})\, q(\mathbf{x}_{t-1}). \tag{11.7}$$

Therefore, the probability of a trajectory from $\mathbf{x}_0$ to $\mathbf{x}_T$ can be expressed as the product

$$q(\mathbf{x}_0, \mathbf{x}_t, \ldots, \mathbf{x}_T) = q(\mathbf{x}_0) \times q(\mathbf{x}_1 | \mathbf{x}_0) \times q(\mathbf{x}_2 | \mathbf{x}_1) \times \cdots \times q(\mathbf{x}_T | \mathbf{x}_{T-1}), \tag{11.8}$$

which can be written succinctly as (Equation 3 in the paper)

$$q(\mathbf{x}_{0\ldots T}) = q(\mathbf{x}_0) \prod_{t=1}^{T} q(\mathbf{x}_t | \mathbf{x}_{t-1}), \tag{11.9}$$

where each term in the product can be evaluated from Equation 11.5. So far, we have merely defined notation for T steps of the forward trajectory of diffusion. Next, we discuss how this can be used to learn the reverse trajectory.

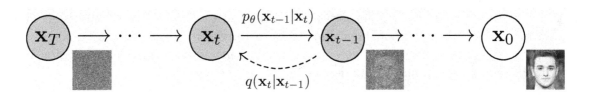

Figure 11.3. Applying the reverse process to images of faces. From Ho, Jain and Abbeel (2020).

11.2. The Reverse Trajectory: Decoding

Like the forward trajectory, the reverse trajectory is defined in terms of the mean and covariance of a multivariate Gaussian distribution. However, in this case, the mean and covariance are estimated using a pair of neural networks. The first network provides an estimate of the mean at time $t-1$, represented by the function

$$\mathbf{x}_{t-1} = \mathbf{f}_{\mu}(\mathbf{x}_t, t), \tag{11.10}$$

where $\mathbf{x}_{t-1}$ is an *estimate* of the image at time step $t-1$ (the paper does not use any special notation to distinguish between the image $\mathbf{x}_{t-1}$ and the estimate of $\mathbf{x}_{t-1}$). The second network provides an estimate of the covariance matrix at time $t-1$, represented by the function

$$\mathbf{\Sigma}_{t-1} = \mathbf{f}_{\Sigma}(\mathbf{x}_t, t). \tag{11.11}$$

The (diagonal) covariance matrix at time step t is parametrised by its variance β_t,

$$\mathbf{\Sigma}_t = \beta_t \mathbf{I}, \tag{11.12}$$

where the variance is learned during training (but the authors note that a fixed schedule also works).

Using the same logic as in the previous section, but now starting from the fully diffused image $\mathbf{x}_T$, the probability of the reverse trajectory from $\mathbf{x}_T$ to $\mathbf{x}_0$ is (Equation 5 in the paper)

$$p(\mathbf{x}_{0...T}) = p(\mathbf{x}_T) \prod_{t=1}^{T} p(\mathbf{x}_{t-1}|\mathbf{x}_t), \tag{11.13}$$

where $p(\mathbf{x}_{t-1}|\mathbf{x}_t)$ is the probability of obtaining $\mathbf{x}_{t-1}$ (by subtracting noise) from $\mathbf{x}_t$. Note that the objective is to find a model for every step $p(\mathbf{x}_{t-1}|\mathbf{x}_t)$ of the reverse trajectory.

11.3. Defining a Lower Bound

Given that the model's estimate of the probability of the forward trajectory is $p(\mathbf{x}_{0...T})$, the model's estimate of the probability of the data $\mathbf{x}_0$ is obtained by marginalising over $p(\mathbf{x}_{0...T})$,

$$p(\mathbf{x}_0) = \int_{\mathbf{x}_{1...T}} p(\mathbf{x}_{0...T}) \, d\mathbf{x}_{1...T}. \tag{11.14}$$

This can be expressed in terms of the forward and reverse trajectories as (Equation 9 in the paper)

$$p(\mathbf{x}_0) = \int_{\mathbf{x}_{1...T}} q(\mathbf{x}_{1...T}|\mathbf{x}_0) \left[p(\mathbf{x}_T) \prod_{t=1}^{T} \frac{p(\mathbf{x}_{t-1}|\mathbf{x}_t)}{q(\mathbf{x}_t|\mathbf{x}_{t-1})} \right] d\mathbf{x}_{1...T}, \tag{11.15}$$

which is the relative probability of the forward and reverse trajectories (the expression in square brackets) averaged over the conditional distribution of the forward trajectories $q(\mathbf{x}_{1...T}|\mathbf{x}_0)$.

At this point, we diverge from the derivation in the paper and use a simpler derivation based on Prince (2023). Taking logarithms of Equation 11.14 defines the log likelihood

$$L = \log \int_{\mathbf{x}_{1...T}} p(\mathbf{x}_{0...T}) \, d\mathbf{x}_{1...T}. \tag{11.16}$$

As with the VAE (Chapter 9), this has a well-defined evidence lower bound (ELBO). To obtain an expression for the ELBO, we first multiply and divide the integrand in Equation 11.16 by $q(\mathbf{x}_{1...T}|\mathbf{x}_0)$,

$$L \;=\; \log \int_{\mathbf{x}_{1...T}} q(\mathbf{x}_{1...T}|\mathbf{x}_0)\, \frac{p(\mathbf{x}_{0...T})}{q(\mathbf{x}_{1...T}|\mathbf{x}_0)}\, d\mathbf{x}_{1...T}. \tag{11.17}$$

Jensen's inequality implies that for any probability distributions $p(x)$ and $q(x)$,

$$\int_x q(x) \log \frac{p(x)}{q(x)}\, dx \;\leq\; \log \int_x q(x) \frac{p(x)}{q(x)}\, dx. \tag{11.18}$$

When Jensen's inequality is applied to Equation 11.17, it defines an ELBO for L as

$$K \;=\; \int q(\mathbf{x}_{1...T}|\mathbf{x}_0) \log \frac{p(\mathbf{x}_{0...T})}{q(\mathbf{x}_{1...T}|\mathbf{x}_0)}\, d\mathbf{x}_{1...T}, \tag{11.19}$$

such that $K \leq L$. Note that K is minus the *Kullback–Leibler divergence* between $p(\mathbf{x}_{0...T})$ and $q(\mathbf{x}_{1...T}|\mathbf{x}_0)$. Training consists of finding neural network weights that maximise K (Equation 15 in the paper),

$$\hat{p}(\mathbf{x}_{t-1}|\mathbf{x}_t) \;=\; \underset{p(\mathbf{x}_{t-1}|\mathbf{x}_t)}{\operatorname{argmax}} K. \tag{11.20}$$

As stated in Section 2.3 of the paper, if the changes in β are sufficiently small then L attains this lower bound, which suggests that a schedule of small, but finite, changes in β ensures that $L \approx K$.

Learning Reverse Trajectories. Next, we express the log ratio in Equation 11.19 in terms of the reverse trajectory $q(\mathbf{x}_{t-1}|\mathbf{x}_t)$ and the model reverse trajectory $p(\mathbf{x}_{t-1}|\mathbf{x}_t)$. We do this because it leads to a measure of the difference between the reverse trajectory $q(\mathbf{x}_{t-1}|\mathbf{x}_t)$ and the model reverse trajectory $p(\mathbf{x}_{t-1}|\mathbf{x}_t)$, and the neural network can then be trained to minimise this difference. In other words, we can train a neural network to transform an image of pure noise into an image that has a statistical structure similar to the images in the training set. For convenience, we write the log ratio as

$$R \;=\; \log \frac{p(\mathbf{x}_{0...T})}{q(\mathbf{x}_{1...T}|\mathbf{x}_0)}. \tag{11.21}$$

Equations 11.9 and 11.13 can be rewritten with lower limits of $t = 2$ as

$$q(\mathbf{x}_{1...T}|\mathbf{x}_0) \;=\; q(\mathbf{x}_1|\mathbf{x}_0) \prod_{t=2}^{T} q(\mathbf{x}_t|\mathbf{x}_{t-1}), \tag{11.22}$$

$$p(\mathbf{x}_{0...T}) \;=\; p(\mathbf{x}_0|\mathbf{x}_1) \prod_{t=2}^{T} p(\mathbf{x}_{t-1}|\mathbf{x}_t) p(\mathbf{x}_T). \tag{11.23}$$

Substituting these into Equation 11.21 yields

$$\begin{aligned}
R \;&=\; \log \frac{p(\mathbf{x}_0|\mathbf{x}_1) \prod_{t=2}^{T} p(\mathbf{x}_{t-1}|\mathbf{x}_t) p(\mathbf{x}_T)}{q(\mathbf{x}_1|\mathbf{x}_0) \prod_{t=2}^{T} q(\mathbf{x}_t|\mathbf{x}_{t-1})} \\
&=\; \log \frac{p(\mathbf{x}_0|\mathbf{x}_1)}{q(\mathbf{x}_1|\mathbf{x}_0)} + \log \frac{\prod_{t=2}^{T} p(\mathbf{x}_{t-1}|\mathbf{x}_t)}{\prod_{t=2}^{T} q(\mathbf{x}_t|\mathbf{x}_{t-1})} + \log p(\mathbf{x}_T).
\end{aligned} \tag{11.24}$$

Notice that this includes the ratio of the *model reverse trajectory* probability divided by the *forward trajectory* probability. Using Bayes' theorem (Appendix D), we can obtain an expression relating the forward trajectory $q(\mathbf{x}_t|\mathbf{x}_{t-1})$ to the reverse trajectory $q(\mathbf{x}_{t-1}|\mathbf{x}_t)$,

$$q(\mathbf{x}_t|\mathbf{x}_{t-1}) \quad = \quad \frac{q(\mathbf{x}_{t-1}|\mathbf{x}_t)q(\mathbf{x}_t)}{q(\mathbf{x}_{t-1})}. \tag{11.25}$$

In the reverse trajectory, $\mathbf{x}_T$ is the initial state and $\mathbf{x}_0$ is the final state. Even though the state $\mathbf{x}_{t-1}$ depends on the initial state $\mathbf{x}_T$, the information about $\mathbf{x}_T$ is implicit in $\mathbf{x}_t$, that is,

$$q(\mathbf{x}_{t-1}|\mathbf{x}_t) \quad = \quad q(\mathbf{x}_{t-1}|\mathbf{x}_t, \mathbf{x}_T). \tag{11.26}$$

In contrast, the *final* state $\mathbf{x}_0$ of the reverse trajectory provides additional information about $\mathbf{x}_{t-1}$, so that

$$q(\mathbf{x}_{t-1}|\mathbf{x}_t) \quad \neq \quad q(\mathbf{x}_{t-1}|\mathbf{x}_t, \mathbf{x}_0), \tag{11.27}$$

where $q(\mathbf{x}_{t-1}|\mathbf{x}_t, \mathbf{x}_0)$ is usually a better estimate than $q(\mathbf{x}_{t-1}|\mathbf{x}_t)$. Accordingly, we can rewrite Equation 11.25 to include $\mathbf{x}_0$:

$$q(\mathbf{x}_t|\mathbf{x}_{t-1}) \quad = \quad q(\mathbf{x}_t|\mathbf{x}_{t-1}, \mathbf{x}_0) \quad = \quad \frac{q(\mathbf{x}_{t-1}|\mathbf{x}_t, \mathbf{x}_0)q(\mathbf{x}_t|\mathbf{x}_0)}{q(\mathbf{x}_{t-1}|\mathbf{x}_0)}. \tag{11.28}$$

Substituting this into Equation 11.24 and rearranging (a lot) gives

$$R \quad \approx \quad \log p(\mathbf{x}_0|\mathbf{x}_1) + \sum_{t=2}^{T} \log \frac{p(\mathbf{x}_{t-1}|\mathbf{x}_t)}{q(\mathbf{x}_{t-1}|\mathbf{x}_t, \mathbf{x}_0)}, \tag{11.29}$$

and then substituting this expression for the log ratio into Equation 11.19 yields

$$K \quad \approx \quad \int_{\mathbf{x}_{1...T}} q(\mathbf{x}_{1...T}|\mathbf{x}_0) \left[\log p(\mathbf{x}_0|\mathbf{x}_1) + \sum_{t=2}^{T} \log \frac{p(\mathbf{x}_{t-1}|\mathbf{x}_t)}{q(\mathbf{x}_{t-1}|\mathbf{x}_t, \mathbf{x}_0)} \right] d\mathbf{x}_{1...T}. \tag{11.30}$$

Finally, we split K into two integrals,

$$K \quad \approx \quad \int q(\mathbf{x}_{1...T}|\mathbf{x}_0) \log p(\mathbf{x}_0|\mathbf{x}_1) \, d\mathbf{x}_{1...T} + \int \sum_{t=2}^{T} q(\mathbf{x}_{1...T}|\mathbf{x}_0) \log \frac{p(\mathbf{x}_{t-1}|\mathbf{x}_t)}{q(\mathbf{x}_{t-1}|\mathbf{x}_t, \mathbf{x}_0)} \, d\mathbf{x}_{1...T}. \tag{11.31}$$

Each term in the summation in Equation 11.31 can be recognised as minus the Kullback–Leibler divergence between (the probability of) a step from $\mathbf{x}_t$ to $\mathbf{x}_{t-1}$ in the reverse trajectory and the same step in the model reverse trajectory, so that

$$K \quad \approx \quad \mathbb{E}_{q(\mathbf{x}_1|\mathbf{x}_0)} \big[\log p(\mathbf{x}_0|\mathbf{x}_1) \big] - \sum_{t=2}^{T} \mathbb{E}_{q(\mathbf{x}_t|\mathbf{x}_0)} \big[D_{\mathrm{KL}} \big(q(\mathbf{x}_{t-1}|\mathbf{x}_t, \mathbf{x}_0) \| p(\mathbf{x}_{t-1}|\mathbf{x}_t) \big) \big]. \tag{11.32}$$

Thus, if the training process maximises K then the reverse trajectory and the model reverse trajectory are forced to be similar.

11.4. Architecture and Training

As can be seen from Figure D.1 in the paper, the network architecture is quite complicated, and it has been modified many times in subsequent papers. However, the core ideas expressed in this paper remain essentially the same.

Without going into details, training the decoder is implemented by gradient ascent on K. This involves several short-cuts covered in this and subsequent papers. After training, the model has an estimate $p(\mathbf{x}_0)$ of the distribution of training images. Generating novel images is achieved by biasing the model reverse trajectory towards a particular type of image $\mathbf{x}_0$.

11.5. Results

Figure 11.2 demonstrates the method. The top row shows the forward trajectory (from left to right) as noise is added to an image $\mathbf{x}_0$, whereas the bottom row shows an estimate of the reverse trajectory (from right to left) as noise is removed from the image.

When the system is trained on handwritten digits from the MNIST data set, the sample images generated are remarkably similar to the types of images in the training set (see Figure App. 1 in the paper). These images are consistent with numerical results reported in Table 2 in the paper.

11.6. List of Mathematical Symbols

β_t parameter that determines the amount of noise at time step t.

ELBO evidence lower bound.

$\mathbf{f}_\mu(\mathbf{x}_t, t)$ neural network used to estimate the image at time step $t - 1$ in the reverse trajectory.

K ELBO for L, such that $K \leq L$.

L log likelihood of the image $\mathbf{x}_0$.

$p(\mathbf{x}_{0...T})$ reverse trajectory from noisy image $\mathbf{x}_T$ to noise-free image $\mathbf{x}_0$; also the model reverse trajectory.

$\hat{p}(\mathbf{x}_{t-1}|\mathbf{x}_t)$ reverse trajectory estimated by maximising K.

$q(\mathbf{x}_t|\mathbf{x}_{t-1})$ probability of $\mathbf{x}_t$ given the immediately preceding image $\mathbf{x}_{t-1}$.

$q(\mathbf{x}_{0...T})$ forward trajectory.

$\mathbf{x}_0$ noise-free image.

$\mathbf{x}_t$ image with some noise added.

$\mathbf{x}_T$ image of almost pure noise.

$\boldsymbol{\Sigma}_{t-1}$ diagonal covariance matrix in which each element is the covariance of one pixel, as estimated by a neural network $\mathbf{f}_\Sigma(\mathbf{x}_t, t)$.

Research Paper: Deep Unsupervised Learning Using Nonequilibrium Thermodynamics

Reference: Sohl-Dickstein, J., Weiss, E., Maheswaranathan, N., and Ganguli, S. (2015). Deep unsupervised learning using nonequilibrium thermodynamics. *Proceedings of Machine Learning Research*, 37:2256–2265. Proceedings of the 32nd International Conference on Machine Learning. https://arxiv.org/abs/1503.03585.
Reproduced with permission.

Deep Unsupervised Learning using Nonequilibrium Thermodynamics

Jascha Sohl-Dickstein   JASCHA@STANFORD.EDU
Stanford University

Eric A. Weiss   EAWEISS@BERKELEY.EDU
University of California, Berkeley

Niru Maheswaranathan   NIRUM@STANFORD.EDU
Stanford University

Surya Ganguli   SGANGULI@STANFORD.EDU
Stanford University

Abstract

A central problem in machine learning involves modeling complex data-sets using highly flexible families of probability distributions in which learning, sampling, inference, and evaluation are still analytically or computationally tractable. Here, we develop an approach that simultaneously achieves both flexibility and tractability. The essential idea, inspired by non-equilibrium statistical physics, is to systematically and slowly destroy structure in a data distribution through an iterative forward diffusion process. We then learn a reverse diffusion process that restores structure in data, yielding a highly flexible and tractable generative model of the data. This approach allows us to rapidly learn, sample from, and evaluate probabilities in deep generative models with thousands of layers or time steps, as well as to compute conditional and posterior probabilities under the learned model. We additionally release an open source reference implementation of the algorithm.

1. Introduction

Historically, probabilistic models suffer from a tradeoff between two conflicting objectives: *tractability* and *flexibility*. Models that are *tractable* can be analytically evaluated and easily fit to data (e.g. a Gaussian or Laplace). However,

Proceedings of the 32nd International Conference on Machine Learning, Lille, France, 2015. JMLR: W&CP volume 37. Copyright 2015 by the author(s).

these models are unable to aptly describe structure in rich datasets. On the other hand, models that are *flexible* can be molded to fit structure in arbitrary data. For example, we can define models in terms of any (non-negative) function $\phi(\mathbf{x})$ yielding the flexible distribution $p(\mathbf{x}) = \frac{\phi(\mathbf{x})}{Z}$, where Z is a normalization constant. However, computing this normalization constant is generally intractable. Evaluating, training, or drawing samples from such flexible models typically requires a very expensive Monte Carlo process.

A variety of analytic approximations exist which ameliorate, but do not remove, this tradeoff–for instance mean field theory and its expansions (T, 1982; Tanaka, 1998), variational Bayes (Jordan et al., 1999), contrastive divergence (Welling & Hinton, 2002; Hinton, 2002), minimum probability flow (Sohl-Dickstein et al., 2011b;a), minimum KL contraction (Lyu, 2011), proper scoring rules (Gneiting & Raftery, 2007; Parry et al., 2012), score matching (Hyvärinen, 2005), pseudolikelihood (Besag, 1975), loopy belief propagation (Murphy et al., 1999), and many, many more. Non-parametric methods (Gershman & Blei, 2012) can also be very effective[1].

1.1. Diffusion probabilistic models

We present a novel way to define probabilistic models that allows:

1. extreme flexibility in model structure,
2. exact sampling,

---

[1] Non-parametric methods can be seen as transitioning smoothly between tractable and flexible models. For instance, a non-parametric Gaussian mixture model will represent a small amount of data using a single Gaussian, but may represent infinite data as a mixture of an infinite number of Gaussians.

3. easy multiplication with other distributions, e.g. in order to compute a posterior, and

4. the model log likelihood, and the probability of individual states, to be cheaply evaluated.

Our method uses a Markov chain to gradually convert one distribution into another, an idea used in non-equilibrium statistical physics (Jarzynski, 1997) and sequential Monte Carlo (Neal, 2001). We build a generative Markov chain which converts a simple known distribution (e.g. a Gaussian) into a target (data) distribution using a diffusion process. Rather than use this Markov chain to approximately evaluate a model which has been otherwise defined, we explicitly define the probabilistic model as the endpoint of the Markov chain. Since each step in the diffusion chain has an analytically evaluable probability, the full chain can also be analytically evaluated.

Learning in this framework involves estimating small perturbations to a diffusion process. Estimating small perturbations is more tractable than explicitly describing the full distribution with a single, non-analytically-normalizable, potential function. Furthermore, since a diffusion process exists for any smooth target distribution, this method can capture data distributions of arbitrary form.

We demonstrate the utility of these *diffusion probabilistic models* by training high log likelihood models for a two-dimensional swiss roll, binary sequence, handwritten digit (MNIST), and several natural image (CIFAR-10, bark, and dead leaves) datasets.

1.2. Relationship to other work

The wake-sleep algorithm (Hinton, 1995; Dayan et al., 1995) introduced the idea of training inference and generative probabilistic models against each other. This approach remained largely unexplored for nearly two decades, though with some exceptions (Sminchisescu et al., 2006; Kavukcuoglu et al., 2010). There has been a recent explosion of work developing this idea. In (Kingma & Welling, 2013; Gregor et al., 2013; Rezende et al., 2014; Ozair & Bengio, 2014) variational learning and inference algorithms were developed which allow a flexible generative model and posterior distribution over latent variables to be directly trained against each other.

The variational bound in these papers is similar to the one used in our training objective and in the earlier work of (Sminchisescu et al., 2006). However, our motivation and model form are both quite different, and the present work retains the following differences and advantages relative to these techniques:

1. We develop our framework using ideas from physics, quasi-static processes, and annealed importance sampling rather than from variational Bayesian methods.

2. We show how to easily multiply the learned distribution with another probability distribution (eg with a conditional distribution in order to compute a posterior)

3. We address the difficulty that training the inference model can prove particularly challenging in variational inference methods, due to the asymmetry in the objective between the inference and generative models. We restrict the forward (inference) process to a simple functional form, in such a way that the reverse (generative) process will have the same functional form.

4. We train models with thousands of layers (or time steps), rather than only a handful of layers.

5. We provide upper and lower bounds on the entropy production in each layer (or time step)

There are a number of related techniques for training probabilistic models (summarized below) that develop highly flexible forms for generative models, train stochastic trajectories, or learn the reversal of a Bayesian network. Reweighted wake-sleep (Bornschein & Bengio, 2015) develops extensions and improved learning rules for the original wake-sleep algorithm. Generative stochastic networks (Bengio & Thibodeau-Laufer, 2013; Yao et al., 2014) train a Markov kernel to match its equilibrium distribution to the data distribution. Neural autoregressive distribution estimators (Larochelle & Murray, 2011) (and their recurrent (Uria et al., 2013a) and deep (Uria et al., 2013b) extensions) decompose a joint distribution into a sequence of tractable conditional distributions over each dimension. Adversarial networks (Goodfellow et al., 2014) train a generative model against a classifier which attempts to distinguish generated samples from true data. A similar objective in (Schmidhuber, 1992) learns a two-way mapping to a representation with marginally independent units. In (Rippel & Adams, 2013; Dinh et al., 2014) bijective deterministic maps are learned to a latent representation with a simple factorial density function. In (Stuhlmüller et al., 2013) stochastic inverses are learned for Bayesian networks. Mixtures of conditional Gaussian scale mixtures (MCGSMs) (Theis et al., 2012) describe a dataset using Gaussian scale mixtures, with parameters which depend on a sequence of causal neighborhoods. There is additionally significant work learning flexible generative mappings from simple latent distributions to data distributions – early examples including (MacKay, 1995) where neural networks are introduced as generative models, and (Bishop et al., 1998) where a stochastic manifold mapping is learned from a latent space to the data space. We will compare experimentally against adversarial networks and MCGSMs.

Related ideas from physics include the Jarzynski equality (Jarzynski, 1997), known in machine learning as An-

$$t = 0 \qquad\qquad t = \frac{T}{2} \qquad\qquad t = T$$

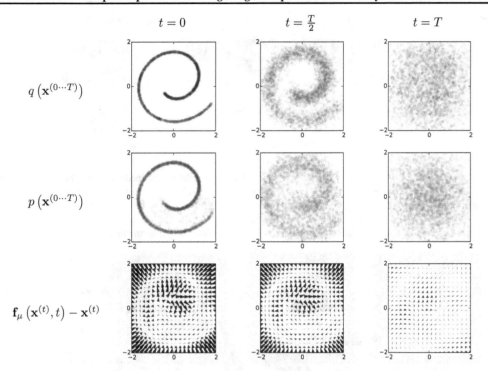

$q\left(\mathbf{x}^{(0\cdots T)}\right)$

$p\left(\mathbf{x}^{(0\cdots T)}\right)$

$\mathbf{f}_\mu\left(\mathbf{x}^{(t)}, t\right) - \mathbf{x}^{(t)}$

Figure 1. The proposed modeling framework trained on 2-d swiss roll data. The top row shows time slices from the forward trajectory $q\left(\mathbf{x}^{(0\cdots T)}\right)$. The data distribution (left) undergoes Gaussian diffusion, which gradually transforms it into an identity-covariance Gaussian (right). The middle row shows the corresponding time slices from the trained reverse trajectory $p\left(\mathbf{x}^{(0\cdots T)}\right)$. An identity-covariance Gaussian (right) undergoes a Gaussian diffusion process with learned mean and covariance functions, and is gradually transformed back into the data distribution (left). The bottom row shows the drift term, $\mathbf{f}_\mu\left(\mathbf{x}^{(t)}, t\right) - \mathbf{x}^{(t)}$, for the same reverse diffusion process.

nealed Importance Sampling (AIS) (Neal, 2001), which uses a Markov chain which slowly converts one distribution into another to compute a ratio of normalizing constants. In (Burda et al., 2014) it is shown that AIS can also be performed using the reverse rather than forward trajectory. Langevin dynamics (Langevin, 1908), which are the stochastic realization of the Fokker-Planck equation, show how to define a Gaussian diffusion process which has any target distribution as its equilibrium. In (Suykens & Vandewalle, 1995) the Fokker-Planck equation is used to perform stochastic optimization. Finally, the Kolmogorov forward and backward equations (Feller, 1949) show that for many forward diffusion processes, the reverse diffusion processes can be described using the same functional form.

2. Algorithm

Our goal is to define a forward (or inference) diffusion process which converts any complex data distribution into a simple, tractable, distribution, and then learn a finite-time reversal of this diffusion process which defines our generative model distribution (See Figure 1). We first describe the forward, inference diffusion process. We then show how the reverse, generative diffusion process can be trained and used to evaluate probabilities. We also derive entropy bounds for the reverse process, and show how the learned distributions can be multiplied by any second distribution (e.g. as would be done to compute a posterior when inpainting or denoising an image).

2.1. Forward Trajectory

We label the data distribution $q\left(\mathbf{x}^{(0)}\right)$. The data distribution is gradually converted into a well behaved (analytically tractable) distribution $\pi(\mathbf{y})$ by repeated application of a Markov diffusion kernel $T_\pi(\mathbf{y}|\mathbf{y}'; \beta)$ for $\pi(\mathbf{y})$, where β is the diffusion rate,

$$\pi(\mathbf{y}) = \int d\mathbf{y}' T_\pi(\mathbf{y}|\mathbf{y}'; \beta)\, \pi(\mathbf{y}') \qquad (1)$$

$$q\left(\mathbf{x}^{(t)}|\mathbf{x}^{(t-1)}\right) = T_\pi\left(\mathbf{x}^{(t)}|\mathbf{x}^{(t-1)}; \beta_t\right). \qquad (2)$$

$$t = 0 \qquad\qquad t = \frac{T}{2} \qquad\qquad t = T$$

$$p\left(\mathbf{x}^{(0\cdots T)}\right)$$

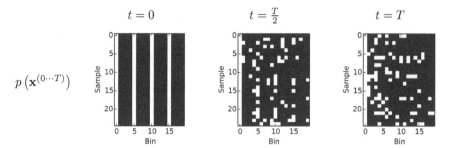

Figure 2. Binary sequence learning via binomial diffusion. A binomial diffusion model was trained on binary 'heartbeat' data, where a pulse occurs every 5th bin. Generated samples (left) are identical to the training data. The sampling procedure consists of initialization at independent binomial noise (right), which is then transformed into the data distribution by a binomial diffusion process, with trained bit flip probabilities. Each row contains an independent sample. For ease of visualization, all samples have been shifted so that a pulse occurs in the first column. In the raw sequence data, the first pulse is uniformly distributed over the first five bins.

(a) (b)

(c) (d)

Figure 3. The proposed framework trained on the CIFAR-10 (Krizhevsky & Hinton, 2009) dataset. *(a)* Example holdout data (similar to training data). *(b)* Holdout data corrupted with Gaussian noise of variance 1 (SNR = 1). *(c)* Denoised images, generated by sampling from the posterior distribution over denoised images conditioned on the images in (b). *(d)* Samples generated by the diffusion model.

The forward trajectory, corresponding to starting at the data distribution and performing T steps of diffusion, is thus

$$q\left(\mathbf{x}^{(0\cdots T)}\right) = q\left(\mathbf{x}^{(0)}\right) \prod_{t=1}^{T} q\left(\mathbf{x}^{(t)}|\mathbf{x}^{(t-1)}\right) \qquad (3)$$

For the experiments shown below, $q\left(\mathbf{x}^{(t)}|\mathbf{x}^{(t-1)}\right)$ corresponds to either Gaussian diffusion into a Gaussian distribution with identity-covariance, or binomial diffusion into an independent binomial distribution. Table App.1 gives the diffusion kernels for both Gaussian and binomial distributions.

2.2. Reverse Trajectory

The generative distribution will be trained to describe the same trajectory, but in reverse,

$$p\left(\mathbf{x}^{(T)}\right) = \pi\left(\mathbf{x}^{(T)}\right) \tag{4}$$

$$p\left(\mathbf{x}^{(0\cdots T)}\right) = p\left(\mathbf{x}^{(T)}\right) \prod_{t=1}^{T} p\left(\mathbf{x}^{(t-1)}|\mathbf{x}^{(t)}\right). \tag{5}$$

For both Gaussian and binomial diffusion, for continuous diffusion (limit of small step size β) the reversal of the diffusion process has the identical functional form as the forward process (Feller, 1949). Since $q\left(\mathbf{x}^{(t)}|\mathbf{x}^{(t-1)}\right)$ is a Gaussian (binomial) distribution, and if β_t is small, then $q\left(\mathbf{x}^{(t-1)}|\mathbf{x}^{(t)}\right)$ will also be a Gaussian (binomial) distribution. The longer the trajectory the smaller the diffusion rate β can be made.

During learning only the mean and covariance for a Gaussian diffusion kernel, or the bit flip probability for a binomial kernel, need be estimated. As shown in Table App.1, $\mathbf{f}_\mu\left(\mathbf{x}^{(t)}, t\right)$ and $\mathbf{f}_\Sigma\left(\mathbf{x}^{(t)}, t\right)$ are functions defining the mean and covariance of the reverse Markov transitions for a Gaussian, and $\mathbf{f}_b\left(\mathbf{x}^{(t)}, t\right)$ is a function providing the bit flip probability for a binomial distribution. The computational cost of running this algorithm is the cost of these functions, times the number of time-steps. For all results in this paper, multi-layer perceptrons are used to define these functions. A wide range of regression or function fitting techniques would be applicable however, including nonparameteric methods.

2.3. Model Probability

The probability the generative model assigns to the data is

$$p\left(\mathbf{x}^{(0)}\right) = \int d\mathbf{x}^{(1\cdots T)} p\left(\mathbf{x}^{(0\cdots T)}\right). \tag{6}$$

Naively this integral is intractable – but taking a cue from annealed importance sampling and the Jarzynski equality, we instead evaluate the relative probability of the forward and reverse trajectories, averaged over forward trajectories,

$$p\left(\mathbf{x}^{(0)}\right) = \int d\mathbf{x}^{(1\cdots T)} p\left(\mathbf{x}^{(0\cdots T)}\right) \frac{q\left(\mathbf{x}^{(1\cdots T)}|\mathbf{x}^{(0)}\right)}{q\left(\mathbf{x}^{(1\cdots T)}|\mathbf{x}^{(0)}\right)} \tag{7}$$

$$= \int d\mathbf{x}^{(1\cdots T)} q\left(\mathbf{x}^{(1\cdots T)}|\mathbf{x}^{(0)}\right) \frac{p\left(\mathbf{x}^{(0\cdots T)}\right)}{q\left(\mathbf{x}^{(1\cdots T)}|\mathbf{x}^{(0)}\right)} \tag{8}$$

$$= \int d\mathbf{x}^{(1\cdots T)} q\left(\mathbf{x}^{(1\cdots T)}|\mathbf{x}^{(0)}\right) \cdot$$
$$p\left(\mathbf{x}^{(T)}\right) \prod_{t=1}^{T} \frac{p\left(\mathbf{x}^{(t-1)}|\mathbf{x}^{(t)}\right)}{q\left(\mathbf{x}^{(t)}|\mathbf{x}^{(t-1)}\right)}. \tag{9}$$

This can be evaluated rapidly by averaging over samples from the forward trajectory $q\left(\mathbf{x}^{(1\cdots T)}|\mathbf{x}^{(0)}\right)$. For infinitesimal β the forward and reverse distribution over trajectories can be made identical (see Section 2.2). If they are identical then only a *single* sample from $q\left(\mathbf{x}^{(1\cdots T)}|\mathbf{x}^{(0)}\right)$ is required to exactly evaluate the above integral, as can be seen by substitution. This corresponds to the case of a quasi-static process in statistical physics (Spinney & Ford, 2013; Jarzynski, 2011).

2.4. Training

Training amounts to maximizing the model log likelihood,

$$L = \int d\mathbf{x}^{(0)} q\left(\mathbf{x}^{(0)}\right) \log p\left(\mathbf{x}^{(0)}\right) \tag{10}$$

$$= \int d\mathbf{x}^{(0)} q\left(\mathbf{x}^{(0)}\right) \cdot$$
$$\log \left[\int d\mathbf{x}^{(1\cdots T)} q\left(\mathbf{x}^{(1\cdots T)}|\mathbf{x}^{(0)}\right) \cdot p\left(\mathbf{x}^{(T)}\right) \prod_{t=1}^{T} \frac{p\left(\mathbf{x}^{(t-1)}|\mathbf{x}^{(t)}\right)}{q\left(\mathbf{x}^{(t)}|\mathbf{x}^{(t-1)}\right)} \right], \tag{11}$$

which has a lower bound provided by Jensen's inequality,

$$L \geq \int d\mathbf{x}^{(0\cdots T)} q\left(\mathbf{x}^{(0\cdots T)}\right) \cdot$$
$$\log \left[p\left(\mathbf{x}^{(T)}\right) \prod_{t=1}^{T} \frac{p\left(\mathbf{x}^{(t-1)}|\mathbf{x}^{(t)}\right)}{q\left(\mathbf{x}^{(t)}|\mathbf{x}^{(t-1)}\right)} \right]. \tag{12}$$

As described in Appendix B, for our diffusion trajectories this reduces to,

$$L \geq K \tag{13}$$

$$K = - \sum_{t=2}^{T} \int d\mathbf{x}^{(0)} d\mathbf{x}^{(t)} q\left(\mathbf{x}^{(0)}, \mathbf{x}^{(t)}\right) \cdot$$
$$D_{KL}\left(q\left(\mathbf{x}^{(t-1)}|\mathbf{x}^{(t)}, \mathbf{x}^{(0)}\right) \middle\| p\left(\mathbf{x}^{(t-1)}|\mathbf{x}^{(t)}\right) \right)$$
$$+ H_q\left(\mathbf{X}^{(T)}|\mathbf{X}^{(0)}\right) - H_q\left(\mathbf{X}^{(1)}|\mathbf{X}^{(0)}\right) - H_p\left(\mathbf{X}^{(T)}\right). \tag{14}$$

where the entropies and KL divergences can be analytically computed. The derivation of this bound parallels the derivation of the log likelihood bound in variational Bayesian methods.

As in Section 2.3 if the forward and reverse trajectories are identical, corresponding to a quasi-static process, then the inequality in Equation 13 becomes an equality.

Training consists of finding the reverse Markov transitions which maximize this lower bound on the log likelihood,

$$\hat{p}\left(\mathbf{x}^{(t-1)}|\mathbf{x}^{(t)}\right) = \underset{p\left(\mathbf{x}^{(t-1)}|\mathbf{x}^{(t)}\right)}{\operatorname{argmax}} K. \tag{15}$$

The specific targets of estimation for Gaussian and binomial diffusion are given in Table App.1.

Thus, the task of estimating a probability distribution has been reduced to the task of performing regression on the functions which set the mean and covariance of a sequence of Gaussians (or set the state flip probability for a sequence of Bernoulli trials).

2.4.1. SETTING THE DIFFUSION RATE β_t

The choice of β_t in the forward trajectory is important for the performance of the trained model. In AIS, the right schedule of intermediate distributions can greatly improve the accuracy of the log partition function estimate (Grosse et al., 2013). In thermodynamics the schedule taken when moving between equilibrium distributions determines how much free energy is lost (Spinney & Ford, 2013; Jarzynski, 2011).

In the case of Gaussian diffusion, we learn[2] the forward diffusion schedule $\beta_{2...T}$ by gradient ascent on K. The variance β_1 of the first step is fixed to a small constant to prevent overfitting. The dependence of samples from $q\left(\mathbf{x}^{(1...T)}|\mathbf{x}^{(0)}\right)$ on $\beta_{1...T}$ is made explicit by using 'frozen noise' – as in (Kingma & Welling, 2013) the noise is treated as an additional auxiliary variable, and held constant while computing partial derivatives of K with respect to the parameters.

For binomial diffusion, the discrete state space makes gradient ascent with frozen noise impossible. We instead choose the forward diffusion schedule $\beta_{1...T}$ to erase a constant fraction $\frac{1}{T}$ of the original signal per diffusion step, yielding a diffusion rate of $\beta_t = (T - t + 1)^{-1}$.

2.5. Multiplying Distributions, and Computing Posteriors

Tasks such as computing a posterior in order to do signal denoising or inference of missing values requires multiplication of the model distribution $p\left(\mathbf{x}^{(0)}\right)$ with a second distribution, or bounded positive function, $r\left(\mathbf{x}^{(0)}\right)$, producing a new distribution $\tilde{p}\left(\mathbf{x}^{(0)}\right) \propto p\left(\mathbf{x}^{(0)}\right) r\left(\mathbf{x}^{(0)}\right)$.

Multiplying distributions is costly and difficult for many techniques, including variational autoencoders, GSNs, NADEs, and most graphical models. However, under a diffusion model it is straightforward, since the second distribution can be treated either as a small perturbation to each step in the diffusion process, or often exactly multiplied into each diffusion step. Figures 3 and 5 demonstrate the use of a diffusion model to perform denoising and inpainting of natural images. The following sections describe how

[2]Recent experiments suggest that it is just as effective to instead use the same fixed β_t schedule as for binomial diffusion.

to multiply distributions in the context of diffusion probabilistic models.

2.5.1. MODIFIED MARGINAL DISTRIBUTIONS

First, in order to compute $\tilde{p}\left(\mathbf{x}^{(0)}\right)$, we multiply each of the intermediate distributions by a corresponding function $r\left(\mathbf{x}^{(t)}\right)$. We use a tilde above a distribution or Markov transition to denote that it belongs to a trajectory that has been modified in this way. $\tilde{p}\left(\mathbf{x}^{(0\cdots T)}\right)$ is the modified reverse trajectory, which starts at the distribution $\tilde{p}\left(\mathbf{x}^{(T)}\right) = \frac{1}{\tilde{Z}_T} p\left(\mathbf{x}^{(T)}\right) r\left(\mathbf{x}^{(T)}\right)$ and proceeds through the sequence of intermediate distributions

$$\tilde{p}\left(\mathbf{x}^{(t)}\right) = \frac{1}{\tilde{Z}_t} p\left(\mathbf{x}^{(t)}\right) r\left(\mathbf{x}^{(t)}\right), \qquad (16)$$

where $\tilde{Z}_t$ is the normalizing constant for the tth intermediate distribution.

2.5.2. MODIFIED DIFFUSION STEPS

The Markov kernel $p\left(\mathbf{x}^{(t)} \mid \mathbf{x}^{(t+1)}\right)$ for the reverse diffusion process obeys the equilibrium condition

$$p\left(\mathbf{x}^{(t)}\right) = \int d\mathbf{x}^{(t+1)} p\left(\mathbf{x}^t \mid \mathbf{x}^{(t+1)}\right) p\left(\mathbf{x}^{(t+1)}\right). \quad (17)$$

We wish the perturbed Markov kernel $\tilde{p}\left(\mathbf{x}^{(t)} \mid \mathbf{x}^{(t+1)}\right)$ to instead obey the equilibrium condition for the perturbed distribution,

$$\tilde{p}\left(\mathbf{x}^{(t)}\right) = \int d\mathbf{x}^{(t+1)} \tilde{p}\left(\mathbf{x}^{(t)} \mid \mathbf{x}^{(t+1)}\right) \tilde{p}\left(\mathbf{x}^{t+1}\right), \tag{18}$$

$$\frac{p\left(\mathbf{x}^{(t)}\right) r\left(\mathbf{x}^{(t)}\right)}{\tilde{Z}_t} = \int d\mathbf{x}^{(t+1)} \tilde{p}\left(\mathbf{x}^{(t)} \mid \mathbf{x}^{(t+1)}\right) \cdot \frac{p\left(\mathbf{x}^{(t+1)}\right) r\left(\mathbf{x}^{(t+1)}\right)}{\tilde{Z}_{t+1}}, \tag{19}$$

$$p\left(\mathbf{x}^{(t)}\right) = \int d\mathbf{x}^{(t+1)} \tilde{p}\left(\mathbf{x}^{(t)} \mid \mathbf{x}^{(t+1)}\right) \cdot \frac{\tilde{Z}_t r\left(\mathbf{x}^{(t+1)}\right)}{\tilde{Z}_{t+1} r\left(\mathbf{x}^{(t)}\right)} p\left(\mathbf{x}^{(t+1)}\right). \tag{20}$$

Equation 20 will be satisfied if

$$\tilde{p}\left(\mathbf{x}^{(t)}|\mathbf{x}^{(t+1)}\right) = p\left(\mathbf{x}^{(t)}|\mathbf{x}^{(t+1)}\right) \frac{\tilde{Z}_{t+1} r\left(\mathbf{x}^{(t)}\right)}{\tilde{Z}_t r\left(\mathbf{x}^{(t+1)}\right)}. \quad (21)$$

Equation 21 may not correspond to a normalized probability distribution, so we choose $\tilde{p}\left(\mathbf{x}^{(t)}|\mathbf{x}^{(t+1)}\right)$ to be the corresponding normalized distribution

$$\tilde{p}\left(\mathbf{x}^{(t)}|\mathbf{x}^{(t+1)}\right) = \frac{1}{\tilde{Z}_t\left(\mathbf{x}^{(t+1)}\right)} p\left(\mathbf{x}^{(t)}|\mathbf{x}^{(t+1)}\right) r\left(\mathbf{x}^{(t)}\right), \tag{22}$$

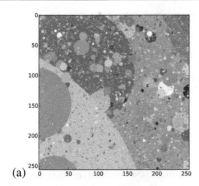 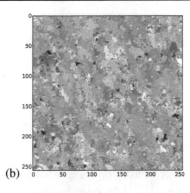 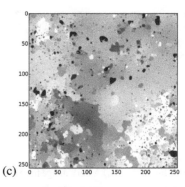

(a)   (b)   (c)

Figure 4. The proposed framework trained on dead leaf images (Jeulin, 1997; Lee et al., 2001). *(a)* Example training image. *(b)* A sample from the previous state of the art natural image model (Theis et al., 2012) trained on identical data, reproduced here with permission. *(c)* A sample generated by the diffusion model. Note that it demonstrates fairly consistent occlusion relationships, displays a multiscale distribution over object sizes, and produces circle-like objects, especially at smaller scales. As shown in Table 2, the diffusion model has the highest log likelihood on the test set.

where $\tilde{Z}_t\left(\mathbf{x}^{(t+1)}\right)$ is the normalization constant.

For a Gaussian, each diffusion step is typically very sharply peaked relative to $r\left(\mathbf{x}^{(t)}\right)$, due to its small variance. This means that $\frac{r\left(\mathbf{x}^{(t)}\right)}{r\left(\mathbf{x}^{(t+1)}\right)}$ can be treated as a small perturbation to $p\left(\mathbf{x}^{(t)}|\mathbf{x}^{(t+1)}\right)$. A small perturbation to a Gaussian effects the mean, but not the normalization constant, so in this case Equations 21 and 22 are equivalent (see Appendix C).

2.5.3. APPLYING $r\left(\mathbf{x}^{(t)}\right)$

If $r\left(\mathbf{x}^{(t)}\right)$ is sufficiently smooth, then it can be treated as a small perturbation to the reverse diffusion kernel $p\left(\mathbf{x}^{(t)}|\mathbf{x}^{(t+1)}\right)$. In this case $\tilde{p}\left(\mathbf{x}^{(t)}|\mathbf{x}^{(t+1)}\right)$ will have an identical functional form to $p\left(\mathbf{x}^{(t)}|\mathbf{x}^{(t+1)}\right)$, but with perturbed mean for the Gaussian kernel, or with perturbed flip rate for the binomial kernel. The perturbed diffusion kernels are given in Table App.1, and are derived for the Gaussian in Appendix C.

If $r\left(\mathbf{x}^{(t)}\right)$ can be multiplied with a Gaussian (or binomial) distribution in closed form, then it can be directly multiplied with the reverse diffusion kernel $p\left(\mathbf{x}^{(t)}|\mathbf{x}^{(t+1)}\right)$ in closed form. This applies in the case where $r\left(\mathbf{x}^{(t)}\right)$ consists of a delta function for some subset of coordinates, as in the inpainting example in Figure 5.

2.5.4. CHOOSING $r\left(\mathbf{x}^{(t)}\right)$

Typically, $r\left(\mathbf{x}^{(t)}\right)$ should be chosen to change slowly over the course of the trajectory. For the experiments in this paper we chose it to be constant,

$$r\left(\mathbf{x}^{(t)}\right) = r\left(\mathbf{x}^{(0)}\right). \qquad (23)$$

Another convenient choice is $r\left(\mathbf{x}^{(t)}\right) = r\left(\mathbf{x}^{(0)}\right)^{\frac{T-t}{T}}$. Under this second choice $r\left(\mathbf{x}^{(t)}\right)$ makes no contribution to the starting distribution for the reverse trajectory. This guarantees that drawing the initial sample from $\tilde{p}\left(\mathbf{x}^{(T)}\right)$ for the reverse trajectory remains straightforward.

2.6. Entropy of Reverse Process

Since the forward process is known, we can derive upper and lower bounds on the conditional entropy of each step in the reverse trajectory, and thus on the log likelihood,

$$H_q\left(\mathbf{X}^{(t)}|\mathbf{X}^{(t-1)}\right) + H_q\left(\mathbf{X}^{(t-1)}|\mathbf{X}^{(0)}\right) - H_q\left(\mathbf{X}^{(t)}|\mathbf{X}^{(0)}\right)$$
$$\le H_q\left(\mathbf{X}^{(t-1)}|\mathbf{X}^{(t)}\right) \le H_q\left(\mathbf{X}^{(t)}|\mathbf{X}^{(t-1)}\right), \qquad (24)$$

where both the upper and lower bounds depend only on $q\left(\mathbf{x}^{(1\cdots T)}|\mathbf{x}^{(0)}\right)$, and can be analytically computed. The derivation is provided in Appendix A.

3. Experiments

We train diffusion probabilistic models on a variety of continuous datasets, and a binary dataset. We then demonstrate sampling from the trained model and inpainting of missing data, and compare model performance against other techniques. In all cases the objective function and gradient were computed using Theano (Bergstra & Breuleux, 2010). Model training was with SFO (Sohl-Dickstein et al., 2014), except for CIFAR-10. CIFAR-10 results used the

---

[3] An earlier version of this paper reported higher log likelihood bounds on CIFAR-10. These were the result of the model learning the 8-bit quantization of pixel values in the CIFAR-10 dataset. The log likelihood bounds reported here are instead for data that has been pre-processed by adding uniform noise to remove pixel quantization, as recommended in (Theis et al., 2015).

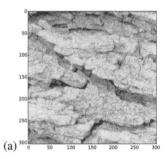 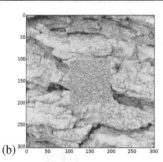 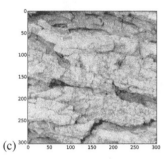

(a)   (b)   (c)

Figure 5. Inpainting. *(a)* A bark image from (Lazebnik et al., 2005). *(b)* The same image with the central 100×100 pixel region replaced with isotropic Gaussian noise. This is the initialization $\tilde{p}\left(\mathbf{x}^{(T)}\right)$ for the reverse trajectory. *(c)* The central 100×100 region has been inpainted using a diffusion probabilistic model trained on images of bark, by sampling from the posterior distribution over the missing region conditioned on the rest of the image. Note the long-range spatial structure, for instance in the crack entering on the left side of the inpainted region. The sample from the posterior was generated as described in Section 2.5, where $r\left(\mathbf{x}^{(0)}\right)$ was set to a delta function for known data, and a constant for missing data.

Dataset	K	$K - L_{null}$
Swiss Roll	2.35 bits	6.45 bits
Binary Heartbeat	-2.414 bits/seq.	12.024 bits/seq.
Bark	-0.55 bits/pixel	1.5 bits/pixel
Dead Leaves	1.489 bits/pixel	3.536 bits/pixel
CIFAR-10[3]	5.4 ± 0.2 bits/pixel	11.5 ± 0.2 bits/pixel
MNIST	See table 2	

Table 1. The lower bound K on the log likelihood, computed on a holdout set, for each of the trained models. See Equation 12. The right column is the improvement relative to an isotropic Gaussian or independent binomial distribution. L_{null} is the log likelihood of $\pi\left(\mathbf{x}^{(0)}\right)$. All datasets except for Binary Heartbeat were scaled by a constant to give them variance 1 before computing log likelihood.

Model	Log Likelihood
Dead Leaves	
MCGSM	1.244 bits/pixel
Diffusion	**1.489 bits/pixel**
MNIST	
Stacked CAE	174 ± 2.3 bits
DBN	199 ± 2.9 bits
Deep GSN	309 ± 1.6 bits
Diffusion	**317 ± 2.7 bits**
Adversarial net	325 ± 2.9 bits
Perfect model	349 ± 3.3 bits

Table 2. Log likelihood comparisons to other algorithms. Dead leaves images were evaluated using identical training and test data as in (Theis et al., 2012). MNIST log likelihoods were estimated using the Parzen-window code from (Goodfellow et al., 2014), with values given in bits, and show that our performance is comparable to other recent techniques. The perfect model entry was computed by applying the Parzen code to samples from the training data.

open source implementation of the algorithm, and RMSprop for optimization. The lower bound on the log likelihood provided by our model is reported for all datasets in Table 1. A reference implementation of the algorithm utilizing Blocks (van Merriënboer et al., 2015) is available at `https://github.com/Sohl-Dickstein/Diffusion-Probabilistic-Models`.

3.1. Toy Problems

3.1.1. SWISS ROLL

A diffusion probabilistic model was built of a two dimensional swiss roll distribution, using a radial basis function network to generate $\mathbf{f}_\mu\left(\mathbf{x}^{(t)}, t\right)$ and $\mathbf{f}_\Sigma\left(\mathbf{x}^{(t)}, t\right)$. As illustrated in Figure 1, the swiss roll distribution was successfully learned. See Appendix Section D.1.1 for more details.

3.1.2. BINARY HEARTBEAT DISTRIBUTION

A diffusion probabilistic model was trained on simple binary sequences of length 20, where a 1 occurs every 5th time bin, and the remainder of the bins are 0, using a multilayer perceptron to generate the Bernoulli rates $\mathbf{f}_b\left(\mathbf{x}^{(t)}, t\right)$ of the reverse trajectory. The log likelihood under the true distribution is $\log_2\left(\frac{1}{5}\right) = -2.322$ bits per sequence. As can be seen in Figure 2 and Table 1 learning was nearly perfect. See Appendix Section D.1.2 for more details.

3.2. Images

We trained Gaussian diffusion probabilistic models on several image datasets. The multi-scale convolutional architecture shared by these experiments is described in Appendix Section D.2.1, and illustrated in Figure D.1.

3.2.1. DATASETS

MNIST In order to allow a direct comparison against previous work on a simple dataset, we trained on MNIST digits (LeCun & Cortes, 1998). Log likelihoods relative to (Bengio et al., 2012; Bengio & Thibodeau-Laufer, 2013; Goodfellow et al., 2014) are given in Table 2. Samples from the MNIST model are given in Appendix Figure App.1. Our training algorithm provides an asymptotically consistent lower bound on the log likelihood. However most previous reported results on continuous MNIST log likelihood rely on Parzen-window based estimates computed from model samples. For this comparison we therefore estimate MNIST log likelihood using the Parzen-window code released with (Goodfellow et al., 2014).

CIFAR-10 A probabilistic model was fit to the training images for the CIFAR-10 challenge dataset (Krizhevsky & Hinton, 2009). Samples from the trained model are provided in Figure 3.

Dead Leaf Images Dead leaf images (Jeulin, 1997; Lee et al., 2001) consist of layered occluding circles, drawn from a power law distribution over scales. They have an analytically tractable structure, but capture many of the statistical complexities of natural images, and therefore provide a compelling test case for natural image models. As illustrated in Table 2 and Figure 4, we achieve state of the art performance on the dead leaves dataset.

Bark Texture Images A probabilistic model was trained on bark texture images (T01-T04) from (Lazebnik et al., 2005). For this dataset we demonstrate that it is straightforward to evaluate or generate from a posterior distribution, by inpainting a large region of missing data using a sample from the model posterior in Figure 5.

4. Conclusion

We have introduced a novel algorithm for modeling probability distributions that enables exact sampling and evaluation of probabilities and demonstrated its effectiveness on a variety of toy and real datasets, including challenging natural image datasets. For each of these tests we used a similar basic algorithm, showing that our method can accurately model a wide variety of distributions. Most existing density estimation techniques must sacrifice modeling power in order to stay tractable and efficient, and sampling or evaluation are often extremely expensive. The core of our algorithm consists of estimating the reversal of a Markov diffusion chain which maps data to a noise distribution; as

the number of steps is made large, the reversal distribution of each diffusion step becomes simple and easy to estimate. The result is an algorithm that can learn a fit to any data distribution, but which remains tractable to train, *exactly* sample from, and evaluate, and under which it is straightforward to manipulate conditional and posterior distributions.

Acknowledgements

We thank Lucas Theis, Subhaneil Lahiri, Ben Poole, Diederik P. Kingma, Taco Cohen, Philip Bachman, and Aäron van den Oord for extremely helpful discussion, and Ian Goodfellow for Parzen-window code. We thank Khan Academy and the Office of Naval Research for funding Jascha Sohl-Dickstein, and we thank the Office of Naval Research and the Burroughs-Wellcome, Sloan, and James S. McDonnell foundations for funding Surya Ganguli.

References

Barron, J. T., Biggin, M. D., Arbelaez, P., Knowles, D. W., Keranen, S. V., and Malik, J. Volumetric Semantic Segmentation Using Pyramid Context Features. In *2013 IEEE International Conference on Computer Vision*, pp. 3448–3455. IEEE, December 2013. ISBN 978-1-4799-2840-8. doi: 10.1109/ICCV.2013.428.

Bengio, Y. and Thibodeau-Laufer, E. Deep generative stochastic networks trainable by backprop. *arXiv preprint arXiv:1306.1091*, 2013.

Bengio, Y., Mesnil, G., Dauphin, Y., and Rifai, S. Better Mixing via Deep Representations. *arXiv preprint arXiv:1207.4404*, July 2012.

Bergstra, J. and Breuleux, O. Theano: a CPU and GPU math expression compiler. *Proceedings of the Python for Scientific Computing Conference (SciPy)*, 2010.

Besag, J. Statistical Analysis of Non-Lattice Data. *The Statistician, 24(3), 179-195*, 1975.

Bishop, C., Svensén, M., and Williams, C. GTM: The generative topographic mapping. *Neural computation*, 1998.

Bornschein, J. and Bengio, Y. Reweighted Wake-Sleep. *International Conference on Learning Representations*, June 2015.

Burda, Y., Grosse, R. B., and Salakhutdinov, R. Accurate and Conservative Estimates of MRF Log-likelihood using Reverse Annealing. *arXiv:1412.8566*, December 2014.

Dayan, P., Hinton, G. E., Neal, R. M., and Zemel, R. S. The helmholtz machine. *Neural computation*, 7(5):889–904, 1995.

Dinh, L., Krueger, D., and Bengio, Y. NICE: Non-linear Independent Components Estimation. *arXiv:1410.8516*, pp. 11, October 2014.

Feller, W. On the theory of stochastic processes, with particular reference to applications. In *Proceedings of the [First] Berkeley Symposium on Mathematical Statistics and Probability*. The Regents of the University of California, 1949.

Gershman, S. J. and Blei, D. M. A tutorial on Bayesian nonparametric models. *Journal of Mathematical Psychology*, 56(1):1–12, 2012.

Gneiting, T. and Raftery, A. E. Strictly proper scoring rules, prediction, and estimation. *Journal of the American Statistical Association*, 102(477):359–378, 2007.

Goodfellow, I. J., Pouget-Abadie, J., Mirza, M., Xu, B., Warde-Farley, D., Ozair, S., Courville, A., and Bengio, Y. Generative Adversarial Nets. *Advances in Neural Information Processing Systems*, 2014.

Gregor, K., Danihelka, I., Mnih, A., Blundell, C., and Wierstra, D. Deep AutoRegressive Networks. *arXiv preprint arXiv:1310.8499*, October 2013.

Grosse, R. B., Maddison, C. J., and Salakhutdinov, R. Annealing between distributions by averaging moments. In *Advances in Neural Information Processing Systems*, pp. 2769–2777, 2013.

Hinton, G. E. Training products of experts by minimizing contrastive divergence. *Neural Computation*, 14(8): 1771–1800, 2002.

Hinton, G. E. The wake-sleep algorithm for unsupervised neural networks). *Science*, 1995.

Hyvärinen, A. Estimation of non-normalized statistical models using score matching. *Journal of Machine Learning Research*, 6:695–709, 2005.

Jarzynski, C. Equilibrium free-energy differences from nonequilibrium measurements: A master-equation approach. *Physical Review E*, January 1997.

Jarzynski, C. Equalities and inequalities: irreversibility and the second law of thermodynamics at the nanoscale. *Annu. Rev. Condens. Matter Phys.*, 2011.

Jeulin, D. Dead leaves models: from space tesselation to random functions. *Proc. of the Symposium on the Advances in the Theory and Applications of Random Sets*, 1997.

Jordan, M. I., Ghahramani, Z., Jaakkola, T. S., and Saul, L. K. An introduction to variational methods for graphical models. *Machine learning*, 37(2):183–233, 1999.

Kavukcuoglu, K., Ranzato, M., and LeCun, Y. Fast inference in sparse coding algorithms with applications to object recognition. *arXiv preprint arXiv:1010.3467*, 2010.

Kingma, D. P. and Welling, M. Auto-Encoding Variational Bayes. *International Conference on Learning Representations*, December 2013.

Krizhevsky, A. and Hinton, G. Learning multiple layers of features from tiny images. *Computer Science Department University of Toronto Tech. Rep.*, 2009.

Langevin, P. Sur la théorie du mouvement brownien. *CR Acad. Sci. Paris*, 146(530-533), 1908.

Larochelle, H. and Murray, I. The neural autoregressive distribution estimator. *Journal of Machine Learning Research*, 2011.

Lazebnik, S., Schmid, C., and Ponce, J. A sparse texture representation using local affine regions. *Pattern Analysis and Machine Intelligence, IEEE Transactions on*, 27 (8):1265–1278, 2005.

LeCun, Y. and Cortes, C. The MNIST database of handwritten digits. 1998.

Lee, A., Mumford, D., and Huang, J. Occlusion models for natural images: A statistical study of a scale-invariant dead leaves model. *International Journal of Computer Vision*, 2001.

Lyu, S. Unifying Non-Maximum Likelihood Learning Objectives with Minimum KL Contraction. *Advances in Neural Information Processing Systems 24*, pp. 64–72, 2011.

MacKay, D. Bayesian neural networks and density networks. *Nuclear Instruments and Methods in Physics Research Section A: Accelerators, Spectrometers, Detectors and Associated Equipment*, 1995.

Murphy, K. P., Weiss, Y., and Jordan, M. I. Loopy belief propagation for approximate inference: An empirical study. In *Proceedings of the Fifteenth conference on Uncertainty in artificial intelligence*, pp. 467–475. Morgan Kaufmann Publishers Inc., 1999.

Neal, R. Annealed importance sampling. *Statistics and Computing*, January 2001.

Ozair, S. and Bengio, Y. Deep Directed Generative Autoencoders. *arXiv:1410.0630*, October 2014.

Parry, M., Dawid, A. P., Lauritzen, S., and Others. Proper local scoring rules. *The Annals of Statistics*, 40(1):561–592, 2012.

Rezende, D. J., Mohamed, S., and Wierstra, D. Stochastic Backpropagation and Approximate Inference in Deep Generative Models. *Proceedings of the 31st International Conference on Machine Learning (ICML-14)*, January 2014.

Rippel, O. and Adams, R. P. High-Dimensional Probability Estimation with Deep Density Models. *arXiv:1410.8516*, pp. 12, February 2013.

Schmidhuber, J. Learning factorial codes by predictability minimization. *Neural Computation*, 1992.

Sminchisescu, C., Kanaujia, A., and Metaxas, D. Learning joint top-down and bottom-up processes for 3D visual inference. In *Computer Vision and Pattern Recognition, 2006 IEEE Computer Society Conference on*, volume 2, pp. 1743–1752. IEEE, 2006.

Sohl-Dickstein, J., Battaglino, P., and DeWeese, M. New Method for Parameter Estimation in Probabilistic Models: Minimum Probability Flow. *Physical Review Letters*, 107(22):11–14, November 2011a. ISSN 0031-9007. doi: 10.1103/PhysRevLett.107.220601.

Sohl-Dickstein, J., Battaglino, P. B., and DeWeese, M. R. Minimum Probability Flow Learning. *International Conference on Machine Learning*, 107(22):11–14, November 2011b. ISSN 0031-9007. doi: 10.1103/PhysRevLett.107.220601.

Sohl-Dickstein, J., Poole, B., and Ganguli, S. Fast large-scale optimization by unifying stochastic gradient and quasi-Newton methods. In *Proceedings of the 31st International Conference on Machine Learning (ICML-14)*, pp. 604–612, 2014.

Spinney, R. and Ford, I. Fluctuation Relations : A Pedagogical Overview. *arXiv preprint arXiv:1201.6381*, pp. 3–56, 2013.

Stuhlmüller, A., Taylor, J., and Goodman, N. Learning stochastic inverses. *Advances in Neural Information Processing Systems*, 2013.

Suykens, J. and Vandewalle, J. Nonconvex optimization using a Fokker-Planck learning machine. In *12th European Conference on Circuit Theory and Design*, 1995.

T, P. Convergence condition of the TAP equation for the infinite-ranged Ising spin glass model. *J. Phys. A: Math. Gen. 15 1971*, 1982.

Tanaka, T. Mean-field theory of Boltzmann machine learning. *Physical Review Letters E*, January 1998.

Theis, L., Hosseini, R., and Bethge, M. Mixtures of conditional Gaussian scale mixtures applied to multiscale image representations. *PloS one*, 7(7):e39857, 2012.

Theis, L., van den Oord, A., and Bethge, M. A note on the evaluation of generative models. *arXiv preprint arXiv:1511.01844*, 2015.

Uria, B., Murray, I., and Larochelle, H. RNADE: The real-valued neural autoregressive density-estimator. *Advances in Neural Information Processing Systems*, 2013a.

Uria, B., Murray, I., and Larochelle, H. A Deep and Tractable Density Estimator. *arXiv:1310.1757*, pp. 9, October 2013b.

van Merriënboer, B., Chorowski, J., Serdyuk, D., Bengio, Y., Bogdanov, D., Dumoulin, V., and Warde-Farley, D. Blocks and Fuel. *Zenodo*, May 2015. doi: 10.5281/zenodo.17721.

Welling, M. and Hinton, G. A new learning algorithm for mean field Boltzmann machines. *Lecture Notes in Computer Science*, January 2002.

Yao, L., Ozair, S., Cho, K., and Bengio, Y. On the Equivalence Between Deep NADE and Generative Stochastic Networks. In *Machine Learning and Knowledge Discovery in Databases*, pp. 322–336. Springer, 2014.

Appendix

A. Conditional Entropy Bounds Derivation

The conditional entropy $H_q\left(\mathbf{X}^{(t-1)}|\mathbf{X}^{(t)}\right)$ of a step in the reverse trajectory is

$$H_q\left(\mathbf{X}^{(t-1)}, \mathbf{X}^{(t)}\right) = H_q\left(\mathbf{X}^{(t)}, \mathbf{X}^{(t-1)}\right) \tag{25}$$

$$H_q\left(\mathbf{X}^{(t-1)}|\mathbf{X}^{(t)}\right) + H_q\left(\mathbf{X}^{(t)}\right) = H_q\left(\mathbf{X}^{(t)}|\mathbf{X}^{(t-1)}\right) + H_q\left(\mathbf{X}^{(t-1)}\right) \tag{26}$$

$$H_q\left(\mathbf{X}^{(t-1)}|\mathbf{X}^{(t)}\right) = H_q\left(\mathbf{X}^{(t)}|\mathbf{X}^{(t-1)}\right) + H_q\left(\mathbf{X}^{(t-1)}\right) - H_q\left(\mathbf{X}^{(t)}\right) \tag{27}$$

An upper bound on the entropy change can be constructed by observing that $\pi(\mathbf{y})$ is the maximum entropy distribution. This holds without qualification for the binomial distribution, and holds for variance 1 training data for the Gaussian case. For the Gaussian case, training data must therefore be scaled to have unit norm for the following equalities to hold. It need not be whitened. The upper bound is derived as follows,

$$H_q\left(\mathbf{X}^{(t)}\right) \geq H_q\left(\mathbf{X}^{(t-1)}\right) \tag{28}$$

$$H_q\left(\mathbf{X}^{(t-1)}\right) - H_q\left(\mathbf{X}^{(t)}\right) \leq 0 \tag{29}$$

$$H_q\left(\mathbf{X}^{(t-1)}|\mathbf{X}^{(t)}\right) \leq H_q\left(\mathbf{X}^{(t)}|\mathbf{X}^{(t-1)}\right). \tag{30}$$

A lower bound on the entropy difference can be established by observing that additional steps in a Markov chain do not increase the information available about the initial state in the chain, and thus do not decrease the conditional entropy of the initial state,

$$H_q\left(\mathbf{X}^{(0)}|\mathbf{X}^{(t)}\right) \geq H_q\left(\mathbf{X}^{(0)}|\mathbf{X}^{(t-1)}\right) \tag{31}$$

$$H_q\left(\mathbf{X}^{(t-1)}\right) - H_q\left(\mathbf{X}^{(t)}\right) \geq H_q\left(\mathbf{X}^{(0)}|\mathbf{X}^{(t-1)}\right) + H_q\left(\mathbf{X}^{(t-1)}\right) - H_q\left(\mathbf{X}^{(0)}|\mathbf{X}^{(t)}\right) - H_q\left(\mathbf{X}^{(t)}\right) \tag{32}$$

$$H_q\left(\mathbf{X}^{(t-1)}\right) - H_q\left(\mathbf{X}^{(t)}\right) \geq H_q\left(\mathbf{X}^{(0)}, \mathbf{X}^{(t-1)}\right) - H_q\left(\mathbf{X}^{(0)}, \mathbf{X}^{(t)}\right) \tag{33}$$

$$H_q\left(\mathbf{X}^{(t-1)}\right) - H_q\left(\mathbf{X}^{(t)}\right) \geq H_q\left(\mathbf{X}^{(t-1)}|\mathbf{X}^{(0)}\right) - H_q\left(\mathbf{X}^{(t)}|\mathbf{X}^{(0)}\right) \tag{34}$$

$$H_q\left(\mathbf{X}^{(t-1)}|\mathbf{X}^{(t)}\right) \geq H_q\left(\mathbf{X}^{(t)}|\mathbf{X}^{(t-1)}\right) + H_q\left(\mathbf{X}^{(t-1)}|\mathbf{X}^{(0)}\right) - H_q\left(\mathbf{X}^{(t)}|\mathbf{X}^{(0)}\right). \tag{35}$$

Combining these expressions, we bound the conditional entropy for a single step,

$$H_q\left(\mathbf{X}^{(t)}|\mathbf{X}^{(t-1)}\right) \geq H_q\left(\mathbf{X}^{(t-1)}|\mathbf{X}^{(t)}\right) \geq H_q\left(\mathbf{X}^{(t)}|\mathbf{X}^{(t-1)}\right) + H_q\left(\mathbf{X}^{(t-1)}|\mathbf{X}^{(0)}\right) - H_q\left(\mathbf{X}^{(t)}|\mathbf{X}^{(0)}\right), \tag{36}$$

where both the upper and lower bounds depend only on the conditional forward trajectory $q\left(\mathbf{x}^{(1\cdots T)}|\mathbf{x}^{(0)}\right)$, and can be analytically computed.

B. Log Likelihood Lower Bound

The lower bound on the log likelihood is

$$L \geq K \tag{37}$$

$$K = \int d\mathbf{x}^{(0\cdots T)} q\left(\mathbf{x}^{(0\cdots T)}\right) \log\left[p\left(\mathbf{x}^{(T)}\right) \prod_{t=1}^{T} \frac{p\left(\mathbf{x}^{(t-1)}|\mathbf{x}^{(t)}\right)}{q\left(\mathbf{x}^{(t)}|\mathbf{x}^{(t-1)}\right)}\right] \tag{38}$$

$$\tag{39}$$

B.1. Entropy of $p\left(\mathbf{X}^{(T)}\right)$

We can peel off the contribution from $p\left(\mathbf{X}^{(T)}\right)$, and rewrite it as an entropy,

$$K = \int d\mathbf{x}^{(0\cdots T)} q\left(\mathbf{x}^{(0\cdots T)}\right) \sum_{t=1}^{T} \log\left[\frac{p\left(\mathbf{x}^{(t-1)}|\mathbf{x}^{(t)}\right)}{q\left(\mathbf{x}^{(t)}|\mathbf{x}^{(t-1)}\right)}\right] + \int d\mathbf{x}^{(T)} q\left(\mathbf{x}^{(T)}\right) \log p\left(\mathbf{x}^{(T)}\right) \tag{40}$$

$$= \int d\mathbf{x}^{(0\cdots T)} q\left(\mathbf{x}^{(0\cdots T)}\right) \sum_{t=1}^{T} \log\left[\frac{p\left(\mathbf{x}^{(t-1)}|\mathbf{x}^{(t)}\right)}{q\left(\mathbf{x}^{(t)}|\mathbf{x}^{(t-1)}\right)}\right] + \int d\mathbf{x}^{(T)} q\left(\mathbf{x}^{(T)}\right) \log \pi\left(\mathbf{x}^{T}\right) \tag{41}$$

$$\tag{42}$$

By design, the cross entropy to $\pi\left(\mathbf{x}^{(t)}\right)$ is constant under our diffusion kernels, and equal to the entropy of $p\left(\mathbf{x}^{(T)}\right)$. Therefore,

$$K = \sum_{t=1}^{T} \int d\mathbf{x}^{(0\cdots T)} q\left(\mathbf{x}^{(0\cdots T)}\right) \log\left[\frac{p\left(\mathbf{x}^{(t-1)}|\mathbf{x}^{(t)}\right)}{q\left(\mathbf{x}^{(t)}|\mathbf{x}^{(t-1)}\right)}\right] - H_p\left(\mathbf{X}^{(T)}\right). \tag{43}$$

B.2. Remove the edge effect at $t = 0$

In order to avoid edge effects, we set the final step of the reverse trajectory to be identical to the corresponding forward diffusion step,

$$p\left(\mathbf{x}^{(0)}|\mathbf{x}^{(1)}\right) = q\left(\mathbf{x}^{(1)}|\mathbf{x}^{(0)}\right) \frac{\pi\left(\mathbf{x}^{(0)}\right)}{\pi\left(\mathbf{x}^{(1)}\right)} = T_\pi\left(\mathbf{x}^{(0)}|\mathbf{x}^{(1)}; \beta_1\right). \tag{44}$$

We then use this equivalence to remove the contribution of the first time-step in the sum,

$$K = \sum_{t=2}^{T} \int d\mathbf{x}^{(0\cdots T)} q\left(\mathbf{x}^{(0\cdots T)}\right) \log\left[\frac{p\left(\mathbf{x}^{(t-1)}|\mathbf{x}^{(t)}\right)}{q\left(\mathbf{x}^{(t)}|\mathbf{x}^{(t-1)}\right)}\right] + \int d\mathbf{x}^{(0)} d\mathbf{x}^{(1)} q\left(\mathbf{x}^{(0)}, \mathbf{x}^{(1)}\right) \log\left[\frac{q\left(\mathbf{x}^{(1)}|\mathbf{x}^{(0)}\right) \pi\left(\mathbf{x}^{(0)}\right)}{q\left(\mathbf{x}^{(1)}|\mathbf{x}^{(0)}\right) \pi\left(\mathbf{x}^{(1)}\right)}\right] - H_p\left(\mathbf{X}^{(T)}\right) \tag{45}$$

$$= \sum_{t=2}^{T} \int d\mathbf{x}^{(0\cdots T)} q\left(\mathbf{x}^{(0\cdots T)}\right) \log\left[\frac{p\left(\mathbf{x}^{(t-1)}|\mathbf{x}^{(t)}\right)}{q\left(\mathbf{x}^{(t)}|\mathbf{x}^{(t-1)}\right)}\right] - H_p\left(\mathbf{X}^{(T)}\right), \tag{46}$$

where we again used the fact that by design $-\int d\mathbf{x}^{(t)} q\left(\mathbf{x}^{(t)}\right) \log \pi\left(\mathbf{x}^{(t)}\right) = H_p\left(\mathbf{X}^{(T)}\right)$ is a constant for all t.

B.3. Rewrite in terms of posterior $q\left(\mathbf{x}^{(t-1)}|\mathbf{x}^{(0)}\right)$

Because the forward trajectory is a Markov process,

$$K = \sum_{t=2}^{T} \int d\mathbf{x}^{(0\cdots T)} q\left(\mathbf{x}^{(0\cdots T)}\right) \log\left[\frac{p\left(\mathbf{x}^{(t-1)}|\mathbf{x}^{(t)}\right)}{q\left(\mathbf{x}^{(t)}|\mathbf{x}^{(t-1)}, \mathbf{x}^{(0)}\right)}\right] - H_p\left(\mathbf{X}^{(T)}\right). \tag{47}$$

Using Bayes' rule we can rewrite this in terms of a posterior and marginals from the forward trajectory,

$$K = \sum_{t=2}^{T} \int d\mathbf{x}^{(0\cdots T)} q\left(\mathbf{x}^{(0\cdots T)}\right) \log\left[\frac{p\left(\mathbf{x}^{(t-1)}|\mathbf{x}^{(t)}\right)}{q\left(\mathbf{x}^{(t-1)}|\mathbf{x}^{(t)}, \mathbf{x}^{(0)}\right)} \frac{q\left(\mathbf{x}^{(t-1)}|\mathbf{x}^{(0)}\right)}{q\left(\mathbf{x}^{(t)}|\mathbf{x}^{(0)}\right)}\right] - H_p\left(\mathbf{X}^{(T)}\right). \tag{48}$$

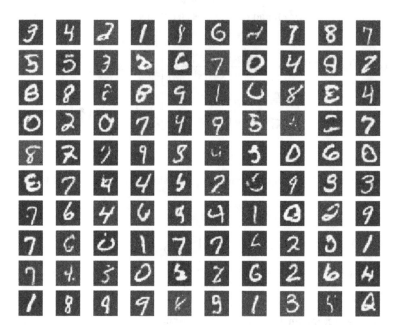

Figure App.1. Samples from a diffusion probabilistic model trained on MNIST digits. Note that unlike many MNIST sample figures, these are true samples rather than the mean of the Gaussian or binomial distribution from which samples would be drawn.

B.4. Rewrite in terms of KL divergences and entropies

We then recognize that several terms are conditional entropies,

$$K = \sum_{t=2}^{T} \int dx^{(0\cdots T)} q\left(x^{(0\cdots T)}\right) \log\left[\frac{p\left(x^{(t-1)}|x^{(t)}\right)}{q\left(x^{(t-1)}|x^{(t)},x^{(0)}\right)}\right] + \sum_{t=2}^{T}\left[H_q\left(X^{(t)}|X^{(0)}\right) - H_q\left(X^{(t-1)}|X^{(0)}\right)\right] - H_p\left(X^{(T)}\right)$$

(49)

$$= \sum_{t=2}^{T} \int dx^{(0\cdots T)} q\left(x^{(0\cdots T)}\right) \log\left[\frac{p\left(x^{(t-1)}|x^{(t)}\right)}{q\left(x^{(t-1)}|x^{(t)},x^{(0)}\right)}\right] + H_q\left(X^{(T)}|X^{(0)}\right) - H_q\left(X^{(1)}|X^{(0)}\right) - H_p\left(X^{(T)}\right).$$

(50)

Finally we transform the log ratio of probability distributions into a KL divergence,

$$K = -\sum_{t=2}^{T} \int dx^{(0)} dx^{(t)} q\left(x^{(0)},x^{(t)}\right) D_{KL}\left(q\left(x^{(t-1)}|x^{(t)},x^{(0)}\right) \middle\| p\left(x^{(t-1)}|x^{(t)}\right)\right)$$

$$+ H_q\left(X^{(T)}|X^{(0)}\right) - H_q\left(X^{(1)}|X^{(0)}\right) - H_p\left(X^{(T)}\right).$$

(51)

Note that the entropies can be analytically computed, and the KL divergence can be analytically computed given $x^{(0)}$ and $x^{(t)}$.

		Gaussian	Binomial
Well behaved (analytically tractable) distribution	$\pi\left(\mathbf{x}^{(T)}\right) =$	$\mathcal{N}\left(\mathbf{x}^{(T)};\mathbf{0},\mathbf{I}\right)$	$\mathcal{B}\left(\mathbf{x}^{(T)};0.5\right)$
Forward diffusion kernel	$q\left(\mathbf{x}^{(t)}|\mathbf{x}^{(t-1)}\right) =$	$\mathcal{N}\left(\mathbf{x}^{(t)};\mathbf{x}^{(t-1)}\sqrt{1-\beta_t},\mathbf{I}\beta_t\right)$	$\mathcal{B}\left(\mathbf{x}^{(t)};\mathbf{x}^{(t-1)}\left(1-\beta_t\right)+0.5\beta_t\right)$
Reverse diffusion kernel	$p\left(\mathbf{x}^{(t-1)}|\mathbf{x}^{(t)}\right) =$	$\mathcal{N}\left(\mathbf{x}^{(t-1)};\mathbf{f}_\mu\left(\mathbf{x}^{(t)},t\right),\mathbf{f}_\Sigma\left(\mathbf{x}^{(t)},t\right)\right)$	$\mathcal{B}\left(\mathbf{x}^{(t-1)};\mathbf{f}_b\left(\mathbf{x}^{(t)},t\right)\right)$
Training targets		$\mathbf{f}_\mu\left(\mathbf{x}^{(t)},t\right),\mathbf{f}_\Sigma\left(\mathbf{x}^{(t)},t\right),\beta_{1\cdots T}$	$\mathbf{f}_b\left(\mathbf{x}^{(t)},t\right)$
Forward distribution	$q\left(\mathbf{x}^{(0\cdots T)}\right) =$	$q\left(\mathbf{x}^{(0)}\right)\prod_{t=1}^{T}q\left(\mathbf{x}^{(t)}|\mathbf{x}^{(t-1)}\right)$	
Reverse distribution	$p\left(\mathbf{x}^{(0\cdots T)}\right) =$	$\pi\left(\mathbf{x}^{(T)}\right)\prod_{t=1}^{T}p\left(\mathbf{x}^{(t-1)}|\mathbf{x}^{(t)}\right)$	
Log likelihood	$L =$	$\int d\mathbf{x}^{(0)}q\left(\mathbf{x}^{(0)}\right)\log p\left(\mathbf{x}^{(0)}\right)$	
Lower bound on log likelihood	$K =$	$-\sum_{t=2}^{T}\mathbb{E}_{q\left(\mathbf{x}^{(0)},\mathbf{x}^{(t)}\right)}\left[D_{KL}\left(q\left(\mathbf{x}^{(t-1)}|\mathbf{x}^{(t)},\mathbf{x}^{(0)}\right)\middle\|p\left(\mathbf{x}^{(t-1)}|\mathbf{x}^{(t)}\right)\right)\right]+H_q\left(\mathbf{X}^{(T)}|\mathbf{X}^{(0)}\right)-H_q\left(\mathbf{X}^{(1)}|\mathbf{X}^{(0)}\right)-H_p\left(\mathbf{X}^{(T)}\right)$	
Perturbed reverse diffusion kernel	$\tilde{p}\left(\mathbf{x}^{(t-1)}|\mathbf{x}^{(t)}\right) =$	$\mathcal{N}\left(x^{(t-1)};\mathbf{f}_\mu\left(\mathbf{x}^{(t)},t\right)+\mathbf{f}_\Sigma\left(\mathbf{x}^{(t)},t\right)\frac{\partial\log r\left(\mathbf{x}^{(t-1)'}\right)}{\partial\mathbf{x}^{(t-1)'}}\bigg|_{\mathbf{x}^{(t-1)'}=f_\mu\left(\mathbf{x}^{(t)},t\right)},\mathbf{f}_\Sigma\left(\mathbf{x}^{(t)},t\right)\right)$	$\mathcal{B}\left(x_i^{(t-1)};\frac{c_i^{t-1}d_i^{t-1}}{x_i^{t-1}d_i^{t-1}+\left(1-c_i^{t-1}\right)\left(1-d_i^{t-1}\right)}\right)$

Table App.1. The key equations in this paper for the specific cases of Gaussian and binomial diffusion processes. $\mathcal{N}\left(u;\mu,\Sigma\right)$ is a Gaussian distribution with mean μ and covariance Σ. $\mathcal{B}\left(u;r\right)$ is the distribution for a single Bernoulli trial, with $u=1$ occurring with probability r, and $u=0$ occurring with probability $1-r$. Finally, for the perturbed Bernoulli trials $b_i^t=\mathbf{x}^{(t-1)}\left(1-\beta_t\right)+0.5\beta_t$, $c_i^t=\left[\mathbf{f}_b\left(\mathbf{x}^{(t+1)},t\right)\right]_i$, and $d_i^t=r\left(x_i^{(t)}=1\right)$, and the distribution is given for a single bit i.

C. Perturbed Gaussian Transition

We wish to compute $\tilde{p}\left(\mathbf{x}^{(t-1)} \mid \mathbf{x}^{(t)}\right)$. For notational simplicity, let $\mu = \mathbf{f}_\mu\left(\mathbf{x}^{(t)}, t\right)$, $\Sigma = \mathbf{f}_\Sigma\left(\mathbf{x}^{(t)}, t\right)$, and $\mathbf{y} = \mathbf{x}^{(t-1)}$. Using this notation,

$$\tilde{p}\left(\mathbf{y} \mid \mathbf{x}^{(t)}\right) \propto p\left(\mathbf{y} \mid \mathbf{x}^{(t)}\right) r\left(\mathbf{y}\right) \tag{52}$$

$$= \mathcal{N}\left(\mathbf{y}; \mu, \Sigma\right) r\left(\mathbf{y}\right). \tag{53}$$

We can rewrite this in terms of energy functions, where $E_r\left(\mathbf{y}\right) = -\log r\left(\mathbf{y}\right)$,

$$\tilde{p}\left(\mathbf{y} \mid \mathbf{x}^{(t)}\right) \propto \exp\left[-E\left(\mathbf{y}\right)\right] \tag{54}$$

$$E\left(\mathbf{y}\right) = \frac{1}{2}\left(\mathbf{y} - \mu\right)^T \Sigma^{-1}\left(\mathbf{y} - \mu\right) + E_r\left(\mathbf{y}\right). \tag{55}$$

If $E_r\left(\mathbf{y}\right)$ is smooth relative to $\frac{1}{2}\left(\mathbf{y} - \mu\right)^T \Sigma^{-1}\left(\mathbf{y} - \mu\right)$, then we can approximate it using its Taylor expansion around μ. One sufficient condition is that the eigenvalues of the Hessian of $E_r\left(\mathbf{y}\right)$ are everywhere much smaller magnitude than the eigenvalues of Σ^{-1}. We then have

$$E_r\left(\mathbf{y}\right) \approx E_r\left(\mu\right) + \left(\mathbf{y} - \mu\right)\mathbf{g} \tag{56}$$

where $\mathbf{g} = \left.\frac{\partial E_r\left(\mathbf{y}'\right)}{\partial \mathbf{y}'}\right|_{\mathbf{y}'=\mu}$. Plugging this in to the full energy,

$$E\left(\mathbf{y}\right) \approx \frac{1}{2}\left(\mathbf{y} - \mu\right)^T \Sigma^{-1}\left(\mathbf{y} - \mu\right) + \left(\mathbf{y} - \mu\right)^T \mathbf{g} + \text{constant} \tag{57}$$

$$= \frac{1}{2}\mathbf{y}^T\Sigma^{-1}\mathbf{y} - \frac{1}{2}\mathbf{y}^T\Sigma^{-1}\mu - \frac{1}{2}\mu^T\Sigma^{-1}\mathbf{y} + \frac{1}{2}\mathbf{y}^T\Sigma^{-1}\Sigma\mathbf{g} + \frac{1}{2}\mathbf{g}^T\Sigma\Sigma^{-1}\mathbf{y} + \text{constant} \tag{58}$$

$$= \frac{1}{2}\left(\mathbf{y} - \mu + \Sigma\mathbf{g}\right)^T \Sigma^{-1}\left(\mathbf{y} - \mu + \Sigma\mathbf{g}\right) + \text{constant}. \tag{59}$$

This corresponds to a Gaussian,

$$\tilde{p}\left(\mathbf{y} \mid \mathbf{x}^{(t)}\right) \approx \mathcal{N}\left(\mathbf{y}; \mu - \Sigma\mathbf{g}, \Sigma\right). \tag{60}$$

Substituting back in the original formalism, this is,

$$\tilde{p}\left(\mathbf{x}^{(t-1)} \mid \mathbf{x}^{(t)}\right) \approx \mathcal{N}\left(x^{(t-1)}; \mathbf{f}_\mu\left(\mathbf{x}^{(t)}, t\right) + \mathbf{f}_\Sigma\left(\mathbf{x}^{(t)}, t\right) \left.\frac{\partial \log r\left(\mathbf{x}^{(t-1)'}\right)}{\partial \mathbf{x}^{(t-1)'}}\right|_{\mathbf{x}^{(t-1)'}=f_\mu\left(\mathbf{x}^{(t)}, t\right)}, \mathbf{f}_\Sigma\left(\mathbf{x}^{(t)}, t\right)\right). \tag{61}$$

D. Experimental Details

D.1. Toy Problems

D.1.1. SWISS ROLL

A probabilistic model was built of a two dimensional swiss roll distribution. The generative model $p\left(\mathbf{x}^{(0\cdots T)}\right)$ consisted of 40 time steps of Gaussian diffusion initialized at an identity-covariance Gaussian distribution. A (normalized) radial basis function network with a single hidden layer and 16 hidden units was trained to generate the mean and covariance functions $\mathbf{f}_\mu\left(\mathbf{x}^{(t)}, t\right)$ and a diagonal $\mathbf{f}_\Sigma\left(\mathbf{x}^{(t)}, t\right)$ for the reverse trajectory. The top, readout, layer for each function was learned independently for each time step, but for all other layers weights were shared across all time steps and both functions. The top layer output of $\mathbf{f}_\Sigma\left(\mathbf{x}^{(t)}, t\right)$ was passed through a sigmoid to restrict it between 0 and 1. As can be seen in Figure 1, the swiss roll distribution was successfully learned.

D.1.2. BINARY HEARTBEAT DISTRIBUTION

A probabilistic model was trained on simple binary sequences of length 20, where a 1 occurs every 5th time bin, and the remainder of the bins are 0. The generative model consisted of 2000 time steps of binomial diffusion initialized at an independent binomial distribution with the same mean activity as the data ($p\left(x_i^{(T)} = 1\right) = 0.2$). A multilayer perceptron with sigmoid nonlinearities, 20 input units and three hidden layers with 50 units each was trained to generate the Bernoulli rates $\mathbf{f}_b\left(\mathbf{x}^{(t)}, t\right)$ of the reverse trajectory. The top, readout, layer was learned independently for each time step, but for all other layers weights were shared across all time steps. The top layer output was passed through a sigmoid to restrict it between 0 and 1. As can be seen in Figure 2, the heartbeat distribution was successfully learned. The log likelihood under the true generating process is $\log_2\left(\frac{1}{5}\right) = -2.322$ bits per sequence. As can be seen in Figure 2 and Table 1 learning was nearly perfect.

D.2. Images

D.2.1. ARCHITECTURE

Readout In all cases, a convolutional network was used to produce a vector of outputs $\mathbf{y}_i \in \mathcal{R}^{2J}$ for each image pixel i. The entries in $\mathbf{y}_i$ are divided into two equal sized subsets, $\mathbf{y}^\mu$ and $\mathbf{y}^\Sigma$.

Temporal Dependence The convolution output $\mathbf{y}^\mu$ is used as per-pixel weighting coefficients in a sum over time-dependent "bump" functions, generating an output $\mathbf{z}_i^\mu \in \mathcal{R}$

for each pixel i,

$$\mathbf{z}_i^\mu = \sum_{j=1}^J \mathbf{y}_{ij}^\mu g_j\left(t\right).\tag{62}$$

The bump functions consist of

$$g_j\left(t\right) = \frac{\exp\left(-\frac{1}{2w^2}\left(t - \tau_j\right)^2\right)}{\sum_{k=1}^J \exp\left(-\frac{1}{2w^2}\left(t - \tau_k\right)^2\right)},\tag{63}$$

where $\tau_j \in (0, T)$ is the bump center, and w is the spacing between bump centers. $\mathbf{z}^\Sigma$ is generated in an identical way, but using y^Σ.

For all image experiments a number of timesteps $T = 1000$ was used, except for the bark dataset which used $T = 500$.

Mean and Variance Finally, these outputs are combined to produce a diffusion mean and variance prediction for each pixel i,

$$\Sigma_{ii} = \sigma\left(z_i^\Sigma + \sigma^{-1}\left(\beta_t\right)\right),\tag{64}$$
$$\mu_i = \left(x_i - z_i^\mu\right)\left(1 - \Sigma_{ii}\right) + z_i^\mu.\tag{65}$$

where both Σ and μ are parameterized as a perturbation around the forward diffusion kernel $T_\pi\left(\mathbf{x}^{(t)}|\mathbf{x}^{(t-1)}; \beta_t\right)$, and z_i^μ is the mean of the equilibrium distribution that would result from applying $p\left(\mathbf{x}^{(t-1)}|\mathbf{x}^{(t)}\right)$ many times. Σ is restricted to be a diagonal matrix.

Multi-Scale Convolution We wish to accomplish goals that are often achieved with pooling networks – specifically, we wish to discover and make use of long-range and multi-scale dependencies in the training data. However, since the network output is a vector of coefficients for every pixel it is important to generate a full resolution rather than down-sampled feature map. We therefore define multi-scale-convolution layers that consist of the following steps:

1. Perform mean pooling to downsample the image to multiple scales. Downsampling is performed in powers of two.
2. Performing convolution at each scale.
3. Upsample all scales to full resolution, and sum the resulting images.
4. Perform a pointwise nonlinear transformation, consisting of a soft relu ($\log\left[1 + \exp\left(\cdot\right)\right]$).

The composition of the first three linear operations resembles convolution by a multiscale convolution kernel, up to blocking artifacts introduced by upsampling. This method of achieving multiscale convolution was described in (Barron et al., 2013).

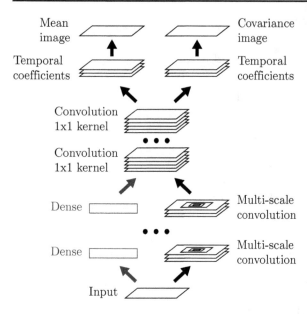

Figure D.1. Network architecture for mean function $\mathbf{f}_\mu\left(\mathbf{x}^{(t)}, t\right)$ and covariance function $\mathbf{f}_\Sigma\left(\mathbf{x}^{(t)}, t\right)$, for experiments in Section 3.2. The input image $\mathbf{x}^{(t)}$ passes through several layers of multi-scale convolution (Section D.2.1). It then passes through several convolutional layers with 1×1 kernels. This is equivalent to a dense transformation performed on each pixel. A linear transformation generates coefficients for readout of both mean $\mu^{(t)}$ and covariance $\Sigma^{(t)}$ for each pixel. Finally, a time dependent readout function converts those coefficients into mean and covariance images, as described in Section D.2.1. For CIFAR-10 a dense (or fully connected) pathway was used in parallel to the multi-scale convolutional pathway. For MNIST, the dense pathway was used to the exclusion of the multi-scale convolutional pathway.

Dense Layers Dense (acting on the full image vector) and kernel-width-1 convolutional (acting separately on the feature vector for each pixel) layers share the same form. They consist of a linear transformation, followed by a tanh nonlinearity.

Chapter 12

Interlude: Learning Sequences

12.1. Introduction

There are two basic approaches to learning sequences, which we refer to as dynamic and static here.

The static approach treats a temporal sequence as if it were laid out in a one-dimensional space, as shown in Figure 12.1. This allows conventional neural networks, which typically have inputs that are vectors of numbers, to process a temporal sequence as if it were a spatial pattern. For text this does not present any major problems, but for speech its effectiveness is limited because (for example) the static input cannot easily accommodate speakers who speak at different rates.

In contrast, the dynamic approach makes use of neural networks with internal feedback connections, so the networks are dynamical systems. This means that, at least in principle, the neural network may be able to learn the dynamical nature of the sequence of words it receives as inputs.

12.2. Static Networks for Sequences

A natural extension of using neural networks to recognise static inputs (e.g. images) is the recognition of sequences[68] (e.g. speech). However, time has an intangible quality that makes it difficult to operationalise in neural networks. Unlike the problem presented by a spatial image, where a unit's weights harvest data from a particular image region, it is not obvious how a network could harvest data from a particular point in time. We therefore begin with networks capable of taking account of events in the recent past.

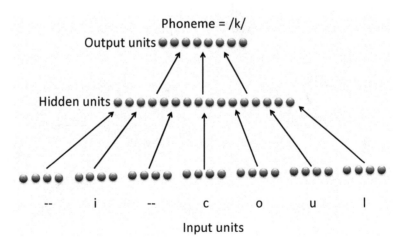

Figure 12.1. The NETtalk neural network used a sliding window of text at its input units, and each output unit represented a different phoneme. Successive phoneme outputs were fed into a speech synthesizer. Reproduced from Stone (2019).

The most obvious method for processing a temporal sequence in a neural network is to represent a series of events as a one-dimensional spatial array. For example, a speech signal can be represented as an array of signal values, and values over the most recent few seconds are presented to the input units of a neural network as a sliding window. This is essentially the strategy adopted for the NETtalk neural network shown in Figure 12.1.

Sejnowski and Rosenberg (1987) used NETtalk to learn the mapping between text and the basic elements of speech, called phonemes. A sliding window of seven consecutive characters was scanned along written text, and the network generated a phoneme code, which allowed a speech generator to produce audible speech. The network architecture comprised 7×29 inputs, 80 hidden units and 26 output units, with one output unit per phoneme, as shown in Figure 12.1. The network was trained on 1,024 words, for which it attained an accuracy of 95%. The ability of this network to generalise beyond the (relatively small) training set was tested by using new text as input, which yielded an accuracy of 78%. A landmark paper by Bengio et al. (2000) is described in the next chapter.

12.3. Dynamic Networks for Sequences

Because temporal sequences are usually generated by dynamical systems, it makes sense to model them using a *recurrent neural network* (RNN) that also behaves like a dynamical system, as in Figure 12.2a. Temporal networks can be represented as equivalent multi-layer networks by *unfolding*, as in Figure 12.2b. Unfolding involves reduplicating the entire network for each time step under consideration, which yields a series of networks in which contiguous networks represent network states at consecutive time steps. Thus, a change in the network state over time is represented as a change from one reduplicated network to the next.

Several variants of the basic backprop network attempt to do this by adding feedback connections between the output layer and special *context units* in the input layer[50], or from the hidden layer to context units in the input layer[24], or within the hidden layer[114], as shown in Figure 12.3. These connections ensure that the input to a unit depends not only on the current input vector from the environment, but also on previous inputs. Thus, the feedback connections implement a form of short-term memory that allows the current input to be interpreted in the context of recent inputs. More generally, the motivation for adding these feedback connections is to allow the network to learn the intrinsic dynamics of a sequence of input and output vectors. Williams and Zipser (1989) provided a backprop rule for online learning of sequences. However, most of the different varieties of RNNs mentioned above are special cases of a general network proposed by Pearlmutter (1988). These early temporal backprop networks have now evolved into temporal deep learning networks.

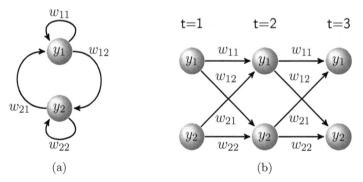

(a)   (b)

Figure 12.2. The recurrent network of two units in (a) can be unfolded to yield the equivalent feedforward network in (b). The recurrent weights (w_{11} and w_{22}) in (a) are represented as the duplicated feedforward weights in (b). Here, the recurrent network is considered over three time steps. Reproduced from Stone (2019).

A problem that plagues networks with many layers, known as the *vanishing gradient problem*, can be particularly acute for temporal RNNs, because the unfolding process portrayed in Figure 12.2 can yield an extremely large number of layers. However, if the feedback connections are set to fixed values (i.e. they do not learn) then the problem of vanishing gradients is avoided [24;43;50]. Finally, Schaffer et al. (2006) proved that RNNs can approximate dynamical systems to arbitrary precision.

12.4. Temporal Deep Neural Networks

A few key developments in temporal neural networks are described briefly here.

Long Short-Term Memory. Long short-term memory (LSTM) networks are essentially a sophisticated form of the RNNs described in Section 12.3. LSTM involves two key innovations that allow the neural network to retain information regarding recent inputs over substantial periods of time [28;43]. First, a *self-loop* implements a recurrent connection that retains information over a period of time determined by the value of a recurrent weight. Second, because the value of this recurrent weight is controlled by the state of a hidden unit, it can be learned according to the current context. In 2015, LSTM networks were used by Google for voice recognition.

Sequence-to-Sequence Learning. In 2014, the ability to translate text from one language to another was dramatically improved based on two papers published almost simultaneously [20;129]. The methods described in these papers also formed the basis of speech-to-text systems used in commercial applications, such as Siri, Cortana and Alexa. Sequence-to-sequence learning is achieved by using two coupled RNNs, an encoder and a decoder. The encoder maps a sequence of inputs $\mathbf{x} = (x_1, \ldots, x_n)$ to

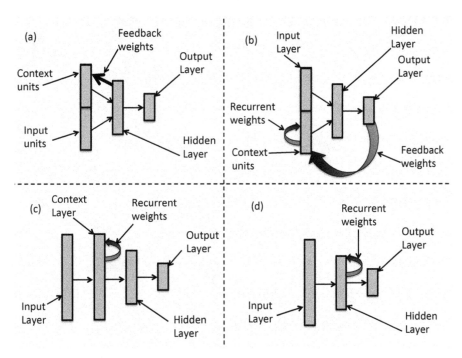

Figure 12.3. Recurrent neural network (RNN) architectures. (a) A set of context units holds a copy of the hidden unit states from the previous time step using fixed feedback weights (Elman, 1990). (b) A set of context units holds information about the output units as a decaying trace of activity using fixed feedback and recurrent weights (Jordan, 1986). (c) Recurrent connections from an extra layer of hidden units ensure that units in the second hidden layer are modulated by previous inputs using learned recurrent weights (Mozer, 1993). (d) Recurrent connections from the hidden units ensure that the hidden unit states are modulated by previous inputs using learned recurrent weights (Stone, 1989). Reproduced from Stone (2019).

277

a final state of its hidden units. If there are k hidden units then the encoder has effectively encoded a sequence of n inputs into a fixed-length k-element vector. The decoder then uses the k hidden unit states as a key to generate an output sequence $\mathbf{y} = (y_1, \ldots, y_m)$ where $m \neq n$.

Temporal Convolutional Networks. Both sequence-to-sequence learning and LSTM networks became standard tools for processing temporal sequences, for example in voice recognition. However, a new form of RNN, called *temporal convolutional networks*, seems to outperform LSTM networks across a range of temporal tasks[6;23].

Pixel Recurrent Neural Networks. Pixel recurrent neural networks (pixelRNNs) are generative models that utilise LSTM to model the long-range statistical structure of images[81]. Strictly speaking, pixelRNNs are not temporal networks, but we mention them here because they make use of RNN architectures and methods.

Chapter 13

A Neural Probabilistic Language Model – 2000

Context

Statistical Language Models

The conventional approach to the statistical modelling of language has its roots in work by Markov (1913) and Shannon (1951). This approach is based on collecting statistics on the co-occurrence of certain word pairs, triplets etc. The basic problem is that the number of possible ordered word pairs in a vocabulary V of size $|V|$ is proportional to $|V|^2$, the number of possible ordered word triplets is proportional to $|V|^3$, and so on. More generally, the number of possible N-grams (word sequences of length N) is proportional to $|V|^N$, so the number of possible N-grams grows exponentially with N. This is known as the *curse of dimensionality*, referred to in the opening sentences of the paper reproduced below and in Bengio's commentary (Section 13.5).

One consequence of this exponential growth is that predicting the next word following a sequence of $N-1$ words requires collecting statistics about sequences of length N, and there are $|V|^N$ such sequences. For example, given a relatively small vocabulary of $|V| = 1,000$ words, the number of N-grams of length $N = 6$ (the value used in the paper) is of the order $1,000^6 = 10^{18}$. Thus, in order to collect statistics on sequences of length $N = 6$, it looks as if we would need to sample a reasonably large subset of 10^{18} sequences. Even with modern computers, and even with the enormous data sets now available, this seems infeasible. Given that the largest data set used in the paper consisted of 'only' 16 million words, with a vocabulary of $|V| \approx 18,000$ words, the task appears almost impossible.

However, an early indication that the task is not impossible was given by the following text generated by Shannon[106] in 1949 using a model based on $N = 2$, which yields a reasonable approximation to English syntax:

> *The head and in frontal attack on an English writer that the character of this point is therefore another method for the letters that the time of who ever told the problem for an unexpected.*

A Neural Probabilistic Language Model

In order for a neural network to process language, each word is usually represented as a list of numbers, called a *feature vector*. The main innovation of the paper discussed in this chapter is that the numerical encoding of words into feature vectors is *learned as part of the overall training procedure*. The idea of encoding words as feature vectors was not new when the paper was published, but earlier methods attempted merely to ensure that 'similar' words have similar feature vectors; however, the definition of similarity had previously been based on sensible, albeit arbitrary, notions of similarity.

The notion of similarity matters because it can be shown that a system which can accurately predict the next word in any sequence from a given language must possess a good model of the overall statistical structure of that language. However, just because two words are deemed similar according to a particular hand-crafted metric, it does not necessarily mean that their (similar) feature vectors will prove useful in predicting the next word in any sequence. In contrast, if the feature vectors assigned to those words are based on a similarity metric that was explicitly designed to maximise prediction accuracy then it should be possible to use those feature vectors to increase prediction accuracy. As demonstrated by the paper, the best similarity metric for predicting the next word in any sequence is obtained not by asking a human expert to hand-craft a similarity metric, but rather by allowing a neural network to define its own metric.

Technical Summary

13.1. Measuring Linguistic Performance

The English language has a statistical structure that can be defined in terms of a probability distribution of word sequences. For example, the word sequence of $T = 5$ words

$$w_1^T = (w_1, w_2, w_3, w_4, w_5) \tag{13.1}$$

$$= (\text{Mary had a little lamb}) \tag{13.2}$$

occurs in English with probability $p(w_1^T)$, where w_i represents a single word and w_i^j represents a sequence that spans the word indices from i to j.

A good model $\hat{P}(w_1^T)$ of English should assign to the sequence w_1^T a probability that is close to the probability $p(w_1^T)$ observed in English, so that

$$\hat{P}(w_1^T) \approx p(w_1^T). \tag{13.3}$$

One measure of the accuracy of a model is its ability to predict the next word in a sequence. A good model should be able to provide an accurate estimate of the probability of the final word in a sequence when it is given the preceding words in that sequence; for example,

$$\hat{P}(w_5|w_1, w_2, w_3, w_4) \approx p(w_5|w_1, w_2, w_3, w_4), \tag{13.4}$$

where the vertical bar | stands for 'given that'. Crucially, such a model contains an implicit estimate of the entire distribution of sequences, because the probability of an entire sequence can be expressed in terms of the conditional probability of each word given previous subsequences[21;104].

Under mild assumptions, Bayes' rule (Appendix D) implies that the probability of the two-word sequence (w_1, w_2) can be expressed as

$$\hat{P}(w_1, w_2) = \hat{P}(w_2|w_1)\hat{P}(w_1). \tag{13.5}$$

By extension, the probability of the three-word sequence (w_1, w_2, w_3) is

$$\hat{P}(w_1, w_2, w_3) = \hat{P}(w_3|w_1, w_2)\hat{P}(w_1, w_2). \tag{13.6}$$

Substituting Equation 13.5 into Equation 13.6 yields the *chain rule of probabilities*,

$$\hat{P}(w_1, w_2, w_3) = \hat{P}(w_3|w_1, w_2)\hat{P}(w_2|w_1)\hat{P}(w_1). \tag{13.7}$$

This can be extended to a sequence of any length T, so that

$$\hat{P}(w_1, \ldots, w_T) = \hat{P}(w_T | w_1, \ldots, w_{T-1}) \, \hat{P}(w_{T-1} | w_1, \ldots, w_{T-2}) \cdots \hat{P}(w_1). \qquad (13.8)$$

Consequently, a model that can accurately predict the probability of the next word in any sequence necessarily possesses an accurate statistical model of the joint distribution of sequences. Rewriting Equation 13.8 in shorthand product notation gives

$$\hat{P}(w_1^T) = \prod_{t=1}^{T} \hat{P}(w_t | w_1, \ldots, w_{t-1}), \qquad (13.9)$$

where the subsequence on the right of the vertical bar can be written as

$$w_1, \ldots, w_{t-1} = w_1^{t-1}. \qquad (13.10)$$

Substituting this into Equation 13.9 yields the first equation in the paper,

$$\hat{P}(w_1^T) = \prod_{t=1}^{T} \hat{P}(w_t | w_1^{t-1}). \qquad (13.11)$$

Perplexity. In the paper, results are reported in terms of *perplexity*, which is the average number of equiprobable words that could follow a given sequence of words, where this average is taken over all the words in a (usually long) sequence. Thus, perplexity is a measure of the unpredictability of words in a language; higher perplexity implies less predictability. Formally, perplexity is the geometric mean of the unpredictability of words in a sequence,

$$P = \left[\prod_{t=1}^{T} \frac{1}{\hat{P}(w_t | w_1^{t-1})} \right]^{1/T}. \qquad (13.12)$$

Therefore, in a good model, the number (and identity) of equiprobable next-word alternatives should match the number of equiprobable next-word alternatives observed in English.

Perplexity and Shannon Entropy. Perplexity can be interpreted in terms of *Shannon entropy* (see Glossary). The paper does not use this interpretation, but it may be helpful to readers familiar with *information theory*[21;106;125].

Taking the logarithm of Equation 13.12 transforms the perplexity into the conditional entropy,

$$H(w_t | w_1^{t-1}) = \frac{1}{T} \sum_{t=1}^{T} \log_2 \frac{1}{\hat{P}(w_t | w_1^{t-1})} \text{ bits.} \qquad (13.13)$$

This implies that when averaged over all words, the uncertainty associated with each next word predicted by the model is H bits. Therefore, in order to predict the next word exactly, we would need (on average) an extra H bits of information. Shannon proved[104] that as $T \to \infty$, the conditional entropy $H(w_t | w_1^{t-1})$ in Equation 13.13 becomes equal to the entropy $H(w_t)$.

Note that it is not possible, in principle, to predict the next word with absolute certainty; there will always be a residual amount of uncertainty in any prediction. Given that English necessarily involves a degree of (average) next-word uncertainty, the objective is to find a model that has a next-word uncertainty H equal to the next-word uncertainty H^* observed in English.

13.2. Architecture and Training

The overall architecture and training method are explained clearly in the paper, so only a brief summary is given here. The model architecture is shown in Figure 1 of the paper, reproduced as Figure 13.1 here for convenience. Each network input is derived from a sequence of $n = 6$ words, $(w_{t-6}, w_{t-5}, w_{t-4}, w_{t-3}, w_{t-2}, w_{t-1})$, which provide context for the next word w_t to be predicted. Each word is one out of $|V|$ words in the vocabulary of $V = (w_1, \ldots, w_{|V|})$ words.

Each word corresponds to a unique word index i, which specifies the ith feature vector (row) of a matrix C of $|V| \times m$ real numbers. The $n = 6$ (learned) feature vectors are concatenated to form a composite input vector of length $n \times m$, so there are $n \times m$ input units. For example, if the first word in the current input sequence of n words is w_{t-n+1}, and if w_{t-n+1} is the ith word in V, then the first m elements of the composite input vector constitute the ith row of C, represented as $C(w_{t-n+1})$.

The network has h hidden units, with sigmoidal (tanh) activation functions. Each hidden unit is connected to the $n \times m$ input units via a set of $n \times m$ weights, which are learned during training. There are $|V|$ output units, each of which corresponds to one word in the vocabulary V. The state of the ith output unit specifies the probability that the next word corresponds to the word represented by that output unit. Thus, each input yields a set of $|V|$ output unit states,

$$\hat{P}(w_1), \ldots, \hat{P}(w_{|V|}), \tag{13.14}$$

where $\hat{P}(w_i)$ is the model's estimate of the probability $\hat{P}(w_i = w_t)$ that the ith word w_i in V is the word w_t that follows the current input $(w_{t-6}, w_{t-5}, w_{t-4}, w_{t-3}, w_{t-2}, w_{t-1})$.

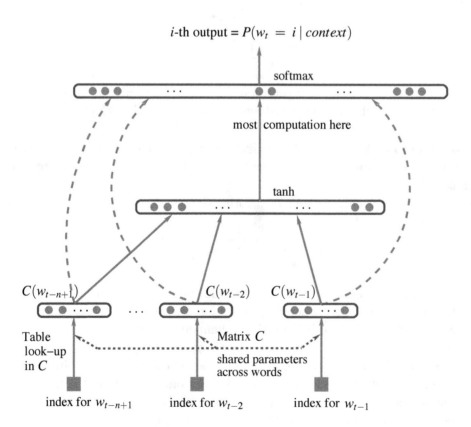

Figure 13.1. Architecture of the model; Figure 1 in the paper.

13.3. Results

The models were tested using two standard data sets: the Brown corpus of about one million words and the Associated Press (AP) corpus of about 14 million words. As an indication of the magnitude of the task of training a neural network on the AP corpus, a total of only five epochs (passes through the data) were accomplished, which took over three weeks using 40 CPUs.

Brown Corpus Results. The Brown corpus consists of about one million words. Of these, approximately 800,000 words were used to train the model, 200,000 words were used as the validation set (e.g. used for early stopping to prevent over-fitting), and 181,041 words were used as the test set. The number of words used in the vocabulary for the training set was $|V| = 16{,}383$. For each predicted word, the number of previous words used as context was $n = 6$. The best results were obtained when each word was encoded as a feature vector of $m = 30$ real numbers, using a neural network with $h = 50$ hidden units and no direct connections from the input to the output layers of units (dashed lines in Figure 13.1). On the test data this model has a perplexity of

$$\hat{P}_{\text{Brown}} = 252 \text{ equiprobable words}, \tag{13.15}$$

as reported in Table 1 of the paper. Comparing with previous methods, the authors show that their model provides a clear advantage in terms of perplexity. Such comparisons are important because they indicate that progress is being made.

However, rather than comparing the results of the proposed model and previous methods, an objective measure of performance can be obtained by comparing the model's next-word Shannon entropy with the observed next-word Shannon entropy of the Brown corpus.

Accordingly, given the model's estimated perplexity of $\hat{P}_{\text{Brown}} = 252$ equiprobable words, the Shannon entropy is

$$\begin{aligned}
\hat{H}_{\text{Brown}} &= \log_2 252 \\
&= 7.98 \text{ bits.}
\end{aligned} \tag{13.16}$$

As an objective benchmark, a statistical estimate of the perplexity of the Brown corpus is

$$P_{\text{Brown}} = 247 \text{ equiprobable words}, \tag{13.17}$$

which corresponds to a Shannon entropy of

$$\begin{aligned}
H_{\text{Brown}} &= \log_2 247 \\
&= 7.95 \text{ bits.}
\end{aligned} \tag{13.18}$$

From this, we can conclude that the best model described in the paper has a next-word entropy close to the theoretical bound imposed by the estimated Shannon entropy of the Brown corpus. More precisely,

$$\begin{aligned}
\frac{\hat{H}_{\text{Brown}} - H_{\text{Brown}}}{H_{\text{Brown}}} \times 100\% &= \frac{7.98 - 7.95}{7.95} \times 100\% \\
&= 0.38\%.
\end{aligned} \tag{13.19}$$

In words, the best model in the paper has a Shannon entropy that is just 0.38% larger than the theoretical bound imposed by the estimated Shannon entropy of the Brown corpus. If we perform the analogous calculation in terms of perplexity then we obtain $(252 - 247)/247 \times 100\% = 2.0\%$.

Associated Press (AP) Corpus Results. The AP corpus consists of about 16 million words. Of these, approximately 14 million (13,994,528) were used to train the model, one million (963,138) were used as the validation set, and one million (963,071) were used as the test set. The number of words used in the vocabulary for the training set was $|V| = 17,964$. Each word was encoded as a feature vector, where each vector consists of $m = 100$ real numbers, using a neural network with $h = 60$ hidden units and with additional direct connections from the input to the output layers of units (dashed lines in Figure 13.1). For each predicted word, the number of previous words used as context was $n = 6$. The best performance of a model in the paper for the AP corpus has a perplexity of

$$\hat{P}_{AP} = 109 \text{ equiprobable words,} \tag{13.20}$$

as reported in Table 2 of the paper, which corresponds to a Shannon entropy of

$$
\begin{aligned}
\hat{H}_{AP} &= \log_2 109 \\
&= 6.77 \text{ bits.}
\end{aligned}
\tag{13.21}
$$

Unfortunately, the perplexity of the AP corpus is not available, so we cannot compare the model perplexity with that of the AP corpus.

It is striking that the model perplexity of the Brown corpus (252, entropy$= 7.98$ bits) is substantially larger than the perplexity of the AP corpus (109, entropy$= 6.77$ bits). This could be because the Brown corpus draws from a more diverse set of sources than the AP corpus, or because the latter is a much larger training set.

13.4. List of Mathematical Symbols

C matrix of $|V| \times m$ real numbers; each row of C is an m-element feature vector for one word.

H Shannon entropy; the Shannon entropy of a variable is the logarithm of the number of equiprobable values. For example, the entropy of an eight-sided die is $H = \log_2 8 = 3$ bits of information.

h number of hidden units.

m number of real numbers per feature vector.

n number of words used as context to predict the next word.

P perplexity, the number of equiprobable next-word predictions.

$\hat{P}(w_i)$ model estimate of the probability that the next word is w_i.

V vocabulary of words.

$|V|$ number of words in the vocabulary V.

w_i ith word in a sequence of words from the vocabulary V.

w_i^j subsequence of words spanning the indices from i to j.

13.5. Comments by the Paper's Author: Y Bengio

I had been obsessed with the idea that neural networks could defeat, or at least greatly reduce, the curse of dimensionality thanks to distributed representations for many years. That idea was certainly derived from Geoff Hinton's excellent earlier work on the notion of distributed representations (1986, 'Learning distributed representations of concepts') and learning to represent symbols with vectors, but I focused on the link to the curse of dimensionality that comes up when trying to represent a rich high-dimensional joint distribution. The easiest case to study was that of a high-dimensional tuple of discrete variables.

My first foray in this question was in my work with my brother Samy Bengio which came out one year earlier at NeurIPS 1999 ('Modeling high-dimensional discrete data with multi-layer neural networks'). We looked at the case of N discrete variables and their joint distribution, and used an auto-regressive architecture to capture the joint distribution as a product of conditional distributions using a special multi-layer perceptron (MLP) architecture.

The step to the NeurIPS 2000 paper[†] involved several changes and novelties:

(a) a linguistically motivated argument for how the curse of dimensionality may be circumvented, specifically because of the notion of word representations, and seeing how these correspond to the input layer of the neural network and generalization across similarly represented words;

(b) compared with the 1999 paper ('Modeling high-dimensional discrete data with multi-layer neural networks'), taking advantage of the stationary nature of text in the neural architecture, i.e., the same conditional distribution (next word given previous ones) can be applied at every word position, along with the word representations;

(c) designing efficient code, running on a parallel computer to run much larger-scale experiments than had been possible in my 1999 work;

(d) introducing an energy-based formulation of the architecture, with the same word vectors used for the context words and the target word.

My obsession with the curse of dimensionality continued in later works, where I came up with arguments showing the mathematical limitations of other machine learning methods due to the high dimensionality of the data, such as kernel methods with Gaussian kernels (with Delalleau, NeurIPS 2005, 'The curse of highly variable functions for local kernel machines) and decision trees ('Decision trees do not generalize to new variations, 2010), which could be bypassed when using appropriate neural network architectures instead.

Research Paper: A Neural Probabilistic Language Model

This paper was originally published as a conference paper[10] in 2000.
Reference: Bengio, Y., Ducharme, R., Vincent, P., and Jauvin, C. (2003). A neural probabilistic language model. *Journal of Machine Learning Research*, 3:1137–1155.
https://jmlr.csail.mit.edu/papers/volume3/bengio03a/bengio03a.pdf.
Reproduced with permission.

[†]The NeurIPS conference paper[10] is a shorter version of the journal paper reproduced below. The author's comments apply to both papers.

Journal of Machine Learning Research 3 (2003) 1137–1155   Submitted 4/02; Published 2/03

A Neural Probabilistic Language Model

Yoshua Bengio   BENGIOY@IRO.UMONTREAL.CA
Réjean Ducharme   DUCHARME@IRO.UMONTREAL.CA
Pascal Vincent   VINCENTP@IRO.UMONTREAL.CA
Christian Jauvin   JAUVINC@IRO.UMONTREAL.CA
Département d'Informatique et Recherche Opérationnelle
Centre de Recherche Mathématiques
Université de Montréal, Montréal, Québec, Canada

Editors: Jaz Kandola, Thomas Hofmann, Tomaso Poggio and John Shawe-Taylor

Abstract

A goal of statistical language modeling is to learn the joint probability function of sequences of words in a language. This is intrinsically difficult because of the **curse of dimensionality**: a word sequence on which the model will be tested is likely to be different from all the word sequences seen during training. Traditional but very successful approaches based on n-grams obtain generalization by concatenating very short overlapping sequences seen in the training set. We propose to fight the curse of dimensionality by **learning a distributed representation for words** which allows each training sentence to inform the model about an exponential number of semantically neighboring sentences. The model learns simultaneously (1) a distributed representation for each word along with (2) the probability function for word sequences, expressed in terms of these representations. Generalization is obtained because a sequence of words that has never been seen before gets high probability if it is made of words that are similar (in the sense of having a nearby representation) to words forming an already seen sentence. Training such large models (with millions of parameters) within a reasonable time is itself a significant challenge. We report on experiments using neural networks for the probability function, showing on two text corpora that the proposed approach significantly improves on state-of-the-art n-gram models, and that the proposed approach allows to take advantage of longer contexts.

Keywords: Statistical language modeling, artificial neural networks, distributed representation, curse of dimensionality

1. Introduction

A fundamental problem that makes language modeling and other learning problems difficult is the *curse of dimensionality*. It is particularly obvious in the case when one wants to model the joint distribution between many discrete random variables (such as words in a sentence, or discrete attributes in a data-mining task). For example, if one wants to model the joint distribution of 10 consecutive words in a natural language with a vocabulary V of size 100,000, there are potentially $100000^{10} - 1 = 10^{50} - 1$ free parameters. When modeling continuous variables, we obtain generalization more easily (e.g. with smooth classes of functions like multi-layer neural networks or Gaussian mixture models) because the function to be learned can be expected to have some local smoothness properties. For discrete spaces, the generalization structure is not as obvious: any change of these discrete variables may have a drastic impact on the value of the function to be estimated, and when the number of values that each discrete variable can take is large, most observed objects are almost maximally far from each other in hamming distance.

A useful way to visualize how different learning algorithms generalize, inspired from the view of non-parametric density estimation, is to think of how probability mass that is initially concentrated on the training points (e.g., training sentences) is distributed in a larger volume, usually in some form of neighborhood around the training points. In high dimensions, it is crucial to distribute probability mass where it matters rather than uniformly in all directions around each training point. We will show in this paper that the way in which the approach proposed here generalizes is fundamentally different from the way in which previous state-of-the-art statistical language modeling approaches are generalizing.

A statistical model of language can be represented by the conditional probability of the next word given all the previous ones, since

$$\hat{P}(w_1^T) = \prod_{t=1}^{T} \hat{P}(w_t|w_1^{t-1}),$$

where w_t is the t-th word, and writing sub-sequence $w_i^j = (w_i, w_{i+1}, \cdots, w_{j-1}, w_j)$. Such statistical language models have already been found useful in many technological applications involving natural language, such as speech recognition, language translation, and information retrieval. Improvements in statistical language models could thus have a significant impact on such applications.

When building statistical models of natural language, one considerably reduces the difficulty of this modeling problem by taking advantage of word order, and the fact that temporally closer words in the word sequence are statistically more dependent. Thus, *n-gram* models construct tables of conditional probabilities for the next word, for each one of a large number of *contexts*, i.e. combinations of the last $n - 1$ words:

$$\hat{P}(w_t|w_1^{t-1}) \approx \hat{P}(w_t|w_{t-n+1}^{t-1}).$$

We only consider those combinations of successive words that actually occur in the training corpus, or that occur frequently enough. What happens when a new combination of n words appears that was not seen in the training corpus? We do not want to assign zero probability to such cases, because such new combinations are likely to occur, and they will occur even more frequently for larger context sizes. A simple answer is to look at the probability predicted using a smaller context size, as done in back-off trigram models (Katz, 1987) or in smoothed (or interpolated) trigram models (Jelinek and Mercer, 1980). So, in such models, how is generalization basically obtained from sequences of words seen in the training corpus to new sequences of words? A way to understand how this happens is to think about a generative model corresponding to these interpolated or back-off n-gram models. Essentially, a new sequence of words is generated by "gluing" very short and overlapping pieces of length 1, 2 ... or up to n words that have been seen frequently in the training data. The rules for obtaining the probability of the next piece are implicit in the particulars of the back-off or interpolated n-gram algorithm. Typically researchers have used $n = 3$, i.e. trigrams, and obtained state-of-the-art results, but see Goodman (2001) for how combining many tricks can yield to substantial improvements. Obviously there is much more information in the sequence that immediately precedes the word to predict than just the identity of the previous couple of words. There are at least two characteristics in this approach which beg to be improved upon, and that we will focus on in this paper. First, it is not taking into account contexts farther than 1 or 2 words,[1] second it is not taking into account the "similarity" between words. For example, having seen the sentence "The cat is walking in the bedroom" in the training corpus should help us generalize to make the sentence "A dog was running in a room" almost as likely, simply because "dog" and "cat" (resp. "the" and "a", "room" and "bedroom", etc...) have similar semantic and grammatical roles.

There are many approaches that have been proposed to address these two issues, and we will briefly explain in Section 1.2 the relations between the approach proposed here and some of these earlier approaches. We will first discuss what is the basic idea of the proposed approach. A more formal presentation will follow in Section 2, using an implementation of these ideas that relies on shared-parameter multi-layer neural networks. Another contribution of this paper concerns the challenge of training such very large neural networks (with millions of parameters) for very large data sets (with millions or tens of millions of examples). Finally, an important contribution of this paper is to show that training such large-scale model is expensive but feasible, scales to large contexts, and yields good comparative results (Section 4).

Many operations in this paper are in matrix notation, with lower case v denoting a column vector and v' its transpose, A_j the j-th row of a matrix A, and $x.y = x'y$.

1.1 Fighting the Curse of Dimensionality with Distributed Representations

In a nutshell, the idea of the proposed approach can be summarized as follows:

---

1. associate with each word in the vocabulary a distributed *word feature vector* (a real-valued vector in $\mathbb{R}^m$),

2. express the joint *probability function* of word sequences in terms of the feature vectors of these words in the sequence, and

3. learn simultaneously the *word feature vectors* and the parameters of that *probability function*.

The feature vector represents different aspects of the word: each word is associated with a point in a vector space. The number of features (e.g. $m = 30$, 60 or 100 in the experiments) is much smaller than the size of the vocabulary (e.g. 17,000). The probability function is expressed as a product of conditional probabilities of the next word given the previous ones, (e.g. using a multi-layer neural network to predict the next word given the previous ones, in the experiments). This function has parameters that can be iteratively tuned in order to **maximize the log-likelihood of the training data** or a regularized criterion, e.g. by adding a weight decay penalty.[2] The feature vectors associated with each word are learned, but they could be initialized using prior knowledge of semantic features.

Why does it work? In the previous example, if we knew that dog and cat played similar roles (semantically and syntactically), and similarly for (the,a), (bedroom,room), (is,was),

---

1. n-grams with n up to 5 (i.e. 4 words of context) have been reported, though, but due to data scarcity, most predictions are made with a much shorter context.

2. Like in ridge regression, the squared norm of the parameters is penalized.

(running,walking), we could naturally generalize (i.e. transfer probability mass) from

> The cat is walking in the bedroom

to

> A dog was running in a room

and likewise to

> The cat is running in a room
>
> A dog is walking in a bedroom
>
> The dog was walking in the room
>
> ...

and many other combinations. In the proposed model, it will so generalize because "similar" words are expected to have a similar feature vector, and because the probability function is a *smooth* function of these feature values, a small change in the features will induce a small change in the probability. Therefore, the presence of only one of the above sentences in the training data will increase the probability, not only of that sentence, but also of its combinatorial number of "neighbors" in sentence space (as represented by sequences of feature vectors).

1.2 Relation to Previous Work

The idea of using neural networks to model high-dimensional discrete distributions has already been found useful to learn the joint probability of $Z_1 \cdots Z_n$, a set of random variables where each is possibly of a different nature (Bengio and Bengio, 2000a,b). In that model, the joint probability is decomposed as a product of conditional probabilities

$$\hat{P}(Z_1 = z_1, \cdots, Z_n = z_n) = \prod_i \hat{P}(Z_i = z_i | g_i(Z_{i-1} = z_{i-1}, Z_{i-2} = z_{i-2}, \cdots, Z_1 = z_1)),$$

where $g(.)$ is a function represented by a neural network with a special left-to-right architecture, with the i-th output block $g_i()$ computing parameters for expressing the conditional distribution of Z_i given the value of the previous Z's, in some arbitrary order. Experiments on four UCI data sets show this approach to work comparatively very well (Bengio and Bengio, 2000a,b). Here we must deal with data of variable length, like sentences, so the above approach must be adapted. Another important difference is that here, all the Z_i (word at i-th position), refer to the same type of object (a word). The model proposed here therefore introduces a sharing of parameters across time – the same g_i is used across time – that is, and across input words at different positions. It is a successful large-scale application of the same idea, along with the (old) idea of learning a distributed representation for symbolic data, that was advocated in the early days of connectionism (Hinton, 1986, Elman, 1990). More recently, Hinton's approach was improved and successfully demonstrated on learning several symbolic relations (Paccanaro and Hinton, 2000). The idea of using neural networks for language modeling is not new either (e.g. Miikkulainen and Dyer, 1991). In contrast, here we push this idea to a **large scale**, and concentrate on learning a **statistical model** of the distribution of word sequences, rather than learning the role of words in a sentence. The approach proposed here is also related to previous proposals of character-based text compression using neural networks to predict the probability of the next character (Schmidhuber, 1996). The idea of using a neural network for language modeling has also been independently proposed by Xu and Rudnicky (2000), although experiments are with networks without hidden units and a single input word, which limit the model to essentially capturing unigram and bigram statistics.

The idea of discovering some similarities between words to obtain generalization from training sequences to new sequences is not new. For example, it is exploited in approaches that are based on learning a clustering of the words (Brown et al., 1992, Pereira et al., 1993, Niesler et al., 1998, Baker and McCallum, 1998): each word is associated deterministically or probabilistically with a discrete class, and words in the same class are similar in some respect. In the model proposed here, instead of characterizing the similarity with a discrete random or deterministic variable (which corresponds to a soft or hard partition of the set of words), we use a continuous real-vector for each word, i.e. a **learned distributed feature vector**, to represent similarity between words. The experimental comparisons in this paper include results obtained with class-based n-grams (Brown et al., 1992, Ney and Kneser, 1993, Niesler et al., 1998).

The idea of using a vector-space representation for words has been well exploited in the area of *information retrieval* (for example see work by Schutze, 1993), where feature vectors for words are learned on the basis of their probability of co-occurring in the same documents (Latent Semantic Indexing, see Deerwester et al., 1990). An important difference is that here we look for a representation for words that is helpful in representing compactly the probability distribution of word sequences from natural language text. Experiments suggest that learning jointly the representation (word features) and the model is very useful. We tried (unsuccessfully) using as fixed word features for each word w the first principal components of the co-occurrence frequencies of w with the words occurring in text around the occurrence of w. This is similar to what has been done with documents for information retrieval with LSI. The idea of using a continuous representation for words has however been exploited successfully by Bellegarda (1997) in the context of an n-gram based statistical language model, using LSI to dynamically identify the topic of discourse.

The idea of a vector-space representation for symbols in the context of neural networks has also previously been framed in terms of a parameter sharing layer, (e.g. Riis and Krogh, 1996) for secondary structure prediction, and for text-to-speech mapping (Jensen and Riis, 2000).

2. A Neural Model

The training set is a sequence $w_1 \cdots w_T$ of words $w_t \in V$, where the vocabulary V is a large but finite set. The objective is to learn a good model $f(w_t, \cdots, w_{t-n+1}) = \hat{P}(w_t|w_1^{t-1})$, in the sense that it gives high out-of-sample likelihood. Below, we report the geometric average of $1/\hat{P}(w_t|w_1^{t-1})$, also known as *perplexity*, which is also the exponential of the average negative log-likelihood. The only constraint on the model is that for any choice of w_1^{t-1}, $\sum_{i=1}^{|V|} f(i, w_{t-1}, \cdots, w_{t-n+1}) = 1$, with $f > 0$. By the product of these conditional probabilities, one obtains a model of the joint probability of sequences of words.

We decompose the function $f(w_t, \cdots, w_{t-n+1}) = \hat{P}(w_t|w_1^{t-1})$ in two parts:

1. A mapping C from any element i of V to a real vector $C(i) \in \mathbb{R}^m$. It represents the *distributed feature vectors* associated with each word in the vocabulary. In practice, C is represented by a $|V| \times m$ matrix of free parameters.

2. The probability function over words, expressed with C: a function g maps an input sequence of feature vectors for words in context, $(C(w_{t-n+1}), \cdots, C(w_{t-1}))$, to a conditional probability distribution over words in V for the next word w_t. The output of g is a vector whose i-th element estimates the probability $\hat{P}(w_t = i|w_1^{t-1})$ as in Figure 1.

$$f(i, w_{t-1}, \cdots, w_{t-n+1}) = g(i, C(w_{t-1}), \cdots, C(w_{t-n+1}))$$

The function f is a composition of these two mappings (C and g), with C being *shared* across all the words in the context. With each of these two parts are associated some parameters. The

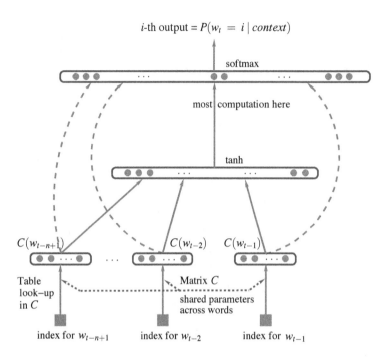

i-th output = $P(w_t = i \mid context)$

softmax

most computation here

tanh

$C(w_{t-n+1})$ $C(w_{t-2})$ $C(w_{t-1})$

Table look-up in C

Matrix C shared parameters across words

index for w_{t-n+1} index for w_{t-2} index for w_{t-1}

Figure 1: Neural architecture: $f(i, w_{t-1}, \cdots, w_{t-n+1}) = g(i, C(w_{t-1}), \cdots, C(w_{t-n+1}))$ where g is the neural network and $C(i)$ is the i-th word feature vector.

parameters of the mapping C are simply the feature vectors themselves, represented by a $|V| \times m$ matrix C whose row i is the feature vector $C(i)$ for word i. The function g may be implemented by a feed-forward or recurrent neural network or another parametrized function, with parameters ω. The overall parameter set is $\theta = (C, \omega)$.

Training is achieved by looking for θ that maximizes the training corpus penalized log-likelihood:

$$L = \frac{1}{T} \sum_t \log f(w_t, w_{t-1}, \cdots, w_{t-n+1}; \theta) + R(\theta),$$

where $R(\theta)$ is a regularization term. For example, in our experiments, R is a weight decay penalty applied only to the weights of the neural network and to the C matrix, not to the biases.[3]

In the above model, the number of free parameters **only scales linearly** with V, the number of words in the vocabulary. It also **only scales linearly** with the order n : the scaling factor could be reduced to sub-linear if more sharing structure were introduced, e.g. using a time-delay neural network or a recurrent neural network (or a combination of both).

In most experiments below, the neural network has one hidden layer beyond the word features mapping, and optionally, direct connections from the word features to the output. Therefore there are really two hidden layers: the shared word features layer C, which has no non-linearity (it would not add anything useful), and the ordinary hyperbolic tangent hidden layer. More precisely, the neural network computes the following function, with a *softmax* output layer, which guarantees positive probabilities summing to 1:

$$\hat{P}(w_t | w_{t-1}, \cdots w_{t-n+1}) = \frac{e^{y_{w_t}}}{\sum_i e^{y_i}}.$$

---

3. The *biases* are the additive parameters of the neural network, such as b and d in equation 1 below.

The y_i are the unnormalized log-probabilities for each output word i, computed as follows, with parameters b, W, U, d and H:

$$y = b + Wx + U \tanh(d + Hx) \tag{1}$$

where the hyperbolic tangent tanh is applied element by element, W is optionally zero (no direct connections), and x is the word features layer activation vector, which is the concatenation of the input word features from the matrix C:

$$x = (C(w_{t-1}), C(w_{t-2}), \cdots, C(w_{t-n+1})).$$

Let h be the number of hidden units, and m the number of features associated with each word. When no direct connections from word features to outputs are desired, the matrix W is set to 0. The free parameters of the model are the output biases b (with $|V|$ elements), the hidden layer biases d (with h elements), the hidden-to-output weights U (a $|V| \times h$ matrix), the word features to output weights W (a $|V| \times (n-1)m$ matrix), the hidden layer weights H (a $h \times (n-1)m$ matrix), and the word features C (a $|V| \times m$ matrix):

$$\theta = (b, d, W, U, H, C).$$

The number of free parameters is $|V|(1 + nm + h) + h(1 + (n-1)m)$. The dominating factor is $|V|(nm + h)$. Note that in theory, if there is a weight decay on the weights W and H but not on C, then W and H could converge towards zero while C would blow up. In practice we did not observe such behavior when training with stochastic gradient ascent.

Stochastic gradient ascent on the neural network consists in performing the following iterative update after presenting the t-th word of the training corpus:

$$\theta \leftarrow \theta + \varepsilon \frac{\partial \log \hat{P}(w_t | w_{t-1}, \cdots w_{t-n+1})}{\partial \theta}$$

where ε is the "learning rate". Note that a large fraction of the parameters needs not be updated or visited after each example: the word features $C(j)$ of all words j that do not occur in the input window.

Mixture of models. In our experiments (see Section 4) we have found improved performance by combining the probability predictions of the neural network with those of an interpolated trigram model, either with a simple fixed weight of 0.5, a learned weight (maximum likelihood on the validation set) or a set of weights that are conditional on the frequency of the context (using the same procedure that combines trigram, bigram, and unigram in the interpolated trigram, which is a mixture).

3. Parallel Implementation

Although the number of parameters scales nicely, i.e. linearly with the size of the input window and linearly with the size of the vocabulary, the amount of computation required for obtaining the output probabilities is much greater than that required from n-gram models. The main reason is that with n-gram models, obtaining a particular $P(w_t | w_{t-1}, \ldots, w_{t-n+1})$ does not require the computation of the probabilities for all the words in the vocabulary, because of the easy normalization (performed when training the model) enjoyed by the linear combinations of relative frequencies. The main computational bottleneck with the neural implementation is the computation of the activations of the output layer.

Running the model (both training and testing) on a parallel computer is a way to reduce computation time. We have explored parallelization on two types of platforms: shared-memory processor machines and Linux clusters with a fast network.

3.1 Data-Parallel Processing

In the case of a shared-memory processor, parallelization is easily achieved, thanks to the very low communication overhead between processors, through the shared memory. In that case we have chosen a data-parallel implementation in which **each processor works on a different subset of the data**. Each processor computes the gradient for its examples, and performs stochastic gradient updates on the parameters of the model, which are simply stored in a shared-memory area. Our first implementation was extremely slow and relied on synchronization commands to make sure that each processor would not write at the same time as another one in one of the above parameter subsets. Most of the cycles of each processor were spent waiting for another processor to release a lock on the write access to the parameters.

Instead we have chosen an **asynchronous implementation** where each processor can write at any time in the shared-memory area. Sometimes, part of an update on the parameter vector by one of the processors is lost, being overwritten by the update of another processor, and this introduces a bit of noise in the parameter updates. However, this noise seems to be very small and did not apparently slow down training.

Unfortunately, large shared-memory parallel computers are very expensive and their processor speed tends to lag behind mainstream CPUs that can be connected in clusters. We have thus been able to obtain much faster training on fast network clusters.

3.2 Parameter-Parallel Processing

If the parallel computer is a network of CPUs, we generally can't afford to frequently exchange all the parameters among the processors, because that represents tens of megabytes (almost 100 megabytes in the case of our largest network), which would take too much time through a local network. Instead we have chosen to **parallelize across the parameters**, in particular the parameters of the output units, because that is where the vast majority of the computation is taking place, in our architecture. Each CPU is responsible for the computation of the unnormalized probability for a subset of the outputs, and performs the updates for the corresponding output unit parameters (weights going into that unit). This strategy allowed us to perform a **parallelized stochastic gradient ascent** with a negligible communication overhead. The CPUs essentially need to communicate two informations: (1) the normalization factor of the output softmax, and (2) the gradients on the hidden layer (denoted a below) and word feature layer (denoted x). All the CPUs duplicate the computations that precede the computation of the output units activations, i.e., the selection of word features and the computation of the hidden layer activation a, as well as the corresponding back-propagation and update steps. However, these computations are a negligible part of the total computation for our networks.

For example, consider the following architecture used in the experiments on the AP (Associated Press) news data: the vocabulary size is $|V| = 17,964$, the number of hidden units is $h = 60$, the order of the model is $n = 6$, the number of word features is $m = 100$. The total number of numerical operations to process a single training example is approximately $|V|(1+nm+h) + h(1+nm) + nm$ (where the terms correspond respectively to the computations of the output units, hidden units, and word feature units). In this example the fraction of the overall computation required for computing the weighted sums of the output units is therefore approximately $\frac{|V|(1+(n-1)m+h)}{|V|(1+(n-1)m+h)+h(1+(n-1)m)+(n-1)m} = 99.7\%$. This calculation is approximate because the actual CPU time associated with different operations differ, but it shows that it is generally advantageous to parallelize the output units computation. The fact that all CPUs will duplicate a very small fraction of the computations is not going to hurt the total computation time for the level of parallelization sought here, i.e. of a few dozen processors. If the number of hidden units was large, parallelizing their computation would also become profitable, but we did not investigate that approach in our experiments.

The implementation of this strategy was done on a cluster of 1.2 GHz clock-speed Athlon processors (32 x 2 CPUs) connected through a Myrinet network (a low-latency Gigabit local area network), using the MPI (Message Passing Interface) library (Dongarra et al., 1995) for the parallelization routines. The parallelization algorithm is sketched below, for a single example $(w_{t-n+1}, \cdots, w_t)$, executed in parallel by CPU i in a cluster of M processors. CPU i (i ranging from 0 to $M-1$) is responsible of a block of output units starting at number $start_i = i \times \lceil |V|/M \rceil$, the block being of length $\min(\lceil |V|/M \rceil, |V| - start_i)$.

COMPUTATION FOR PROCESSOR i, example t

1. **FORWARD PHASE**

    (a) Perform forward computation for the word features layer:
    $x(k) \leftarrow C(w_{t-k})$,
    $x = (x(1), x(2), \cdots, x(n-1))$

    (b) Perform forward computation for the hidden layer:
    $o \leftarrow d + Hx$
    $a \leftarrow \tanh(o)$

    (c) Perform forward computation for output units in the i-th block:
    $s_i \leftarrow 0$
    Loop over j in the i-th block

        i. $y_j \leftarrow b_j + a.U_j$
        ii. If (direct connections) $y_j \leftarrow y_j + x.W_j$
        iii. $p_j \leftarrow e^{y_j}$
        iv. $s_i \leftarrow s_i + p_j$

    (d) Compute and share $S = \sum_i s_i$ among the processors. This can easily be achieved with an MPI `Allreduce` operation, which can efficiently compute and share this sum.

    (e) Normalize the probabilities:
    Loop over j in the i-th block, $p_j \leftarrow p_j/S$.

    (f) Update the log-likelihood. If w_t falls in the block of CPU $i > 0$, then CPU i sends p_{w_t} to CPU 0. CPU 0 computes $L = \log p_{w_t}$ and keeps track of the total log-likelihood.

2. **BACKWARD/UPDATE PHASE**, with learning rate ε.

    (a) Perform backward gradient computation for output units in the i-th block:
    clear gradient vectors $\frac{\partial L}{\partial a}$ and $\frac{\partial L}{\partial x}$.
    Loop over j in the i-th block

   i. $\frac{\partial L}{\partial y_j} \leftarrow 1_{j==w_t} - p_j$

   ii. $b_j \leftarrow b_j + \varepsilon \frac{\partial L}{\partial y_j}$

      If (direct connections) $\frac{\partial L}{\partial x} \leftarrow \frac{\partial L}{\partial x} + \frac{\partial L}{\partial y_j} W_j$

      $\frac{\partial L}{\partial a} \leftarrow \frac{\partial L}{\partial a} + \frac{\partial L}{\partial y_j} U_j$

      If (direct connections) $W_j \leftarrow W_j + \varepsilon \frac{\partial L}{\partial y_j} x$

      $U_j \leftarrow U_j + \varepsilon \frac{\partial L}{\partial y_j} a$

(b) Sum and share $\frac{\partial L}{\partial x}$ and $\frac{\partial L}{\partial a}$ across processors. This can easily be achieved with an MPI `Allreduce` operation.

(c) Back-propagate through and update hidden layer weights:
Loop over k between 1 and h,

      $\frac{\partial L}{\partial o_k} \leftarrow (1 - a_k^2) \frac{\partial L}{\partial a_k}$

   $\frac{\partial L}{\partial x} \leftarrow \frac{\partial L}{\partial x} + H' \frac{\partial L}{\partial o}$

   $d \leftarrow d + \varepsilon \frac{\partial L}{\partial o}$

   $H \leftarrow H + \varepsilon \frac{\partial L}{\partial o} x'$

(d) Update word feature vectors for the input words:
Loop over k between 1 and $n - 1$

     $C(w_{t-k}) \leftarrow C(w_{t-k}) + \varepsilon \frac{\partial L}{\partial x(k)}$

     where $\frac{\partial L}{\partial x(k)}$ is the k-th block (of length m) of the vector $\frac{\partial L}{\partial x}$.

The weight decay regularization was not shown in the above implementation but can easily be put in (by subtracting the weight decay factor times the learning rate times the value of the parameter, from each parameter, at each update). Note that parameter updates are done directly rather than through a parameter gradient vector, to increase speed, a limiting factor in computation speed being the access to memory, in our experiments.

There could be a numerical problem in the computation of the exponentials in the forward phase, whereby all the p_j could be numerically zero, or one of them could be too large for computing the exponential (step 1(c)ii above). To avoid this problem, the usual solution is to subtract the maximum of the y_j's before taking the exponentials in the *softmax*. Thus we have added an extra `Allreduce` operation to share among the M processors the maximum of the y_j's, before computing the exponentials in p_j. Let q_i be the maximum of the y_j's in block i. Then the overall maximum $Q = \max_i q_i$ is collectively computed and shared among the M processors. The exponentials are then computed as follows: $p_j \leftarrow e^{y_j - Q}$ (instead of step 1(c)ii) to guarantee that at least one of the p_j's will be numerically non-zero, and the maximum of the exponential's argument is 1.

By comparing clock time of the parallel version with clock time on a single processor, we found that the communication overhead was only 1/15th of the total time (for one training epoch): thus we get an almost perfect speed-up through parallelization, using this algorithm on a fast network.

On clusters with a slow network, it might be possible to still obtain an efficient parallelization by performing the communications every K examples (a *mini-batch*) rather than for each example. This requires storing K versions of the activities and gradients of the neural network in each processor. After the forward phase on the K examples, the probability sums must be shared among the processors. Then the K backward phases are initiated, to obtain the K partial gradient vectors $\frac{\partial L}{\partial a}$ and $\frac{\partial L}{\partial x}$. After exchanging these gradient vectors among the processors, each processor can complete the backward phase and update parameters. This method mainly saves time because of the savings in network communication latency (the amount of data transferred is the same). It may lose in convergence time if K is too large, for the same reason that batch gradient descent is generally much slower than stochastic gradient descent (LeCun et al., 1998).

4. Experimental Results

Comparative experiments were performed on the Brown corpus which is a stream of 1,181,041 words, from a large variety of English texts and books. The first 800,000 words were used for training, the following 200,000 for validation (model selection, weight decay, early stopping) and the remaining 181,041 for testing. The number of different words is $47,578$ (including punctuation, distinguishing between upper and lower case, and including the syntactical marks used to separate texts and paragraphs). Rare words with frequency ≤ 3 were merged into a single symbol, reducing the vocabulary size to $|V| = 16,383$.

An experiment was also run on text from the Associated Press (AP) News from 1995 and 1996. The training set is a stream of about 14 million (13,994,528) words, the validation set is a stream of about 1 million (963,138) words, and the test set is also a stream of about 1 million (963,071) words. The original data has 148,721 different words (including punctuation), which was reduced to $|V| = 17964$ by keeping only the most frequent words (and keeping punctuation), mapping upper case to lower case, mapping numeric forms to special symbols, mapping rare words to a special symbol and mapping proper nouns to another special symbol.

For training the neural networks, the initial learning rate was set to $\varepsilon_o = 10^{-3}$ (after a few trials with a tiny data set), and gradually decreased according to the following schedule: $\varepsilon_t = \frac{\varepsilon_o}{1+rt}$ where t represents the number of parameter updates done and r is a decrease factor that was heuristically chosen to be $r = 10^{-8}$.

4.1 N-Gram Models

The first benchmark against which the neural network was compared is an interpolated or smoothed trigram model (Jelinek and Mercer, 1980). Let $q_t = l(freq(w_{t-1}, w_{t-2}))$ represents the discretized frequency of occurrence of the input context (w_{t-1}, w_{t-2}).[4] Then the conditional probability estimates have the form of a conditional mixture:

$$\hat{P}(w_t|w_{t-1}, w_{t-2}) = \alpha_0(q_t)p_0 + \alpha_1(q_t)p_1(w_t) + \alpha_2(q_t)p_2(w_t|w_{t-1}) + \alpha_3(q_t)p_3(w_t|w_{t-1}, w_{t-2})$$

with conditional weights $\alpha_i(q_t) \geq 0, \sum_i \alpha_i(q_t) = 1$. The base predictors are the following: $p_0 = 1/|V|$, $p_1(i)$ is a unigram (relative frequency of word i in the training set), $p_2(i|j)$ is the bigram (relative frequency of word i when the previous word is j), and $p_3(i|j,k)$ is the trigram (relative frequency of word i when the previous 2 words are j and k). The motivation is that when the frequency of (w_{t-1}, w_{t-2}) is large, p_3 is most reliable, whereas when it is lower, the lower-order statistics of p_2, p_1, or even p_0 are more reliable. There is a different set of mixture weights α for each of the discrete values of q_t (which are context frequency bins). They can be easily estimated with

---

4. We used $l(x) = \lceil -\log((1+x)/T) \rceil$ where $freq(w_{t-1}, w_{t-2})$ is the frequency of occurrence of the input context and T is the size of the training corpus.

the EM algorithm in about 5 iterations, on a set of data (the validation set) not used for estimating the unigram, bigram and trigram relative frequencies. The interpolated n-gram was used to form a mixture with the MLPs since they appear to make "errors" in very different ways.

Comparisons were also made with other state-of-the-art n-gram models: back-off n-gram models with the *Modified Kneser-Ney* algorithm (Kneser and Ney, 1995, Chen and Goodman., 1999), as well as class-based n-gram models (Brown et al., 1992, Ney and Kneser, 1993, Niesler et al., 1998). The validation set was used to choose the order of the n-gram and the number of word classes for the class-based models. We used the implementation of these algorithms in the SRI Language Modeling toolkit, described by Stolcke (2002) and in www.speech.sri.com/projects/srilm/. They were used for computing the back-off models perplexities reported below, noting that we did not give a special status to end-of-sentence tokens in the accounting of the log-likelihood, just as for our neural network perplexity. All tokens (words and punctuation) were treated the same in averaging the log-likelihood (hence in obtaining the perplexity).

4.2 Results

Below are measures of test set perplexity (geometric average of $1/\hat{P}(w_t|w_1^{t-1})$) for different models $\hat{P}$. Apparent convergence of the stochastic gradient ascent procedure was obtained after around 10 to 20 epochs for the Brown corpus. On the AP News corpus we were not able to see signs of overfitting (on the validation set), possibly because we ran only 5 epochs (over 3 weeks using 40 CPUs). Early stopping on the validation set was used, but was necessary only in our Brown experiments. A weight decay penalty of 10^{-4} was used in the Brown experiments and a weight decay of 10^{-5} was used in the APNews experiments (selected by a few trials, based on validation set perplexity). Table 1 summarizes the results obtained on the Brown corpus. All the back-off models of the table are modified Kneser-Ney n-grams, which worked significantly better than standard back-off models. When m is specified for a back-off model in the table, a class-based n-gram is used (m is the number of word classes). Random initialization of the word features was done (similarly to initialization of neural network weights), but we suspect that better results might be obtained with a knowledge-based initialization.

The **main result** is that significantly better results can be obtained when using the neural network, in comparison with the best of the n-grams, with a test perplexity difference of about 24% on Brown and about 8% on AP News, when taking the MLP versus the n-gram that worked best on the validation set. The table also suggests that the neural network was able to take advantage of more context (on Brown, going from 2 words of context to 4 words brought improvements to the neural network, not to the n-grams). It also shows that the hidden units are useful (MLP3 vs MLP1 and MLP4 vs MLP2), and that mixing the output probabilities of the neural network with the interpolated trigram always helps to reduce perplexity. The fact that simple averaging helps suggests that the neural network and the trigram make errors (i.e. low probability given to an observed word) in different places. The results do not allow to say whether the direct connections from input to output are useful or not, but suggest that on a smaller corpus at least, better generalization can be obtained without the direct input-to-output connections, at the cost of longer training: without direct connections the network took twice as much time to converge (20 epochs instead of 10), albeit to a slightly lower perplexity. A reasonable interpretation is that direct input-to-output connections provide a bit more capacity and faster learning of the "linear" part of the mapping from word features to log-

	n	c	h	m	direct	mix	train.	valid.	test.
MLP1	5		50	60	yes	no	182	284	268
MLP2	5		50	60	yes	yes		275	257
MLP3	5		0	60	yes	no	201	327	310
MLP4	5		0	60	yes	yes		286	272
MLP5	5		50	30	yes	no	209	296	279
MLP6	5		50	30	yes	yes		273	259
MLP7	3		50	30	yes	no	210	309	293
MLP8	3		50	30	yes	yes		284	270
MLP9	5		100	30	no	no	175	280	276
MLP10	5		100	30	no	yes		265	**252**
Del. Int.	3						31	352	336
Kneser-Ney back-off	3							334	323
Kneser-Ney back-off	4							332	321
Kneser-Ney back-off	5							332	321
class-based back-off	3	150						348	334
class-based back-off	3	200						354	340
class-based back-off	3	500						326	**312**
class-based back-off	3	1000						335	319
class-based back-off	3	2000						343	326
class-based back-off	4	500						327	312
class-based back-off	5	500						327	312

Table 1: Comparative results on the Brown corpus. The deleted interpolation trigram has a test perplexity that is 33% above that of the neural network with the lowest validation perplexity. The difference is 24% in the case of the best n-gram (a class-based model with 500 word classes). n : order of the model. c : number of word classes in class-based n-grams. h : number of hidden units. m : number of word features for MLPs, number of classes for class-based n-grams. *direct*: whether there are direct connections from word features to outputs. *mix*: whether the output probabilities of the neural network are mixed with the output of the trigram (with a weight of 0.5 on each). The last three columns give perplexity on the training, validation and test sets.

probabilities. On the other hand, without those connections the hidden units form a tight bottleneck which might force better generalization.

Table 2 gives similar results on the larger corpus (AP News), albeit with a smaller difference in perplexity (8%). Only 5 epochs were performed (in approximately three weeks with 40 CPUs). The class-based model did not appear to help the n-gram models in this case, but the high-order modified Kneser-Ney back-off model gave the best results among the n-gram models.

5. Extensions and Future Work

In this section, we describe extensions to the model described above, and directions for future work.

	n	h	m	direct	mix	train.	valid.	test.
MLP10	6	60	100	yes	yes		104	**109**
Del. Int.	3						126	132
Back-off KN	3						121	127
Back-off KN	4						113	119
Back-off KN	5						112	**117**

Table 2: Comparative results on the AP News corpus. See the previous table for the column labels.

5.1 An Energy Minimization Network

A variant of the above neural network can be interpreted as an energy minimization model following Hinton's recent work on products of experts (Hinton, 2000). In the neural network described in the previous sections the distributed word features are used only for the "input" words and not for the "output" word (next word). Furthermore, a very large number of parameters (the majority) are expanded in the output layer: the semantic or syntactic similarities between output words are not exploited. In the variant described here, the output word is also represented by its feature vector. The network takes in input a sub-sequence of words (mapped to their feature vectors) and outputs an energy function E which is low when the words form a likely sub-sequence, high when it is unlikely. For example, the network outputs an "energy" function

$$E(w_{t-n+1}, \cdots, w_t) = v.\tanh(d + Hx) + \sum_{i=0}^{n-1} b_{w_{t-i}}$$

where b is the vector of biases (which correspond to unconditional probabilities), d is the vector of hidden units biases, v is the output weight vector, and H is the hidden layer weight matrix, and unlike in the previous model, input and output words contribute to x:

$$x = (C(w_t), C(w_{t-1}), C(w_{t-2}), \cdots, C(w_{t-n+1})).$$

The energy function $E(w_{t-n+1}, \cdots, w_t)$ can be interpreted as an unnormalized log-probability for the joint occurrence of $(w_{t-n+1}, \cdots, w_t)$. To obtain a conditional probability $\hat{P}(w_t | w_{t-n+1}^{t-1})$ it is enough (but costly) to normalize over the possible values of w_t, as follows:

$$\hat{P}(w_t | w_{t-1}, \cdots, w_{t-n+1}) = \frac{e^{-E(w_{t-n+1}, \cdots, w_t)}}{\sum_i e^{-E(w_{t-n+1}, \cdots, w_{t-1}, i)}}$$

Note that the total amount of computation is comparable to the architecture presented earlier, and the number of parameters can also be matched if the v parameter is indexed by the identity of the target word (w_t). Note that only b_{w_t} remains after the above softmax normalization (any linear function of the w_{t-i} for $i > 0$ is canceled by the softmax normalization). As before, the parameters of the model can be tuned by stochastic gradient ascent on $\log \hat{P}(w_t | w_{t-1}, \cdots, w_{t-n+1})$, using similar computations.

In the products-of-experts framework, the hidden units can be seen as the experts: the joint probability of a sub-sequence $(w_{t-n+1}, \cdots, w_t)$ is proportional to the exponential of a sum of terms associated with each hidden unit j, $v_j \tanh(d_j + H_j x)$. Note that because we have chosen to decompose the probability of a whole sequence in terms of conditional probabilities for each element, the computation of the gradient is tractable. This is not the case for example with products-of-HMMs (Brown and Hinton, 2000), in which the product is over experts that view the whole sequence, and which can be trained with approximate gradient algorithms such as the contrastive divergence algorithm (Brown and Hinton, 2000). Note also that this architecture and the products-of-experts formulation can be seen as extensions of the very successful **Maximum Entropy** models (Berger et al., 1996), but where the basis functions (or "features", here the hidden units activations) are learned by penalized maximum likelihood at the same time as the parameters of the features linear combination, instead of being learned in an outer loop, with greedy feature subset selection methods.

We have implemented and experimented with the above architecture, and have developed a speed-up technique for the neural network training, based on importance sampling and yielding a 100-fold speed-up (Bengio and Senécal, 2003).

Out-of-vocabulary words. An advantage of this architecture over the previous one is that it can easily deal with out-of-vocabulary words (and even assign them a probability!). The main idea is to first guess an initial feature vector for such a word, by taking a weighted convex combination of the feature vectors of other words that could have occurred in the same context, with weights proportional to their conditional probability. Suppose that the network assigned a probability $\hat{P}(i|w_{t-n+1}^{t-1})$ to words $i \in V$ in context w_{t-n+1}^{t-1}, and that in this context we observe a new word $j \notin V$. We initialize the feature vector $C(j)$ for j as follows: $C(j) \leftarrow \sum_{i \in V} C(i)\hat{P}(i|w_{t-n+1}^{t-1})$. We can then incorporate j in V and re-compute probabilities for this slightly larger set (which only requires a renormalization for all the words, except for word i, which requires a pass through the neural network). This feature vector $C(i)$ can then be used in the input context part when we try to predict the probabilities of words that follow word i.

5.2 Other Future Work

There are still many challenges ahead to follow-up on this work. In the short term, methods to speed-up training and recognition need to be designed and evaluated. In the longer term, more ways to generalize should be introduced, in addition to the two main ways exploited here. Here are some ideas that we intend to explore:

1. Decomposing the network in sub-networks, for example using a clustering of the words. Training many smaller networks should be easier and faster.

2. Representing the conditional probability with a tree structure where a neural network is applied at each node, and each node represents the probability of a word class given the context and the leaves represent the probability of words given the context. This type of representation has the potential to reduce computation time by a factor $|V|/\log|V|$ (see Bengio, 2002).

3. Propagating gradients only from a subset of the output words. It could be the words that are conditionally most likely (based on a faster model such as a trigram, see Schwenk and Gauvain, 2002, for an application of this idea), or it could be a subset of the words for which the trigram has been found to perform poorly. If the language model is coupled to a speech recognizer, then only the scores (unnormalized probabilities) of the acoustically ambiguous words need to be computed. See also Bengio and Senécal (2003) for a new accelerated training method using importance sampling to select the words.

4. Introducing a-priori knowledge. Several forms of such knowledge could be introduced, such as: semantic information (e.g., from WordNet, see Fellbaum, 1998), low-level grammatical information (e.g., using parts-of-speech), and high-level grammatical information, e.g., coupling the model to a stochastic grammar, as suggested in Bengio (2002). The effect of longer term context could be captured by introducing more structure and parameter sharing in the neural network, e.g. using time-delay or recurrent neural networks. In such a multi-layered network the computation that has been performed for small groups of consecutive words does not need to be redone when the network input window is shifted. Similarly, one could use a recurrent network to capture potentially even longer term information about the subject of the text.

5. Interpreting (and possibly using) the word feature representation learned by the neural network. A simple first step would start with $m = 2$ features, which can be more easily displayed. We believe that more meaningful representations will require large training corpora, especially for larger values of m.

6. Polysemous words are probably not well served by the model presented here, which assigns to each word a single point in a continuous semantic space. We are investigating extensions of this model in which each word is associated with multiple points in that space, each associated with the different senses of the word.

6. Conclusion

The experiments on two corpora, one with more than a million examples, and a larger one with above 15 million words, have shown that the proposed approach yields much better perplexity than a state-of-the-art method, the smoothed trigram, with differences between 10 and 20% in perplexity.

We believe that the main reason for these improvements is that the proposed approach allows to take advantage of the learned distributed representation to fight the curse of dimensionality with its own weapons: each training sentence informs the model about a combinatorial number of other sentences.

There is probably much more to be done to improve the model, at the level of architecture, computational efficiency, and taking advantage of prior knowledge. An important priority of future research should be to improve speed-up techniques[5] as well as ways to increase capacity without increasing training time too much (to deal with corpora with hundreds of millions of words or more). A simple idea to take advantage of temporal structure and extend the size of the input window to include possibly a whole paragraph (without increasing too much the number of parameters or computation time) is to use a time-delay and possibly recurrent neural networks. Evaluations of the type of models presented here in applicative contexts would also be useful, but see work already done by Schwenk and Gauvain (2002) for improvements in speech recognition word error rate.

More generally, the work presented here opens the door to improvements in statistical language models brought by replacing "tables of conditional probabilities" by more compact and smoother representations based on distributed representations that can accommodate far more conditioning variables. Whereas much effort has been spent in statistical language models (e.g. stochastic grammars) to restrict or summarize the conditioning variables in order to avoid overfitting, the type of

---

5. See work by Bengio and Senécal (2003) for a 100-fold speed-up technique.

models described here shifts the difficulty elsewhere: many more computations are required, but computation and memory requirements scale linearly, not exponentially with the number of conditioning variables.

ACKNOWLEDGMENTS

The authors would like to thank Léon Bottou, Yann Le Cun and Geoffrey Hinton for useful discussions. This research was made possible by funding from the NSERC granting agency, as well as the MITACS and IRIS networks.

References

D. Baker and A. McCallum. Distributional clustering of words for text classification. In *SIGIR'98*, 1998.

J.R. Bellegarda. A latent semantic analysis framework for large–span language modeling. In *Proceedings of Eurospeech 97*, pages 1451–1454, Rhodes, Greece, 1997.

S. Bengio and Y. Bengio. Taking on the curse of dimensionality in joint distributions using neural networks. *IEEE Transactions on Neural Networks, special issue on Data Mining and Knowledge Discovery*, 11(3):550–557, 2000a.

Y. Bengio. New distributed probabilistic language models. Technical Report 1215, Dept. IRO, Université de Montréal, 2002.

Y. Bengio and S. Bengio. Modeling high-dimensional discrete data with multi-layer neural networks. In S. A. Solla, T. K. Leen, and K-R. Müller, editors, *Advances in Neural Information Processing Systems*, volume 12, pages 400–406. MIT Press, 2000b.

Y. Bengio and J-S. Senécal. Quick training of probabilistic neural nets by importance sampling. In *AISTATS*, 2003.

A. Berger, S. Della Pietra, and V. Della Pietra. A maximum entropy approach to natural language processing. *Computational Linguistics*, 22:39–71, 1996.

A. Brown and G.E. Hinton. Products of hidden markov models. Technical Report GCNU TR 2000-004, Gatsby Unit, University College London, 2000.

P.F. Brown, V.J. Della Pietra, P.V. DeSouza, J.C. Lai, and R.L. Mercer. Class-based n-gram models of natural language. *Computational Linguistics*, 18:467–479, 1992.

S.F. Chen and J.T. Goodman. An empirical study of smoothing techniques for language modeling. *Computer, Speech and Language*, 13(4):359–393, 1999.

S. Deerwester, S.T. Dumais, G.W. Furnas, T.K. Landauer, and R. Harshman. Indexing by latent semantic analysis. *Journal of the American Society for Information Science*, 41(6):391–407, 1990.

J. Dongarra, D. Walker, and The Message Passing Interface Forum. MPI: A message passing interface standard. Technical Report http://www-unix.mcs.anl.gov/mpi, University of Tenessee, 1995.

J.L. Elman. Finding structure in time. *Cognitive Science*, 14:179–211, 1990.

C. Fellbaum. *WordNet: An Electronic Lexical Database*. MIT Press, 1998.

J .Goodman. A bit of progress in language modeling. Technical Report MSR-TR-2001-72, Microsoft Research, 2001.

G.E. Hinton. Learning distributed representations of concepts. In *Proceedings of the Eighth Annual Conference of the Cognitive Science Society*, pages 1–12, Amherst 1986, 1986. Lawrence Erlbaum, Hillsdale.

G.E. Hinton. Training products of experts by minimizing contrastive divergence. Technical Report GCNU TR 2000-004, Gatsby Unit, University College London, 2000.

F. Jelinek and R. L. Mercer. Interpolated estimation of Markov source parameters from sparse data. In E. S. Gelsema and L. N. Kanal, editors, *Pattern Recognition in Practice*. North-Holland, Amsterdam, 1980.

K.J. Jensen and S. Riis. Self-organizing letter code-book for text-to-phoneme neural network model. In *Proceedings ICSLP*, 2000.

S.M. Katz. Estimation of probabilities from sparse data for the language model component of a speech recognizer. *IEEE Transactions on Acoustics, Speech, and Signal Processing*, ASSP-35 (3):400–401, March 1987.

R. Kneser and H. Ney. Improved backing-off for m-gram language modeling. In *International Conference on Acoustics, Speech and Signal Processing*, pages 181–184, 1995.

Y. LeCun, L. Bottou, G.B. Orr, and K.-R. Müller. Efficient backprop. In G.B. Orr and K.-R. Müller, editors, *Neural Networks: Tricks of the Trade*, pages 9–50. Springer, 1998.

R. Miikkulainen and M.G. Dyer. Natural language processing with modular neural networks and distributed lexicon. *Cognitive Science*, 15:343–399, 1991.

H. Ney and R. Kneser. Improved clustering techniques for class-based statistical language modelling. In *European Conference on Speech Communication and Technology (Eurospeech)*, pages 973–976, Berlin, 1993.

T.R. Niesler, E.W.D. Whittaker, and P.C. Woodland. Comparison of part-of-speech and automatically derived category-based language models for speech recognition. In *International Conference on Acoustics, Speech and Signal Processing*, pages 177–180, 1998.

A. Paccanaro and G.E. Hinton. Extracting distributed representations of concepts and relations from positive and negative propositions. In *Proceedings of the International Joint Conference on Neural Network, IJCNN'2000*, Como, Italy, 2000. IEEE, New York.

F. Pereira, N. Tishby, and L. Lee. Distributional clustering of english words. In *30th Annual Meeting of the Association for Computational Linguistics*, pages 183–190, Columbus, Ohio, 1993.

S. Riis and A. Krogh. Improving protein secondary structure prediction using structured neural networks and multiple sequence profiles. *Journal of Computational Biology*, pages 163–183, 1996.

J. Schmidhuber. Sequential neural text compression. *IEEE Transactions on Neural Networks*, 7(1): 142–146, 1996.

H. Schutze. Word space. In S. J. Hanson, J. D. Cowan, and C. L. Giles, editors, *Advances in Neural Information Processing Systems 5*, pages pp. 895–902, San Mateo CA, 1993. Morgan Kaufmann.

H. Schwenk and J-L. Gauvain. Connectionist language modeling for large vocabulary continuous speech recognition. In *International Conference on Acoustics, Speech and Signal Processing*, pages 765–768, Orlando, Florida, 2002.

A. Stolcke. SRILM - an extensible language modeling toolkit. In *Proceedings of the International Conference on Statistical Language Processing*, Denver, Colorado, 2002.

W. Xu and A. Rudnicky. Can artificial neural network learn language models. In *International Conference on Statistical Language Processing*, pages M1–13, Beijing, China, 2000.

Chapter 14

Transformer Networks – 2017

Context

This paper[134] by Vaswani et al., titled 'Attention is all you need', presented a significant improvement in machine translation, demonstrated on English-to-German and English-to-French translation of text. However, the main reason for the paper's impact is that it provides an unambiguous demonstration that the complex computational machinery required for recurrent neural networks (RNNs) can be discarded (as indicated in the first sentence of the abstract).

A major component of the model is *attention*, which is essentially a mechanism for learning the extent to which different words are related to each other. Although the idea of attention was not new, the main innovation introduced in the paper is the use of parallel streams of attention (*multi-head attention*) in a static neural network.

The main reason for the complexity of the architecture is to enable the neural network to learn to summarise all the information regarding the position of each word in a sentence, the relationship between words, and the next-word probability of each word in German (for example) given a sequence of words in English. This information summary is represented as a set of *self-attention head matrices*. Each head matrix is based on three learned matrices, each of which captures a different aspect of the data. The head matrices form the basis of the next-word prediction in German given a sequence of words in English. Loosely speaking, the encoding process is like a form of entanglement (in quantum mechanics), such that all the information required to perform next-word (English-to-German) prediction is implicit in the matrices, and the job of the neural network is to disentangle, or make explicit, the information in these matrices.

Unlike models from the early years of AI, which usually had simple architectures based on a single principle (e.g. Hopfield nets), the complexity of the transformer network architecture and the multiplicity of the mechanisms included seem hard to justify – aside from the brutal fact that the architecture works. It is as if someone had invented an aeroplane with four wings stacked on top each other (Figure 14.1), but no one could figure out how to reduce the number of wings without the aeroplane crashing (in fairness, this general point applies to many modern AI systems). One particularly valiant attempt at reducing the 'number of wings' was made by Yu et al. (2023). It is also noteworthy that Oren et al. (2024) proved that transformers are equivalent to RNNs.

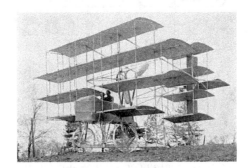

Figure 14.1. Jacob's multi-plane from 1910.

Technical Summary

Notation. In the following account, the notation is consistent with that used in the paper. However, the brevity required for research papers means that several intermediate quantities are not represented by any variable in the paper. We therefore introduce a few new variables below, and these, along with the variable names used in the paper, are *italicised*.

14.1. The Short Version

The objective of the transformer is to translate each *input sequence* of n English words,

$$\mathbf{w}^E = (w_1^E, \ldots, w_n^E), \tag{14.1}$$

into a *target sequence* of words in another language (e.g. German),

$$\mathbf{w}^G = (w_1^G, \ldots, w_m^G), \tag{14.2}$$

where m is not necessarily equal to n. For convenience, here we assume that English gets translated into German, but the paper uses both German and French as the target language.

The transformer consists of two main modules, an *encoder* and a *decoder*, as shown in Figure 14.2. The job of the encoder is to repackage each input sequence of n English words into a succinct representation that carries enough information (e.g. about the position and identity of words) for the

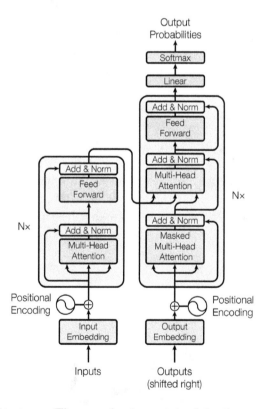

Figure 14.2. Transformer architecture. The encoder is contained in the grey rectangle on the left, and the decoder is in the grey rectangle on the right. All parameters used to transform input word sequences into output sequences can be learned (however, the positional encoding was not learned in the paper). This is Figure 1 in the paper.

decoder to generate a translation of $\mathbf{w}^E$ into German,

$$\hat{\mathbf{w}}^G \;=\; (\hat{w}_1^G, \ldots, \hat{w}_m^G). \tag{14.3}$$

After training, the decoder generates one predicted next word in German $\hat{w}_{t+1}^G$ at each time step t, and this is added to the decoder input at the next time step, so the prediction $\hat{w}_{t+1}^G$ contributes to the predictions of subsequent German words $\hat{w}_{t+2}^G, \hat{w}_{t+3}^G, \ldots, \hat{w}_m^G$.

14.2. The Long Version

The transformer includes several large matrices, and it can be difficult to keep track of which quantities are represented in the rows and columns of each matrix. To circumvent this problem, we begin by focusing on a single word in a single sequence of n words. Later, we will apply the same reasoning to multiple words, which will result in the matrix formulation given in the paper.

Encoder Scaled Dot Product Attention. This section takes us up to Equation 1 in the paper. Referring to Figure 14.2 (or Figure 1 in the paper), this corresponds to a transition from the encoder input to a box labelled 'Add & Norm' (we omit the optional 'masking' procedure for now).

Given a vocabulary V of $|V|$ words, the encoding process for each of the n words in a word sequence $w_1, \ldots, w_n$ can be summarised in the following 13 steps:

1. The input sequence of n words is split into *tokens*. For simplicity, we assume that each word is represented by one token.

2. Each token is assigned a unique identity (ID) number i, which corresponds to its position amongst the list of $|V|$ words in V.

3. The ID number is used to form a one-hot row vector $\mathbf{x}_i^{\text{hot}}$ of $|V|$ elements in which every element is zero except the ith one. For example, if we had a small vocabulary of only $|V| = 5$ words then the second word in this vocabulary would be represented by the one-hot row vector

$$\mathbf{x}_2^{\text{hot}} \;=\; (0, 1, 0, 0, 0). \tag{14.4}$$

4. Next, we define a learned embedding matrix $W_{|V| \times d_{\text{model}}}^{\text{embed}}$ with $|V|$ rows and d_{model} columns. For example, if $|V| = 5$ then the embedding matrix is of the form

$$W^{\text{embed}} \;=\; \begin{pmatrix} a_{11} & a_{12} & \cdots & a_{1d_{\text{model}}} \\ a_{21} & a_{22} & \cdots & a_{2d_{\text{model}}} \\ a_{31} & a_{32} & \cdots & a_{3d_{\text{model}}} \\ a_{41} & a_{42} & \cdots & a_{4d_{\text{model}}} \\ a_{51} & a_{52} & \cdots & a_{5d_{\text{model}}} \end{pmatrix}. \tag{14.5}$$

5. The embedding matrix projects the one-hot row vector $\mathbf{x}_i^{\text{hot}}$ to a *embedded token vector* $\mathbf{x}_i^{\text{token}}$,

$$\mathbf{x}_i^{\text{token}} \;=\; \mathbf{x}_i^{\text{hot}} W^{\text{embed}}, \tag{14.6}$$

which is a row vector of d_{model} real numbers. Because each column of W^{embed} gets multiplied by $\mathbf{x}_i^{\text{hot}}$ and the only non-zero element in $\mathbf{x}_i^{\text{hot}}$ is its ith element, the $1 \times d_{\text{model}}$ embedded token vector $\mathbf{x}_i^{\text{token}}$ is the same as the ith row of the embedding matrix W^{embed}. The number d_{model} of columns in W^{embed} is called the *embedding dimension* because each word is represented as an embedded token vector $\mathbf{x}_i^{\text{token}}$ in d_{model}-dimensional space $\mathbb{R}^{d_{\text{model}}}$. For example, if we focus on

$i = 2$ then Equations 14.4–14.6 give

$$\mathbf{x}_2^{\text{token}} = (0, 1, 0, 0, 0) \begin{pmatrix} a_{11} & a_{12} & \cdots & a_{1d_{\text{model}}} \\ a_{21} & a_{22} & \cdots & a_{2d_{\text{model}}} \\ a_{31} & a_{32} & \cdots & a_{3d_{\text{model}}} \\ a_{41} & a_{42} & \cdots & a_{4d_{\text{model}}} \\ a_{51} & a_{52} & \cdots & a_{5d_{\text{model}}} \end{pmatrix}$$

$$= (a_{21}, a_{22}, \ldots, a_{2d_{\text{model}}}). \tag{14.7}$$

Since the only non-zero element in $\mathbf{x}_i^{\text{hot}}$ is the second element, the $1 \times d_{\text{model}}$ vector $\mathbf{x}_2^{\text{token}}$ is the same as the second row of the embedding matrix W^{embed}.

The point of learning the word-embedding matrix is to ensure that vectors corresponding to similar words (e.g. 'apple' and 'pear') get projected to embedded token vectors at similar locations in the space $\mathbb{R}^{d_{\text{model}}}$ inhabited by embedded token vectors. Two vectors having similar locations in $\mathbb{R}^{d_{\text{model}}}$ means that the angle between them is small, and it is this angle that is used as the measure of similarity. In contrast, the one-hot vectors that represent the ID numbers of words are all mutually orthogonal (i.e. the angle between any pair of one-hot vectors is 90°). Note that the encoder and decoder have different embedded token vectors because they have independent embedding matrices, which are learned as part of the training process.

6. The embedded vector $\mathbf{x}_i^{\text{token}}$ for the word w_i carries information about the identity of that word, but it provides no information regarding the position of w_i in the input sequence of n words. To add this positional information, both the encoder and the decoder modules represent the position of the ith word in the input sequence by a *positional vector* $\mathbf{x}_i^{\text{pos}}$ of d_{model} elements, which is added to the embedded token vector to yield the *word+position vector*

$$\mathbf{x}_i = \mathbf{x}_i^{\text{token}} + \mathbf{x}_i^{\text{pos}}. \tag{14.8}$$

Specifically, if *pos* is the position of a word within a sequence of n words, the following *positional encodings* are used in the paper:

$$PE_{(pos,2i)} = \sin(pos/10000^{2i/d_{\text{model}}}), \tag{14.9}$$
$$PE_{(pos,2i+1)} = \cos(pos/10000^{2i/d_{\text{model}}}), \tag{14.10}$$

where $i = 1, \ldots, d_{\text{model}}/2$. For a given word position *pos* there are $d_{\text{model}}/2$ sine values, each of which is associated with a different wavelength λ (between $\lambda = 2\pi$ and $\lambda = 10{,}000 \times 2\pi$), and $d_{\text{model}}/2$ corresponding cosine values[†]. Notice that the set of wavelengths for a given word position *pos* shrinks as *pos* increases, which ensures that each word is assigned a unique position vector.

The input sequence of n words has now been transformed into n embedded vectors $\mathbf{x}_1, \ldots, \mathbf{x}_n$ in $\mathbb{R}^{d_{\text{model}}}$.

[†]The use of the index i here is easily confused with the index used for word position in other parts of the paper, but that is the notation used in the paper, so we follow it here.

7. The word+position row vector $\mathbf{x}_i$ is transformed by three different (learned) embedding matrices into three corresponding row vectors,

$$\text{a query vector} \quad \mathbf{q}_i = \mathbf{x}_i W^Q, \tag{14.11}$$

$$\text{a key vector} \quad \mathbf{k}_i = \mathbf{x}_i W^K, \tag{14.12}$$

$$\text{a value vector} \quad \mathbf{v}_i = \mathbf{x}_i W^V. \tag{14.13}$$

Both the *embedding query matrix* W^Q and the *embedding key matrix* W^K have d_{model} rows, and each of these rows has d_k elements, so both $\mathbf{q}_i$ and $\mathbf{k}_i$ have d_k elements,

$$
\begin{aligned}
\mathbf{q}_i &= (x_{i1}, x_{i2}, \ldots, x_{id_{\text{model}}})
\begin{pmatrix}
b_{11} & b_{12} & \cdots & b_{1d_k} \\
b_{21} & b_{22} & \cdots & b_{2d_k} \\
\vdots & \vdots & \ddots & \vdots \\
b_{d_{\text{model}}1} & b_{d_{\text{model}}2} & \cdots & b_{d_{\text{model}}d_k}
\end{pmatrix} \\
&= (q_{i1}, q_{i2}, \ldots, q_{id_k})
\end{aligned}
\tag{14.14}
$$

and

$$
\begin{aligned}
\mathbf{k}_i &= (x_{i1}, x_{i2}, \ldots, x_{id_{\text{model}}})
\begin{pmatrix}
c_{11} & c_{12} & \cdots & c_{1d_k} \\
c_{21} & c_{22} & \cdots & c_{2d_k} \\
\vdots & \vdots & \ddots & \vdots \\
c_{d_{\text{model}}1} & c_{d_{\text{model}}2} & \cdots & c_{d_{\text{model}}d_k}
\end{pmatrix} \\
&= (k_{i1}, k_{i2}, \ldots, k_{id_k}).
\end{aligned}
\tag{14.15}
$$

The value of d_k was set to $d_{\text{model}} = 512$ in the first part of the paper, but this was reduced to $d_k = 64$ in the part of the paper that describes multi-head attention. The embedding value matrix W^V also has d_{model} rows but can have a different number d_v of columns, so that the value vector $\mathbf{v}_i$ has d_v elements.

8. At this point, the relationship between the ith and jth words in the sequence can be summarised as the *dot product attention*

$$\mathbf{q}_i \cdot \mathbf{k}_j. \tag{14.16}$$

However, under the plausible assumption that the vectors have Gaussian distributions, the dot product increases with $\sqrt{d_k}$, so we normalise it with $\sqrt{d_k}$ and use the normalised dot product of the ith query vector $\mathbf{q}_i$ and the jth key vector $\mathbf{k}_j$ as a measure of the proximity of w_i to w_j,

$$b_{ij} = \frac{\mathbf{q}_i \cdot \mathbf{k}_j}{\sqrt{d_k}}, \tag{14.17}$$

where $d_k = d_{\text{model}} = 512$ here (later we set $d_k = 64$). Similarly, we can summarise the relationship between the ith word and all of the n words in the input sequence as the ith proximity vector

$$
\begin{aligned}
\mathbf{b}_i &= (b_{i1}, b_{i2}, \ldots, b_{in}) \\
&= (\mathbf{q}_i \cdot \mathbf{k}_1, \mathbf{q}_i \cdot \mathbf{k}_2, \ldots, \mathbf{q}_i \cdot \mathbf{k}_n)/\sqrt{d_k}.
\end{aligned}
\tag{14.18}
$$

Crucially, the embedding matrices W^Q, W^K and W^V are learned as part of the training process, so these matrices produce embedding vectors optimised for translating words from one language to another. This learning means that the embedding vectors produced by W^Q, W^K and W^V for the words *lamb* and *fleece* (for example) can be made similar to each other if doing so enhances the ability of the transformer to translate between languages.

The notion of similarity is well defined for vectors, and is usually expressed in terms of the dot or inner product (see Appendix B). For example, given an embedding vector $\mathbf{q}_i$ for the word $w_i = lamb$ and an embedding vector $\mathbf{k}_j$ for a different word $w_j = fleece$, their similarity (as determined by W^Q and W^K) is a monotonic (cosine) function of the angle θ between $\mathbf{q}_i$ and $\mathbf{k}_j$:

$$\text{Similarity}(\mathbf{q}_i, \mathbf{k}_j) \quad = \quad \mathbf{q}_i \cdot \mathbf{k}_j \tag{14.19}$$

$$= \quad c_{ij} \cos\theta, \tag{14.20}$$

where $c_{ij} = 1/(|\mathbf{q}_i|\,|\mathbf{k}_j|)$ is a constant. For example, if $\theta \approx 0$ then $\text{Similarity}(\mathbf{q}_i, \mathbf{k}_j) \approx c_{ij}$, whereas if $\theta \approx 90°$ then $\text{Similarity}(\mathbf{q}_i, \mathbf{k}_j) \approx 0$. Notice that the similarity measure defined here is the same as the dot product attention defined in Equation 14.16.

The similarity measure defined above determines the amount of attention that the (embedded representation of the) word w_i pays to the word w_j (which depends on their similarity measure). However, this is almost always different from the amount of attention that the word w_j pays to the word w_i.

9. If we normalise the elements of the proximity vector $\mathbf{b}_i = (b_{i1}, b_{i2}, \ldots, b_{in})$ using the *softmax function* then we obtain a *scaled dot product attention* value

$$p_{ij} \quad = \quad \frac{e^{b_{ij}}}{\sum_{j=1}^{n} e^{b_{ij}}}, \tag{14.21}$$

where the p_{ij} values for the word w_i sum to unity,

$$\sum_{j=1}^{n} p_{ij} \quad = \quad 1. \tag{14.22}$$

Loosely speaking, this allows the system to treat p_{ij} as the probability that the words w_i and w_j occur in their respective positions in the current input sequence. For each input word w_i, we now have a summary of the probabilities for that word in the context of the input sequence of n words,

$$\mathbf{p}_i \quad = \quad (p_{i1}, p_{i2}, \ldots, p_{in}). \tag{14.23}$$

10. Next, we combine the query, key and value vectors (n of each) to encapsulate the relationships between the words in the input sequence:

a) The n row vectors $\mathbf{q}_i$, each with d_k elements, are used to form the rows of the $n \times d_k$ *query matrix*

$$Q \quad = \quad \begin{pmatrix} \mathbf{q}_1 \\ \mathbf{q}_2 \\ \vdots \\ \mathbf{q}_n \end{pmatrix} = \begin{pmatrix} q_{11} & q_{12} & \cdots & q_{1d_k} \\ q_{21} & q_{22} & \cdots & q_{2d_k} \\ \vdots & \vdots & \ddots & \vdots \\ q_{n1} & q_{n2} & \cdots & q_{nd_k} \end{pmatrix}. \tag{14.24}$$

b) The n row vectors $\mathbf{k}_i$, each with d_k elements, are used to form the rows of the $n \times d_k$ *key matrix*

$$K = \begin{pmatrix} \mathbf{k}_1 \\ \mathbf{k}_2 \\ \vdots \\ \mathbf{k}_n \end{pmatrix} = \begin{pmatrix} k_{11} & k_{12} & \cdots & k_{1d_k} \\ k_{21} & k_{22} & \cdots & k_{2d_k} \\ \vdots & \vdots & \ddots & \vdots \\ k_{n1} & k_{n2} & \cdots & k_{nd_k} \end{pmatrix}. \tag{14.25}$$

c) The n row vectors $\mathbf{v}_i$, each with d_v elements, are used to form the rows of the $n \times d_v$ *value matrix*

$$V = \begin{pmatrix} \mathbf{v}_1 \\ \mathbf{v}_2 \\ \vdots \\ \mathbf{v}_n \end{pmatrix} = \begin{pmatrix} v_{11} & v_{12} & \cdots & v_{1d_v} \\ v_{21} & v_{22} & \cdots & v_{2d_v} \\ \vdots & \vdots & \ddots & \vdots \\ v_{n1} & v_{n2} & \cdots & v_{nd_v} \end{pmatrix}. \tag{14.26}$$

11. We can now obtain the *self-attention matrix* given in Equation 1 of the paper,

$$\text{Attention}(Q, K, V) = \text{softmax}\left(\frac{QK^{\mathsf{T}}}{\sqrt{d_k}}\right) V, \tag{14.27}$$

where the scaled product $QK^{\mathsf{T}}/\sqrt{d_k}$ is an $n \times n$ matrix in which each element b_{ij} (defined in Equation 14.17) is a measure of the closeness between two words in the input sequence of n words. Multiplying the $n \times n$ matrix that results from applying the softmax function to $QK^{\mathsf{T}}/\sqrt{d_k}$ by the $n \times d_v$ value matrix V yields the $n \times d_v$ self-attention matrix.

12. **Multi-Head Attention.** The steps leading to Equation 14.27 result in a single *head*. In contrast, *multi-head attention* involves using the embedded vectors $\mathbf{x}_1, \ldots, \mathbf{x}_n$ obtained from the input sequence of n words to form $N = 6$ different versions of the query, key and value matrices at the same time, which results in N corresponding versions of the self-attention matrix. The exact form of each different self-attention matrix depends on the (random) initial state of each learned matrix before training begins. The stated reason for employing multi-head attention is to give the system the opportunity to emphasize different relationships between words in each of the N versions of the self-attention matrix.

13. **Add & Norm.** This refers to the yellow boxes in Figure 1 of the paper. The outputs of multi-head attention have a copy of the vectors $\mathbf{x}_1, \ldots, \mathbf{x}_n$ added to them. This is to ensure that any information lost between the inputs and outputs of the multi-head attention module remains available to subsequent stages of processing. The word *norm* is short for normalisation, which ensures that final outputs have a mean and a variance learned during training.

The penultimate module in the encoder and decoder in Figure 1 of the paper is labelled 'Feed Forward', which refers to a fully connected feedforward (backpropagation) network.

The Decoder. To simplify the account, we define an *MA-module* as any combination of two boxes in Figure 1 of the paper labelled '(Masked) Multi-Head Attention' and 'Add & Norm'; the term MA-module is not used in the paper.

Whereas the encoder's input is a word sequence in English, the decoder's input sequence is in a different language (German or French). Referring to Figure 1 in the paper, the lower MA-module of the decoder is almost identical to the lower MA-module in the encoder, the main difference being the addition of *masking*. This ensures that each predicted German word w_{t+1}^G is based only on the subsequence $w_1^G, w_2^G, \ldots, w_t^G$ of words that precede w_{t+1}^G in the input sequence. Masking is implemented with an additional matrix in which the upper triangular elements are set to zero. The upper MA-module in the decoder has inputs consisting of a pair of query and key matrices from the encoder, plus a value matrix from the lower MA-module of the decoder. Note that the transformer has access to the full sequence of n English words (via the encoder) at all times.

The transformer output is fairly complicated, and is different before and after training, where post-training is referred to as inference. During training, the output of the transformer takes the form of a probability distribution of $|V|$ elements (i.e. over the vocabulary) for each word in the decoder's German target sequence. Specifically, the output for each German word in the target sequence is a vector of $|V|$ probabilities, where the ith element is the probability that the next German word w_{t+1}^G in the sequence is w_i^G.

During inference (or testing/prediction), the output of the transformer takes a different form. Instead of producing a probability distribution for every word in the target sequence simultaneously, the model generates a probability distribution for every word in the target sequence one word at a time. A common decoding strategy is the *greedy decoding* approach, where at each decoding step the model selects the word with the highest probability as being the predicted next word. The predicted word is then fed back into the model as input for the next decoding step. In practice, *beam search* can be used to explore multiple possible sequences to select the most likely one.

14.3. Results

The best results were achieved by the *big model* in the paper, which has $N = 6$ versions of the self-attention matrix within each MA-module and a total of 213 million parameters (e.g. in the embedding matrices and feedforward neural networks). The results are presented in terms of perplexity and a metric called the *bilingual evaluation understudy* (BLEU). Here, we concentrate on BLEU because the perplexity measure used in the paper refers to word-pieces, so it is not comparable to conventional measures of perplexity. The main results are neatly summarised in two paragraphs, paraphrased here:

On the WMT 2014 English-to-German translation task, the big transformer model outperformed the best previously reported models by more than 2.0 BLEU, establishing a new state-of-the-art BLEU score of 28.4. Training took 3.5 days on eight P100 GPUs.

On the WMT 2014 English-to-French translation task, the big model achieved a BLEU score of 41.0, outperforming all of the previously published single models, at less than a quarter of the training cost of the previous state-of-the-art model.

14.4. List of Mathematical Symbols

b_{ij} normalised dot product attention, a measure of proximity between the query vector $\mathbf{q}_i$ and the key vector $\mathbf{k}_j$.

h number of heads in multi-head attention ($h = 8$ in the paper).

K key matrix.

$\mathbf{k}_i$ key vector for the ith word in the input sequence.

N number of layers of encoders and decoders ($N = 6$ in the paper).

n number of words or tokens in an input sequence.

p_{ij} scaled dot product attention, a measure of the probability that the words w_i and w_j occur in their respective positions in the current input sequence.

Q query matrix.

$\mathbf{q}_i$ query vector for the ith word in the input sequence.

V a vocabulary of $|V|$ words; also the value matrix.

$\mathbf{v}_i$ value vector for the ith word in the input sequence.

W^{embed} a learned embedding matrix with $|V|$ rows and d_{model} columns.

W^K embedding key matrix.

W^Q embedding query matrix.

W^V embedding value matrix.

$\mathbf{w}^E$ input sequence of n English words, $\mathbf{w}^E = (w_1^E, \ldots, w_n^E)$.

$\mathbf{w}^G$ a target sequence of m German words, $\mathbf{w}^G = (w_1^G, \ldots, w_m^G)$.

$\hat{\mathbf{w}}^G$ a translation of $\mathbf{w}^E$ into German.

$\mathbf{x}_i^{\mathrm{hot}}$ a one-hot row vector $\mathbf{x}_i^{\mathrm{hot}}$ of $|V|$ elements in which every element is zero except for the ith one, where i is the ID number of the word in V.

$\mathbf{x}_i^{\mathrm{pos}}$ positional vector, a row vector of d_{model} elements that represents the position of the ith word in the input sequence $\mathbf{w}^E$.

$\mathbf{x}_i^{\mathrm{token}}$ embedded token vector, which is a row vector of d_{model} elements.

Research Paper: Attention Is All You Need

Note that there are several versions of this paper available on the internet (e.g. `https://arxiv.org/abs/1706.03762`). The version reproduced here is from the Neural Information Processing Systems conference website: `https://papers.nips.cc/paper_files/paper/2017/file/3f5ee243547dee91fbd053c1c4a845aa-Paper.pdf`.

Reference: Vaswani, A., Shazeer, N., Parmar, N., Uszkoreit, J., Jones, L., Gomez, A. N., Kaiser, L., and Polosukhin, I. (2017). Attention is all you need. In *Proceedings of the 31st International Conference on Neural Information Processing Systems (NIPS'17)*, pages 6000–6010.

Reproduced with permission.

Attention Is All You Need

Ashish Vaswani[*]
Google Brain
avaswani@google.com

Noam Shazeer[*]
Google Brain
noam@google.com

Niki Parmar[*]
Google Research
nikip@google.com

Jakob Uszkoreit[*]
Google Research
usz@google.com

Llion Jones[*]
Google Research
llion@google.com

Aidan N. Gomez[* †]
University of Toronto
aidan@cs.toronto.edu

Łukasz Kaiser[*]
Google Brain
lukaszkaiser@google.com

Illia Polosukhin[* ‡]
illia.polosukhin@gmail.com

Abstract

The dominant sequence transduction models are based on complex recurrent or convolutional neural networks that include an encoder and a decoder. The best performing models also connect the encoder and decoder through an attention mechanism. We propose a new simple network architecture, the Transformer, based solely on attention mechanisms, dispensing with recurrence and convolutions entirely. Experiments on two machine translation tasks show these models to be superior in quality while being more parallelizable and requiring significantly less time to train. Our model achieves 28.4 BLEU on the WMT 2014 English-to-German translation task, improving over the existing best results, including ensembles, by over 2 BLEU. On the WMT 2014 English-to-French translation task, our model establishes a new single-model state-of-the-art BLEU score of 41.0 after training for 3.5 days on eight GPUs, a small fraction of the training costs of the best models from the literature.

1 Introduction

Recurrent neural networks, long short-term memory [12] and gated recurrent [7] neural networks in particular, have been firmly established as state of the art approaches in sequence modeling and transduction problems such as language modeling and machine translation [29, 2, 5]. Numerous efforts have since continued to push the boundaries of recurrent language models and encoder-decoder architectures [31, 21, 13].

---

[*]Equal contribution. Listing order is random. Jakob proposed replacing RNNs with self-attention and started the effort to evaluate this idea. Ashish, with Illia, designed and implemented the first Transformer models and has been crucially involved in every aspect of this work. Noam proposed scaled dot-product attention, multi-head attention and the parameter-free position representation and became the other person involved in nearly every detail. Niki designed, implemented, tuned and evaluated countless model variants in our original codebase and tensor2tensor. Llion also experimented with novel model variants, was responsible for our initial codebase, and efficient inference and visualizations. Lukasz and Aidan spent countless long days designing various parts of and implementing tensor2tensor, replacing our earlier codebase, greatly improving results and massively accelerating our research.

[†]Work performed while at Google Brain.

[‡]Work performed while at Google Research.

Recurrent models typically factor computation along the symbol positions of the input and output sequences. Aligning the positions to steps in computation time, they generate a sequence of hidden states h_t, as a function of the previous hidden state h_{t-1} and the input for position t. This inherently sequential nature precludes parallelization within training examples, which becomes critical at longer sequence lengths, as memory constraints limit batching across examples. Recent work has achieved significant improvements in computational efficiency through factorization tricks [18] and conditional computation [26], while also improving model performance in case of the latter. The fundamental constraint of sequential computation, however, remains.

Attention mechanisms have become an integral part of compelling sequence modeling and transduction models in various tasks, allowing modeling of dependencies without regard to their distance in the input or output sequences [2, 16]. In all but a few cases [22], however, such attention mechanisms are used in conjunction with a recurrent network.

In this work we propose the Transformer, a model architecture eschewing recurrence and instead relying entirely on an attention mechanism to draw global dependencies between input and output. The Transformer allows for significantly more parallelization and can reach a new state of the art in translation quality after being trained for as little as twelve hours on eight P100 GPUs.

2 Background

The goal of reducing sequential computation also forms the foundation of the Extended Neural GPU [20], ByteNet [15] and ConvS2S [8], all of which use convolutional neural networks as basic building block, computing hidden representations in parallel for all input and output positions. In these models, the number of operations required to relate signals from two arbitrary input or output positions grows in the distance between positions, linearly for ConvS2S and logarithmically for ByteNet. This makes it more difficult to learn dependencies between distant positions [11]. In the Transformer this is reduced to a constant number of operations, albeit at the cost of reduced effective resolution due to averaging attention-weighted positions, an effect we counteract with Multi-Head Attention as described in section 3.2.

Self-attention, sometimes called intra-attention is an attention mechanism relating different positions of a single sequence in order to compute a representation of the sequence. Self-attention has been used successfully in a variety of tasks including reading comprehension, abstractive summarization, textual entailment and learning task-independent sentence representations [4, 22, 23, 19].

End-to-end memory networks are based on a recurrent attention mechanism instead of sequence-aligned recurrence and have been shown to perform well on simple-language question answering and language modeling tasks [28].

To the best of our knowledge, however, the Transformer is the first transduction model relying entirely on self-attention to compute representations of its input and output without using sequence-aligned RNNs or convolution. In the following sections, we will describe the Transformer, motivate self-attention and discuss its advantages over models such as [14, 15] and [8].

3 Model Architecture

Most competitive neural sequence transduction models have an encoder-decoder structure [5, 2, 29]. Here, the encoder maps an input sequence of symbol representations $(x_1, ..., x_n)$ to a sequence of continuous representations $\mathbf{z} = (z_1, ..., z_n)$. Given $\mathbf{z}$, the decoder then generates an output sequence $(y_1, ..., y_m)$ of symbols one element at a time. At each step the model is auto-regressive [9], consuming the previously generated symbols as additional input when generating the next.

The Transformer follows this overall architecture using stacked self-attention and point-wise, fully connected layers for both the encoder and decoder, shown in the left and right halves of Figure 1, respectively.

3.1 Encoder and Decoder Stacks

Encoder: The encoder is composed of a stack of $N = 6$ identical layers. Each layer has two sub-layers. The first is a multi-head self-attention mechanism, and the second is a simple, position-

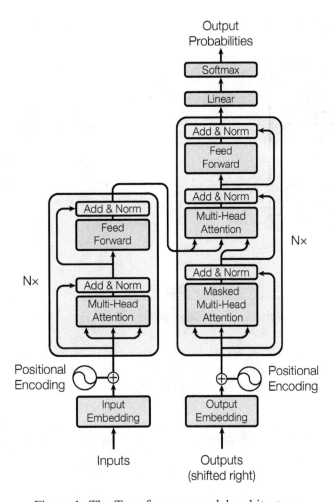

Figure 1: The Transformer - model architecture.

wise fully connected feed-forward network. We employ a residual connection [10] around each of the two sub-layers, followed by layer normalization [1]. That is, the output of each sub-layer is LayerNorm(x + Sublayer(x)), where Sublayer(x) is the function implemented by the sub-layer itself. To facilitate these residual connections, all sub-layers in the model, as well as the embedding layers, produce outputs of dimension $d_{\text{model}} = 512$.

Decoder: The decoder is also composed of a stack of $N = 6$ identical layers. In addition to the two sub-layers in each encoder layer, the decoder inserts a third sub-layer, which performs multi-head attention over the output of the encoder stack. Similar to the encoder, we employ residual connections around each of the sub-layers, followed by layer normalization. We also modify the self-attention sub-layer in the decoder stack to prevent positions from attending to subsequent positions. This masking, combined with fact that the output embeddings are offset by one position, ensures that the predictions for position i can depend only on the known outputs at positions less than i.

3.2 Attention

An attention function can be described as mapping a query and a set of key-value pairs to an output, where the query, keys, values, and output are all vectors. The output is computed as a weighted sum of the values, where the weight assigned to each value is computed by a compatibility function of the query with the corresponding key.

3.2.1 Scaled Dot-Product Attention

We call our particular attention "Scaled Dot-Product Attention" (Figure 2). The input consists of queries and keys of dimension d_k, and values of dimension d_v. We compute the dot products of the

Scaled Dot-Product Attention

Multi-Head Attention

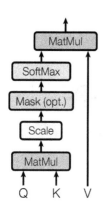
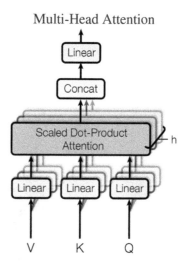

Figure 2: (left) Scaled Dot-Product Attention. (right) Multi-Head Attention consists of several attention layers running in parallel.

query with all keys, divide each by $\sqrt{d_k}$, and apply a softmax function to obtain the weights on the values.

In practice, we compute the attention function on a set of queries simultaneously, packed together into a matrix Q. The keys and values are also packed together into matrices K and V. We compute the matrix of outputs as:

$$\text{Attention}(Q, K, V) = \text{softmax}(\frac{QK^T}{\sqrt{d_k}})V \tag{1}$$

The two most commonly used attention functions are additive attention [2], and dot-product (multiplicative) attention. Dot-product attention is identical to our algorithm, except for the scaling factor of $\frac{1}{\sqrt{d_k}}$. Additive attention computes the compatibility function using a feed-forward network with a single hidden layer. While the two are similar in theoretical complexity, dot-product attention is much faster and more space-efficient in practice, since it can be implemented using highly optimized matrix multiplication code.

While for small values of d_k the two mechanisms perform similarly, additive attention outperforms dot product attention without scaling for larger values of d_k [3]. We suspect that for large values of d_k, the dot products grow large in magnitude, pushing the softmax function into regions where it has extremely small gradients [4]. To counteract this effect, we scale the dot products by $\frac{1}{\sqrt{d_k}}$.

3.2.2 Multi-Head Attention

Instead of performing a single attention function with d_{model}-dimensional keys, values and queries, we found it beneficial to linearly project the queries, keys and values h times with different, learned linear projections to d_k, d_k and d_v dimensions, respectively. On each of these projected versions of queries, keys and values we then perform the attention function in parallel, yielding d_v-dimensional output values. These are concatenated and once again projected, resulting in the final values, as depicted in Figure 2.

Multi-head attention allows the model to jointly attend to information from different representation subspaces at different positions. With a single attention head, averaging inhibits this.

---

[4]To illustrate why the dot products get large, assume that the components of q and k are independent random variables with mean 0 and variance 1. Then their dot product, $q \cdot k = \sum_{i=1}^{d_k} q_i k_i$, has mean 0 and variance d_k.

$$\text{MultiHead}(Q, K, V) = \text{Concat}(\text{head}_1, ..., \text{head}_\text{h})W^O$$
$$\text{where head}_\text{i} = \text{Attention}(QW_i^Q, KW_i^K, VW_i^V)$$

Where the projections are parameter matrices $W_i^Q \in \mathbb{R}^{d_{\text{model}} \times d_k}$, $W_i^K \in \mathbb{R}^{d_{\text{model}} \times d_k}$, $W_i^V \in \mathbb{R}^{d_{\text{model}} \times d_v}$ and $W^O \in \mathbb{R}^{hd_v \times d_{\text{model}}}$.

In this work we employ $h = 8$ parallel attention layers, or heads. For each of these we use $d_k = d_v = d_{\text{model}}/h = 64$. Due to the reduced dimension of each head, the total computational cost is similar to that of single-head attention with full dimensionality.

3.2.3 Applications of Attention in our Model

The Transformer uses multi-head attention in three different ways:

- In "encoder-decoder attention" layers, the queries come from the previous decoder layer, and the memory keys and values come from the output of the encoder. This allows every position in the decoder to attend over all positions in the input sequence. This mimics the typical encoder-decoder attention mechanisms in sequence-to-sequence models such as [31, 2, 8].

- The encoder contains self-attention layers. In a self-attention layer all of the keys, values and queries come from the same place, in this case, the output of the previous layer in the encoder. Each position in the encoder can attend to all positions in the previous layer of the encoder.

- Similarly, self-attention layers in the decoder allow each position in the decoder to attend to all positions in the decoder up to and including that position. We need to prevent leftward information flow in the decoder to preserve the auto-regressive property. We implement this inside of scaled dot-product attention by masking out (setting to $-\infty$) all values in the input of the softmax which correspond to illegal connections. See Figure 2.

3.3 Position-wise Feed-Forward Networks

In addition to attention sub-layers, each of the layers in our encoder and decoder contains a fully connected feed-forward network, which is applied to each position separately and identically. This consists of two linear transformations with a ReLU activation in between.

$$\text{FFN}(x) = \max(0, xW_1 + b_1)W_2 + b_2 \tag{2}$$

While the linear transformations are the same across different positions, they use different parameters from layer to layer. Another way of describing this is as two convolutions with kernel size 1. The dimensionality of input and output is $d_{\text{model}} = 512$, and the inner-layer has dimensionality $d_{ff} = 2048$.

3.4 Embeddings and Softmax

Similarly to other sequence transduction models, we use learned embeddings to convert the input tokens and output tokens to vectors of dimension d_{model}. We also use the usual learned linear transformation and softmax function to convert the decoder output to predicted next-token probabilities. In our model, we share the same weight matrix between the two embedding layers and the pre-softmax linear transformation, similar to [24]. In the embedding layers, we multiply those weights by $\sqrt{d_{\text{model}}}$.

3.5 Positional Encoding

Since our model contains no recurrence and no convolution, in order for the model to make use of the order of the sequence, we must inject some information about the relative or absolute position of the tokens in the sequence. To this end, we add "positional encodings" to the input embeddings at the

Table 1: Maximum path lengths, per-layer complexity and minimum number of sequential operations for different layer types. n is the sequence length, d is the representation dimension, k is the kernel size of convolutions and r the size of the neighborhood in restricted self-attention.

Layer Type	Complexity per Layer	Sequential Operations	Maximum Path Length
Self-Attention	$O(n^2 \cdot d)$	$O(1)$	$O(1)$
Recurrent	$O(n \cdot d^2)$	$O(n)$	$O(n)$
Convolutional	$O(k \cdot n \cdot d^2)$	$O(1)$	$O(log_k(n))$
Self-Attention (restricted)	$O(r \cdot n \cdot d)$	$O(1)$	$O(n/r)$

bottoms of the encoder and decoder stacks. The positional encodings have the same dimension d_{model} as the embeddings, so that the two can be summed. There are many choices of positional encodings, learned and fixed [8].

In this work, we use sine and cosine functions of different frequencies:

$$PE_{(pos,2i)} = sin(pos/10000^{2i/d_{\text{model}}})$$
$$PE_{(pos,2i+1)} = cos(pos/10000^{2i/d_{\text{model}}})$$

where pos is the position and i is the dimension. That is, each dimension of the positional encoding corresponds to a sinusoid. The wavelengths form a geometric progression from 2π to $10000 \cdot 2\pi$. We chose this function because we hypothesized it would allow the model to easily learn to attend by relative positions, since for any fixed offset k, PE_{pos+k} can be represented as a linear function of PE_{pos}.

We also experimented with using learned positional embeddings [8] instead, and found that the two versions produced nearly identical results (see Table 3 row (E)). We chose the sinusoidal version because it may allow the model to extrapolate to sequence lengths longer than the ones encountered during training.

4   Why Self-Attention

In this section we compare various aspects of self-attention layers to the recurrent and convolutional layers commonly used for mapping one variable-length sequence of symbol representations $(x_1, ..., x_n)$ to another sequence of equal length $(z_1, ..., z_n)$, with $x_i, z_i \in \mathbb{R}^d$, such as a hidden layer in a typical sequence transduction encoder or decoder. Motivating our use of self-attention we consider three desiderata.

One is the total computational complexity per layer. Another is the amount of computation that can be parallelized, as measured by the minimum number of sequential operations required.

The third is the path length between long-range dependencies in the network. Learning long-range dependencies is a key challenge in many sequence transduction tasks. One key factor affecting the ability to learn such dependencies is the length of the paths forward and backward signals have to traverse in the network. The shorter these paths between any combination of positions in the input and output sequences, the easier it is to learn long-range dependencies [11]. Hence we also compare the maximum path length between any two input and output positions in networks composed of the different layer types.

As noted in Table 1, a self-attention layer connects all positions with a constant number of sequentially executed operations, whereas a recurrent layer requires $O(n)$ sequential operations. In terms of computational complexity, self-attention layers are faster than recurrent layers when the sequence length n is smaller than the representation dimensionality d, which is most often the case with sentence representations used by state-of-the-art models in machine translations, such as word-piece [31] and byte-pair [25] representations. To improve computational performance for tasks involving very long sequences, self-attention could be restricted to considering only a neighborhood of size r in the input sequence centered around the respective output position. This would increase the maximum path length to $O(n/r)$. We plan to investigate this approach further in future work.

A single convolutional layer with kernel width $k < n$ does not connect all pairs of input and output positions. Doing so requires a stack of $O(n/k)$ convolutional layers in the case of contiguous kernels, or $O(log_k(n))$ in the case of dilated convolutions [15], increasing the length of the longest paths between any two positions in the network. Convolutional layers are generally more expensive than recurrent layers, by a factor of k. Separable convolutions [6], however, decrease the complexity considerably, to $O(k \cdot n \cdot d + n \cdot d^2)$. Even with $k = n$, however, the complexity of a separable convolution is equal to the combination of a self-attention layer and a point-wise feed-forward layer, the approach we take in our model.

As side benefit, self-attention could yield more interpretable models. We inspect attention distributions from our models and present and discuss examples in the appendix. Not only do individual attention heads clearly learn to perform different tasks, many appear to exhibit behavior related to the syntactic and semantic structure of the sentences.

5 Training

This section describes the training regime for our models.

5.1 Training Data and Batching

We trained on the standard WMT 2014 English-German dataset consisting of about 4.5 million sentence pairs. Sentences were encoded using byte-pair encoding [3], which has a shared source-target vocabulary of about 37000 tokens. For English-French, we used the significantly larger WMT 2014 English-French dataset consisting of 36M sentences and split tokens into a 32000 word-piece vocabulary [31]. Sentence pairs were batched together by approximate sequence length. Each training batch contained a set of sentence pairs containing approximately 25000 source tokens and 25000 target tokens.

5.2 Hardware and Schedule

We trained our models on one machine with 8 NVIDIA P100 GPUs. For our base models using the hyperparameters described throughout the paper, each training step took about 0.4 seconds. We trained the base models for a total of 100,000 steps or 12 hours. For our big models,(described on the bottom line of table 3), step time was 1.0 seconds. The big models were trained for 300,000 steps (3.5 days).

5.3 Optimizer

We used the Adam optimizer [17] with $\beta_1 = 0.9$, $\beta_2 = 0.98$ and $\epsilon = 10^{-9}$. We varied the learning rate over the course of training, according to the formula:

$$lrate = d_{\text{model}}^{-0.5} \cdot \min(step\_num^{-0.5}, step\_num \cdot warmup\_steps^{-1.5}) \tag{3}$$

This corresponds to increasing the learning rate linearly for the first $warmup\_steps$ training steps, and decreasing it thereafter proportionally to the inverse square root of the step number. We used $warmup\_steps = 4000$.

5.4 Regularization

We employ three types of regularization during training:

Residual Dropout   We apply dropout [27] to the output of each sub-layer, before it is added to the sub-layer input and normalized. In addition, we apply dropout to the sums of the embeddings and the positional encodings in both the encoder and decoder stacks. For the base model, we use a rate of $P_{drop} = 0.1$.

Table 2: The Transformer achieves better BLEU scores than previous state-of-the-art models on the English-to-German and English-to-French newstest2014 tests at a fraction of the training cost.

Model	BLEU		Training Cost (FLOPs)	
	EN-DE	EN-FR	EN-DE	EN-FR
ByteNet [15]	23.75			
Deep-Att + PosUnk [32]		39.2		$1.0 \cdot 10^{20}$
GNMT + RL [31]	24.6	39.92	$2.3 \cdot 10^{19}$	$1.4 \cdot 10^{20}$
ConvS2S [8]	25.16	40.46	$9.6 \cdot 10^{18}$	$1.5 \cdot 10^{20}$
MoE [26]	26.03	40.56	$2.0 \cdot 10^{19}$	$1.2 \cdot 10^{20}$
Deep-Att + PosUnk Ensemble [32]		40.4		$8.0 \cdot 10^{20}$
GNMT + RL Ensemble [31]	26.30	41.16	$1.8 \cdot 10^{20}$	$1.1 \cdot 10^{21}$
ConvS2S Ensemble [8]	26.36	**41.29**	$7.7 \cdot 10^{19}$	$1.2 \cdot 10^{21}$
Transformer (base model)	27.3	38.1	$\mathbf{3.3 \cdot 10^{18}}$	
Transformer (big)	**28.4**	**41.0**	$2.3 \cdot 10^{19}$	

Label Smoothing   During training, we employed label smoothing of value $\epsilon_{ls} = 0.1$ [30]. This hurts perplexity, as the model learns to be more unsure, but improves accuracy and BLEU score.

6   Results

6.1   Machine Translation

On the WMT 2014 English-to-German translation task, the big transformer model (Transformer (big) in Table 2) outperforms the best previously reported models (including ensembles) by more than 2.0 BLEU, establishing a new state-of-the-art BLEU score of 28.4. The configuration of this model is listed in the bottom line of Table 3. Training took 3.5 days on 8 P100 GPUs. Even our base model surpasses all previously published models and ensembles, at a fraction of the training cost of any of the competitive models.

On the WMT 2014 English-to-French translation task, our big model achieves a BLEU score of 41.0, outperforming all of the previously published single models, at less than $1/4$ the training cost of the previous state-of-the-art model. The Transformer (big) model trained for English-to-French used dropout rate $P_{drop} = 0.1$, instead of 0.3.

For the base models, we used a single model obtained by averaging the last 5 checkpoints, which were written at 10-minute intervals. For the big models, we averaged the last 20 checkpoints. We used beam search with a beam size of 4 and length penalty $\alpha = 0.6$ [31]. These hyperparameters were chosen after experimentation on the development set. We set the maximum output length during inference to input length + 50, but terminate early when possible [31].

Table 2 summarizes our results and compares our translation quality and training costs to other model architectures from the literature. We estimate the number of floating point operations used to train a model by multiplying the training time, the number of GPUs used, and an estimate of the sustained single-precision floating-point capacity of each GPU [5].

6.2   Model Variations

To evaluate the importance of different components of the Transformer, we varied our base model in different ways, measuring the change in performance on English-to-German translation on the development set, newstest2013. We used beam search as described in the previous section, but no checkpoint averaging. We present these results in Table 3.

In Table 3 rows (A), we vary the number of attention heads and the attention key and value dimensions, keeping the amount of computation constant, as described in Section 3.2.2. While single-head attention is 0.9 BLEU worse than the best setting, quality also drops off with too many heads.

---

[5]We used values of 2.8, 3.7, 6.0 and 9.5 TFLOPS for K80, K40, M40 and P100, respectively.

Table 3: Variations on the Transformer architecture. Unlisted values are identical to those of the base model. All metrics are on the English-to-German translation development set, newstest2013. Listed perplexities are per-wordpiece, according to our byte-pair encoding, and should not be compared to per-word perplexities.

	N	d_{model}	d_{ff}	h	d_k	d_v	P_{drop}	ϵ_{ls}	train steps	PPL (dev)	BLEU (dev)	params $\times 10^6$
base	6	512	2048	8	64	64	0.1	0.1	100K	4.92	25.8	65
(A)				1	512	512				5.29	24.9	
				4	128	128				5.00	25.5	
				16	32	32				4.91	25.8	
				32	16	16				5.01	25.4	
(B)					16					5.16	25.1	58
					32					5.01	25.4	60
(C)	2									6.11	23.7	36
	4									5.19	25.3	50
	8									4.88	25.5	80
		256			32	32				5.75	24.5	28
		1024			128	128				4.66	26.0	168
			1024							5.12	25.4	53
			4096							4.75	26.2	90
(D)							0.0			5.77	24.6	
							0.2			4.95	25.5	
								0.0		4.67	25.3	
								0.2		5.47	25.7	
(E)		positional embedding instead of sinusoids								4.92	25.7	
big	6	1024	4096	16			0.3		300K	**4.33**	**26.4**	213

In Table 3 rows (B), we observe that reducing the attention key size d_k hurts model quality. This suggests that determining compatibility is not easy and that a more sophisticated compatibility function than dot product may be beneficial. We further observe in rows (C) and (D) that, as expected, bigger models are better, and dropout is very helpful in avoiding over-fitting. In row (E) we replace our sinusoidal positional encoding with learned positional embeddings [8], and observe nearly identical results to the base model.

7   Conclusion

In this work, we presented the Transformer, the first sequence transduction model based entirely on attention, replacing the recurrent layers most commonly used in encoder-decoder architectures with multi-headed self-attention.

For translation tasks, the Transformer can be trained significantly faster than architectures based on recurrent or convolutional layers. On both WMT 2014 English-to-German and WMT 2014 English-to-French translation tasks, we achieve a new state of the art. In the former task our best model outperforms even all previously reported ensembles.

We are excited about the future of attention-based models and plan to apply them to other tasks. We plan to extend the Transformer to problems involving input and output modalities other than text and to investigate local, restricted attention mechanisms to efficiently handle large inputs and outputs such as images, audio and video. Making generation less sequential is another research goals of ours.

The code we used to train and evaluate our models is available at https://github.com/tensorflow/tensor2tensor.

Acknowledgements   We are grateful to Nal Kalchbrenner and Stephan Gouws for their fruitful comments, corrections and inspiration.

References

[1] Jimmy Lei Ba, Jamie Ryan Kiros, and Geoffrey E Hinton. Layer normalization. *arXiv preprint arXiv:1607.06450*, 2016.

[2] Dzmitry Bahdanau, Kyunghyun Cho, and Yoshua Bengio. Neural machine translation by jointly learning to align and translate. *CoRR*, abs/1409.0473, 2014.

[3] Denny Britz, Anna Goldie, Minh-Thang Luong, and Quoc V. Le. Massive exploration of neural machine translation architectures. *CoRR*, abs/1703.03906, 2017.

[4] Jianpeng Cheng, Li Dong, and Mirella Lapata. Long short-term memory-networks for machine reading. *arXiv preprint arXiv:1601.06733*, 2016.

[5] Kyunghyun Cho, Bart van Merrienboer, Caglar Gulcehre, Fethi Bougares, Holger Schwenk, and Yoshua Bengio. Learning phrase representations using rnn encoder-decoder for statistical machine translation. *CoRR*, abs/1406.1078, 2014.

[6] Francois Chollet. Xception: Deep learning with depthwise separable convolutions. *arXiv preprint arXiv:1610.02357*, 2016.

[7] Junyoung Chung, Çaglar Gülçehre, Kyunghyun Cho, and Yoshua Bengio. Empirical evaluation of gated recurrent neural networks on sequence modeling. *CoRR*, abs/1412.3555, 2014.

[8] Jonas Gehring, Michael Auli, David Grangier, Denis Yarats, and Yann N. Dauphin. Convolutional sequence to sequence learning. *arXiv preprint arXiv:1705.03122v2*, 2017.

[9] Alex Graves. Generating sequences with recurrent neural networks. *arXiv preprint arXiv:1308.0850*, 2013.

[10] Kaiming He, Xiangyu Zhang, Shaoqing Ren, and Jian Sun. Deep residual learning for image recognition. In *Proceedings of the IEEE Conference on Computer Vision and Pattern Recognition*, pages 770–778, 2016.

[11] Sepp Hochreiter, Yoshua Bengio, Paolo Frasconi, and Jürgen Schmidhuber. Gradient flow in recurrent nets: the difficulty of learning long-term dependencies, 2001.

[12] Sepp Hochreiter and Jürgen Schmidhuber. Long short-term memory. *Neural computation*, 9(8):1735–1780, 1997.

[13] Rafal Jozefowicz, Oriol Vinyals, Mike Schuster, Noam Shazeer, and Yonghui Wu. Exploring the limits of language modeling. *arXiv preprint arXiv:1602.02410*, 2016.

[14] Łukasz Kaiser and Ilya Sutskever. Neural GPUs learn algorithms. In *International Conference on Learning Representations (ICLR)*, 2016.

[15] Nal Kalchbrenner, Lasse Espeholt, Karen Simonyan, Aaron van den Oord, Alex Graves, and Koray Kavukcuoglu. Neural machine translation in linear time. *arXiv preprint arXiv:1610.10099v2*, 2017.

[16] Yoon Kim, Carl Denton, Luong Hoang, and Alexander M. Rush. Structured attention networks. In *International Conference on Learning Representations*, 2017.

[17] Diederik Kingma and Jimmy Ba. Adam: A method for stochastic optimization. In *ICLR*, 2015.

[18] Oleksii Kuchaiev and Boris Ginsburg. Factorization tricks for LSTM networks. *arXiv preprint arXiv:1703.10722*, 2017.

[19] Zhouhan Lin, Minwei Feng, Cicero Nogueira dos Santos, Mo Yu, Bing Xiang, Bowen Zhou, and Yoshua Bengio. A structured self-attentive sentence embedding. *arXiv preprint arXiv:1703.03130*, 2017.

[20] Samy Bengio Łukasz Kaiser. Can active memory replace attention? In *Advances in Neural Information Processing Systems, (NIPS)*, 2016.

[21] Minh-Thang Luong, Hieu Pham, and Christopher D Manning. Effective approaches to attention-based neural machine translation. *arXiv preprint arXiv:1508.04025*, 2015.

[22] Ankur Parikh, Oscar Täckström, Dipanjan Das, and Jakob Uszkoreit. A decomposable attention model. In *Empirical Methods in Natural Language Processing*, 2016.

[23] Romain Paulus, Caiming Xiong, and Richard Socher. A deep reinforced model for abstractive summarization. *arXiv preprint arXiv:1705.04304*, 2017.

[24] Ofir Press and Lior Wolf. Using the output embedding to improve language models. *arXiv preprint arXiv:1608.05859*, 2016.

[25] Rico Sennrich, Barry Haddow, and Alexandra Birch. Neural machine translation of rare words with subword units. *arXiv preprint arXiv:1508.07909*, 2015.

[26] Noam Shazeer, Azalia Mirhoseini, Krzysztof Maziarz, Andy Davis, Quoc Le, Geoffrey Hinton, and Jeff Dean. Outrageously large neural networks: The sparsely-gated mixture-of-experts layer. *arXiv preprint arXiv:1701.06538*, 2017.

[27] Nitish Srivastava, Geoffrey E Hinton, Alex Krizhevsky, Ilya Sutskever, and Ruslan Salakhutdinov. Dropout: a simple way to prevent neural networks from overfitting. *Journal of Machine Learning Research*, 15(1):1929–1958, 2014.

[28] Sainbayar Sukhbaatar, arthur szlam, Jason Weston, and Rob Fergus. End-to-end memory networks. In C. Cortes, N. D. Lawrence, D. D. Lee, M. Sugiyama, and R. Garnett, editors, *Advances in Neural Information Processing Systems 28*, pages 2440–2448. Curran Associates, Inc., 2015.

[29] Ilya Sutskever, Oriol Vinyals, and Quoc VV Le. Sequence to sequence learning with neural networks. In *Advances in Neural Information Processing Systems*, pages 3104–3112, 2014.

[30] Christian Szegedy, Vincent Vanhoucke, Sergey Ioffe, Jonathon Shlens, and Zbigniew Wojna. Rethinking the inception architecture for computer vision. *CoRR*, abs/1512.00567, 2015.

[31] Yonghui Wu, Mike Schuster, Zhifeng Chen, Quoc V Le, Mohammad Norouzi, Wolfgang Macherey, Maxim Krikun, Yuan Cao, Qin Gao, Klaus Macherey, et al. Google's neural machine translation system: Bridging the gap between human and machine translation. *arXiv preprint arXiv:1609.08144*, 2016.

[32] Jie Zhou, Ying Cao, Xuguang Wang, Peng Li, and Wei Xu. Deep recurrent models with fast-forward connections for neural machine translation. *CoRR*, abs/1606.04199, 2016.

Chapter 15

GPT-2 – 2019

Context

Generative pre-trained transformer (GPT) models belong to a class of *large language models* (LLMs), which have revolutionised the field of *natural language processing* (NLP). At the time of writing, there are four GPT models, which depend heavily on transformer networks (Chapter 14). In fact, GPT models are essentially scaled-up versions of transformer networks, with increasing numbers of parameters (learnable weights), trained on increasingly large data sets, which include internet sources and books. The first model[88], which has since become known as GPT-1, was introduced in 2018 and has 117 million parameters. To get some of idea of the impact of each GPT model, it is worth noting the number of times the paper that introduced the model has been cited; in the case of GPT-1, the number of citations was 8,500 (according to Google Scholar). GPT-2 (2019) has 1.5 billion parameters, and the paper[89] has been cited 9,000 times. GPT-3 (2020) has a staggering 175 billion parameters and can perform an array of tasks, from language translation to code generation; the paper[17] been cited 24,407 times. GPT-4 (2023) has 1.7 trillion (1.7×10^{12}) parameters, and the (recently published) paper[1] has been cited 786 times.

Even though neural network connections (parameters) in GPT models have little in common with the synaptic connections between biological neurons, the two are often compared. The human brain has 86 billion neurons (86×10^9), each of which has about 7,000 synaptic connections, making a total of around 6×10^{14} or 600 trillion synapses. If, for the sake of argument, we assume that synapses can be compared to neural network parameters, this suggests that the brain has about 300 times as many connections as GPT-4. Of course, it will soon become possible to build LLMs with more than 600 trillion parameters. However, it is unlikely that the competence of LLMs is purely a matter of how many parameters they possess. For example, a bee's brain has only about 100,000 neurons, with less than one billion synapses; but despite their relatively small brains, bees can perform complex tasks[15].

The comments in Chapter 14 regarding the arbitrary nature of design choices for transformers apply equally well to GPT models, which is perhaps unsurprising given that GPTs rely heavily on transformers. Additionally, it should be borne in mind that a GPT is, in essence, a sophisticated algorithm for generating the next word in a sequence, and that its training data are derived from the internet. Insofar as a GPT has a model of the world, that model is based entirely on the sequences of words it has encountered on the internet. Accordingly, we should not expect GPTs to possess a general ability to know, or to reason about, the physical world.

For example, when asked to "Explain Heisenberg's uncertainty principle in terms of measurement", GPT-3.5 replied with a long, sometimes contradictory, answer, which included these words: "Heisenberg's uncertainty principle arises from the inherent disturbance caused by the measurement process in quantum mechanics." This is plain wrong, but it is a common mistake, so GPT-3 is simply expressing the error it has encountered on the internet. And GPT-4.0 did not do any better: "It tells us that the act of measuring affects the system in such a way that certain pairs of properties cannot both be known to arbitrary precision simultaneously." (In fact, Heisenberg's uncertainty principle is a consequence of the intrinsic joint uncertainty of certain pairs of physical quantities, but it does not depend on the disturbance caused by measurement.)

Despite examples like this, recent LLMs (e.g. GPT-3 and Gemini) are much better than their predecessors at answering questions. And, in fairness, these 'trick' questions were formulated with the knowledge that LLMs are essentially 'next-word predictors' and that their predictions are limited precisely because they are based solely on word sequences encountered on the internet. In most other respects, the most recent generation of LLMs is impressive.

Technical Summary

Of the four currently available versions of GPT, the 2019 paper introducing GPT-2 was chosen for inclusion here because it strikes a reasonable balance between technical detail and discussion. GPT-2 was trained on 8 million documents, comprising 40 Gb of text. Its results are impressive with respect to a wide variety of linguistic benchmarks, such as comprehension and translation.

Strikingly, the problem faced by LLMs is formulated in the most general probabilistic terms, such as $p(\text{output} \,|\, \text{input}, \text{task})$. Thus, instead of designing networks for specific tasks, it is intended that a sufficiently powerful LLM will be capable of estimating $p(\text{output} \,|\, \text{input}, \text{task})$ for any number of different tasks; in other words, it should be capable of performing a wide range of tasks. It is noteworthy that GPT-2, and later incarnations of LLMs, seem to have the potential to estimate such probabilities.

The paper includes four versions of GPT-2, with increasing numbers of parameters N (specifically, $N = 124$ million, 355 million, 762 million and 1542 million)*. When plotted against various measures of performance, the scaling characteristics observed in all of the graphs in the paper suggest a law of diminishing returns. For example, Figure 4 in the paper shows that reducing the perplexity by a factor of 2 requires N to increase by a factor of about 14, and the trend line in that figure suggests that the perplexity decreases approximately linearly with $\log N$. Other studies[128] suggest a similar law of diminishing returns for neural networks as the size of the training set is increased†. Taken together, these results indicate that the benefits of increasing the number of parameters and the size of training set may soon be out-weighed by the cost of training LLMs, which currently runs to millions of dollars. Indeed, there is a growing trend to reduce costs by designing modular LLMs, which consist of several interconnected small LLMs.

It is worth noting that the term *zero-shot task transfer* (ZSTT), which occurs frequently in the paper, is just another term for *generalisation*; this means the ability of a model to tackle new tasks without having been trained on those tasks.

Because the paper is reproduced here without colour, the plotted lines in Figures 3 and 4 are hard to distinguish. In Figure 3, the lower curve represents 'Full Scoring' and the upper curve represents 'Partial Scoring'. In Figure 4, the lower curve represents 'WebText train' and the upper curve represents 'WebText test'. These figures can be viewed in colour in the original paper available at `https://cdn.openai.com/better-language-models/language_models_are_unsupervised_multitask_learners.pdf`.

Research Paper:
Language Models Are Unsupervised Multitask Learners

Reference: Radford, A., Wu, J., Child, R., Luan, D., Amodei, D., and Sutskever, I. (2019). Language models are unsupervised multitask learners. *OpenAI Blog*, 1(8):9. `https://cdn.openai.com/better-language-models/language_models_are_unsupervised_multitask_learners.pdf`. Reproduced with permission.

---

*According to OpenAI's github repository, the correct values of $N = 124$ million and $N = 355$ million were quoted incorrectly as $N = 117$ million and 345 million (respectively) in the paper (see `https://github.com/openai/gpt-2`).

†Counter-intuitively, *double descent*[79] suggests that test performance actually *increases* if the number of parameters exceeds the number of training items.

Language Models are Unsupervised Multitask Learners

Alec Radford [* 1]   Jeffrey Wu [* 1]   Rewon Child [1]   David Luan [1]   Dario Amodei [** 1]   Ilya Sutskever [** 1]

Abstract

Natural language processing tasks, such as question answering, machine translation, reading comprehension, and summarization, are typically approached with supervised learning on task-specific datasets. We demonstrate that language models begin to learn these tasks without any explicit supervision when trained on a new dataset of millions of webpages called WebText. When conditioned on a document plus questions, the answers generated by the language model reach 55 F1 on the CoQA dataset - matching or exceeding the performance of 3 out of 4 baseline systems without using the 127,000+ training examples. The capacity of the language model is essential to the success of zero-shot task transfer and increasing it improves performance in a log-linear fashion across tasks. Our largest model, GPT-2, is a 1.5B parameter Transformer that achieves state of the art results on 7 out of 8 tested language modeling datasets in a zero-shot setting but still underfits WebText. Samples from the model reflect these improvements and contain coherent paragraphs of text. These findings suggest a promising path towards building language processing systems which learn to perform tasks from their naturally occurring demonstrations.

1. Introduction

Machine learning systems now excel (in expectation) at tasks they are trained for by using a combination of large datasets, high-capacity models, and supervised learning (Krizhevsky et al., 2012) (Sutskever et al., 2014) (Amodei et al., 2016). Yet these systems are brittle and sensitive to slight changes in the data distribution (Recht et al., 2018) and task specification (Kirkpatrick et al., 2017). Current systems are better characterized as narrow experts rather than

---
[*, **]Equal contribution   [1]OpenAI, San Francisco, California, United States.   Correspondence to: Alec Radford <alec@openai.com>.

competent generalists. We would like to move towards more general systems which can perform many tasks – eventually without the need to manually create and label a training dataset for each one.

The dominant approach to creating ML systems is to collect a dataset of training examples demonstrating correct behavior for a desired task, train a system to imitate these behaviors, and then test its performance on independent and identically distributed (IID) held-out examples. This has served well to make progress on narrow experts. But the often erratic behavior of captioning models (Lake et al., 2017), reading comprehension systems (Jia & Liang, 2017), and image classifiers (Alcorn et al., 2018) on the diversity and variety of possible inputs highlights some of the shortcomings of this approach.

Our suspicion is that the prevalence of single task training on single domain datasets is a major contributor to the lack of generalization observed in current systems. Progress towards robust systems with current architectures is likely to require training and measuring performance on a wide range of domains and tasks. Recently, several benchmarks have been proposed such as GLUE (Wang et al., 2018) and decaNLP (McCann et al., 2018) to begin studying this.

Multitask learning (Caruana, 1997) is a promising framework for improving general performance. However, multitask training in NLP is still nascent. Recent work reports modest performance improvements (Yogatama et al., 2019) and the two most ambitious efforts to date have trained on a total of 10 and 17 (`dataset, objective`) pairs respectively (McCann et al., 2018) (Bowman et al., 2018). From a meta-learning perspective, each (`dataset, objective`) pair is a single training example sampled from the distribution of datasets and objectives. Current ML systems need hundreds to thousands of examples to induce functions which generalize well. This suggests that multitask training many need just as many effective training pairs to realize its promise with current approaches. It will be very difficult to continue to scale the creation of datasets and the design of objectives to the degree that may be required to brute force our way there with current techniques. This motivates exploring additional setups for performing multitask learning.

The current best performing systems on language tasks

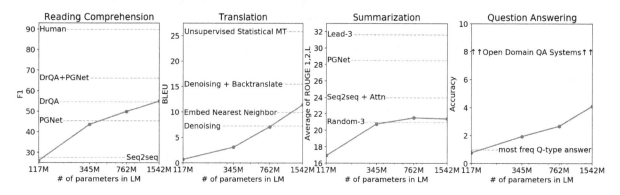

Figure 1. Zero-shot task performance of WebText LMs as a function of model size on many NLP tasks. Reading Comprehension results are on CoQA (Reddy et al., 2018), translation on WMT-14 Fr-En (Artetxe et al., 2017), summarization on CNN and Daily Mail (See et al., 2017), and Question Answering on Natural Questions (Kwiatkowski et al., 2019). Section 3 contains detailed descriptions of each result.

utilize a combination of pre-training and supervised fine-tuning. This approach has a long history with a trend towards more flexible forms of transfer. First, word vectors were learned and used as inputs to task-specific architectures (Mikolov et al., 2013) (Collobert et al., 2011), then the contextual representations of recurrent networks were transferred (Dai & Le, 2015) (Peters et al., 2018), and recent work suggests that task-specific architectures are no longer necessary and transferring many self-attention blocks is sufficient (Radford et al., 2018) (Devlin et al., 2018).

These methods still require supervised training in order to perform a task. When only minimal or no supervised data is available, another line of work has demonstrated the promise of language models to perform specific tasks, such as commonsense reasoning (Schwartz et al., 2017) and sentiment analysis (Radford et al., 2017).

In this paper, we connect these two lines of work and continue the trend of more general methods of transfer. We demonstrate language models can perform down-stream tasks in a zero-shot setting – without any parameter or architecture modification. We demonstrate this approach shows potential by highlighting the ability of language models to perform a wide range of tasks in a zero-shot setting. We achieve promising, competitive, and state of the art results depending on the task.

2. Approach

At the core of our approach is language modeling. Language modeling is usually framed as unsupervised distribution estimation from a set of examples $(x_1, x_2, ..., x_n)$ each composed of variable length sequences of symbols $(s_1, s_2, ..., s_n)$. Since language has a natural sequential ordering, it is common to factorize the joint probabilities over

symbols as the product of conditional probabilities (Jelinek & Mercer, 1980) (Bengio et al., 2003):

$$p(x) = \prod_{i=1}^{n} p(s_n | s_1, ..., s_{n-1}) \qquad (1)$$

This approach allows for tractable sampling from and estimation of $p(x)$ as well as any conditionals of the form $p(s_{n-k}, ..., s_n | s_1, ..., s_{n-k-1})$. In recent years, there have been significant improvements in the expressiveness of models that can compute these conditional probabilities, such as self-attention architectures like the Transformer (Vaswani et al., 2017).

Learning to perform a single task can be expressed in a probabilistic framework as estimating a conditional distribution $p(output|input)$. Since a general system should be able to perform many different tasks, even for the same input, it should condition not only on the input but also on the task to be performed. That is, it should model $p(output|input, task)$. This has been variously formalized in multitask and meta-learning settings. Task conditioning is often implemented at an architectural level, such as the task specific encoders and decoders in (Kaiser et al., 2017) or at an algorithmic level such as the inner and outer loop optimization framework of MAML (Finn et al., 2017). But as exemplified in McCann et al. (2018), language provides a flexible way to specify tasks, inputs, and outputs all as a sequence of symbols. For example, a translation training example can be written as the sequence (translate to french, english text, french text). Likewise, a reading comprehension training example can be written as (answer the question, document, question, answer). McCann et al. (2018) demonstrated it was possible to train a single model, the MQAN, to infer and perform many different tasks on examples with this type of format.

Language modeling is also able to, in principle, learn the tasks of McCann et al. (2018) without the need for explicit supervision of which symbols are the outputs to be predicted. Since the supervised objective is the the same as the unsupervised objective but only evaluated on a subset of the sequence, the global minimum of the unsupervised objective is also the global minimum of the supervised objective. In this slightly toy setting, the concerns with density estimation as a principled training objective discussed in (Sutskever et al., 2015) are side stepped. The problem instead becomes whether we are able to, in practice, optimize the unsupervised objective to convergence. Preliminary experiments confirmed that sufficiently large language models are able to perform multitask learning in this toy-ish setup but learning is much slower than in explicitly supervised approaches.

While it is a large step from the well-posed setup described above to the messiness of "language in the wild", Weston (2016) argues, in the context of dialog, for the need to develop systems capable of learning from natural language directly and demonstrated a proof of concept – learning a QA task without a reward signal by using forward prediction of a teacher's outputs. While dialog is an attractive approach, we worry it is overly restrictive. The internet contains a vast amount of information that is passively available without the need for interactive communication. Our speculation is that a language model with sufficient capacity will begin to learn to infer and perform the tasks demonstrated in natural language sequences in order to better predict them, regardless of their method of procurement. If a language model is able to do this it will be, in effect, performing unsupervised multitask learning. We test whether this is the case by analyzing the performance of language models in a zero-shot setting on a wide variety of tasks.

2.1. Training Dataset

Most prior work trained language models on a single domain of text, such as news articles (Jozefowicz et al., 2016), Wikipedia (Merity et al., 2016), or fiction books (Kiros et al., 2015). Our approach motivates building as large and diverse a dataset as possible in order to collect natural language demonstrations of tasks in as varied of domains and contexts as possible.

A promising source of diverse and nearly unlimited text is web scrapes such as Common Crawl. While these archives are many orders of magnitude larger than current language modeling datasets, they have significant data quality issues. Trinh & Le (2018) used Common Crawl in their work on commonsense reasoning but noted a large amount of documents "whose content are mostly unintelligible". We observed similar data issues in our initial experiments with

"I'm not the cleverest man in the world, but like they say in French: **Je ne suis pas un imbecile [I'm not a fool].**

In a now-deleted post from Aug. 16, Soheil Eid, Tory candidate in the riding of Joliette, wrote in French: **"Mentez mentez, il en restera toujours quelque chose,"** which translates as, **"Lie lie and something will always remain."**

"I hate the word '**perfume**,'" Burr says. 'It's somewhat better in French: '**parfum**.'

If listened carefully at 29:55, a conversation can be heard between two guys in French: "**-Comment on fait pour aller de l'autre coté? -Quel autre coté?**", which means "**- How do you get to the other side? - What side?**".

If this sounds like a bit of a stretch, consider this question in French: **As-tu aller au cinéma?**, or **Did you go to the movies?**, which literally translates as Have-you to go to movies/theater?

"**Brevet Sans Garantie Du Gouvernement**", translated to English: "**Patented without government warranty**".

Table 1. Examples of naturally occurring demonstrations of English to French and French to English translation found throughout the WebText training set.

Common Crawl. Trinh & Le (2018)'s best results were achieved using a small subsample of Common Crawl which included only documents most similar to their target dataset, the Winograd Schema Challenge. While this is a pragmatic approach to improve performance on a specific task, we want to avoid making assumptions about the tasks to be performed ahead of time.

Instead, we created a new web scrape which emphasizes document quality. To do this we only scraped web pages which have been curated/filtered by humans. Manually filtering a full web scrape would be exceptionally expensive so as a starting point, we scraped all outbound links from Reddit, a social media platform, which received at least 3 karma. This can be thought of as a heuristic indicator for whether other users found the link interesting, educational, or just funny.

The resulting dataset, WebText, contains the text subset of these 45 million links. To extract the text from HTML responses we use a combination of the Dragnet (Peters & Lecocq, 2013) and Newspaper[1] content extractors. All results presented in this paper use a preliminary version of WebText which does not include links created after Dec 2017 and which after de-duplication and some heuristic based cleaning contains slightly over 8 million documents for a total of 40 GB of text. We removed all Wikipedia documents from WebText since it is a common data source for other datasets and could complicate analysis due to over-

---

[1] https://github.com/codelucas/newspaper

lapping training data with test evaluation tasks.

2.2. Input Representation

A general language model (LM) should be able to compute the probability of (and also generate) any string. Current large scale LMs include pre-processing steps such as lower-casing, tokenization, and out-of-vocabulary tokens which restrict the space of model-able strings. While processing Unicode strings as a sequence of UTF-8 bytes elegantly fulfills this requirement as exemplified in work such as Gillick et al. (2015), current byte-level LMs are not competitive with word-level LMs on large scale datasets such as the One Billion Word Benchmark (Al-Rfou et al., 2018). We observed a similar performance gap in our own attempts to train standard byte-level LMs on WebText.

Byte Pair Encoding (BPE) (Sennrich et al., 2015) is a practical middle ground between character and word level language modeling which effectively interpolates between word level inputs for frequent symbol sequences and character level inputs for infrequent symbol sequences. Despite its name, reference BPE implementations often operate on Unicode code points and not byte sequences. These implementations would require including the full space of Unicode symbols in order to model all Unicode strings. This would result in a base vocabulary of over 130,000 before any multi-symbol tokens are added. This is prohibitively large compared to the 32,000 to 64,000 token vocabularies often used with BPE. In contrast, a byte-level version of BPE only requires a base vocabulary of size 256. However, directly applying BPE to the byte sequence results in suboptimal merges due to BPE using a greedy frequency based heuristic for building the token vocabulary. We observed BPE including many versions of common words like `dog` since they occur in many variations such as `dog.` `dog!` `dog?` . This results in a sub-optimal allocation of limited vocabulary slots and model capacity. To avoid this, we prevent BPE from merging across character categories for any byte sequence. We add an exception for spaces which significantly improves the compression efficiency while adding only minimal fragmentation of words across multiple vocab tokens.

This input representation allows us to combine the empirical benefits of word-level LMs with the generality of byte-level approaches. Since our approach can assign a probability to any Unicode string, this allows us to evaluate our LMs on any dataset regardless of pre-processing, tokenization, or vocab size.

2.3. Model

We use a Transformer (Vaswani et al., 2017) based architecture for our LMs. The model largely follows the details of the OpenAI GPT model (Radford et al., 2018) with a

Parameters	Layers	d_{model}
117M	12	768
345M	24	1024
762M	36	1280
1542M	48	1600

Table 2. Architecture hyperparameters for the 4 model sizes.

few modifications. Layer normalization (Ba et al., 2016) was moved to the input of each sub-block, similar to a pre-activation residual network (He et al., 2016) and an additional layer normalization was added after the final self-attention block. A modified initialization which accounts for the accumulation on the residual path with model depth is used. We scale the weights of residual layers at initialization by a factor of $1/\sqrt{N}$ where N is the number of residual layers. The vocabulary is expanded to 50,257. We also increase the context size from 512 to 1024 tokens and a larger batchsize of 512 is used.

3. Experiments

We trained and benchmarked four LMs with approximately log-uniformly spaced sizes. The architectures are summarized in Table 2. The smallest model is equivalent to the original GPT, and the second smallest equivalent to the largest model from BERT (Devlin et al., 2018). Our largest model, which we call GPT-2, has over an order of magnitude more parameters than GPT. The learning rate of each model was manually tuned for the best perplexity on a 5% held-out sample of WebText. All models still underfit WebText and held-out perplexity has as of yet improved given more training time.

3.1. Language Modeling

As an initial step towards zero-shot task transfer, we are interested in understanding how WebText LM's perform at zero-shot domain transfer on the primary task they are trained for – language modeling. Since our model operates on a byte level and does not require lossy pre-processing or tokenization, we can evaluate it on any language model benchmark. Results on language modeling datasets are commonly reported in a quantity which is a scaled or exponentiated version of the average negative log probability per canonical prediction unit - usually a character, a byte, or a word. We evaluate the same quantity by computing the log-probability of a dataset according to a WebText LM and dividing by the number of canonical units. For many of these datasets, WebText LMs would be tested significantly out-of-distribution, having to predict aggressively standardized text, tokenization artifacts such as disconnected punctuation and contractions, shuffled sentences, and even the string

	LAMBADA (PPL)	LAMBADA (ACC)	CBT-CN (ACC)	CBT-NE (ACC)	WikiText2 (PPL)	PTB (PPL)	enwik8 (BPB)	text8 (BPC)	WikiText103 (PPL)	1BW (PPL)
SOTA	99.8	59.23	85.7	82.3	39.14	46.54	0.99	1.08	18.3	**21.8**
117M	**35.13**	45.99	**87.65**	**83.4**	**29.41**	65.85	1.16	1.17	37.50	75.20
345M	**15.60**	55.48	**92.35**	**87.1**	**22.76**	47.33	1.01	**1.06**	26.37	55.72
762M	**10.87**	**60.12**	**93.45**	**88.0**	**19.93**	40.31	0.97	**1.02**	22.05	44.575
1542M	**8.63**	**63.24**	**93.30**	**89.05**	**18.34**	35.76	0.93	0.98	17.48	42.16

Table 3. Zero-shot results on many datasets. No training or fine-tuning was performed for any of these results. PTB and WikiText-2 results are from (Gong et al., 2018). CBT results are from (Bajgar et al., 2016). LAMBADA accuracy result is from (Hoang et al., 2018) and LAMBADA perplexity result is from (Grave et al., 2016). Other results are from (Dai et al., 2019).

<UNK> which is extremely rare in WebText - occurring only 26 times in 40 billion bytes. We report our main results in Table 3 using invertible de-tokenizers which remove as many of these tokenization / pre-processing artifacts as possible. Since these de-tokenizers are invertible, we can still calculate the log probability of a dataset and they can be thought of as a simple form of domain adaptation. We observe gains of 2.5 to 5 perplexity for GPT-2 with these de-tokenizers.

WebText LMs transfer well across domains and datasets, improving the state of the art on 7 out of the 8 datasets in a zero-shot setting. Large improvements are noticed on small datasets such as Penn Treebank and WikiText-2 which have only 1 to 2 million training tokens. Large improvements are also noticed on datasets created to measure long-term dependencies like LAMBADA (Paperno et al., 2016) and the Children's Book Test (Hill et al., 2015). Our model is still significantly worse than prior work on the One Billion Word Benchmark (Chelba et al., 2013). This is likely due to a combination of it being both the largest dataset and having some of the most destructive pre-processing - 1BW's sentence level shuffling removes all long-range structure.

3.2. Children's Book Test

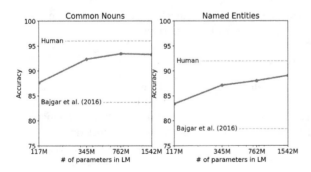

Figure 2. Performance on the Children's Book Test as a function of model capacity. Human performance are from Bajgar et al. (2016), instead of the much lower estimates from the original paper.

The Children's Book Test (CBT) (Hill et al., 2015) was created to examine the performance of LMs on different categories of words: named entities, nouns, verbs, and prepositions. Rather than reporting perplexity as an evaluation metric, CBT reports accuracy on an automatically constructed cloze test where the task is to predict which of 10 possible choices for an omitted word is correct. Following the LM approach introduced in the original paper, we compute the probability of each choice and the rest of the sentence conditioned on this choice according to the LM, and predict the one with the highest probability. As seen in Figure 2 performance steadily improves as model size is increased and closes the majority of the gap to human performance on this test. Data overlap analysis showed one of the CBT test set books, The Jungle Book by Rudyard Kipling, is in WebText, so we report results on the validation set which has no significant overlap. GPT-2 achieves new state of the art results of 93.3% on common nouns and 89.1% on named entities. A de-tokenizer was applied to remove PTB style tokenization artifacts from CBT.

3.3. LAMBADA

The LAMBADA dataset (Paperno et al., 2016) tests the ability of systems to model long-range dependencies in text. The task is to predict the final word of sentences which require at least 50 tokens of context for a human to successfully predict. GPT-2 improves the state of the art from 99.8 (Grave et al., 2016) to 8.6 perplexity and increases the accuracy of LMs on this test from 19% (Dehghani et al., 2018) to 52.66%. Investigating GPT-2's errors showed most predictions are valid continuations of the sentence, but are not valid final words. This suggests that the LM is not using the additional useful constraint that the word must be the final of the sentence. Adding a stop-word filter as an approximation to this further increases accuracy to 63.24%, improving the overall state of the art on this task by 4%. The previous state of the art (Hoang et al., 2018) used a different restricted prediction setting where the outputs of the model were constrained to only words that appeared in the context. For GPT-2, this restriction is harmful rather than helpful since 19% of answers are not in context. We use a version of the dataset without preprocessing.

3.4. Winograd Schema Challenge

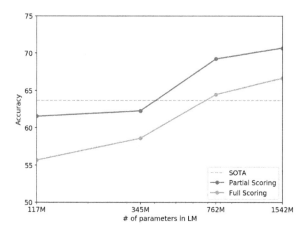

Figure 3. Performance on the Winograd Schema Challenge as a function of model capacity.

The Winograd Schema challenge (Levesque et al., 2012) was constructed to measure the capability of a system to perform commonsense reasoning by measuring its ability to resolve ambiguities in text. Recently Trinh & Le (2018) demonstrated significant progress on this challenge using LMs, by predicting the resolution of the ambiguity with higher probability. We follow their problem formulation and visualize the performance of our models with both full and partial scoring techniques in Figure 3. GPT-2 improves state of the art accuracy by 7%, achieving 70.70%. The dataset is quite small with only 273 examples so we recommend reading Trichelair et al. (2018) to help contextualize this result.

3.5. Reading Comprehension

The Conversation Question Answering dataset (CoQA) Reddy et al. (2018) consists of documents from 7 different domains paired with natural language dialogues between a question asker and a question answerer about the document. CoQA tests reading comprehension capabilities and also the ability of models to answer questions that depend on conversation history (such as "Why?").

Greedy decoding from GPT-2 when conditioned on a document, the history of the associated conversation, and a final token A: achieves 55 F1 on the development set. This matches or exceeds the performance of 3 out of 4 baseline systems without using the 127,000+ manually collected question answer pairs those baselines were trained on. The supervised SOTA, a BERT based system (Devlin et al.,

	R-1	R-2	R-L	R-AVG
Bottom-Up Sum	**41.22**	**18.68**	**38.34**	**32.75**
Lede-3	40.38	17.66	36.62	31.55
Seq2Seq + Attn	31.33	11.81	28.83	23.99
GPT-2 `TL;DR:`	29.34	8.27	26.58	21.40
Random-3	28.78	8.63	25.52	20.98
GPT-2 no hint	21.58	4.03	19.47	15.03

Table 4. Summarization performance as measured by ROUGE F1 metrics on the CNN and Daily Mail dataset. Bottom-Up Sum is the SOTA model from (Gehrmann et al., 2018)

2018), is nearing the 89 F1 performance of humans. While GPT-2's performance is exciting for a system without any supervised training, some inspection of its answers and errors suggests GPT-2 often uses simple retrieval based heuristics such as *answer with a name from the document in response to a who question.*

3.6. Summarization

We test GPT-2's ability to perform summarization on the CNN and Daily Mail dataset (Nallapati et al., 2016). To induce summarization behavior we add the text `TL;DR:` after the article and generate 100 tokens with Top-k random sampling (Fan et al., 2018) with $k = 2$ which reduces repetition and encourages more abstractive summaries than greedy decoding. We use the first 3 generated sentences in these 100 tokens as the summary. While qualitatively the generations resemble summaries, as shown in Table 14, they often focus on recent content from the article or confuse specific details such as how many cars were involved in a crash or whether a logo was on a hat or shirt. On the commonly reported ROUGE 1,2,L metrics the generated summaries only begin to approach the performance of classic neural baselines and just barely outperforms selecting 3 random sentences from the article. GPT-2's performance drops by 6.4 points on the aggregate metric when the task hint is removed which demonstrates the ability to invoke task specific behavior in a language model with natural language.

3.7. Translation

We test whether GPT-2 has begun to learn how to translate from one language to another. In order to help it infer that this is the desired task, we condition the language model on a context of example pairs of the format `english sentence = french sentence` and then after a final prompt of `english sentence =` we sample from the model with greedy decoding and use the first generated sentence as the translation. On the WMT-14 English-French test set, GPT-2 gets 5 BLEU, which is slightly worse than a word-by-word substitution with a bilingual lexicon inferred in previous work on unsupervised word translation

Question	Generated Answer	Correct	Probability
Who wrote the book the origin of species?	Charles Darwin	✓	83.4%
Who is the founder of the ubuntu project?	Mark Shuttleworth	✓	82.0%
Who is the quarterback for the green bay packers?	Aaron Rodgers	✓	81.1%
Panda is a national animal of which country?	China	✓	76.8%
Who came up with the theory of relativity?	Albert Einstein	✓	76.4%
When was the first star wars film released?	1977	✓	71.4%
What is the most common blood type in sweden?	A	✗	70.6%
Who is regarded as the founder of psychoanalysis?	Sigmund Freud	✓	69.3%
Who took the first steps on the moon in 1969?	Neil Armstrong	✓	66.8%
Who is the largest supermarket chain in the uk?	Tesco	✓	65.3%
What is the meaning of shalom in english?	peace	✓	64.0%
Who was the author of the art of war?	Sun Tzu	✓	59.6%
Largest state in the us by land mass?	California	✗	59.2%
Green algae is an example of which type of reproduction?	parthenogenesis	✗	56.5%
Vikram samvat calender is official in which country?	India	✓	55.6%
Who is mostly responsible for writing the declaration of independence?	Thomas Jefferson	✓	53.3%
What us state forms the western boundary of montana?	Montana	✗	52.3%
Who plays ser davos in game of thrones?	Peter Dinklage	✗	52.1%
Who appoints the chair of the federal reserve system?	Janet Yellen	✗	51.5%
State the process that divides one nucleus into two genetically identical nuclei?	mitosis	✓	50.7%
Who won the most mvp awards in the nba?	Michael Jordan	✗	50.2%
What river is associated with the city of rome?	the Tiber	✓	48.6%
Who is the first president to be impeached?	Andrew Johnson	✓	48.3%
Who is the head of the department of homeland security 2017?	John Kelly	✓	47.0%
What is the name given to the common currency to the european union?	Euro	✓	46.8%
What was the emperor name in star wars?	Palpatine	✓	46.5%
Do you have to have a gun permit to shoot at a range?	No	✓	46.4%
Who proposed evolution in 1859 as the basis of biological development?	Charles Darwin	✓	45.7%
Nuclear power plant that blew up in russia?	Chernobyl	✓	45.7%
Who played john connor in the original terminator?	Arnold Schwarzenegger	✗	45.2%

Table 5. The 30 most confident answers generated by GPT-2 on the development set of Natural Questions sorted by their probability according to GPT-2. None of these questions appear in WebText according to the procedure described in Section 4.

(Conneau et al., 2017b). On the WMT-14 French-English test set, GPT-2 is able to leverage its very strong English language model to perform significantly better, achieving 11.5 BLEU. This outperforms several unsupervised machine translation baselines from (Artetxe et al., 2017) and (Lample et al., 2017) but is still much worse than the 33.5 BLEU of the current best unsupervised machine translation approach (Artetxe et al., 2019). Performance on this task was surprising to us, since we deliberately removed non-English webpages from WebText as a filtering step. In order to confirm this, we ran a byte-level language detector[2] on WebText which detected only 10MB of data in the French language which is approximately 500x smaller than the monolingual French corpus common in prior unsupervised machine translation research.

3.8. Question Answering

A potential way to test what information is contained within a language model is to evaluate how often it generates the correct answer to factoid-style questions. Previous showcasing of this behavior in neural systems where all information is stored in parameters such as *A Neural Conversational Model* (Vinyals & Le, 2015) reported qualitative results due to the lack of high-quality evaluation datasets. The recently introduced Natural Questions dataset (Kwiatkowski et al.,

2019) is a promising resource to test this more quantitatively. Similar to translation, the context of the language model is seeded with example question answer pairs which helps the model infer the short answer style of the dataset. GPT-2 answers 4.1% of questions correctly when evaluated by the exact match metric commonly used on reading comprehension datasets like SQUAD.[3] As a comparison point, the smallest model does not exceed the 1.0% accuracy of an incredibly simple baseline which returns the most common answer for each question type (who, what, where, etc...). GPT-2 answers 5.3 times more questions correctly, suggesting that model capacity has been a major factor in the poor performance of neural systems on this kind of task as of yet. The probability GPT-2 assigns to its generated answers is well calibrated and GPT-2 has an accuracy of 63.1% on the 1% of questions it is most confident in. The 30 most confident answers generated by GPT-2 on development set questions are shown in Table 5. The performance of GPT-2 is still much, much, worse than the 30 to 50% range of open domain question answering systems which hybridize information retrieval with extractive document question answering (Alberti et al., 2019).

---

[2] https://github.com/CLD2Owners/cld2

[3] Alec, who previously thought of himself as good at random trivia, answered 17 of 100 randomly sampled examples correctly when tested in the same setting as GPT-2. He actually only got 14 right but he should have gotten those other 3

	PTB	WikiText-2	enwik8	text8	Wikitext-103	1BW
Dataset train	**2.67%**	0.66%	**7.50%**	2.34%	**9.09%**	**13.19%**
WebText train	0.88%	**1.63%**	6.31%	**3.94%**	2.42%	3.75%

Table 6. Percentage of test set 8 grams overlapping with training sets.

4. Generalization vs Memorization

Recent work in computer vision has shown that common image datasets contain a non-trivial amount of near-duplicate images. For instance CIFAR-10 has 3.3% overlap between train and test images (Barz & Denzler, 2019). This results in an over-reporting of the generalization performance of machine learning systems. As the size of datasets increases this issue becomes increasingly likely which suggests a similar phenomena could be happening with WebText. Therefore it is important to analyze how much test data also shows up in the training data.

To study this we created Bloom filters containing 8-grams of WebText training set tokens. To improve recall, strings were normalized to contain only lower-cased alphanumeric words with a single space as a delimiter. The Bloom filters were constructed such that the false positive rate is upper bounded by $\frac{1}{10^8}$. We further verified the low false positive rate by generating 1M strings, of which zero were found by the filter.

These Bloom filters let us calculate, given a dataset, the percentage of 8-grams from that dataset that are also found in the WebText training set. Table 6 shows this overlap analysis for the test sets of common LM benchmarks. Common LM datasets' test sets have between 1-6% overlap with WebText train, with an average of overlap of 3.2%. Somewhat surprisingly, many datasets have larger overlaps with their own training splits, with an average of 5.9% overlap.

Our approach optimizes for recall, and while manual inspection of the overlaps shows many common phrases, there are many longer matches that are due to duplicated data. This is not unique to WebText. For instance, we discovered that the test set of WikiText-103 has an article which is also in the training dataset. Since there are only 60 articles in the test set there is at least an overlap of 1.6%.[4] Potentially more worryingly, 1BW has an overlap of nearly 13.2% with its own training set according to our procedure.

For the Winograd Schema Challenge, we found only 10 schemata which had any 8-gram overlaps with the WebText training set. Of these, 2 were spurious matches. Of the remaining 8, only 1 schema appeared in any contexts that

gave away the answer.

For CoQA, about 15% of documents in the news domain are already in WebText and the model performs about 3 F1 better on these. CoQA's development set metric reports the average performance over 5 different domains and we measure a gain of about 0.5-1.0 F1 due to overlap across the various domains. However, no actual training questions or answers are in WebText since CoQA was released after the cutoff date for links in WebText.

On LAMBADA, the average overlap is 1.2%. GPT-2 performs about 2 perplexity better on examples with greater than 15% overlap. Recalculating metrics when excluding all examples with any overlap shifts results from 8.6 to 8.7 perplexity and reduces accuracy from 63.2% to 62.9%. This very small change in overall results is likely due to only 1 in 200 examples having significant overlap.

Overall, our analysis suggests that data overlap between WebText training data and specific evaluation datasets provides a small but consistent benefit to reported results. However, for most datasets we do not notice significantly larger overlaps than those already existing between standard training and test sets, as Table 6 highlights.

Understanding and quantifying how highly similar text impacts performance is an important research question. Better de-duplication techniques such as scalable fuzzy matching could also help better answer these questions. For now, we recommend the use of n-gram overlap based de-duplication as an important verification step and sanity check during the creation of training and test splits for new NLP datasets.

Another potential way of determining whether the performance of WebText LMs is attributable to memorization is inspecting their performance on their own held-out set. As shown in Figure 4, performance on both the training and test sets of WebText are similar and improve together as model size is increased. This suggests even GPT-2 is still underfitting on WebText in many ways.

GPT-2 is also able to write news articles about the discovery of talking unicorns. An example is provided in Table 13.

5. Related Work

A significant portion of this work measured the performance of larger language models trained on larger datasets. This

---

[4]A significant portion of additional overlap is due to editors reusing some paragraphs across multiple articles with a shared theme such as various battles in the Korean War.

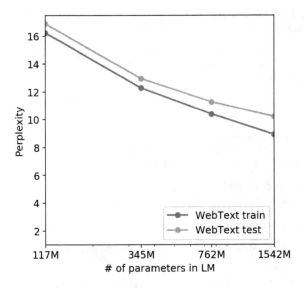

Figure 4. The performance of LMs trained on WebText as a function of model size.

is similar to the work of Jozefowicz et al. (2016) which scaled RNN based language models on the 1 Billion Word Benchmark. Bajgar et al. (2016) also previously improved results on the Children's Book Test by creating a much larger training dataset out of Project Gutenberg to supplement the standard training dataset. Hestness et al. (2017) conducted a thorough analysis of how the performance of various deep learning models changes as a function of both model capacity and dataset size. Our experiments, while much noisier across tasks, suggest similar trends hold for sub-tasks of an objective and continue into the 1B+ parameter regime.

Interesting learned functionality in generative models has been documented before such as the cells in an RNN language model performing line-width tracking and quote/comment detection Karpathy et al. (2015). More inspirational to our work was the observation of Liu et al. (2018) that a model trained to generate Wikipedia articles also learned to translate names between languages.

Previous work has explored alternative approaches to filtering and constructing a large text corpus of web pages, such as the iWeb Corpus (Davies, 2018).

There has been extensive work on pre-training methods for language tasks. In addition to those mentioned in the introduction, GloVe (Pennington et al., 2014) scaled word vector representation learning to all of Common Crawl. An influential early work on deep representation learning for text was *Skip-thought Vectors* (Kiros et al., 2015). McCann et al. (2017) explored the use of representations derived from machine translation models and Howard & Ruder (2018)

improved the RNN based fine-tuning approaches of (Dai & Le, 2015). (Conneau et al., 2017a) studied the transfer performance of representations learned by natural language inference models and (Subramanian et al., 2018) explored large-scale multitask training.

(Ramachandran et al., 2016) demonstrated that seq2seq models benefit from being initialized with pre-trained language models as encoders and decoders. More recent work has shown that LM pre-training is helpful when fine-tuned for difficult generation tasks like chit-chat dialog and dialog based question answering systems as well (Wolf et al., 2019) (Dinan et al., 2018).

6. Discussion

Much research has been dedicated to learning (Hill et al., 2016), understanding (Levy & Goldberg, 2014), and critically evaluating (Wieting & Kiela, 2019) the representations of both supervised and unsupervised pre-training methods. Our results suggest that unsupervised task learning is an additional promising area of research to explore. These findings potentially help explain the widespread success of pre-training techniques for down-stream NLP tasks as we show that, in the limit, one of these pre-training techniques begins to learn to perform tasks directly without the need for supervised adaption or modification.

On reading comprehension the performance of GPT-2 is competitive with supervised baselines in a zero-shot setting. However, on other tasks such as summarization, while it is qualitatively performing the task, its performance is still only rudimentary according to quantitative metrics. While suggestive as a research result, in terms of practical applications, the zero-shot performance of GPT-2 is still far from use-able.

We have studied the zero-shot performance of WebText LMs on many canonical NLP tasks, but there are many additional tasks that could be evaluated. There are undoubtedly many practical tasks where the performance of GPT-2 is still no better than random. Even on common tasks that we evaluated on, such as question answering and translation, language models only begin to outperform trivial baselines when they have sufficient capacity.

While zero-shot performance establishes a baseline of the potential performance of GPT-2 on many tasks, it is not clear where the ceiling is with finetuning. On some tasks, GPT-2's fully abstractive output is a significant departure from the extractive pointer network (Vinyals et al., 2015) based outputs which are currently state of the art on many question answering and reading comprehension datasets. Given the prior success of fine-tuning GPT, we plan to investigate fine-tuning on benchmarks such as decaNLP and GLUE, especially since it is unclear whether the additional training data and capacity of GPT-2 is sufficient to overcome the inefficiencies of uni-directional representations demonstrated by BERT (Devlin et al., 2018).

7. Conclusion

When a large language model is trained on a sufficiently large and diverse dataset it is able to perform well across many domains and datasets. GPT-2 zero-shots to state of the art performance on 7 out of 8 tested language modeling datasets. The diversity of tasks the model is able to perform in a zero-shot setting suggests that high-capacity models trained to maximize the likelihood of a sufficiently varied text corpus begin to learn how to perform a surprising amount of tasks without the need for explicit supervision.[5]

Acknowledgements

Thanks to everyone who wrote the text, shared the links, and upvoted the content in WebText. Many millions of people were involved in creating the data that GPT-2 was trained on. Also thanks to all the Googlers who helped us with training infrastructure, including Zak Stone, JS Riehl, Jonathan Hseu, Russell Power, Youlong Cheng, Noam Shazeer, Solomon Boulos, Michael Banfield, Aman Gupta, Daniel Sohn, and many more. Finally thanks to the people who gave feedback on drafts of the paper: Jacob Steinhardt, Sam Bowman, Geoffrey Irving, and Madison May.

References

Al-Rfou, R., Choe, D., Constant, N., Guo, M., and Jones, L. Character-level language modeling with deeper self-attention. *arXiv preprint arXiv:1808.04444*, 2018.

Alberti, C., Lee, K., and Collins, M. A bert baseline for the natural questions. *arXiv preprint arXiv:1901.08634*, 2019.

Alcorn, M. A., Li, Q., Gong, Z., Wang, C., Mai, L., Ku, W.-S., and Nguyen, A. Strike (with) a pose: Neural networks are easily fooled by strange poses of familiar objects. *arXiv preprint arXiv:1811.11553*, 2018.

Amodei, D., Ananthanarayanan, S., Anubhai, R., Bai, J., Battenberg, E., Case, C., Casper, J., Catanzaro, B., Cheng, Q., Chen, G., et al. Deep speech 2: End-to-end speech recognition in english and mandarin. In *International Conference on Machine Learning*, pp. 173–182, 2016.

Artetxe, M., Labaka, G., Agirre, E., and Cho, K. Unsupervised neural machine translation. *arXiv preprint arXiv:1710.11041*, 2017.

Artetxe, M., Labaka, G., and Agirre, E. An effective approach to unsupervised machine translation. *arXiv preprint arXiv:1902.01313*, 2019.

Ba, J. L., Kiros, J. R., and Hinton, G. E. Layer normalization. *arXiv preprint arXiv:1607.06450*, 2016.

Bajgar, O., Kadlec, R., and Kleindienst, J. Embracing data abundance: Booktest dataset for reading comprehension. *arXiv preprint arXiv:1610.00956*, 2016.

Barz, B. and Denzler, J. Do we train on test data? purging cifar of near-duplicates. *arXiv preprint arXiv:1902.00423*, 2019.

Bengio, Y., Ducharme, R., Vincent, P., and Jauvin, C. A neural probabilistic language model. *Journal of machine learning research*, 3(Feb):1137–1155, 2003.

Bowman, S. R., Pavlick, E., Grave, E., Van Durme, B., Wang, A., Hula, J., Xia, P., Pappagari, R., McCoy, R. T., Patel, R., et al. Looking for elmo's friends: Sentence-level pretraining beyond language modeling. *arXiv preprint arXiv:1812.10860*, 2018.

Caruana, R. Multitask learning. *Machine learning*, 28(1):41–75, 1997.

Chelba, C., Mikolov, T., Schuster, M., Ge, Q., Brants, T., Koehn, P., and Robinson, T. One billion word benchmark for measuring progress in statistical language modeling. *arXiv preprint arXiv:1312.3005*, 2013.

Collobert, R., Weston, J., Bottou, L., Karlen, M., Kavukcuoglu, K., and Kuksa, P. Natural language processing (almost) from scratch. *Journal of Machine Learning Research*, 12(Aug):2493–2537, 2011.

Conneau, A., Kiela, D., Schwenk, H., Barrault, L., and Bordes, A. Supervised learning of universal sentence representations from natural language inference data. *arXiv preprint arXiv:1705.02364*, 2017a.

Conneau, A., Lample, G., Ranzato, M., Denoyer, L., and Jégou, H. Word translation without parallel data. *arXiv preprint arXiv:1710.04087*, 2017b.

Dai, A. M. and Le, Q. V. Semi-supervised sequence learning. In *Advances in neural information processing systems*, pp. 3079–3087, 2015.

Dai, Z., Yang, Z., Yang, Y., Cohen, W. W., Carbonell, J., Le, Q. V., and Salakhutdinov, R. Transformer-xl: Attentive language models beyond a fixed-length context. *arXiv preprint arXiv:1901.02860*, 2019.

Davies, M. The 14 billion word iweb corpus. *https://corpus.byu.edu/iWeb/*, 2018.

Dehghani, M., Gouws, S., Vinyals, O., Uszkoreit, J., and Kaiser, Ł. Universal transformers. *arXiv preprint arXiv:1807.03819*, 2018.

Devlin, J., Chang, M.-W., Lee, K., and Toutanova, K. Bert: Pre-training of deep bidirectional transformers for language understanding. *arXiv preprint arXiv:1810.04805*, 2018.

Dinan, E., Roller, S., Shuster, K., Fan, A., Auli, M., and Weston, J. Wizard of wikipedia: Knowledge-powered conversational agents. *arXiv preprint arXiv:1811.01241*, 2018.

Fan, A., Lewis, M., and Dauphin, Y. Hierarchical neural story generation. *arXiv preprint arXiv:1805.04833*, 2018.

---

[5]Preliminary code for downloading and using the small model is available at https://github.com/openai/gpt-2

Finn, C., Abbeel, P., and Levine, S. Model-agnostic meta-learning for fast adaptation of deep networks. *arXiv preprint arXiv:1703.03400*, 2017.

Gehrmann, S., Deng, Y., and Rush, A. M. Bottom-up abstractive summarization. *arXiv preprint arXiv:1808.10792*, 2018.

Gillick, D., Brunk, C., Vinyals, O., and Subramanya, A. Multilingual language processing from bytes. *arXiv preprint arXiv:1512.00103*, 2015.

Gong, C., He, D., Tan, X., Qin, T., Wang, L., and Liu, T.-Y. Frage: frequency-agnostic word representation. In *Advances in Neural Information Processing Systems*, pp. 1341–1352, 2018.

Grave, E., Joulin, A., and Usunier, N. Improving neural language models with a continuous cache. *arXiv preprint arXiv:1612.04426*, 2016.

He, K., Zhang, X., Ren, S., and Sun, J. Identity mappings in deep residual networks. In *European conference on computer vision*, pp. 630–645. Springer, 2016.

Hestness, J., Narang, S., Ardalani, N., Diamos, G., Jun, H., Kianinejad, H., Patwary, M., Ali, M., Yang, Y., and Zhou, Y. Deep learning scaling is predictable, empirically. *arXiv preprint arXiv:1712.00409*, 2017.

Hill, F., Bordes, A., Chopra, S., and Weston, J. The goldilocks principle: Reading children's books with explicit memory representations. *arXiv preprint arXiv:1511.02301*, 2015.

Hill, F., Cho, K., and Korhonen, A. Learning distributed representations of sentences from unlabelled data. *arXiv preprint arXiv:1602.03483*, 2016.

Hoang, L., Wiseman, S., and Rush, A. M. Entity tracking improves cloze-style reading comprehension. *arXiv preprint arXiv:1810.02891*, 2018.

Howard, J. and Ruder, S. Universal language model fine-tuning for text classification. In *Proceedings of the 56th Annual Meeting of the Association for Computational Linguistics (Volume 1: Long Papers)*, volume 1, pp. 328–339, 2018.

Jelinek, F. and Mercer, R. L. Interpolated estimation of markov source parameters from sparse data. *In Proceedings of the Workshop on Pattern Recognition in Practice, Amsterdam, The Netherlands: North-Holland, May.*, 1980.

Jia, R. and Liang, P. Adversarial examples for evaluating reading comprehension systems. *arXiv preprint arXiv:1707.07328*, 2017.

Jozefowicz, R., Vinyals, O., Schuster, M., Shazeer, N., and Wu, Y. Exploring the limits of language modeling. *arXiv preprint arXiv:1602.02410*, 2016.

Kaiser, L., Gomez, A. N., Shazeer, N., Vaswani, A., Parmar, N., Jones, L., and Uszkoreit, J. One model to learn them all. *arXiv preprint arXiv:1706.05137*, 2017.

Karpathy, A., Johnson, J., and Fei-Fei, L. Visualizing and understanding recurrent networks. *arXiv preprint arXiv:1506.02078*, 2015.

Kirkpatrick, J., Pascanu, R., Rabinowitz, N., Veness, J., Desjardins, G., Rusu, A. A., Milan, K., Quan, J., Ramalho, T., Grabska-Barwinska, A., et al. Overcoming catastrophic forgetting in neural networks. *Proceedings of the national academy of sciences*, pp. 201611835, 2017.

Kiros, R., Zhu, Y., Salakhutdinov, R. R., Zemel, R., Urtasun, R., Torralba, A., and Fidler, S. Skip-thought vectors. In *Advances in neural information processing systems*, pp. 3294–3302, 2015.

Krizhevsky, A., Sutskever, I., and Hinton, G. E. Imagenet classification with deep convolutional neural networks. In *Advances in neural information processing systems*, pp. 1097–1105, 2012.

Kwiatkowski, T., Palomaki, J., Rhinehart, O., Collins, M., Parikh, A., Alberti, C., Epstein, D., Polosukhin, I., Kelcey, M., Devlin, J., et al. Natural questions: a benchmark for question answering research. 2019.

Lake, B. M., Ullman, T. D., Tenenbaum, J. B., and Gershman, S. J. Building machines that learn and think like people. *Behavioral and Brain Sciences*, 40, 2017.

Lample, G., Conneau, A., Denoyer, L., and Ranzato, M. Unsupervised machine translation using monolingual corpora only. *arXiv preprint arXiv:1711.00043*, 2017.

Levesque, H., Davis, E., and Morgenstern, L. The winograd schema challenge. In *Thirteenth International Conference on the Principles of Knowledge Representation and Reasoning*, 2012.

Levy, O. and Goldberg, Y. Neural word embedding as implicit matrix factorization. In *Advances in neural information processing systems*, pp. 2177–2185, 2014.

Liu, P. J., Saleh, M., Pot, E., Goodrich, B., Sepassi, R., Kaiser, L., and Shazeer, N. Generating wikipedia by summarizing long sequences. *arXiv preprint arXiv:1801.10198*, 2018.

McCann, B., Bradbury, J., Xiong, C., and Socher, R. Learned in translation: Contextualized word vectors. In *Advances in Neural Information Processing Systems*, pp. 6294–6305, 2017.

McCann, B., Keskar, N. S., Xiong, C., and Socher, R. The natural language decathlon: Multitask learning as question answering. *arXiv preprint arXiv:1806.08730*, 2018.

Merity, S., Xiong, C., Bradbury, J., and Socher, R. Pointer sentinel mixture models. *arXiv preprint arXiv:1609.07843*, 2016.

Mikolov, T., Sutskever, I., Chen, K., Corrado, G. S., and Dean, J. Distributed representations of words and phrases and their compositionality. In *Advances in neural information processing systems*, pp. 3111–3119, 2013.

Nallapati, R., Zhou, B., Gulcehre, C., Xiang, B., et al. Abstractive text summarization using sequence-to-sequence rnns and beyond. *arXiv preprint arXiv:1602.06023*, 2016.

Paperno, D., Kruszewski, G., Lazaridou, A., Pham, Q. N., Bernardi, R., Pezzelle, S., Baroni, M., Boleda, G., and Fernández, R. The lambada dataset: Word prediction requiring a broad discourse context. *arXiv preprint arXiv:1606.06031*, 2016.

Pennington, J., Socher, R., and Manning, C. Glove: Global vectors for word representation. In *Proceedings of the 2014 conference on empirical methods in natural language processing (EMNLP)*, pp. 1532–1543, 2014.

Peters, M. E. and Lecocq, D. Content extraction using diverse feature sets. In *Proceedings of the 22nd International Conference on World Wide Web*, pp. 89–90. ACM, 2013.

Peters, M. E., Neumann, M., Iyyer, M., Gardner, M., Clark, C., Lee, K., and Zettlemoyer, L. Deep contextualized word representations. *arXiv preprint arXiv:1802.05365*, 2018.

Radford, A., Jozefowicz, R., and Sutskever, I. Learning to generate reviews and discovering sentiment. *arXiv preprint arXiv:1704.01444*, 2017.

Radford, A., Narasimhan, K., Salimans, T., and Sutskever, I. Improving language understanding by generative pre-training. 2018.

Ramachandran, P., Liu, P. J., and Le, Q. V. Unsupervised pre-training for sequence to sequence learning. *arXiv preprint arXiv:1611.02683*, 2016.

Recht, B., Roelofs, R., Schmidt, L., and Shankar, V. Do cifar-10 classifiers generalize to cifar-10? *arXiv preprint arXiv:1806.00451*, 2018.

Reddy, S., Chen, D., and Manning, C. D. Coqa: A conversational question answering challenge. *arXiv preprint arXiv:1808.07042*, 2018.

Schwartz, R., Sap, M., Konstas, I., Zilles, L., Choi, Y., and Smith, N. A. Story cloze task: Uw nlp system. In *Proceedings of the 2nd Workshop on Linking Models of Lexical, Sentential and Discourse-level Semantics*, pp. 52–55, 2017.

See, A., Liu, P. J., and Manning, C. D. Get to the point: Summarization with pointer-generator networks. *arXiv preprint arXiv:1704.04368*, 2017.

Sennrich, R., Haddow, B., and Birch, A. Neural machine translation of rare words with subword units. *arXiv preprint arXiv:1508.07909*, 2015.

Subramanian, S., Trischler, A., Bengio, Y., and Pal, C. J. Learning general purpose distributed sentence representations via large scale multi-task learning. *arXiv preprint arXiv:1804.00079*, 2018.

Sutskever, I., Vinyals, O., and Le, Q. V. Sequence to sequence learning with neural networks. In *Advances in neural information processing systems*, pp. 3104–3112, 2014.

Sutskever, I., Jozefowicz, R., Gregor, K., Rezende, D., Lillicrap, T., and Vinyals, O. Towards principled unsupervised learning. *arXiv preprint arXiv:1511.06440*, 2015.

Trichelair, P., Emami, A., Cheung, J. C. K., Trischler, A., Suleman, K., and Diaz, F. On the evaluation of common-sense reasoning in natural language understanding. *arXiv preprint arXiv:1811.01778*, 2018.

Trinh, T. H. and Le, Q. V. A simple method for commonsense reasoning. *arXiv preprint arXiv:1806.02847*, 2018.

Vaswani, A., Shazeer, N., Parmar, N., Uszkoreit, J., Jones, L., Gomez, A. N., Kaiser, Ł., and Polosukhin, I. Attention is all you need. In *Advances in Neural Information Processing Systems*, pp. 5998–6008, 2017.

Vinyals, O. and Le, Q. A neural conversational model. *arXiv preprint arXiv:1506.05869*, 2015.

Vinyals, O., Fortunato, M., and Jaitly, N. Pointer networks. In *Advances in Neural Information Processing Systems*, pp. 2692–2700, 2015.

Wang, A., Singh, A., Michael, J., Hill, F., Levy, O., and Bowman, S. R. Glue: A multi-task benchmark and analysis platform for natural language understanding. *arXiv preprint arXiv:1804.07461*, 2018.

Weston, J. E. Dialog-based language learning. In *Advances in Neural Information Processing Systems*, pp. 829–837, 2016.

Wieting, J. and Kiela, D. No training required: Exploring random encoders for sentence classification. *arXiv preprint arXiv:1901.10444*, 2019.

Wolf, T., Sanh, V., Chaumond, J., and Delangue, C. Transfertransfo: A transfer learning approach for neural network based conversational agents. *arXiv preprint arXiv:1901.08149*, 2019.

Yogatama, D., d'Autume, C. d. M., Connor, J., Kocisky, T., Chrzanowski, M., Kong, L., Lazaridou, A., Ling, W., Yu, L., Dyer, C., et al. Learning and evaluating general linguistic intelligence. *arXiv preprint arXiv:1901.11373*, 2019.

Chapter 16

Conclusion

16.1. Steam-Powered AI

In the early days of steam engines, efforts to increase their power output relied on tinkering with various components and knowledge of the resultant steam engine lore. But that was before Carnot (1796–1832) placed the physics of steam engines on a proper scientific footing and provided a clear account of how steam engines work. Once the underlying mechanisms were understood, it became clear how to improve the engines' efficiency, and so tinkering and steam engine lore became redundant.

At the present time, AI is in a pre-Carnot stage of development (or in a multi-wing stage of flight; see Chapter 14). We can see that AI systems work, but we don't really understand how they work, nor why they work as well as they do[5;42]. Consequently, there is a great deal of neural network lore, which is the result of historical tinkering with things such as the number of units per layer, layer normalisation, unit activation functions, learning rate schedules, and number of layers.

If we had a proper understanding of AI systems then two things would happen. First, we would be able to identify (and discard) aspects of AI models that have arisen from past tinkering but which are merely useless baggage. Second, and more importantly, we would understand what the theoretical limits of present-day systems are and how to build AI systems that could exceed those limits.

It is tempting to think that we can treat the problem of how a brain or a neural network works just like any other scientific problem, by reducing it to its fundamental elements. Indeed, this approach has a long tradition, as enunciated by Nicolaus Steno (1638–1686) in 1669:

> The brain being indeed a machine, we must not hope to find its artifice through other ways than those which are used to find the artifice of the other machines. It thus remains to do what we would do for any other machine; I mean to dismantle it piece by piece and to consider what these can do separately and together.

But, as has been rediscovered many times in the intervening centuries, an overly reductionist approach is doomed to failure. By analogy to research on the human brain, the successes of AI systems have been tempered by the growing realisation that understanding how the components (e.g. neurons) of a system work is not the same as understanding how the whole system works. This realisation was foreshadowed by Gottfried Leibniz (1646–1716), the philosopher and co-inventor (with Isaac Newton) of calculus:

> Let us pretend that there was a machine, which was constructed in such a way as to give rise to thinking, sensing, and having perceptions. You could imagine it expanded in size ..., so that you could go inside it, like going into a mill. On this assumption, your tour inside it would show you the working parts pushing each other, but never anything which would explain a perception.

Thus, relating high-order concepts such as perception to particular mechanisms or components may be unrealistic for any intelligent system, whether it is biological or not. However, that does not mean there is no point in striving to understand the principles which govern how such systems work. Just as the route to understanding steam engines does not depend on studying the structure of water molecules, so the route to understanding AI systems does not depend on studying individual connections. And, just as steam engines were ultimately understood by analysing them in terms of collective abstract physical quantities such as entropy, so AI systems will ultimately be understood by studying the collective properties of neural networks in a principled manner.

16.2. Black Boxes

A common criticism of AI systems based on neural networks is that they are essentially *black box* systems. In other words, because a neural network learns from data, and because the knowledge gained by the neural network is implicit in the trillions of connections between millions of units, it is extremely difficult to know exactly how the network makes decisions regarding any particular input. There are two broad responses to this problem of *interpretability*. The first argues that it *is* possible[53] to interpret the decisions made by a neural network.

The second response is that such a criticism is simply unfair. When humans become expert in a particular domain, they can rarely explain why they took a particular decision[110]. Or rather, they can explain, but their explanations may have little to do with the actual process by which they made that decision[13]. This is because knowledge gained over many years becomes like intuition, so that experts have little insight into why they took certain decisions.

In fact, even though we do not recognise it as such, we all have expertise in everyday domains such as face recognition, language and social interactions (well, most of us do). Despite the wealth of knowledge implicit in such expertise, we cannot easily articulate how we recognise a friend, or know when (or if) to laugh at a joke, any more than we can explain how to ride a bike or how to chew gum without biting our own tongue. Thus doctors, pilots, car drivers, and anyone with expertise in any domain are, to all intents and purposes, as opaque as AI systems.

16.3. AI: Back to the Future

The future of AI has been compared to the future of nuclear fusion. The claim that nuclear fusion would provide free, unlimited power within ten years was first made soon after it was discovered in 1930, and that claim has been renewed about every ten years since then. Similarly, a promise that AI would exist within ten years was first made in the initial wave of AI research in the 1950s, and that promise has been renewed on a fairly regular basis ever since. For example, in 1961, Claude Shannon (the inventor of *information theory*[106;120]) said:

> I confidently expect that, within 10 to 15 years, we will find emerging from the laboratory something not too far from the robots of science fiction fame.

If a genuinely intelligent AI will one day become a reality then continually renewing a prediction that such an AI will exist within ten years is a safe bet, because at some point that prediction must come true. However, before we consider the question of whether computers will ever be intelligent, we should address the following fundamental question: is AI possible, even in principle? The answer is *yes*, and here's why.

Before Orville and Wilbur Wright flew the first aeroplane in 1903, sceptics declared that a machine could never fly like a bird (in 1901, Wilbur had predicted powered flight would take another 50 years). Today, many of us are like those sceptics, doubting that a machine can ever achieve human levels of intelligence. But for a compelling counter-argument to such scepticism, consider the fact that birds and brains are physical devices that must obey the same universal laws of physics. In other words, a bird is a flying machine that happens to be made of organic matter, and a brain is a computational machine that happens to be made of neurons.

Finally, irrespective of the arguments for or against this or that aspect of AI, the final sentence of Turing's 1950 paper[133] provides a timely reminder:

> We can only see a short distance ahead, but we can see plenty there that needs to be done.

Appendix A

Glossary

activation function A function that maps unit inputs to output values.

autoencoder A network that maps input vectors to themselves as output vectors via one or more hidden layers.

backprop Abbreviation for backpropagation, the principal method for training multi-layer networks.

Bayes' rule/theorem A probability rule that provides a rigorous method for interpreting evidence (data) in the context of previous experience. See Appendix D.

bias unit A unit with a fixed state (usually $+1$ or -1) that supplies input to other units in a network via learned bias weights.

connectionism A theory which seeks to explain how the human brain works, expressed in terms of the interactions of interconnected networks of simple processing units. These units are often likened to neurons in the brain, and the connections between them represent synapses.

content addressable memory If an association between an input and an output vector is learned, and if the input vector evokes that output, then the memory is said to be content addressable.

convolution A set of connection weights defined over a small region in an image defines a filter. Applying these weights across all regions in an image yields the convolution of the image with the filter.

convolutional neural network A neural network in which each hidden unit in a layer L_i has a receptive field defined over a small region of units in the previous layer L_{i-1}.

deep neural network A network with more than one hidden layer of units between the input and output layers.

early stopping A method used to reduce over-fitting by monitoring progress on a validation set while training the network on a training set.

ELBO Evidence lower bound. Rather than maximising the evidence (likelihood) L during training, a (simpler) variational function that provides a lower bound on L is maximised instead.

energy function A function defined in terms of the states of units in a network. A set of network states such that connected units have mutually compatible states has a low energy value.

error Difference between the output of a unit and the target (desired) output.

generalisation The ability to extrapolate beyond items in the training data.

generative model A neural network that models the distribution $p(\mathbf{x})$ of input vectors $\mathbf{x}$ as $\hat{p}(\mathbf{x}) \approx p(\mathbf{x})$, which can be used to generate new instances of $\mathbf{x}$ by sampling from $\hat{p}(\mathbf{x})$.

GOFAI Good old fashioned AI, an approach that dominated AI research in the latter part of the 20th century. It assumes that intelligence requires only a purely symbolic or logic-based representation of the physical world.

gradient ascent/descent If performance is measured in terms of an error function $E(\mathbf{w})$, the optimal weights can be found by iteratively changing the network weight vector $\mathbf{w}$ in a direction that is effectively downhill with respect to E. Gradient ascent just means performing gradient descent with respect to $-E$.

inference Using data to estimate the posterior distribution of unknown parameters (e.g. weights or latent variables).

Kullback–Leibler divergence A measure of the difference between two probability distributions $p(x)$ and $q(x)$:

$$D(q(x)\|p(x)) \;=\; \int_x q(x) \log_2 \frac{q(x)}{p(x)} \, dx \text{ bits.}$$

Also called the relative entropy, D is a measure of how surprising the distribution $p(x)$ is, if values of x are chosen with probability $q(x)$. Note that $D(q(x)\|p(x)) \neq D(p(x)\|q(x))$, and $D(q(x)\|p(x)) = 0$ if $q(x) = p(x)$.

likelihood Given data $\mathbf{x}$ generated by a process with parameters θ, the probability of the data is $p(\mathbf{x}|\theta)$, which is also known as the likelihood of θ. See Appendix C.

maximum likelihood estimation Method of using data to estimate the value of parameters. See Appendix C.

MNIST data set A data set containing handwritten images of digits from 0 to 9. It consists of 60,000 images in a training set and 10,000 images in a test set. Each image is 28×28 pixels, with grey-levels between 0 and 255.

multi-layer perceptron A neural network with multiple layers; usually refers to a backprop network.

one-hot vector A binary vector used to indicate one out of n classes. For example, given 10 digit classes (0–9) the class for the digit 3 would be represented as the one-hot vector $[0\ 0\ 0\ 1\ 0\ 0\ 0\ 0\ 0\ 0]$. Used for conditional variational autoencoders and transformer networks.

orthogonal If two vectors $\mathbf{x}_1 = (x_{1,1}, \ldots, x_{1,n})$ and $\mathbf{x}_2 = (x_{2,1}, \ldots, x_{2,n})$ are orthogonal then they are perpendicular to each other, which means that their inner product equals zero: $\sum_{i=1}^{n} x_{1,i} x_{2,i} = 0$.

over-fitting Given noisy data with an underlying trend, an ideal model would fit a function to the trend. Over-fitting occurs when the model fits a function to the noise, or when the fitted function does not interpolate smoothly between widely separated data points.

receptive field A spatially coherent set of photoreceptors connected to a single neuron, with connection strengths that usually fall as a function of distance from the centre of the receptive field.

rectified linear unit (ReLU) A unit with an output of 0 unless the input is positive ($net > 0$), in which case the output equals the input ($y = net$).

regularisation A method to reduce over-fitting, usually by adding constraint terms to a cost function.

Shannon entropy Given a variable X with probability distribution $p(X)$, if that distribution is divided into N intervals with centre values $x_1, x_2, \ldots, x_N$ then the entropy of X is

$$H(X) \;=\; \sum_{i=1}^{N} p(x_i) \log_2 \frac{1}{p(x_i)} \text{ bits.}$$

supervised learning Learning to map input vectors to corresponding outputs, where the differences between the network outputs and the correct outputs are used to train the network.

top-n classification performance Classification performance is often reported as a percentage given a top-n threshold. If inputs are classified as belonging to the top five classes 80% of the time, this is reported as a top-5 performance of 80%. A top-1 performance specifies how often the network classification is the correct class label.

training data Data used to train a neural network, usually split into three sets.

unit Artificial neuron.

validation set Part of the training data used to monitor progress on the training set. The validation set is not used for updating weights.

weight Value representing the strength of the connection between two units.

Appendix B

A Vector and Matrix Tutorial

The single key fact to know about vectors and matrices is that each vector represents a point located in space, and a matrix moves that point to a different location. Everything else is just details.

Vectors. A number, such as 1.234, is known as a *scalar*, and a *vector* is an ordered list of scalars. A vector with two components w_1 and w_2 is written as $\mathbf{w} = (w_1, w_2)$. Note that vectors are written in bold type. The vector $\mathbf{w}$ can be represented as a single point in a 2D graph; the location of this point is by convention a distance of w_1 from the origin along the horizontal axis and a distance of w_2 from the origin along the vertical axis.

Adding Vectors. The *vector sum* of two vectors is obtained by adding their corresponding elements. Adding two ordered pairs of scalars (w_1, w_2) and (x_1, x_2) gives

$$(w_1 + x_1), \quad (w_2 + x_2). \tag{B.1}$$

Writing (w_1, w_2), (x_1, x_2) and their sum as vectors:

$$\begin{aligned} \mathbf{z} &= \big((w_1 + x_1), (w_2 + x_2)\big) & \text{(B.2)} \\ &= (x_1, x_2) + (w_1, w_2) \\ &= \mathbf{x} + \mathbf{w}. & \text{(B.3)} \end{aligned}$$

Thus the sum of two vectors is another vector, which is known as the *resultant* of those two vectors.

Subtracting Vectors. Subtracting vectors is implemented similarly by subtracting corresponding elements, so that

$$\begin{aligned} \mathbf{z} &= \mathbf{x} - \mathbf{w} & \text{(B.4)} \\ &= \big((x_1 - w_1), (x_2 - w_2)\big). & \text{(B.5)} \end{aligned}$$

Multiplying Vectors. Consider the result of multiplying the corresponding elements of two pairs of scalars (x_1, x_2) and (w_1, w_2) and adding the two products together:

$$y = w_1 x_1 + w_2 x_2. \tag{B.6}$$

We can write (x_1, x_2) and (w_1, w_2) as vectors and express y as

$$\begin{aligned} y &= (x_1, x_2) \cdot (w_1, w_2) \\ &= \mathbf{x} \cdot \mathbf{w}, & \text{(B.7)} \end{aligned}$$

where Equation B.7 is to be interpreted as Equation B.6. This sum of the products of corresponding vector elements is known as the *inner, scalar* or *dot* product and is often denoted by a dot, as here.

Vector Length. As each vector represents a point in space, it must have a distance from the origin, and this distance is known as the vector's length or *modulus*, denoted by $|\mathbf{x}|$ for a vector $\mathbf{x}$. For a vector $\mathbf{x} = (x_1, x_2)$ with two components, this distance is given by the length of the hypotenuse of a right-angled triangle with sides x_1 and x_2, so

$$|\mathbf{x}| = \sqrt{x_1^2 + x_2^2}. \tag{B.8}$$

Angle between Vectors. The angle θ between two vectors $\mathbf{x}$ and $\mathbf{w}$ is given by

$$\cos \theta = \frac{\mathbf{x} \cdot \mathbf{w}}{|\mathbf{x}| \, |\mathbf{w}|}. \tag{B.9}$$

Crucially, if $\theta = 90°$ then the inner product in the numerator is zero, because $\cos 90° = 0$ irrespective of the lengths of the vectors. Vectors at 90 degrees to each other are known as *orthogonal vectors*.

Row and Column Vectors. Vectors come in two basic flavours, *row vectors* and *column vectors*. A simple notational device to transform a row vector (x_1, x_2) into a column vector (or vice versa) is the *transpose operator*, T:

$$(x_1, x_2)^{\mathsf{T}} = \begin{pmatrix} x_1 \\ x_2 \end{pmatrix}. \tag{B.10}$$

The reason for having row and column vectors is that it is often necessary to combine several vectors into a single *matrix*, which is then used to multiply a single column vector $\mathbf{x}$, defined here as

$$\mathbf{x} = (x_1, x_2)^{\mathsf{T}}. \tag{B.11}$$

In such cases, we need to keep track of which vectors are row vectors and which are column vectors. If we redefine $\mathbf{w}$ as a column vector, $\mathbf{w} = (w_1, w_2)^{\mathsf{T}}$, then the inner product $\mathbf{w} \cdot \mathbf{x}$ can be written as

$$y = \mathbf{w}^{\mathsf{T}} \mathbf{x} \tag{B.12}$$

$$= (w_1, w_2) \begin{pmatrix} x_1 \\ x_2 \end{pmatrix} \tag{B.13}$$

$$= w_1 x_1 + w_2 x_2. \tag{B.14}$$

Here, each element of the row vector $\mathbf{w}^{\mathsf{T}}$ is multiplied by the corresponding element of the column vector $\mathbf{x}$, and the results are summed. Writing the inner product in this way allows us to simultaneously specify many pairs of such products as a vector–matrix product. For example, if x_1 and x_2 of the vector variable $\mathbf{x}$ have been measured n times (e.g. at n consecutive time steps), then taking the inner product of these pairs of measurements with $\mathbf{w}$ yields n values of the variable y:

$$(y_1, y_2, \ldots, y_n) = (w_1, w_2) \begin{pmatrix} x_{11} & x_{12} & \cdots & x_{1n} \\ x_{21} & x_{22} & \cdots & x_{2n} \end{pmatrix}. \tag{B.15}$$

Here, each (single-element) column y_t is given by the inner product of the corresponding column in $\mathbf{x}$ with the row vector $\mathbf{w}^{\mathsf{T}}$, so that

$$y = \mathbf{w}^{\mathsf{T}} \mathbf{x}.$$

Vector–Matrix Multiplication. If we reset the number of times $\mathbf{x}$ has been measured to $n = 1$ for now, we can consider the simple case of how two scalar values y_1 and y_2 are given by the inner products

$$y_1 = \mathbf{w}_1^{\mathsf{T}} \mathbf{x} \tag{B.16}$$

$$y_2 = \mathbf{w}_2^{\mathsf{T}} \mathbf{x}, \tag{B.17}$$

where $\mathbf{w}_1 = (w_1, w_2)^\mathsf{T}$ and $\mathbf{w}_2 = (w_3, w_4)^\mathsf{T}$. If we consider the pair of values y_1 and y_2 as a vector $\mathbf{y} = (y_1, y_2)^\mathsf{T}$ then we can rewrite Equations B.16 and B.17 as

$$(y_1, y_2)^\mathsf{T} = (\mathbf{w}_1^\mathsf{T}\mathbf{x}, \mathbf{w}_2^\mathsf{T}\mathbf{x})^\mathsf{T}. \tag{B.18}$$

Combining the column vectors $\mathbf{w}_1$ and $\mathbf{w}_2$ defines a *matrix* W:

$$W = (\mathbf{w}_1, \mathbf{w}_2)^\mathsf{T} = \begin{pmatrix} w_1 & w_2 \\ w_3 & w_4 \end{pmatrix}. \tag{B.19}$$

We can now rewrite equation B.18 as

$$(y_1, y_2)^\mathsf{T} = \begin{pmatrix} w_1 & w_2 \\ w_3 & w_4 \end{pmatrix} (x_1, x_2)^\mathsf{T}, \tag{B.20}$$

or more succinctly as $\mathbf{y} = W\mathbf{x}$. This defines the standard syntax for vector–matrix multiplication. Note that the column vector $(x_1, x_2)^\mathsf{T}$ is multiplied by the first row in W to obtain the row y_1 and is multiplied by the second row in W to obtain the row y_2. Just as the vector $\mathbf{x}$ represents a point on a plane, so the point $\mathbf{y}$ represents a (usually different) point on the plane. Thus *the matrix W implements a linear geometric transformation of points from $\mathbf{x}$ to $\mathbf{y}$.*

If $n > 1$ then the tth column $(y_{1t}, y_{2t})^\mathsf{T}$ in $\mathbf{y}$ is obtained as the product of tth column $(x_{1t}, x_{2t})^\mathsf{T}$ in $\mathbf{x}$ with the row vectors in W:

$$\mathbf{y} = \begin{pmatrix} y_{11} & \cdots & y_{1n} \\ y_{21} & \cdots & y_{2n} \end{pmatrix} \tag{B.21}$$

$$= \begin{pmatrix} w_1 & w_2 \\ w_3 & w_4 \end{pmatrix} \begin{pmatrix} x_{11} & \cdots & x_{1n} \\ x_{21} & \cdots & x_{2n} \end{pmatrix} \tag{B.22}$$

$$= W\mathbf{x}. \tag{B.23}$$

Note that $\mathbf{y}$ has the same number of rows as W and the same number of columns as $\mathbf{x}$.

The Outer Product. If an m-element column vector $\mathbf{x}$ is multiplied by an n-element row vector $\mathbf{y}^\mathsf{T}$ then the result is a matrix W with m rows and n columns. For example, if $m = 2$ and $n = 3$ then

$$W = \mathbf{x}\mathbf{y}^\mathsf{T} \tag{B.24}$$

$$= \begin{pmatrix} x_1 \\ x_2 \end{pmatrix} (y_1, y_2, y_3) \tag{B.25}$$

$$= \begin{pmatrix} x_1 y_1 & x_1 y_2 & x_1 y_3 \\ x_2 y_1 & x_2 y_2 & x_2 y_3 \end{pmatrix}. \tag{B.26}$$

Transpose of Vector–Matrix Product. If $\mathbf{y} = W\mathbf{x}$ then the transpose of $\mathbf{y}$ is

$$\mathbf{y}^\mathsf{T} = (W\mathbf{x})^\mathsf{T} = \mathbf{x}^\mathsf{T}W^\mathsf{T}, \tag{B.27}$$

where W^T is obtained by swapping off-diagonal elements:

$$W^\mathsf{T} = \begin{pmatrix} w_1 & w_2 \\ w_3 & w_4 \end{pmatrix}^\mathsf{T} = \begin{pmatrix} w_1 & w_3 \\ w_2 & w_4 \end{pmatrix}. \tag{B.28}$$

Matrix Inverse. By analogy with scalar algebra, if $\mathbf{y} = W\mathbf{x}$ then $\mathbf{x} = W^{-1}\mathbf{y}$, where W^{-1} is the inverse of W. If the columns of a matrix are orthogonal then $W^{-1} = W^\mathsf{T}$.

Appendix C

Maximum Likelihood Estimation

A popular method for estimating the values of parameters is maximum likelihood estimation (MLE). Because it is often assumed that data have a Gaussian distribution, we use MLE to estimate the mean of Gaussian data here. Given Gaussian data, the probability (density) of observing a value x_i is

$$p(x_i|\mu, \sigma) = c\, e^{-(\mu-x_i)^2/(2\sigma^2)}, \qquad (C.1)$$

where μ is the mean (centre) of the Gaussian distribution, σ is the standard deviation, which is a measure of the spread of the distribution, and $c = 1/\sqrt{2\pi\sigma^2}$ is a constant which ensures that the distribution has a total value (integral) of 1. A graph of a typical Gaussian distribution is shown in Figure C.1, which has a mean of μ and a standard deviation of σ. We wish to estimate the mean μ.

A sample of N values can be represented as a vector $\mathbf{x} = (x_1, \ldots, x_N)$. If these N values are mutually independent then

$$p(\mathbf{x}|\mu) = \prod_{i=1}^{N} p(x_i|\mu), \qquad (C.2)$$

where we have omitted σ for simplicity.

Maximum likelihood estimation consists in finding an estimate $\hat{\mu}$ of μ that is maximally consistent with the data $\mathbf{x}$. More precisely, $p(\mathbf{x}|\hat{\mu})$ is the probability of the observed data if the mean were equal

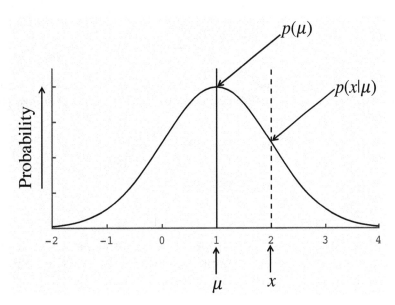

Figure C.1. Gaussian distribution with mean $\mu = 1$ and standard deviation σ. The probability $p(x|\mu)$ of observing the value x is given by Equation C.1.

to $\hat{\mu}$, even though the true (unknown) mean is μ. However, because we are interested in finding the optimal value of $\hat{\mu}$, we consider $p(\mathbf{x}|\hat{\mu})$ to be a function of μ, called the *likelihood function*. Consequently, even though $p(\mathbf{x}|\hat{\mu})$ is the probability of observing the data $\mathbf{x}$, it is called the likelihood of μ,

$$p(\mathbf{x}|\hat{\mu}) = \prod_{i=1}^{N} p(x_i|\hat{\mu}). \tag{C.3}$$

In the process of estimating μ, we could simply 'try out' different values of $\hat{\mu}$ in Equation C.3 to see which value is most consistent with the data $\mathbf{x}$, that is, which value $\hat{\mu}$ of μ makes the data most probable. The result represents our best estimate of μ. Note that a different set of data would yield a different likelihood function, with a different value for the maximum likelihood estimate $\hat{\mu}$.

At first, this way of thinking about the data seems odd. It just sounds wrong to speak of the probability of the data, which are the things we have already observed – so why would we care how probable they are? In fact, we do not care about the probability of the data *per se*, but we do care how probable those data are in the context of the parameters we wish to estimate.

In practice, we usually choose to deal with log likelihoods rather than likelihoods. Because the log function is monotonic, the estimate $\hat{\mu}$ of μ that maximises $p(\mathbf{x}|\hat{\mu})$ also maximises $\log p(\mathbf{x}|\hat{\mu})$, where

$$\log p(\mathbf{x}|\hat{\mu}) = \log \prod_{i=1}^{N} p(x_i|\hat{\mu}) \tag{C.4}$$

$$= \sum_{i=1}^{N} \log p(x_i|\hat{\mu}). \tag{C.5}$$

Substituting Equation C.1 into Equation C.5 gives

$$\log p(\mathbf{x}|\hat{\mu}) = k - \sum_{i=1}^{N} \frac{(\hat{\mu} - x_i)^2}{2\sigma^2}, \tag{C.6}$$

where $k = -N \log c$. A common notation for the log likelihood is $L(\hat{\mu})$. If the standard deviation $\sigma = 1$ then this becomes

$$L(\hat{\mu}) = k - \frac{1}{2} \sum_{i=1}^{N} (\hat{\mu} - x_i)^2. \tag{C.7}$$

Maximising $L(\hat{\mu})$ is equivalent to minimising $-L(\hat{\mu})$, which resembles the equation used for *least-squares estimation* (LSE), so LSE is a form of MLE. At a maximum, $dL(\hat{\mu})/d\hat{\mu} = 0$ and so (because $dk/d\hat{\mu} = 0$)

$$\frac{dL(\hat{\mu})}{d\hat{\mu}} = -\frac{1}{2} \sum_{i=1}^{N} \frac{d(\hat{\mu} - x_i)^2}{d\hat{\mu}} \tag{C.8}$$

$$= -\sum_{i=1}^{N} (\hat{\mu} - x_i). \tag{C.9}$$

Thus, if we set $dL(\hat{\mu})/d\hat{\mu}$ to zero and solve for $\hat{\mu}$, then we obtain the maximum likelihood estimate $\hat{\mu}$ of μ, where $\hat{\mu}$ turns out to be the mean of the sample, that is, $\hat{\mu} = (1/N) \sum x_i$. Finally, note that MLE corresponds to Bayesian estimation with a uniform prior distribution[119].

Appendix D

Bayes' Theorem

Bayes' theorem is essentially a rigorous method for interpreting evidence (data) in the context of previous experience or knowledge. Bayes' theorem is also known as *Bayes' rule*[109;119].

If we construe each possible value of a parameter as one of many possible answers then Bayesian inference is not guaranteed to provide the correct answer. Instead, it provides the probability that each of a number of alternative answers is true, and these probabilities can then be used to find the answer that is most probably true. In other words, Bayesian inference provides an informed guess. While this may not sound like much, it is far from random guessing. Indeed, it can be shown that no other procedure can provide a better guess, so that Bayesian inference can be justifiably interpreted as the output of a perfect guessing machine – a perfect inference engine. This perfect inference engine is fallible, but it is provably less fallible than any other.

For notational convenience, we denote a variable by x, and a particular value of x is indicated with a subscript (e.g. x_t).

Conditional Probability. The conditional probability that $y = y_t$ given that $x = x_t$ is defined as

$$p(y_t|x_t) = \frac{p(x_t, y_t)}{p(x_t)}, \tag{D.1}$$

where the vertical bar stands for 'given that' and $p(x_t, y_t)$ is the *joint probability* that $x = x_t$ and $y = y_t$. Equation D.1 holds true for all values of y, so the *conditional distribution* is

$$p(y|x_t) = \frac{p(x_t, y)}{p(x_t)}, \tag{D.2}$$

where $p(x_t, y)$ is a cross-section of the *joint probability distribution* $p(x, y)$ at $x = x_t$. When considered over all values of x, this yields a family of conditional distributions (one member for each value of x)

$$p(y|x) = \frac{p(x, y)}{p(x)}. \tag{D.3}$$

The Product Rule: Multiplying both sides of Equation D.3 by $p(x)$ yields the *product rule* (also known as the *chain rule*)

$$p(x, y) = p(y|x)p(x). \tag{D.4}$$

The Sum Rule and Marginalisation: The *sum rule* is also known as the *law of total probability*. In the case of a discrete variable, given a joint distribution $p(x, y)$, the *marginal distribution* $p(y)$ is obtained by *marginalisation* (summing) over values of x:

$$p(y) \;=\; \sum_t p(x_t, y). \tag{D.5}$$

If we apply the product rule then we obtain

$$p(y) \;=\; \sum_t p(y|x_t)p(x_t). \tag{D.6}$$

In the case of a continuous variable, the sum and product rules become

$$p(y) \;=\; \int_x p(x, y)\, dx \tag{D.7}$$

$$\;=\; \int_x p(y|x)p(x)\, dx. \tag{D.8}$$

Bayes' Theorem: If we swap x and y in Equation D.4 then

$$p(y, x) = p(x|y)p(y). \tag{D.9}$$

Because $p(x, y) = p(y, x)$, it follows that

$$p(y|x)p(x) \;=\; p(x|y)p(y). \tag{D.10}$$

Dividing both sides by $p(x)$ yields Bayes' theorem:

$$p(y|x) \;=\; \frac{p(x|y)p(y)}{p(x)}. \tag{D.11}$$

If we have an observed value x_t of x, then Bayes' theorem can be used to infer the *posterior probability distribution* of y:

$$p(y|x_t) \;=\; \frac{p(x_t|y)p(y)}{p(x_t)}, \tag{D.12}$$

where $p(y)$ is the *prior probability distribution* of y, $p(x_t|y)$ is the conditional probability that $x = x_t$, which is defined by the *likelihood function* of y, and $p(x_t)$ is the *marginal likelihood* or *evidence*. If $p(x_t)$ is not known then it can be obtained by marginalising the numerator over $p(y)$:

$$p(x_t) \;=\; \int_y p(x_t|y)p(y)\, dy. \tag{D.13}$$

Therefore, Bayes' theorem in Equation D.12 can be written as

$$p(y|x_t) \;=\; \frac{p(x_t|y)p(y)}{\int_y p(x_t|y)p(y)\, dy}. \tag{D.14}$$

References

[1] Achiam, J., Adler, S., Agarwal, S., Ahmad, L., Akkaya, I., Aleman, F. L., Almeida, D., Altenschmidt, J., Altman, S., Anadkat, S., et al. (2023). GPT-4 technical report. *arXiv e-prints*. arXiv:2303.08774.

[2] Ackley, D. H., Hinton, G. E., and Sejnowski, T. J. (1985). A learning algorithm for Boltzmann machines. *Cognitive Science*, 9:147–169.

[3] Alemi, A. A., Fischer, I., Dillon, J. V., and Murphy, K. (2017). Deep variational information bottleneck. *arXiv e-prints*. arXiv:1612.00410.

[4] Anderson, J. (1968). A model for memory using spatial correlation functions. *Kybernetik*, 5:113–119.

[5] Bahri, Y., Kadmon, J., Pennington, J., Schoenholz, S. S., Sohl-Dickstein, J., and Ganguli, S. (2020). Statistical mechanics of deep learning. *Annual Review of Condensed Matter Physics*, 11:501–528.

[6] Bai, S., Kolter, Z., and Koltun, V. (2018). An empirical evaluation of generic convolutional and recurrent networks for sequence modeling. *arXiv e-prints*. arXiv:1803.01271.

[7] Bain, A. (1873). *Mind and Body: The Theories of Their Relation*. Henry S. King & Co.

[8] Barto, A., Sutton, R., and Anderson, C. (1983). Neuronlike adaptive elements that can solve difficult learning control problems. *IEEE Transactions on Systems, Man, and Cybernetics*, 13(5):834–846.

[9] Barto, A. G., Sutton, R. S., and Anderson, C. W. (2021). Looking back on the actor–critic architecture. *IEEE Transactions on Systems, Man, and Cybernetics: Systems*, 51(1):40–50.

[10] Bengio, Y., Ducharme, R., and Vincent, P. (2000). A neural probabilistic language model. In *Advances in Neural Information Processing Systems 13*.

[11] Bengio, Y., Ducharme, R., Vincent, P., and Jauvin, C. (2003). A neural probabilistic language model. *Journal of Machine Learning Research*, 3:1137–1155.

[12] Berlucchi, G. and Buchtel, H. A. (2009). Neuronal plasticity: historical roots and evolution of meaning. *Experimental Brain Research*, 192:307–319.

[13] Berry, D. C. and Broadbent, D. E. (1988). Interactive tasks and the implicit-explicit distinction. *British Journal of Psychology*, 79(2):251–272.

[14] Bishop, C. M. (1996). *Neural Networks for Pattern Recognition*. Oxford University Press.

[15] Bridges, A. D., Royka, A., Wilson, T., Lockwood, C., Richter, J., Juusola, M., and Chittka, L. (2024). Bumblebees socially learn behaviour too complex to innovate alone. *Nature*, 627:572–578.

[16] Brinkmann, L., Baumann, F., Bonnefon, J.-F., Derex, M., Müller, T. F., Nussberger, A.-M., Czaplicka, A., Acerbi, A., Griffiths, T. L., Henrich, J., Leibo, J. Z., McElreath, R., Oudeyer, P.-Y., Stray, J., and Rahwan, I. (2023). Machine culture. *Nature Human Behaviour*, 7(11):1855–1868.

[17] Brown, T., Mann, B., Ryder, N., Subbiah, M., Kaplan, J. D., Dhariwal, P., Neelakantan, A., Shyam, P., Sastry, G., Askell, A., et al. (2020). Language models are few-shot learners. In *Proceedings of the 34th Conference on Neural Information Processing Systems (NeurIPS 2020)*, pages 1877–1901. Curran Associates.

[18] Burda, Y., Grosse, R., and Salakhutdinov, R. (2015). Importance weighted autoencoders. *arXiv e-prints*. arXiv:1509.00519.

[19] Burgess, C. P., Higgins, I., Pal, A., Matthey, L., Watters, N., Desjardins, G., and Lerchner, A. (2018). Understanding disentangling in β-VAE. *arXiv e-prints*. arXiv:1804.03599.

[20] Cho, K., van Merrienboer, B., Gulcehre, C., Bahdanau, D., Bougares, F., Schwenk, H., and Bengio, Y. (2014). Learning phrase representations using RNN encoder-decoder for statistical machine translation. *arXiv e-prints*. arXiv:1406.1078.

[21] Cover, T. M. and Thomas, J. A. (1991). *Elements of Information Theory*. John Wiley and Sons.

[22] Cybenko, G. (1989). Approximation by superposition of a sigmoidal function. *Mathematics of Control, Signals and Systems*, 2:303–314.

[23] Elbayad, M., Besacier, L., and Verbeek, J. (2018). Pervasive attention: 2D convolutional neural networks for sequence-to-sequence prediction. *arXiv e-prints.* arXiv:1808.03867.

[24] Elman, J. L. (1990). Finding structure in time. *Cognitive Science*, 14(2):179–211.

[25] Florian, R. V. (2007). Correct equations for the dynamics of the cart-pole system. Technical report, Center for Cognitive and Neural Studies (Coneural), Romania.

[26] Frisby, J. P. and Stone, J. V. (2010). *Seeing: The Computational Approach to Biological Vision.* MIT Press.

[27] Geman, S. and Geman, D. (1993). Stochastic relaxation, Gibbs distributions and the Bayesian restoration of images. *Journal of Applied Statistics*, 20:25–62.

[28] Gers, F. A., Schmidhuber, J., and Cummins, F. (2000). Learning to forget: continual prediction with LSTM. *Neural Computation*, 12(10):2451–2471.

[29] Goodfellow, I. J., Pouget-Abadie, J., Mirza, M., Xu, B., Warde-Farley, D., Ozair, S., Courville, A., and Bengio, Y. (2014). Generative adversarial networks. *arXiv e-prints.* arXiv:1406.2661.

[30] Gorman, R. P. and Sejnowski, T. J. (1988). Analysis of hidden units in a layered network trained to classify sonar targets. *Neural Networks*, 1(1):75–89.

[31] Greener, J. G., Moffat, L., and Jones, D. T. (2018). Design of metalloproteins and novel protein folds using variational autoencoders. *Scientific Reports*, 8(1):16189.

[32] Guilliard, I., Rogahn, R., Piavis, J., and Kolobov, A. (2018). Autonomous thermalling as a partially observable Markov decision process. *arXiv e-prints.* arXiv:1805.09875.

[33] Haarnoja, T., Zhou, A., Ha, S., Tan, J., Tucker, G., and Levine, S. (2018). Learning to walk via deep reinforcement learning. *arXiv e-prints.* arXiv:1812.11103.

[34] Harvey, I. and Stone, J. V. (1996). Unicycling helps your French: spontaneous recovery of associations by learning unrelated tasks. *Neural Computation*, 8:697–704.

[35] Hay, J. C., Lynch, B. E., and Smith, D. R. (1960). *Mark I Perceptron Operators' Manual.* Cornell Aeronautical Laboratory Inc., Buffalo, NY.

[36] Hebb, D. O. (1949). *The Organization of Behavior: A Neuropsychological Theory.* Wiley.

[37] Helmbold, D. and Long, P. (2017). Surprising properties of dropout in deep networks. *Journal of Machine Learning Research*, 18(1):7284–7311.

[38] Higgins, I., Matthey, L., Glorot, X., Pal, A., Uria, B., Blundell, C., Mohamed, S., and Lerchner, A. (2016). Early visual concept learning with unsupervised deep learning. *arXiv e-prints.* arXiv:1606.05579.

[39] Hinton, G. E. and Salakhutdinov, R. R. (2006). Reducing the dimensionality of data with neural networks. *Science*, 313(5786):504–507.

[40] Hinton, G. E., Sejnowski, T. J., and Ackley, D. H. (1984). Boltzmann machines: Constraint satisfaction networks that learn. Technical Report CMU-CS-84-119, Department of Computer Science, Carnegie-Mellon University.

[41] Ho, J., Jain, A., and Abbeel, P. (2020). Denoising diffusion probabilistic models. In *Proceedings of the 34th Conference on Neural Information Processing Systems (NeurIPS 2020)*, pages 6840–6851.

[42] Hoang, L. and Guerraoui, R. (2018). Deep learning works in practice. But does it work in theory? *arXiv e-prints.* arXiv:1801.10437.

[43] Hochreiter, S. and Schmidhuber, J. (1997). Long short-term memory. *Neural Computation*, 9(8):1735–1780.

[44] Hopfield, J. J. (1982). Neural networks and physical systems with emergent collective computational abilities. *Proc. Nat. Acad. Sci. USA*, 79(8):2554–2558.

[45] Hopfield, J. J. (1984). Neurons with graded response have collective computational properties like those of two-state neurons. *Proc. Nat. Acad. Sci. USA*, 81:3088–3092.

[46] Hopfield, J. J. and Tank, D. W. (1985). "Neural" computation of decisions in optimization problems. *Biological Cybernetics*, 52(3):141–152.

[47] Hsu, W., Zhang, Y., and Glass, J. (2017). Unsupervised domain adaptation for robust speech recognition via variational autoencoder-based data augmentation. *arXiv e-prints.* arXiv:1707.06265.

[48] Hubel, D. H. and Wiesel, T. N. (1962). Receptive fields, binocular interaction, and functional architecture in the cat's visual cortex. *Journal of Physiology*, 160:160–154.

[49] Hwangbo, J., Lee, J., Dosovitskiy, A., Bellicoso, D., Tsounis, V., Kolton, V., and Hutter, M. (2019). Learning agile and dynamic motor skills for legged robots. *Science Robotics*, 4(26):eaau5872.

[50] Jordan, M. I. (1986). Serial order: a parallel distributed approach. Technical Report 8604, Institute for Cognitive Science, University of California, San Diego.

[51] Karras, T., Aila, T., Laine, S., and Lehtinen, J. (2017). Progressive growing of GANs for improved quality, stability, and variation. *arXiv e-prints*. arXiv:1710.10196.

[52] Kavasidis, I., Palazzo, S., Spampinato, C., Giordano, D., and Shah, M. (2017). Brain2Image: converting brain signals into images. In *Proceedings of the 25th ACM International Conference on Multimedia (MM'17)*, pages 1809–1817. Association for Computing Machinery.

[53] Kim, B., Varshney, K. R., and Weller, A. (2018a). *Proceedings of the 2018 ICML Workshop on Human Interpretability in Machine Learning (WHI 2018)*. arXiv:1807.01308.

[54] Kim, Y., Wiseman, S., and Rush, A. M. (2018b). A tutorial on deep latent variable models of natural language. *arXiv e-prints*. arXiv:1812.06834.

[55] Kingma, D. P. (2017). *Variational Inference & Deep Learning*. PhD thesis, University of Amsterdam.

[56] Kingma, D. P. and Welling, M. (2013). Auto-encoding variational Bayes. *arXiv e-prints*. arXiv:1312.6114.

[57] Kirkpatrick, S., Gelat, C., and Vecchi, M. (1983). Optimization by simulated annealing. *Science*, 220(4598):671–680.

[58] Kohonen, T. (1972). Correlation matrix memories. *IEEE Transactions on Computers*, C-21(4):353–359.

[59] Krizhevsky, A., Sutskever, I., and Hinton, G. E. (2012). ImageNet classification with deep convolutional neural networks. In *Advances in Neural Information Processing Systems 25 (NIPS 2012)*, pages 1097–1105.

[60] Krotov, D. (2023). A new frontier for Hopfield networks. *Nature Reviews Physics*, 5:366–367.

[61] Larsen, A., Sønderby, S., Larochelle, H., and Winther, O. (2015). Autoencoding beyond pixels using a learned similarity metric. *arXiv e-prints*. arXiv:1512.09300.

[62] LeCun, Y., Boser, B., Denker, J. S., Henderson, D., Howard, R. E., Hubbard, W., and Jackel, L. D. (1989a). Handwritten digit recognition with a back-propagation network. In *Advances in Neural Information Processing Systems 2*, pages 396–404. Morgan Kaufman Publishers.

[63] LeCun, Y., Boser, B., Denker, J. S., Henderson, R. E., Hubbard, W., and Jackel, L. D. (1989b). Backpropagation applied to handwritten ZIP code recognition. *Neural Computation*, 1:541–551.

[64] LeCun, Y., Bottou, L., Bengio, Y., and Haffner, P. (1998). Gradient-based learning applied to document recognition. *Proceedings of the IEEE*, 86(11):2278–2324.

[65] Lettvin, J. V., Maturana, H. R., McCulloch, W. S., and Pitts, W. H. (1959). What the frog's eye tells the frog's brain. *Proceedings of the Institute of Radio Engineers*, 47(11):1940–1951.

[66] Lim, J., Ryu, S., Kim, J. W., and Kim, W. Y. (2018). Molecular generative model based on conditional variational autoencoder for de novo molecular design. *arXiv e-prints*. arXiv:1806.05805.

[67] Lister, R. and Stone, J. V. (1995). An empirical study of the time complexity of various error functions with conjugate gradient back propagation. In *Proceedings of ICNN'95: International Conference on Neural Networks*. IEEE.

[68] Longuet-Higgins, H. C. (1968). The non-local storage of temporal information. *Proceedings of the Royal Society B*, 171(1024):327–334.

[69] Longuet-Higgins, H. C., Willshaw, D. J., and Buneman, O. P. (1970). Theories of associative recall. *Quarterly Reviews of Biophysics*, 3(2):223–244.

[70] Maaløe, L., Fraccaro, M., Liévin, V., and Winther, O. (2019). BIVA: a very deep hierarchy of latent variables for generative modeling. *arXiv e-prints*. arXiv:1902.02102.

[71] MacKay, D. J. C. (2003). *Information Theory, Inference, and Learning Algorithms*. Cambridge University Press.

[72] Markov, A. A. (2006). An example of statistical investigation of the text *Eugene Onegin* concerning the connection of samples in chains. *Science in Context*, 19(4):591–600.

[73] McCulloch, W. S. and Pitts, W. (1943). A logical calculus of the ideas immanent in nervous activity. *Bulletin of Mathematical Biophysics*, 5:115–133.

[74] Michie, D. and Chambers, R. A. (1968). BOXES: an experiment in adaptive control. *Machine Intelligence*, 2(2):137–152.

[75] Minsky, M. and Papert, S. (1969). *Perceptrons: An Introduction to Computational Geometry*. MIT Press.

[76] Mnih, V., Kavukcuoglu, K., Silver, D., Graves, A., Antonoglou, I., Wierstra, D., and Riedmiller, M. (2013). Playing Atari with deep reinforcement learning. *arXiv e-prints*. arXiv:1312.5602.

[77] Mnih, V., Kavukcuoglu, K., Silver, D., Rusuand, A. A., Veness, J., Bellemare, M. G., Graves, A., Riedmiller, M., Fidjeland, A. K., Ostrovski, G., et al. (2015). Human-level control through deep reinforcement learning. *Nature*, 518(7540):529.

[78] Mozer, M. C. (1993). Neural net architectures for temporal sequence processing. In *Santa Fe Institute Studies in the Sciences of Complexity*, volume 15, pages 243–243. Addison-Wesley.

[79] Nakkiran, P., Kaplun, G., Bansal, Y., Yang, T., Barak, B., and Sutskever, I. (2021). Deep double descent: where bigger models and more data hurt. *Journal of Statistical Mechanics: Theory and Experiment*, 2021(12):124003.

[80] Novikoff, A. B. J. (1962). On convergence proofs on perceptrons. In *Proceedings of the Symposium on the Mathematical Theory of Automata*, volume 12, pages 615–622.

[81] Oord, A., Kalchbrenner, N., and Kavukcuoglu, K. (2016). Pixel recurrent neural networks. *arXiv e-prints*. arXiv:1601.06759.

[82] Oren, M., Hassid, M., Adi, Y., and Schwartz, R. (2024). Transformers are multi-state RNNs. *arXiv e-prints*. arXiv:2401.06104.

[83] Pearlmutter, B. A. (1989). Learning state space trajectories in recurrent neural networks. *Neural Computation*, 1(2):263–269.

[84] Polykovskiy, D., Zhebrak, A., Vetrov, D., Ivanenkov, Y., Aladinskiy, V., Mamoshina, P., Bozdaganyan, M., Aliper, A., Zhavoronkov, A., and Kadurin, A. (2018). Entangled conditional adversarial autoencoder for de novo drug discovery. *Molecular Pharmaceutics*, 15(10):4398–4405.

[85] Pomerleau, D. A. (1989). ALVINN: an autonomous land vehicle in a neural network. In *Advances in Neural Information Processing Systems 1 (NIPS 1988)*, pages 305–313.

[86] Prince, S. J. D. (2023). *Understanding Deep Learning*. MIT Press.

[87] Radford, A., Metz, L., and Chintala, S. (2015). Unsupervised representation learning with deep convolutional generative adversarial networks. *arXiv e-prints*. arXiv:1511.06434.

[88] Radford, A., Narasimhan, K., Salimans, T., and Sutskever, I. (2018). Improving language understanding by generative pre-training. Preprint, OpenAI.

[89] Radford, A., Wu, J., Child, R., Luan, D., Amodei, D., and Sutskever, I. (2019). Language models are unsupervised multitask learners. *OpenAI Blog*, 1(8):9.

[90] Randløv, J. and Alstrøm, P. (1998). Learning to drive a bicycle using reinforcement learning and shaping. In *Proceedings of the 15th International Conference on Machine Learning (ICML'98)*, pages 463–471. Morgan Kaufmann Publishers.

[91] Reddy, G., Celani, A., Sejnowski, T., and Vergassola, M. (2016). Learning to soar in turbulent environments. *Proc. Nat. Acad. Sci. USA*, 113(33):E4877–E4884.

[92] Reddy, G., Wong-Ng, J., Celani, A., Sejnowski, T., and Vergassola, M. (2018). Glider soaring via reinforcement learning in the field. *Nature*, 562(7726):236.

[93] Rezende, D. J., Mohamed, S., and Wierstra, D. (2014). Stochastic backpropagation and approximate inference in deep generative models. *arXiv e-prints*. arXiv:1401.4082.

[94] Rosenblatt, F. (1958). The perceptron: a probabilistic model for information storage and organization in the brain. *Psychological Review*, 65(6):386–408.

[95] Rumelhart, D. E., Hinton, G. E., and Williams, R. J. (1985). Learning internal representations by error propagation. Technical Report ICS 8506, Institute for Cognitive Science, University of California, San Diego.

[96] Rumelhart, D. E., Hinton, G. E., and Williams, R. J. (1986). Learning representations by back-propagating errors. *Nature*, 323:533–536.

[97] Samuel, A. L. (1959). Some studies in machine learning using the game of checkers. *IBM Journal of Research and Development*, 3(3):210–229.

[98] Schäfer, A. M. and Zimmermann, H. G. (2006). Recurrent neural networks are universal approximators. In *Proceedings of the International Conference on Artificial Neural Networks (ICANN 2006)*, pages 632–640. Springer.

[99] Schmidhuber, J. (2015). Deep learning in neural networks: an overview. *Neural Networks*, 61:85–117.

[100] Sejnowski, T. J. (1977). Storing covariance with nonlinearly interacting neurons. *Journal of Mathematical Biology*, 4(4):303–321.

[101] Sejnowski, T. J. (2018). *The Deep Learning Revolution*. MIT Press.

[102] Sejnowski, T. J. and Rosenberg, C. R. (1987). NETtalk: parallel networks that learn to pronounce English text. *Complex Systems*, 1(1):145–168.

[103] Selfridge, O. G. (1958). Pandemonium: a paradigm for learning. In *The Mechanisation of Thought Processes: Proceedings of a Symposium Held at the National Physical Laboratory*, pages 513–526.

[104] Shannon, C. (1951). Prediction and entropy of printed English. *Bell System Technical Journal*, 30:47–51.

[105] Shannon, C. E. (1950). Programming a computer for playing chess. *The London, Edinburgh, and Dublin Philosophical Magazine and Journal of Science*, 41(314):256–275.

[106] Shannon, C. E. and Weaver, W. (1949). *The Mathematical Theory of Communication*. University of Illinois Press.

[107] Silver, D., Huang, A., Maddison, C. J., Guez, A., Sifre, L., van den Driessche, G., Schrittwieser, J., Antonoglou, I., Panneershelvam, V., Lanctot, M., et al. (2016). Mastering the game of Go with deep neural networks and tree search. *Nature*, 529:484–503.

[108] Silver, D., Schrittwieser, J., Simonyan, K., Antonoglou, I., Huang, A., Guez, A., Hubert, T., Baker, L., Lai, M., Bolton, A., et al. (2017). Mastering the game of Go without human knowledge. *Nature*, 550(7676):354–359.

[109] Sivia, D. (1996). *Data Analysis: A Bayesian Tutorial*. Oxford University Press.

[110] Smith, J. D., Berg, M. E., Cook, R. G., Murphy, M. S., Crossley, M. J., Boomer, J., Spiering, B., Beran, M. J., Church, B. A., Ashby, F. G., et al. (2012). Implicit and explicit categorization: a tale of four species. *Neuroscience & Biobehavioral Reviews*, 36(10):2355–2369.

[111] Sohl-Dickstein, J., Weiss, E., Maheswaranathan, N., and Ganguli, S. (2015). Deep unsupervised learning using nonequilibrium thermodynamics. *Proceedings of Machine Learning Research*, 37:2256–2265. Proceedings of the 32nd International Conference on Machine Learning.

[112] Stone, J. (1996). Learning perceptually salient visual parameters through spatiotemporal smoothness constraints. *Neural Computation*, 8(7):1463–1492.

[113] Stone, J. V. (1988). Connectionist models: theoretical status, form, and function. *Artificial Intelligence and Simulation of Behaviour*, 66.

[114] Stone, J. V. (1989). Learning sequences with a connectionist model. Invited paper presented to the *Rank Prize Funds Symposium on Neural Networks*.

[115] Stone, J. V. (1992). The optimal elastic net: finding solutions to the travelling salesman problem. In *Proceedings of the 1992 International Conference on Artificial Neural Networks (ICANN-92)*, pages 170–174.

[116] Stone, J. V. (2004). *Independent Component Analysis: A Tutorial Introduction*. MIT Press.

[117] Stone, J. V. (2007). Distributed representations accelerate evolution of adaptive behaviours. *PLoS Computational Biology*, 3(8):e147.

[118] Stone, J. V. (2012). *Vision and Brain*. MIT Press.

[119] Stone, J. V. (2013). *Bayes' Rule: A Tutorial Introduction to Bayesian Analysis*. Sebtel Press.

[120] Stone, J. V. (2015). *Information Theory: A Tutorial Introduction*. Sebtel Press.

[121] Stone, J. V. (2018). *Principles of Neural Information Theory: Computational Neuroscience and Metabolic Efficiency*. Sebtel Press.

[122] Stone, J. V. (2019). *Artificial Intelligence Engines: A Tutorial Introduction to the Mathematics of Deep Learning*. Sebtel Press.

[123] Stone, J. V. (2020a). *A Brief Guide to Artificial Intelligence*. Sebtel Press.

[124] Stone, J. V. (2020b). *The Quantum Menagerie: A Tutorial Introduction to the Mathematics of Quantum Mechanics*. Sebtel Press.

[125] Stone, J. V. (2022). *Information Theory: A Tutorial Introduction*. Sebtel Press, 2nd edition.

[126] Stone, J. V. and Harper, N. (1999). Temporal constraints on visual learning. *Perception*, 28:1089–1104.

[127] Stone, J. V., Hunkin, N. M., and Hornby, A. (2001). Predicting spontaneous recovery of memory. *Nature*, 414:167–168.

[128] Sun, C., Shrivastava, A., Singh, S., and Gupta, A. (2017). Revisiting unreasonable effectiveness of data in deep learning era. In *Proceedings of the IEEE International Conference on Computer Vision (ICCV)*, pages 843–852.

[129] Sutskever, I., Vinyals, O., and Le, Q. V. (2014). Sequence to sequence learning with neural networks. *arXiv e-prints*. arXiv:1409.3215.

[130] Sutton, R. S. and Barto, A. G. (2018). *Reinforcement Learning: An Introduction*. MIT Press.

[131] Tesauro, G. (1995). Temporal difference learning and TD-Gammon. *Communications of the ACM*, 38(3):58–68.

[132] Tishby, N., Pereira, F. C., and Bialek, W. (2000). The information bottleneck method. *arXiv e-prints*. arXiv:physics/0004057.

[133] Turing, A. M. (1950). Computing machinery and intelligence. *Mind*, 49:433–460.

[134] Vaswani, A., Shazeer, N., Parmar, N., Uszkoreit, J., Jones, L., Gomez, A. N., Kaiser, L., and Polosukhin, I. (2017). Attention is all you need. In *Proceedings of the 31st International Conference on Neural Information Processing Systems (NIPS'17)*, pages 6000–6010. Curran Associates.

[135] Vincent, P., Larochelle, H., Lajoie, I., Bengio, Y., and Manzagol, P. (2010). Stacked denoising autoencoders: learning useful representations in a deep network with a local denoising criterion. *Journal of Machine Learning Research*, 11:3371–3408.

[136] Werbos, P. (1974). *Beyond Regression: New Tools for Prediction and Analysis in the Behavioral Sciences*. PhD thesis, Harvard University.

[137] Widrow, B. and Hoff, M. E. (1960). Adaptive switching circuits. *1960 WESCON Convention Record, Part IV*, pages 96–104.

[138] Williams, R. J. and Zipser, D. (1989). A learning algorithm for continually running fully recurrent neural networks. *Neural Computation*, 1(2):270–280.

[139] Winograd, T. (1987). Thinking machines: Can there be? Are we? Department of Computer Science, Stanford University.

[140] Yu, Y., Buchanan, S., Pai, D., Chu, T., Wu, Z., Tong, S., Haeffele, B. D., and Ma, Y. (2023). White-box transformers via sparse rate reduction. *arXiv e-prints*. arXiv:2306.01129.

[141] Yuret, D. (2016). Knet: beginning deep learning with 100 lines of Julia. In *Machine Learning Systems Workshop at the International Conference on Neural Information Processing Systems*.

Index

Printed in the USA
CPSIA information can be obtained
at www.ICGtesting.com
LVHW081730201024
794332LV00005B/520